Digital Seismology and Fine Modeling of the Lithosphere

ETTORE MAJORANA
INTERNATIONAL SCIENCE SERIES
Series Editor:
Antonino Zichichi
European Physical Society
Geneva, Switzerland

(PHYSICAL SCIENCES)

Recent volumes in the series:

A Continuation Order Plan is available for this series. A continuation order will bring delivery of each new volume immediately upon publication. Volumes are billed only upon actual shipment. For further information please contact the publisher.

Digital Seismology and Fine Modeling of the Lithosphere

Edited by

R. Cassinis
University of Milan
Milan, Italy

G. Nolet
University of Utrecht
Utrecht, The Netherlands

and

G. F. Panza
University of Trieste
Trieste, Italy

Springer Science+Business Media, LLC

Library of Congress Cataloging in Publication Data

Digital seismology and fine modeling of the lithosphere / edited by R. Cassinis, G. Nolet, and G. F. Panza.

    p.   cm. — (Ettore Majorana international science series. Physical sciences; v. 42)

    "Proceedings of the International School of Applied Geophysics. sixth course on digital seismology and fine modeling of the lithosphere, held March 14–23, 1987, in Erice, Italy"—T.p. verso.

    Includes bibliographies and index.

    1. Seismometry. 2. Earth—Crust. 3. Earth—Mantle. I Cassinis, R. (Roberto) II. Nolet, Guust, 1945–   . III. Panza, Giuliano F. IV. International School of Applied Geophysics. V. Series.

QE541.D48 1989   89-8435
551.2′2—dc20   CIP

ISBN 978-1-4899-6761-9   ISBN 978-1-4899-6759-6 (eBook)
DOI 10.1007/978-1-4899-6759-6

Proceedings of the International School of Applied Geophysics,
Sixth Course on Digital Seismology and Fine Modeling of the
Lithosphere, held March 14–23, 1987, in Erice, Italy

© 1989 Springer Science+Business Media New York
Originally published by Plenum Plenum Press, New York in 1989.
Softcover reprint of the hardcover 1st edition 1989

PREFACE

During last years the seismological science experienced an explosive growth. The development was probably the fastest one among Earth Sciences, the contribution of seismology to the knowledge of the Earth's interior being outstanding. Both the scientific and the economic communities realized that, among the objectives of seismology, besides the scientific speculations there are also medium and short term applications like the surveys for energy source and mineral deposits or the evaluation of the earthquake hazard. Even the activities that are defined as "basic studies" can be considered as compulsory steps to prepare the applications and, therefore, as "long term applications".

The development was also accelerated by the contribution of exploration geophysicists, namely the seismic reflection specialists, who, so far had proceeded as a separate body. The common point of interest is the exploration of the Earth's crust and especially of the continental crust. In order to aim this objective the seismic techniques of exploration were modified: on one side it required the introduction in the research domain of the expensive and sophisticated equipments as well as of the processing methods employed so far for the surveys of hydrocarbons in sedimentary basins. "Active" and "passive" seismology are now being integrated for the fine exploration of the Lithosphere and Upper Mantle. On the other hand, several exploration seismologists are becoming interested in more general problems and are operating in close cooperation with the classical seismologists.

These joint activities have accelerated the "digital revolution" in seismological science: the "tomography" of the Mantle, the extraction from the noise of the feeble reflected signals from the inner discontinuities of the Earth and the fine exploration of the Lithospheric structure would not be possible without the contribution of the experience gathered by the oil explorationists.

The present difficulties of the oil industry (although temporary) will probably make easier this process of unification.

Another integration which is almost complete is the one between earthquake engineers and seismologists; the common points being the strong motion seismology and the seismic risk. However, there is a well definite boundary between the two categories that is represented by the modifications of the terrain shaking induced by the load of a building. Finally, the cooperation with geologists has been strongly improved.

The seismic exploration of the Lithosphere is probably the field where the achievements of the seismological science became more apparent to the Earth Science's community. Two reasons can explain this interest: first, the lithospheric structure and, especially, the structure of the continental crust, became of practical interest as a future reserve of natural resources; second, new and exciting information was given on the superficial part of the Earth, where the Geologists can compare and try to fit to the physical data their models based on outcrops, that is on visible evidences.

It seems correct to say that the geophysical discoveries on Lithosphere have transformed the Geological science, opening a new phase of research.

While during the sixties and the early seventies the major achievements were obtained on the oceanic crust and lithosphere, since the late seventies the exploration has been concentrated on the continental crust.

The purpose of the 6th Course of the Int'l School of Applied Geophysics was the presentation of the updated seismic techniques for the exploration of the shallower structure of our Planet as well as for the understanding of the dynamic processes taking place in the Crust and Upper Mantle. Also the theoretical background leading to the technical developments were discussed. Of course, a special emphasis was put in the analysis of the consequences, implications and potential of the digital techniques in every phase of the research and exploration work, from instrumentation and data gathered to processing and interpretation.

Almost all the lectures presented at the Course are reproduced here in full length. While being tutorial, they contain original views and treatments.

Some research papers presented by participants are also published. They witness only in part the lively discussions following the lectures that, unfortunately, were not recorded like the other forms of tutorial and seminarial activity carried out during the Course.

R. Cassinis, G. F. Panza and G. Nolet

CONTENTS

VERY-BROAD-BAND SEISMOMETRY

E. Wielandt

Institute of Geophysics
Stuttgart University
Richard-Wagner-Str. 44, D-7000 Stuttgart 1

INTRODUCTION

Broad-band seismic recording means recording all seismic signals of interest in a single seismogram (or in a single three-component set of seismograms), as opposed to recording different kinds of seismic signals in separate narrow-band channels as it is done, e.g., in the world-wide standard seismograph network (WWSSN). Although broad-band seismometry may appear to be a modern trend, it actually formed the very beginning of instrumental seismology. The first known mechanical seismograph was constructed in Italy around 1875; the first reasonably clear seismograms were obtained by English scientists in Japan in 1880. Most of their instruments were designed to record the largest possible range of periods. The Japanese seismologist Omori recorded beautiful broad-band seismograms of large earthquakes around 1900 which until recently one could not have obtained with modern seismographs.

The magnification of mechanical seismographs, especially at long periods, is limited by friction. This problem does not exist in the electromagnetic seismograph (invented by Galitzin in 1904) which permits an almost arbitrary magnification. However, there is another phenomenon that limits the useful gain at long periods: the marine microseisms, at periods around 6 sec, obscure the records as soon as the magnification is increased to more than a few hundred. One then has the choice of either keeping the gain below that limit, or choosing the response so that the seismograph acts as a band-pass filter that suppresses the microseisms. Seismologists of the old generation were always aware that the new narrow-band instruments could not replace the old mechanical seismographs and therefore kept a few of these going; nevertheless the historical development until about 1970 has been constantly away from broad-band recording, and towards specialized seismographs permitting large magnifications in a narrow bandwidth. The "high-gain long-period" type of seismographs is at the end of this development; its photographic records are good for the detection of surface waves but hardly for anything else.

The trend was reversed with the advent of magnetic tape recording, especially digital recording, first used in 1962 in the "CALTECH digital seismograph". It now became possible to record the seismic signals as they come out of the seismometer, and to do the necessary filtration in a flexible manner at playback time. However, digital recording had first to

1

reach a certain technical standard before digital broad-band recording could become a practical alternative to the traditional visual techniques. The first permanent digital broad-band installation, the German GRF array, became operational in 1976.

TECHNICAL CONSIDERATIONS

It is now time to discuss some general technical aspects of broad-band seismic recording. In order to be able to record seismic signals of different character equally well, and occasionally at the same time, a broad-band seismic system must have three specific technical qualities:

1. The system must have a large relative bandwidth. The lowest frequencies of interest are in the order of 0.3 mHz while the highest are about 10 Hz for teleseismic investigations and 100 Hz or more in local studies.

2. The system must have a large dynamic range because seismic signals occur in a wide range of amplitudes. A dynamic range of 140 decibels is considered desirable, i.e., the strongest signal that can be recorded unclipped should be 10,000,000 times larger than the faintest signal that can be detected.

3. The system must have an excellent linearity since signals of vastly different amplitude may be present at the same time in different parts of the spectrum. For example, the accelerations caused by marine microseisms, although minute, may still be 200,000 times larger than those due to a free oscillation of the earth; nevertheless we want to detect this oscillation in the presence of microseisms.

Analog methods of recording are limited to a dynamic range in the order of 60 dB. Broad-band seismic recording therefore usually implies digital recording. Even for digital standards, the requirements to the hardware, especially to the digitizer, are high; fully satisfactory technical solutions, which we call very-broad-band seismographs, have emerged only in the past few years.

It is however remarkable that once the technical solution exists, digital very-broad-band seismographs are now not only technically superior but also simpler than digital narrow-band systems. That is, except for applications limited to the short-period band, it is now more economic to install a broad-band seismograph and filter its output signal digitally to the desired bandwidth, than it would be to install a conventional narrow-band system with analog filters.

The main advantage offered by a very-broad-band seismograph is that it does not predetermine the kind of research for which its data can be used. Narrow-band data tend to become obsolete when the scientific interest shifts to a different part of the seismic spectrum. A library of digital broad-band data can be expected to retain its value even if the research objectives change. The broad-band concept also offers a number of other, less obvious advantages:

- easy specification of the response
- better stability versus time
- easier absolute calibration (e.g., at zero frequency)
- easy restitution of the true ground motion
- better use of the dynamic range of the digitizer
- transient disturbances can be removed from the record before filtration.

SENSOR DESIGN

The resolution and linearity required for VBB seismic recording can at present only be achieved with small, hermetically sealed, and well-isolated (or borehole mounted) instruments with negative force feedback. Since design considerations for such instruments have been discussed elsewhere [1,2,3], we give here only a brief summary.

General Design Principles

All inertial seismometers measure the seismic motion of the ground against an inertial reference, i.e., against an elastically suspended mass which does not fully participate in the motion to be observed. The conventional approach is to make the restoring force of the suspension as small as possible so that the mass can be used as an inertial frame against which the motion of the ground can directly be measured. When the period of the ground motion is longer than the free period, the suspended mass will follow the ground motion as the rest of the instrument does, so the differential motion becomes very small; the passband of the seismometer is naturally limited by the free period of the suspension.

Instruments with negative force feedback, or force-balance systems, sense the ground motion after the same inertial principle, but provide an additional electrostatic or electromagnetic restoring force that causes the mass to follow the motion of the ground (Figure 1). This eliminates most problems associated with the mechanical properties of the suspension. As long as the feedback system is efficient in keeping the relative motion of the mass small, the driving force is nearly proportional to ground acceleration and nearly independent of the circuit that controls it. The electric voltage or current that generates the force is used as the output signal. The precision of the sensor now essentially depends on only one component, the force transducer, in the same way as the precision of a potentiometric recorder depends on the quality of the potentiometer. Electromagnetic force transducers, as passive components, have an almost unlimited dynamic range and can be made very stable and linear. By charging a capacitor with the current that drives the transducer, we can obtain an output voltage that is proportional to ground velocity.

With modern electronics, it is not difficult to resolve extremely small relative movements of the seismic mass and to obtain any desired gain. The main problem in the design of broad-band sensors is to protect the sensor from environmental and internal disturbances to such a degree that the extremely small inertial forces in the order of 10^{-11} g which are associated with free oscillations of the earth are not masked by other forces acting on the mass, such as caused by thermal effects, buoyancy, magnetic fields, ageing of the spring or corrosion.

The dynamic range of a force-balance system depends on that of its feedback path, i.e., on the dynamic range of the components inserted

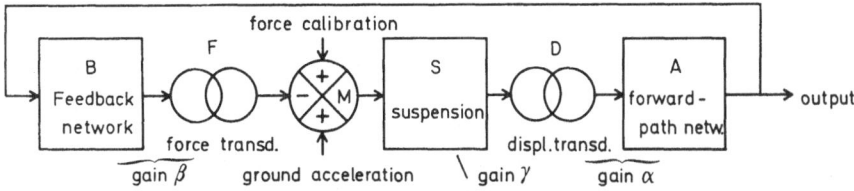

Fig. 1. Block diagram of a force-balance seismometer.

between the output and the force transducer. For highest dynamic range, active components should be avoided there. Unfortunately, active components are usually required in the feedback loop in order to shape the response.

Specific Considerations for VBB Instruments

The large relative bandwidth of very-broad-band sensors (four to five decades of frequency) requires a careful choice of the response. The most critical part, in terms of linearity and dynamic range, in a digital VBB seismograph is presently not the sensor but the digitizer. The sensor response must be chosen so that the requirements to the digitizer are minimized. A frequently used, although somewhat too sloppy, formulation is that the sensor must "pre-whiten" the seismic noise, i.e., the latter should be white when it enters the digitizer. The matter is discussed in more detail in paper [3].

HIGH-RESOLUTION DIGITIZERS

The digitizer of a VBB seismograph should cover a dynamic range of at least 140 dB, and it must have an excellent linearity and resolution because signals of greatly different amplitude may simultaneously be present in different parts of the VBB seismic band. Gain-ranging digitizers can easily provide the required dynamic range but are notorious for nonlinear distortions. If such a digitizer is used, care must be taken that the fundamental range is large enough so that microseisms and local noise do not force gain-ranging. Even then, nonlinear distortion of larger earthquake signals may occasionally be noticeable.

A perfect solution to the problem of digitizing VBB seismic signals has recently emerged in the form of fixed-range, 24-bit digitizers. Within a few years, 24-bit digitizers should become less expensive than present gain-ranging systems, and still offer a linearity and resolution two orders of magnitude better than these.

STATION PROCESSORS

A station processor is not strictly an essential part of a digital seismograph, yet seismic data acquisition systems are normally designed around a microcomputer for greater flexibility. The rapid developments in the field of storage media and telecommunications would otherwise make the system obsolete in a few years. Also, the amount of data generated by a VBB seismic system is considerable; a station processor incorporating data compression, intelligent event selection, and/or a signal-dependent choice of the sampling rate could greatly reduce the data volume.

In any case, it is desirable to derive from the primary data stream at (say) 20 Hz sampling one or two additional data streams at, say, 1 Hz and 0.05 Hz rates by digital low-pass filtration and decimation in the station processor, and to record these "long-period" and "very-long-period" data in separate blocks on the same medium as the primary stream. This will only insignificantly increase the data volume but greatly simplify data screening and the extraction of long-period signals. The low-rate data should be continuous even if the high-rate data are kept only for events.

VBB station processors and recording systems are being developed in various places, and at least one of them is now commercially available. It appears also possible to use one of the better personal computers as a

VBB station processor, although the task of writing the necessary software should not be underestimated. A recently proposed solution is that seismometers, digitizers, a timing system and a microcomputer for data formatting and buffering are combined to form a peripheral device to a personal computer, which then takes care of event detection, data storage and communications. The advantage of this concept would be that the latter functions, which are likely to evolve rapidly, are not tied to the specific hardware and software of the data acquisition system and can be modified by the user.

RECORDING MEDIA

A continuously recording three-component VBB station with a sampling rate of 20 Hz will produce between 40 and 100 Mbyte of data per week (depending on the amount of data compression). These data can at present be recorded on one C-300 data cartridge, or data from two weeks on a 6250 bpi computer tape. Developments in this field are however so rapid that it is impossible to recommend a specific solution. It is expected that the technology of optical discs will soon provide an alternative to magnetic recording. The small volume and greater durability of these media, as well as the rapid access to the data, would in fact suit seismological needs very well.

CALIBRATION

Modern feedback seismometers are so stable in their transfer properties that they need not be calibrated by the user; nevertheless it is desirable to be able to verify the transfer function. Conventional methods of calibration, such as analyzing the pulse or step response, are not always practical when the system has a large bandwidth, and are normally not precise enough to identify imperfections of the system, although the result would be good enough for most seismological applications.

Any method for the calibration of digital seismographs should have the following properties:

1. The analysis should be free of systematic errors, i.e., it should give exact results when applied to exact synthetic signals of finite length and finite sampling rate.

2. The method should not require that certain properties of the test signal, such as waveform, amplitude, frequency, or timing of transitions, are known with great precision or kept constant. Ideally, the method should work with any input signal that has sufficient energy in the frequency band of interest.

3. The method should be as insensitive as possible to quantization errors, seismic and other noise, instrument drift, and nonlinear behavior of the instrument; it should however permit the identification of such effects.

The following general concept seems to comply best with these requirements. Both the test signal and the output signal of the seismometer are digitally recorded. The seismometer is modelled in the computer as a linear filter whose form and parametrization can be chosen according to the problem. If realized in time domain, the filter could even include nonlinear elements. A synthetic output signal is constructed by applying that filter to the recorded input signal. The rms deviation between the synthetic and the observed output signal is determined and iteratively minimized by adjusting the model, using one out of a number of well-known inversion algorithms. The result of the calibration procedure is thus a computer model for the transfer properties of the seismometer. Its accuracy can be assessed by plotting the residual error and inspecting it for nonrandom components, such as linear and nonlinear distortions.

This concept is obviously much more flexible than other methods where predefined input signals are used and where the accuracy of the calibration depends on the precision of a specific signal generator, or on its synchronization with the digitizer. It must also be noted that the most common calibration methods rely on step functions as test signals. These functions, and the resulting output signals, tend to generate serious nonrandom quantization errors. Our concept permits the selection of test waveforms that randomize the quantization error, as well as the duplication of other calibration methods that require specific waveforms.

TECHNICAL LIMITATIONS

In this final section we will outline some technical limitations encountered in the realization of broad-band seismographs. We consider only the case that the resolution of small signals has priority over the ability to record large signals. In other words, we assume that the output signal of the sensor is so large that no small signals are lost in the digitizer.

We use a two-dimensional representation of signal ranges in terms of amplitude and frequency (Figure 2), noting however that such a representation has no exact meaning because noise levels must be measured in frequency domain and clip levels in time domain. As a compromise, the figure and the discussion refer to peak-to-peak signals in a constant relative bandwidth of one-sixth decade.

The figure illustrates the range of signals that can be handled with different broad-band systems. The minimum system would correspond to a combination of standard (WWSSN) short-period and long-period seismographs; this is considered as a minimum requirement for a system to be in the

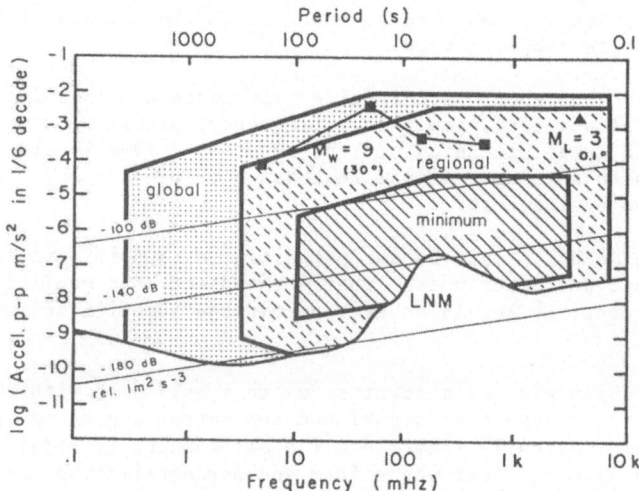

Fig. 2. Range of signals that can be handled with different broad-band systems. See text for explanation.

broad-band category. For today's technical standard, this would however be a poorly performing system. The area labelled "regional" indicates a signal range that can be handled without major problems. The "global" range illustrates the design goals for the planned IRIS global network.

The proposed signal ranges are limited partly by seismological considerations and partly by technical constraints. Seismologically, it seems not to make sense to extend the signal range below 0.3 mHz (left side) or below minimum ground noise (bottom). The ability to record large signals (top) is limited by the dynamic ranges of the sensor and of the digitizer, which depend on the state of semiconductor technology. The limitation at high frequencies (right side) has mainly economical reasons - the cost of data storage and processing increases with the sampling rate; there is also the technical limitation that 24-bit digitizers are presently not available for sampling rates higher than 20 Hz.

There are additional technical constraints that require a compromise between two desiderata. The resolution of small signals at high frequencies is limited by the resolution of the displacement transducer. At low frequencies, the resolution is usually limited by intrinsic and ambient noise affecting the mechanical sensor. Large signals at high frequencies require a large feedback force, and if we want to improve the performance there, power consumption and the generation of heat in the seismometer become a problem. For large signals at low frequencies, a limit is presently set by the dynamic range of the semiconductors used in the feedback circuit.

REFERENCES

1. E. Wielandt, Design principles of electronic inertial seismometers, in: "Earthquakes - Observation, Theory and Interpretation", Proc. Scool of Physics Enrico Fermi, Varenna, 85. Corso, Soc. Italiana di Fisica, Bologna, pp 354-365 (1983).
2. E. Wielandt and G. Streckeisen, The leaf-spring seismometer: design and performance, BSSA 72 No. 6, pp 2349-2367 (1982).
3. E. Wielandt and J. M. Steim, A digital very-broad-band seismograph, Annales Geophysicae, B4, No. 3, pp 227-232 (1986).

PORTABLE SHORT PERIOD VERTICAL SEISMIC STATIONS

TRANSMITTING VIA TELEPHONE OR SATELLITE

G. Poupinet, J. Fréchet and F. Thouvenot

LGIT, Observatoire de Grenoble
BP 53X, F 38041 Grenoble
France

INTRODUCTION

Short period seismology is the main technique for the investigation of the seismic activity and for imaging the lithosphere. Natural earthquakes are an efficient tool to investigate the earth: they provide a way to understand how rocks break through the accumulation of stress in some particular places, as well as a tool to map the P- and S-velocity structure in the medium in between sources and receivers. The wealth of information accessible by recording earthquakes is limited by our ability to deploy dense arrays of sensors in geologically interesting regions.

We briefly review the principal data acquisition techniques used in reflection and refraction seismology and for monitoring local or distant seismicity. Then, we present the dilemma that arises when designing a multipurpose portable seismological network. We describe the simplified digital stations which have been built within the framework of the LITHOSCOPE project, a French program for the study of the lithosphere by means of a short period portable array. These stations are modular: they run independently, storing data in a large memory, or are connected to a phone modem or to a METEOSAT satellite transmitter. Our design resembles recorders recently proposed by several manufacturers and universities.

I.   COLLECTING DATA FOR SHORT PERIOD SEISMOLOGY STUDIES

I.1. Arrays for Reflection and Refraction Seismology

In both reflection and refraction experiments, the time of occurrence of the source and thus of the recording windows is preset. Seismic vertical reflection experiments, like COCORP or the Canadian and European equivalents, investigate hectometric features and describe the geometry of faults within the crust and exceptionally in the upper lithosphere. The possibility to track major faults from the surface to deep in the earth helps to understand how nappes have been displaced during continental collision or the mechanism of formation of extensive features like grabens or sedimentary basins. COCORP uses the standard oil industry equipment, but records longer time windows, to investigate the deep crust. Usually five trucks generate vibration sweeps that are recorded on a linear array composed of several hundred 10 Hz geophones, aligned on a few

kilometers. All geophones are connected via cables to a laboratory truck
and vibration sequences are stacked and recorded on 9-tracks magnetic
tapes, which are later on processed in a laboratory equipped with
mainframe computers. A seismic section is produced: it is a plot of
seismic impedance as a function of depth and distance along a profile.

In 3-dimensional studies, geophones are spread on a surface and
groups of geophones are gathered on cassette recorders, which are
triggered by a radio signal telemetered from the laboratory truck. Radio
trigger is a function that should be incorporated in a multipurpose array.

On a larger scale, in refraction and wide angle reflection, a few
large explosive shots are recorded by about a hundred stations ranging at
distances up to 300 kilometers. The seismometer period is usually 0.5 s.
Preset shot time windows turn on independently each recorder: the vertical
and eventually the two horizontal ground velocities are stored on a FM or
digital magnetic tape as well as a time code. Cassettes are played-back
and merged later on. Seismograms are plotted as a function of distance.
Hodochrones are interpreted in terms of a layered structure and of lateral
heterogeneities. These techniques have been extensively applied in
Western Europe where many Universities bought identical FM refraction
stations, manufactured by Lennartz, a German firm; this de facto
standardization greatly simplified the necessary exchange of analog data.
Europe has been crisscrossed by refraction profiles and the depth of the
Moho has been mapped beneath all the continent.

The USGS, Menlo Park, has built 100 refraction recorders that are
easier to operate than previous equipment (Healy and Hill, 1984). The
array is deployed by 5 observers, each one driving a pick-up truck
containing 20 stations. Many interesting regions have been investigated
with this array, mainly in Western United States, Saudia Arabia and
Alaska. This equipment prefigures the type of operation that should be
retained for a large array. The Earth Physics Branch of the Canadian
Department of Energy, Mines and Resources has developed a similar
instrument, incorporating more recent electronics components: it is
manufactured by EDA, a Canadian firm. This new refraction instrument,
called PRS 1, is a field microcomputer with a large dynamic memory: the
data acquisition program is loaded in the computer just before the
deployment in the field. Data are stored in the large memory and
retrieved by a portable computer. This new concept eliminates any
mechanical device, like a tape drive.

I.2. Temporary Stations for Seismicity Studies

When recording natural seismicity, the time of occurrence of
earthquakes is not known in advance. Therefore there are two strategies:
continuous recording or event detection.

Continuous recording is the best technique for studying low magnitude
events; the Sprengnether MEQ 800 smoke paper recorder or similar
instruments have been extensively used for monitoring seismicity. Between
10 to 40 smoke paper recorders are deployed in a region and each station
is serviced every one or two days. Paper is replaced and a master clock
comparison is recorded. These experiments are very heavy in manpower and
cannot be planned for periods longer than one month.

In his study of Yellowstone and in later work, Iyer (1975)
continuously recorded the ground motion on tape drives with a 5 days
autonomy. A small German firm, Reimer, manufactures a continuously
recording station with an autonomy of 20 days. Tapes are then played back
on paper or digitized in order to pick first arrival times. Continuous

recording in direct mode is common in ocean bottom seismology and is a robust technique (Prothero, 1984). The main drawback is a low dynamic range and a high cost for handling tapes when they are returned to the laboratory. Tapes have to be played back at a speed higher than the recording speed, then the parts containing events are digitized. The data from each station are read on a minicomputer and merged to display them on a graphic terminal, where P-phases are picked by an operator. Previously, a paper record was produced in the time windows of interest and times were visually picked. The main interest of the continuously recording system is that it is always possible to return to the original data and to retrieve seismograms that would not have been detected by a trigger. However the system, as it is used now, may not be practicable for a large number of sensors, as it would ask a very large manpower to be operated. An upgrading of continuous recording may soon benefit from the improvements in portable microcomputer storage capacities. However, the operation of mechanical devices like tape drives, is difficult in harsh environments, when temperature and humidity are extreme.

Another possibility is to continuously radio telemeter to a central site and to record all traces on a multichannel analog or digital tape recorder. Digital telemetry gives a better dynamic range than analog telemetry but uses a larger bandwidth. Topography limits the geographical coverage of ground telemetered network; many relays have to be installed in mountainous regions. A standard radio telemetered network with more than 100 digital stations, uses a wide frequency band; a frequency and time multiplexed system has to be considered.

Event triggered magnetic recorders are often used in seismicity surveys: a threshold detection level is set to start storage on a cassette with a small time lag, which allows to record the beginning of the P-wave train. Lennartz, Kinemetrics, Sprengnether and most manufacturers build 3 or 6 components event recorders. The generalization of low-power digital chips facilitates the set up of triggering algorithms in electronics as simple as refraction stations. EDA proposes an event recorder, the PRS 4, derived from their refraction recorder: it stores 3 components and a time signal on a dynamic memory of up to 6 Megabytes.

I.3. Permanent Microearthquakes and Large Aperture Arrays

Lee and Stewart (1981) describe permanent microearthquake networks in detail. They present the Central California Network which is presently the largest array installed with the purpose of monitoring earthquakes on a regional scale. About 400 stations are continuously telemetered to a recording and processing center in Menlo Park, California.

The field package is composed of a 1 Hz vertical seismometer, an amplifier, a voltage-controlled oscillator, an automatic calibration system, a telemetry interface connected to a phone line and a power supply; the electronics are compact and very low-power. Seismic signals are carried along dedicated phone lines by frequency multiplexing 8 channels. Many small telemetered networks were based on frequency modulation, as it provides a very large storage or transmission capacity. In the USGS network, multiplexed signals are recorded on 3 video analog tape-recorders in direct mode, 14 tracks (i.e., 112 stations) on a one-inch tape. A radio time code (WWVB) and a reference sinusoid are simultaneously recorded. The analog tapes used to be processed off lines through a very complex operation, involving manual selection of events, creation of events tapes, and for special studies digitization on a minicomputer. Routine P arrival times and magnitude were read from microfilm records. Recently the USGS network has been upgraded: all traces are digitized by a dedicated computer, which performs an automatic

P-picking in real time (Allen, 1982). Earthquakes are located in real-time and the long and tedious phase reading work is avoided. Data are stored on 9-tracks tapes. The limitation of the system comes from the low dynamic range of the transmission, close to 50 dB, due to the FM transmission. The cost of operation is high because all lines are permanently rented from phone companies.

Another example of a dense network is the NORSAR array: 40 stations (it started with 120) are located in Southern Norway with the purpose to monitor nuclear explosions. Data are transmitted on permanent phone lines to a computer that detects tiny distant events through multiple channels processing techniques; they are stored continuously on 9-tracks tapes, with a digitization rate of 20 points per second. The NORSAR array is particularly adapted for teleseismic studies.

The USGS CALNET and NORSAR networks provided the P-residuals sets for the first three-dimensional inversions of crustal and lithospheric structures (Aki and Lee, 1976; Aki et al., 1977).

II. A SEISMOLOGICAL MODULAR SYSTEM DEVELOPED FOR LITHOSCOPE PROJECT

II.1. Choice of a Short Period Portable Multipurpose Array

Our objective is to study seismicity and the 3-dimensional structure of the lithosphere beneath any geologic object, like a mountain range, a rift, a craton, a sedimentary basin ... A region has to be covered by a dense array of seismographs. A distance between stations of the order of 10 kilometers is adapted to most regional experiments. Long time periods of observation are needed to get a reliable estimate of seismicity and a proper azimuthal and incidence coverage for distant earthquakes. In volcanology, a network should be installed on a surface of the order of a few hundred square kilometers to record a large number of local events, during an eruption. On the other hand, the same array has to be easy to operate for refraction and wide-angle reflection experiments.

Several research groups are planning to deploy dense short period arrays. The choice of a universal station is difficult because an instrument that would cover all seismological techniques, from vertical reflection to teleseismic studies, is very complex and expensive. Thus, the limited resources devoted to seismicity and structural studies make it impossible to build a large number of such stations. Spatial aliasing remains the principal limit in most seismological studies and can only be counterbalanced by increasing the number and density of sensors.

PASSCAL, a group of American Universities (Program for Array Seismological Studies of the Continental Lithosphere, IRIS, 1985) is developing a universal instrument with impressive technical characteristics (Williams, 1987). It is composed of:

- a Data Acquisition Subsystem that digitizes up to 6 channels at 1000 points per second. Each channel is then digitally filtered in order to store it at a sampling rate between 1 and 1000 samples per second. Several event detection algorithms are possible, because the CPU is a powerful Motorola 68HC000 microprocessor. A 4 Megabytes CMOS battery backed-up static memory allows a temporary storage of the data.
- a Time Keeping Subsystem will maintain a precision less than 1 ms. It will be radio synchronized
- an Auxiliary Recording Subsystem will store 60 Megabytes and soon 120 Megabytes on a portable tape drive.

12

- a Field Set-Up Terminal will set up parameters, communicate and visualize the ground motion.

PASSCAL intends to buy 1000 recorders. For budgetary reasons another simpler recorder, retaining the philosophy of the EDA PRS 4 or any similar instrument, may be chosen in complement to the previously described one.

The best array for long duration experiments would be one that can be installed rapidly in the field and would send data regularly to the investigator's laboratory. The field maintenance would be reduced to one visit per year. Such equipment would require a link between any place on earth and laboratories. Satellite telemetry is adapted for monitoring remote and difficult access zones and phone telemetry for long duration studies close to a permanent or temporary laboratory. The basic design of satellite or phone telemetered stations are similar. When transmission is not permitted, on the site recording has to be envisaged and the autonomy of each station should be months.

II.2. Description of the Electronics and the Basic Software

A basic seismological electronics has been designed within the framework of the LITHOSCOPE project. This equipment is built by an electronic firm, specialized in satellite telemetry, CEIS-ESPACE. It is intended to be a one component digital electronic module interfaced to different peripheral devices. Our Seismological Acquisition Module (SAM) is a 1 Hz seismometer, an analog board and a microprocessor-controlled board (Figure 1). A single board design is possible (e.g., Hayman, 1988).

Analog board. A 1-Hz seismometer, usually the Mark Product L4 or any equivalent light and easy to install geophone, is connected to a preamplifier-amplifier board. The gain-ranging amplifier has a maximum

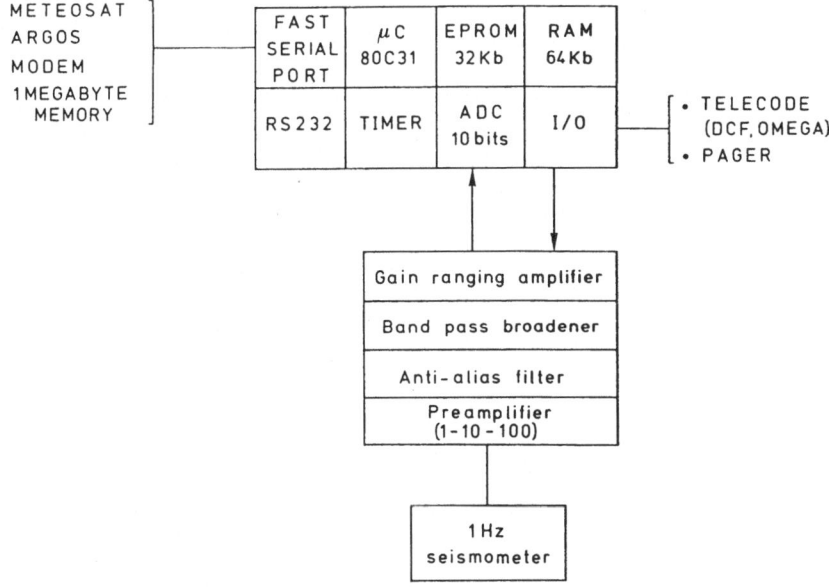

Fig. 1. Functional diagram of a basic Seismological Acquisition Module, composed of an analog board and a microprocessor board. This unit digitizes and processes the seismic signal; it communicates with other boards like a 1 Megabytes CMOS battery backed-up memory, a phone modem, a METEOSAT transmitter or a radio synchronization receiver.

gain of 96 dB, with 6 dB steps. A 6-poles anti-aliasing Butterworth filter cuts off at 25 Hz. The band pass of the geophone is broadened electronically from 1.0 to about 0.2 Hz. This device is implemented to improve teleseismic detection with a cheap seismometer and for future developments to detect teleseismic S-waves. This filter is by-passed in local seismicity applications. The signal is then digitized by the microprocessing board.

Digital microprocessor board. A low-power processing board has the following characteristics:

- Intel 80C31 microprocessor
- up to 32 Kbytes EPROM
- up to 64 Kbytes RAM
- a calendar-clock chip
- Analog to Digital Conversion on 10 bits on 3 channels (we only use one)
- RS 232 communication with an external computer
- fast speed serial communication through the 80C31 with other boards. This fast link replaces the usual bus around which any electronic device is built.
- a watchdog which resets the microprocessor in case of a spurious behavior, for example, due to electromagnetic perturbations.

The software has the following functions:

- clock. The quartz of the time-chip is not temperature compensated but is a good quality clock-making quartz. Up to 14 turn-on times are stored in the clock-chip and can start digitization for refraction.

- digitization and gain-ranging. The amplified seismometer output is digitized. In the absence of an event, the gain is set according to the long term average (LTA), which is a low-pass filter on the absolute value of the noise. The microprocessor board sets the gain of the analog board.

- detection of seismic events. In the present version, a short term average over long term average-STA/LTA-algorithm is applied to trigger. The constants of the STA and LTA and the threshold value are adjustable. When an event is declared, the time, 256 samples of noise and 768 or more samples of signal are stored in RAM. A pseudo-magnitude is computed, by summing the absolute values of the event. The event is declared over when the noise level returns below the LTA value for more than a given number of seconds. A Walsh transform algorithm (Goforth and Herrin, 1981) has been written and tested in the laboratory. It will be implemented on future EPROMs and we intend to apply a frequency criterion for the recognition of distant earthquakes (Evans and Allen, 1983).

- sorting of events according to pseudo-magnitude. The last largest events are stored in memory. The smallest event in the buffer is removed if its magnitude is smaller than that of the last detection.

- RS 232 software for communication with a field terminal or a portable computer to set the detection algorithm parameters and to check or retrieve data.

- fast communication software for the 80C31 serial output.

By itself our basic module has two major drawbacks:

- its time base is not precise enough for a long duration stand-alone experiment.

- 64 Kbytes of memory is at most 10 minutes of signal at 100 points
  per second.

A quartz clock radio synchronized on DCF, HBG or OMEGA is
manufactured by the Observatoire de Neuchâtel (Switzerland) (Bonanomi,
1986). It maintains a precision better than 5 ms and can drive the clock
of the acquisition module. When the basic module is connected to a modem
or satellite emitter its clock is periodically compared to a master clock.

1 Megabyte battery backed-up static memory board. A 1 Megabyte solid
state static memory board has been built to store data. It communicates
with the basic unit through the 80C31 serial link. Data are first sorted
as a function of mean magnitude and stored in the 64 Kilobytes RAM. Then,
they are transferred in the 1 Megabyte board at regular time intervals,
for example, every day. The station is filled with the largest events of
the last week.

The data are retrieved from the memory at a fast speed through the
microprocessor serial link and temporarily stored on a retrieving board,
similar to the memory board, except for its communication software.
This retrieving board plays also the role of a master clock. The data
are then transferred onto a PC compatible computer. The microcomputer
can also be directly connected to the 1 Megabyte memory to unload it at
a speed of 19200 bauds. A portable microcomputer with a disk of 20
Megabytes is used to set-up the station and to unload the data.

III. CONFIGURATIONS FOR A PORTABLE SEISMOLOGICAL ARRAY

Our system is modular and the basic unit can be connected to
different types of receiving, emitting and storage devices. We present
four configurations:

- Seismological Acquisition Module + 1 Megabyte memory + Telecode
  (DCF or OMEGA)
- Seismological Acquisition Module + 1 Megabyte + pager (local radio
  receiver)
- Seismological Acquisition Module + phone modem
- Seismological Acquisition Module + METEOSAT satellite transmitter

Other configurations are possible.

III.1. Independent Seismic Recorders Array

A station composed of the basic electronic unit and the 1 Megabyte
board can operate independently. For short duration experiments, for
example, in refraction, the clocks can be time synchronized by a single
master clock comparison at the installation and when data are recovered.
For long duration, we attach a Telecode radio synchronized clock (DCF in
Europe and OMEGA elsewhere). In refraction experiments, the system is
activated at preset times. For seismicity studies, the event detection
algorithm records earthquakes and stores the largest in the mass storage
at regular time intervals.

III.2. Radio-Triggered Network

A local radio receiver (pager or beeper) is connected to a basic unit
and a 1 Megabyte memory. For array deployment on the 10-kilometers scale
or larger in flat regions, radio triggering is an efficient solution.
Commercial pagers are easily adapted to seismic surveying. Hirn et al.
(1987) recorded a seismo-volcanic crisis on Etna with an array of 30

three-components stations, originally designed for refraction. A small telemetered array detected earthquakes and triggered the FM recorders. Magnetic tapes were replaced every day, because of the large number of events. For teleseismic studies on a kilometric scale, stations could be serviced every few weeks. The triggering radio signal also time synchronizes the clocks. All stations record the same events: this is important during a volcanic crisis, with thousands of microearthquakes and tremor sequences. This type of telemetry works only in direct view, but could be experimented on longer distances: HBG, operated by the Observatoire de Neuchâtel, transmits a code, that could command each recorder, when a distant earthquake is detected by a pilot observatory.

III.3. A Phone-Interrogated Array: SISMALP

SISMALP is a permanent network of 30 stations which is being installed in the Western Alps for a study of the regional seismicity as well as for the investigation of the lithospheric structure. 4 stations have been in operation since August 1986 (Figure 2). Each station is composed of the basic module plus a modem.

Usually permanent phone lines are rented to transmit seismic signals continuously. This solution is expensive when the number of stations is large. The use of non-dedicated dial-up phone lines decreases the cost by a large factor. Such dial-up networks are used in hydrology and meteorology.

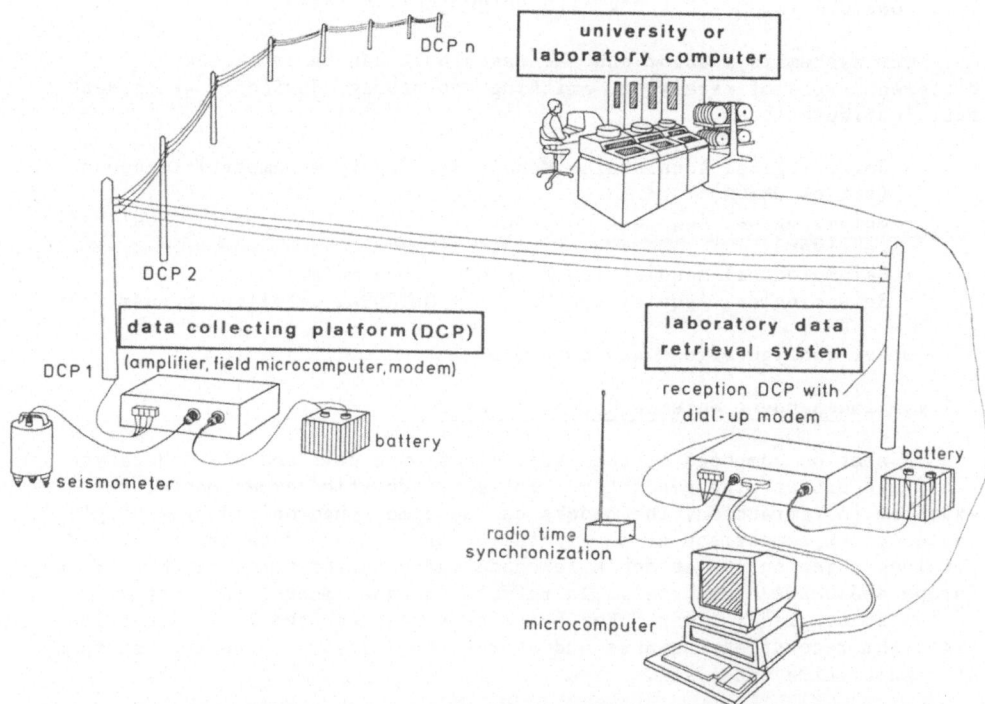

Fig. 2. Functional diagram of a phone interrogated seismological network. Each event recorder stores the last largest events in the static RAM. A central station, piloted by a microcomputer, dials up alternatively the phone number of each station, then time synchronizes it and collects seismograms. The operational cost is limited to the duration of communications with the slave DCPs.

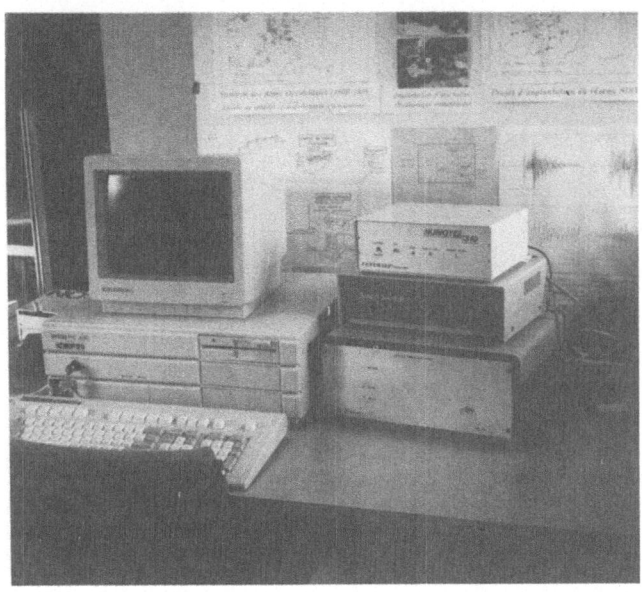

Fig. 3. View of the microcomputer that automatically controls the
operation of a dial-up phone array. On the bottom right of the
picture, we see the master Data Collecting Platform, which is
radio time synchronized and which dialogs with slave DCPs, spread
in the field. The computer calls the slave DCP phone number
through an automatic dial-up system (NUMOTEL) and then recovers
the data transferred to the master DCP.

We have developed a low-cost battery-powered digital seismic station
transmitting its data through a non-dedicated phone line using a 1200
bauds field modem. This modem, specially signed for hydrological
applications, is activated only during the transmission. The data
collection is made at periodic intervals, ranging from hours to days,
depending on the seismicity level. Each Data Collecting Platform (DCP)
dialogs with a central station built around a microcomputer (IBM-PC
compatible) connected to a 1200 bauds modem (Figure 3). The dialling is
initiated either by the laboratory station, or by the slave field
platforms. The dialog software performs the following tasks:

- synchronizing the slave and the master clock (which is referenced
  to an accurate radio time signal like DCF), and measuring the time
  delay on the phone line.
- transmitting the (up to 64 Kilobytes) data stored in the DCP
- modifying the functional parameters of the DCP, as the sampling
  interval, the memory partition, the gain level, the sensitivity,
  etc.

III.4. Satellite Transmitted Seismological Networks

Our basic module is connected to a METEOSAT (or ARGOS) emitter. It
transmits essential data from the field to our laboratory. Figure 4 is a
schematic diagram of a METEOSAT seismological network. We review the
characteristics of the satellites available for collecting environmental
data, as they are seldom used in seismology.

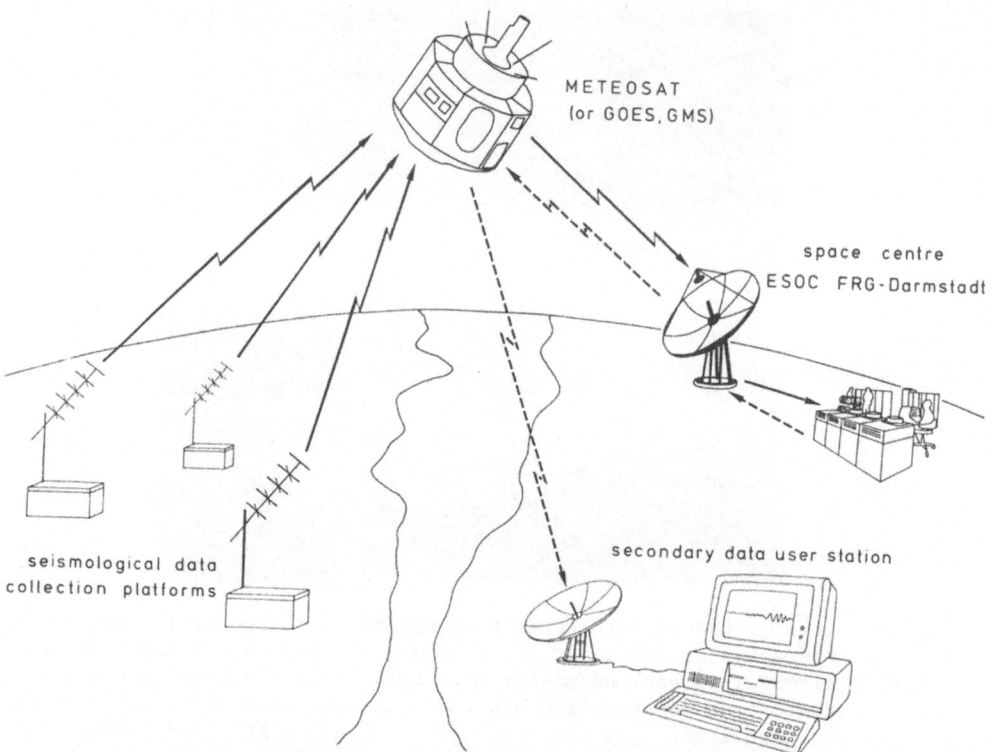

METEOSAT
(or GOES,GMS)

space centre
ESOC FRG·Darmstadt

seismological data
collection platforms

secondary data user station

Fig. 4.   Functional diagram of a METEOSAT seismological network.
Seismological Data Collecting Platforms send data to the
geostationary satellite that relays them to the Space Center.
The Space center stores the data and sends them back to the
satellite which dispatches them to users equipped with low cost
secondary user stations. A portable seismological network can
work anywhere on the globe, except close to the poles, where
ARGOS is the only system able to collect data. The main drawback
of the system is the actual limitation in transmission capacity
of the satellite; 1200 bauds transmission will soon be
implemented on several GOES channels.

A.   Satellite Links

    Phone satellites of the INTELSAT or INMARSAT series are accessible
from many countries; it would be an interesting solution, if it was
possible to transmit with low-cost emitters. For permanent stations, the
wide spectrum technique gives access to INTELSAT with a low-cost terminal
(IRIS, 1985; Berger and Lusignan, 1986). Phone satellites are used in
seismology to transfer high data flow already multiplexed, for instance in
the RSTN network or from NORESS in Southern Norway to the United States.

    Environmental satellite data collection systems have been built for
meteorology and hydrology. They transmit a small number of bytes at a low
speed. A tsunami alert system (Clark and Medina, 1976), in which P
arrival-times are transmitted on GOES and a volcano monitoring system
(Endo et al., 1974) were experimented by the USGS more than 10 years ago.

    Satellites are expensive to build and launch. A data collecting
system is expected to last many years and is planned during at least 5

years. There are two main systems of environmental data collecting systems which were designed 10 years ago (WMO, 1982):

- geostationary satellites of the GOES, METEOSAT, GMS series
- polar orbiting satellites of the NOAA-N series

a. Geostationary satellites. Five geostationary satellites are positioned above the equator at an altitude of 36000 km. They transmit pictures of the cloud cover of the earth and collect data from environmental sensors. A publication from the World Meteorological Organization (1985) describes their technical features. The five geostationary satellites are compatible and provide so-called regional and international data links. 2 GOES are covering North America and the pacific, METEOSAT the Atlantic, Africa and Europe and GMS the Far East. Each satellite sees about one third of the surface of the earth (see Figure 5).

METEOSAT has 66 channels on which messages of at most 5192 bits are transmitted every 3 hours. The European Space Agency allows to collect more messages, so that a capacity of 1.8 Megabyte per year per DCP is a minimum. An alert channel is reserved for short messages of less than 64 bytes, when a threshold is overpassed. Each self timed DCP is attributed a frequency between 402-402.2 Mhz and a time window. Data are collected by the European Space Operating Center In Darmstadt (FRG) and dispatched in nearly real time on the WEFAX channel of the satellite, which disseminates images. Data are received on a Secondary Data User Station (SDUS), built with a 1.2 meter antenna, a low cost receiver, a bit synchronizer and a microcomputer. In 1988, the precise date of reception of messages will be transmitted on the SDUS: DCPs will then be time synchronized by comparing the DCP clock at emission with the satellite clock at reception. Geostationary satellites are extremely interesting

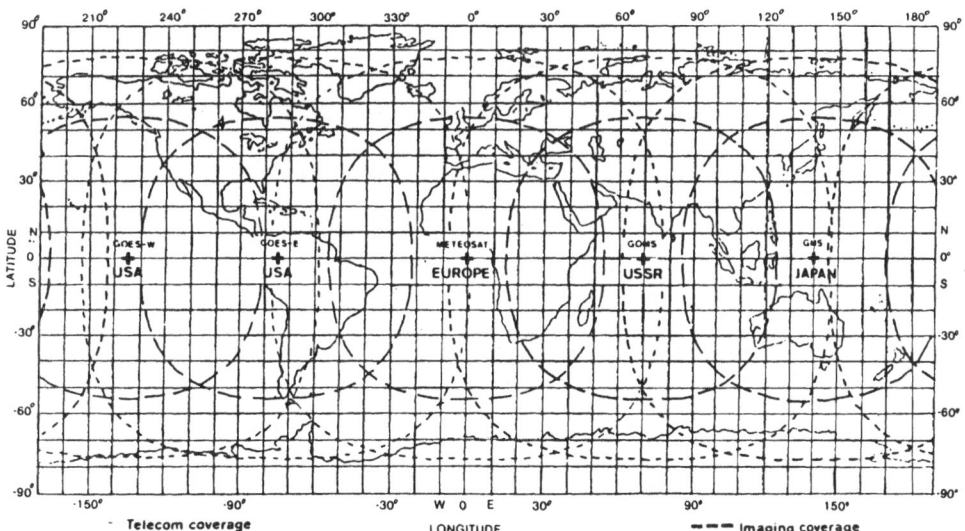

Fig. 5. Visibility zones of the 5 Meteorological Geostationary satellites. 2 GOES, GMS and METEOSAT are in operation. Their data collecting system are compatible, despite some small differences in the format accepted for collect and distribution. A seismological DCP working on one satellite can work on the others with minor software modifications.

for seismology, particularly for mantle tomography and for monitoring regional earthquakes: their visibility zone is large and they handle several thousands DCPs. Distant users have fast access to the data through the SDUS for a small investment.

b. ARGOS. ARGOS is a localization and data collection system. It is installed on two NOAA spacecraft which are orbiting on polar sun synchroneous orbits at an altitude of about 800 km. ARGOS DCPs are simple compared to METEOSAT: they emit on a single frequency (401.65 Mhz) and do not need a precise timing. They transmit in the so-called random access mode, 32 bytes every 150 seconds, but the DCP does not know if it has been received. For normal visibility conditions, an average of 300-400 bytes is transmitted daily at latitudes close to 45 degrees. This capacity decreases on the equator and increases near the poles, where ARGOS is the only collecting system: geostationary satellites do not see the poles. On the average, 100 Kilobytes are transmitted per year and per DCP. The world-wide coverage of the system may be important for some applications. ARGOS handles more than 1000 DCPs, but performances decrease when too many DCPs interfere in the same geographical area. Reception of messages in the satellite are clocked with a 20 milliseconds precision and this time is transmitted with the collected data. Messages arriving in the satellite are relayed on a VHF down link. A direct read out station receives about 50% of the messages within a circle of 1000 kilometers. The data are also stored on board and then transferred to the ARGOS processing Centres in Toulouse (France) and in the United States; they are archived on tapes and distributed through telex and phone lines. A delay of about 3 hours is required to access data from any point on the globe. Monthly tapes are handy for experiments that do not require real time observations.

c. Future requirements for seismology-oriented satellite data collecting systems. The main functions of a seismological data collecting system are:

- to collect data
- to dispatch them as fast as possible.

Meteorological satellites have actually two drawbacks for seismology:

- they have not been conceived to time synchronize DCPs, through emission
- their transmission capacity and speed is very low.

Time synchronization has been implemented on various satellites (GOES) but it requires one radio receiver per station. We need:

- a time synchronization through emission better than 1/1000 s. The datation of the time of reception of messages on board the satellite is a reliable time synchronization technique when DCPs remain unattended for several months
- an increase in transmission capacities and the possibility to collect a large number of DCPs.

METEOSAT receives data at 100 bauds, the speed of TELEX. The next GOES will allow transmission at 1200 bauds, a speed comparable to the actual non-dedicated phone network.

ARGOS uses very simple and light emitters and is accessible worldwide. Its time synchronization could be improved just by changing the satellite clock time increment. Its transmission capacity could be enhanced by using several frequencies.

B. <u>Our Experiments with Seismological Satellite Transmitting DCPs</u>

a. <u>ARGOS seismological Data Collecting Platforms</u>. ARGOS has a very low data transmission capacity, so that its use in seismology is limited to:

- time synchronize DCPs
- transmit P-picks (about 100 per day) or P-wave forms (about 6-10 per day).

There is no possibility to transmit a complete seismogram with ARGOS, but a local storage on the 1 Megabytes RAM can complement the ARGOS DCP.

Our first experiment with ARGOS was related to volcanology: we wanted to monitor microearthquakes on a volcano from large distances. We started with an analog P-picker connected to ARGOS (Poupinet and Glot, 1982; Glot and Poupinet, 1987). P-arrivals are coded on 4-bytes, with the time of the P-pick, the polarity and quality of the signal and the first zero-crossing. The purpose is to locate earthquakes, obtain fault plane solutions and eventually get some idea on the frequency of the onset. Up to 14 events are stored and transmitted in two consecutive messages. 10 ARGOS P-pickers have been successfully operated on Mount Etna (Italy) in complement to the University of Catania telemetered network. Our minicomputer gets data automatically every night from the ARGOS Centre in Toulouse and computes hypocenter locations.

The P-picker in its original form was not reliable for low-frequency events, like distant earthquakes. For teleseismic applications, we need to correlate wave forms from one site to the other, in order to obtain reliable relative arrival times. A microprocessor system transmitting P-wave forms, was built and tested in the French Alps (Poupinet et al., 1985). To improve on our first design, the gain-ranging amplifier and the seismological board described in II-1 were built. They are interfaced to both ARGOS and METEOSAT.

ARGOS use is limited for seismology but may find applications for experiments near the poles, at sea and in locations where a geostationary satellite is difficult to access.

b. <u>METEOSAT seismological Data Collecting Platform</u>. GOES, METEOSAT or GMS have a transmission capacity that allows to construct a general purpose short period seismic station, sorting and storing events before it transmits them.

A basic seismological acquisition module is connected to a frequency synthetized METEOSAT emitter. Self-timed messages are transmitted on a frequency different from alert messages. In the METEOSAT seismological DCP, each event is digitized at 100 points per second and stored on 1024 points with a variable time rate. The last 63 largest events are buffered. We transmit 1 or 2 triggers every hour. Each message of 638 bytes includes:

- the time of emission on 5 bytes
- the voltage of the battery
- a status byte giving the number of events transmitted and the gain of the LTA
- the number of events recorded in RAM
- 2 blank bytes.

Then 1 or 2 identical zones formatted as follows:

Fig. 6.   a. Global view of a seismological METEOSAT Data Collecting
          Platform. The antenna is the large black tube seen in the
          foreground.
          b. Electronic box of the METEOSAT DCP. On the left, we see the
          METEOSAT frequency synthetized emitter and on the right, the
          analog board in its metal casing and the digital board. An
          internal lithium-cadmium battery is recharged by a small solar
          panel.

- the time of the event coded on 5 bytes
- a status byte for the transmission of: calibration test of the seismometer; the quality of the reception of the radio time synchronization top; the digitization rate; and the value of the gain before and after detection
- 1 correction code for the previous 6 bytes
- 76 bytes of data preceding triggering
- 230 bytes or 550 bytes of signal.

The largest events are transmitted first. When the number of events in memory is less than 2, a sequence of noise is transmitted. Every n hours a calibration test is performed and transmitted. Differential data compaction is easy to install on the DCP but the gain is of the order of 2.

In the METEOSAT prototype, a radio receiver synchronizes the clock; this will not be needed in 1988 when ESA will dispatch the time of reception of messages with a precision of one millisecond. Our first METEOSAT prototype DCP was installed in the Pyrenees in July 1986. Figure 6 presents the METEOSAT DCP composed of an electronics case, an antenna, a solar panel and a seismometer.

c. Data retrieval and Secondary Data User Station. For METEOSAT, seismograms are received:

- in ESOC which stores them and sends magnetic tapes
- on a SDUS, or direct reception station, where they are decoded and plotted in near real-time. A network can work in the entire visibility zone of METEOSAT. Figure 7 shows a sequence of seismograms and noise windows collected during 2 days. The numbers on this diagram are transmission and detection hours.

Large magnitude earthquakes coded on short messages are transmitted in real time on the alert channel, on a frequency different from the self timed transmission channel. The SDUS is built around a WEFAX meteorological image receiver and a personal microcomputer. Seismograms are decoded and plotted on the microcomputer and earthquakes can be located automatically in nearly real-time.

CONCLUSION

Research projects in seismological tomography are the incentive for the development of new short period portable instruments. The goal is to deploy several hundred sensors on any interesting geological feature and to record shots and earthquakes. Two types of instruments are proposed:

- very sophisticated ones able to record 6 components in vertical reflection as well as in teleseismic tomography (PASSCAL choice).
- more simple, one or three components event recorders.

Both instruments are versatile because they are controlled by a microprocessor.

We have presented a version developed within the framework of the LITHOSCOPE project, a French cooperative group, devoted to the study of the deep structure of the continental lithosphere by means of a short period seismology array. The LITHOSCOPE instrument is composed of a basic

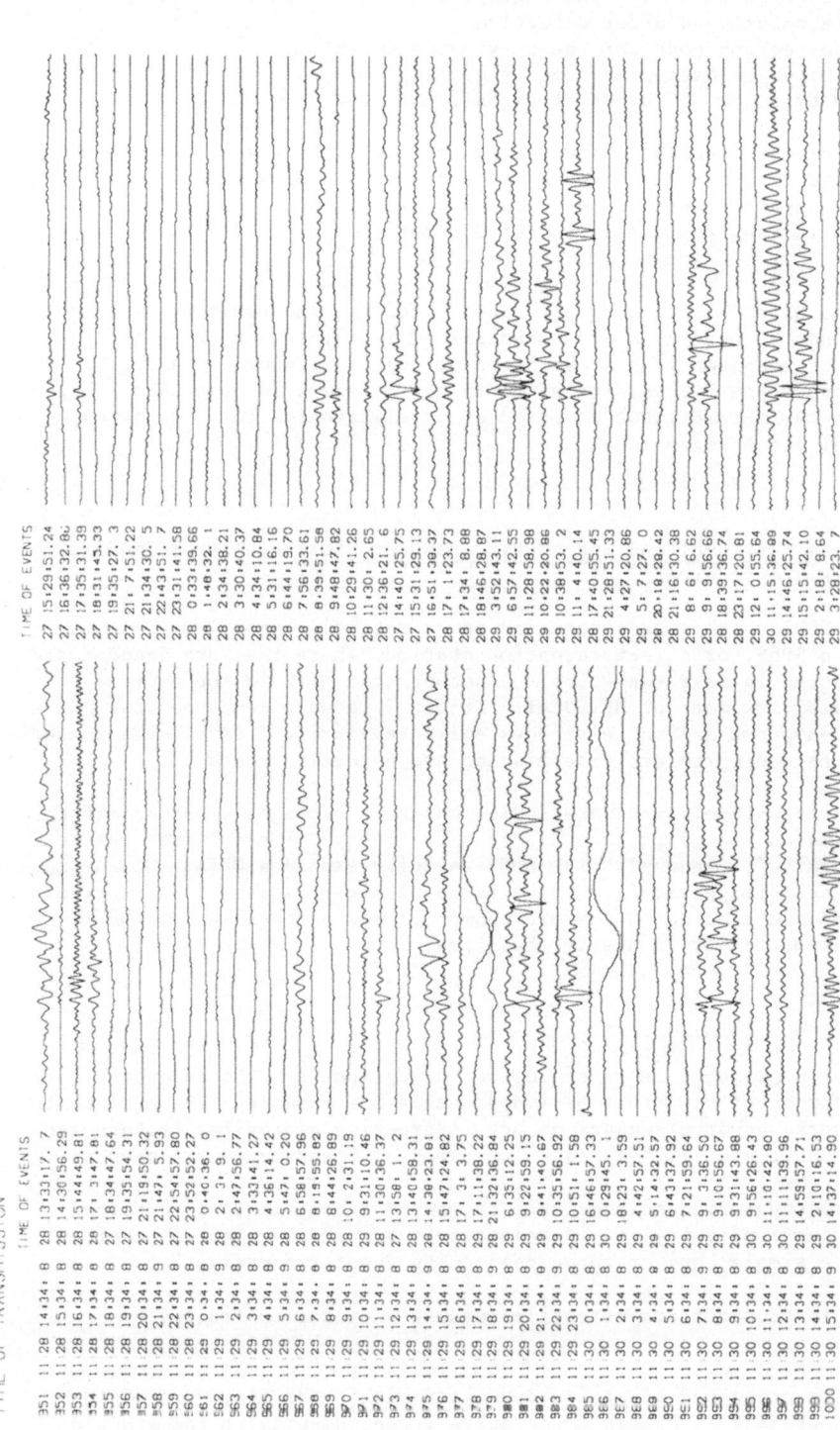

Fig. 7. Meteosat Seismological Platform. Messages transmitted by the prototype METEOSAT DCP during two days. Numbers are transmission and detection hours. The digitization rate is 100 points per second and each seismogram contains 307 points. The format of the data transmitted can be modified: this figure illustrates the amount of data that can be collected by the actual METEOSAT system.

24

data acquisition module (1 vertical component) connected to peripherals chosen according to the type of experiment:

- a 1 Megabyte static memory and a time code receiver for short duration deployments in refraction or earthquakes studies.
- a 1 Megabyte memory + a pager radio receiver for local monitoring for instance on a volcano, during a seismic crisis.
- a modem for long duration monitoring seismicity around a laboratory.
- a METEOSAT or GOES emitter for earthquakes studies in remote sites.

REFERENCES

Aki, K., Christoffersson, A., and Husebye, E., 1977, Determination of the three-dimensional seismic structure of the lithosphere, J. Geophys. Res., 82:277-296.

Aki, K., and Lee, W. H. K., 1976, Determination of three-dimensional velocity anomalies under a seismic array using first P arrivals times from local earthquakes, Part 1. A homogeneous initial model, J. Geophys. Res., 81:4381-4399.

Allen, R. V., 1982, Automatic phase pickers: their present use and future prospects, Bull. Seism. Soc. Am., 72:S225-S242.

Berger, J., and Lusignan, B. B., 1986, Satellite telemetry of seismic data status report, Reprint, Scripps Institution of Oceanology, La Jolla, California, USA.

Bonanomi, J., 1986, Telecode. Unité électronique à quartz radio-synchronisée, Observatoire de Neuchâtel, CH.

Clark, H. E., and Medina, E. S., 1976, Tsunami seismic system, USGS Albuquerque Seism. Lab., Rept 77-777.

Endo, E. T., Ward., P. L., Harlow, D. H., Allen, R. V., and Eaton, J. P., 1974, A prototype global volcano surveillance system monitoring seismic activity and tilt, Bull. Volcanol., 38:315-344.

Evans, J. R., and Allen, S. S., 1983, A teleseismic-specific detection algorithm for single short period traces, Bull. Seismol. Soc. Am., 71:1173-1186.

Glot, J. P., and Poupinet, G., 1987, ARGOS and seismic monitoring of volcanoes, ARGOS Newsletter, 30:4-7.

Goforth, T., and Herrin, E., 1981, An automatic seismic signal detection algorithm based on the Walsh transform, Bull. Seismol. Soc. Am., 71:1351-1360.

Healy, J. H., and Hill, D. P., 1984, A new system for seismic-refraction studies of active faults, in: "Earthquake Prediction, Proceedings of the International Symposium on earthquake prediction", Terra Scientific Publishing Company, Tokyo, UNESCO, Paris, pp 421-423.

Hirn, A., Nercessian, A., Sapin, M., Ferruci, F., and Wittlinger, G., 1987, Relation between seismicity, eruptive activity and structural heterogeneity at Mount Etna volcano, Sicily, submitted to J. Geophys. Res.

IRIS - Incorporated Research Institutions for Seismology, 1985, The program for Seismic Studies of the Continental Lithosphere (PASSCAL), Washington DC, USA.

Iyer, H. M., 1975, Anomalous delays of teleseismic P-waves in Yellowstone National Park, Nature, 253:425-427.

Lee, W. H. K., and Stewart, S. W., 1981, Principles and applications of microearthquake networks, Advances in Geophysics, Vol. 23, Academic Press.

Meyer, R. P., and Mereu, R. F., 1983, Proceedings of the Workshop on Portable seismograph development, International Association of Seismology and Physics of the Earth's Interior, Commission on

Controlled Source Seismology, University of Wisconsin, Madison, WI, USA.

Poupinet, G., and Glot, J. P., 1982, A low power seismic event detector transmitting data via ARGOS satellite data collecting system, EOS, 63:1266.

Poupinet, G., Bresson, A., Goula, X., and Coutant, O., 1985, A test of SEISPACE seismological DCPs in Ubaye (French Alps), Proceedings of the 10th ARGOS users Conference, Kiel (FRG), 21-23 May 1985, CNES, Toulouse, France.

Poupinet, G., 1987, Seismic data collection platforms for satellite transmission, in: "Seismic Tomography", G. Nolet, ed., D. Reidel Publishing Company, pp 239-250.

Prothero, W. A., 1984, Ocean bottom seismometer technology, EOS, 65-13, 113-116.

Williams, R. T., 1987, PASSCAL Newsletter No. 4, University of South Carolina, Columbia, USA.

World Meteorological Organization, 1982, Satellites in Meteorology, Oceanography and Hydrology, WMO, Geneva, CH.

World Meteorological Organization, 1985, Information on meteorological satellite programmes operated by Members and Organizations, WMO, publication No. 411, Geneva, CH.

IMAGING WITH TELESEISMIC DATA

Guust Nolet and Berend Scheffers

Department of Theoretical Geophysics
PO Box 80.021
3508 TA Utrecht
The Netherlands

INTRODUCTION

For the first time in history, seismological data are becoming available in sufficient quantity, and with sufficient accuracy, to attempt a realistic, three dimensional modeling of the elastic properties of the deep Earth's interior.

Until recently, Earth scientists where severly hampered by lack of knowledge about the structure of the Earth's mantle. Seismological data have provided us one-dimensional Earth models - essentially density and the seismic velocities as a function of depth - since the early 20th century. Although it became soon evident that there exist lateral variations in seismic travel times, much effort has been spent to find the *best* one-dimensional Earth model. One reason for this is that the resolution of lateral heterogeneity asks for much more than just a marginal increase in data quantity. A simple calculation illustrates this: if we disregard the crust, we can specify the seismic velocity and the density with sufficient resolution at some 100-200 depth nodes in a 1D model. With reasonably independent, accurate data, we need some 10^3 data to resolve this. In fact, model 1066B from Gilbert and Dziewonski (1975), which in our view still represents one of the best approaches to the average Earth, was derived from 1066 'gross Earth' data, mostly normal mode frequencies. If we want to obtain a similar resolution in a 3D Earth, we must specify the three parameters at more than 10^6 spatial nodes, giving us, say 10^7 unknowns to be resolved with at least as many data. Thus, imaging of the whole Earth asks for an increase of the data quantity by *four orders of magnitude!*

We have not yet reached that stage. One dimensional Earth models continue to exist for that reason (the PREM model by Dziewonski and Anderson, 1981, is the most recent example), but also because they serve a very useful role as a global reference, in earthquake location studies and in more general petrological modelling of the deep rock properties. On the other hand, local studies have been made with a 3D resolution of some 100 km or better, whereas global 3D models are available with a resolution of several 1000 km (see Nolet, 1987a, for references). In this paper we shall review the basic techniques that have been developed recently to image the Earth with the type of seismic data now available.

SEISMIC DATA

Although the Earth's surface is covered with thousands of seismographs, the distribution of this instrumentation is rather uneven (mainly covering the continents) and the quality of the records varies strongly. The first attempt at standardizing seismological recordings was made when the WWSSN network was installed in the early 1960's. This network, which counted at some time more than 150 high quality stations, is now fast becoming obsolete: the analog recording forces separate measurement of the short and long period signals and suppression of the noisy intermediate periods (5-10 seconds). Digital instrumentation, with a high dynamic range and a broad frequency band sensitivity, are now replacing the older instrumentation.

One of the most powerful data bases that has been growing in the course of the past 20 years, are the arrival times of seismic waves as reported by local station operators to the International Seismological Centre (ISC). These data are available on tape, and soon on compact disk as well. In fact, the total number of arrival time readings in the ISC data base is estimated to come close to 10^7. Unfortunately, these are not all independent, and a high degree of averaging and data winnowing is needed to reduce the effect of - often large - errors. It is obvious that with such large data bases, data selection must be done in some automatic way.

Accurate travel time readings from local networks are often used to image local structures such as volcanoes and their magma system. Travel time readings do have some obvious disadvantages, however. The identification of a particular phase may be uncertain. For example, if we read the arrival time of Sp (the S wave converted to a P wave at the Moho discontinuity) but identify it as S, we introduce an error of some 5 seconds. Another source of error is the ray approximation. Wielandt (1987) has shown that diffracted waves may overtake the direct wave if it is slowed down - thus masking heterogeneities with a negative velocity contrast. Finally, some regions in the Earth are badly covered by seismic rays, such as the Upper Mantle under the oceans, and rays provide only a very crude resolution when low velocity areas are present, since Fermat's Principle, which tells them to find the fastest ray path, forced them to avoid such regions as much as possible.

But there is much more information in the seismogram record than just the arrival time. Again, because of the large quantity of data needed for imaging, digital recording with some form of highly automatic processing is called for. The low-frequency part of the record, with the surface waves and their higher modes, is the best candidate for such an analysis, but in the near future it will perhaps also be possible to use synthetic bodywaves for modelling of the Earth. If we use the waveforms of the record or part of the record, we shall use the term 'waveform modelling' for the imaging technique. It is common practice also to use the term 'seismic tomography' for these kind of imaging techniques.

In the following sections we shall briefly sketch the theoretical background which is needed for a thorough understanding of the imaging problem in seismology.

SEISMIC RAYS

As a starting point we take the elastodynamic equations in a solid:

$$\rho \partial_{tt} u_i = \partial_j \sigma_{ij} + f_i \tag{1}$$

where u_i is the component of a (small) displacement in direction i due to incremental stresses σ_{ij} and the body force component f_i. The stress is linearly related to the strain ε_{kl} by Hooke's law:

$$\sigma_{ij} = c_{ijkl} \varepsilon_{kl} = c_{ijkl} \partial_k u_l \tag{2}$$

where the last term can easily be derived from the definition of the strain ($\varepsilon_{kl} = \frac{1}{2}\partial_k u_l + \frac{1}{2}\partial_l u_k$) and the symmetry properties of the *elasticity tensor*

$$c_{ijkl} = c_{jikl} = c_{ijlk} = c_{lkij}$$

If we insert (2) into (1) we find:

$$\rho \partial_{tt} u_i = \partial_j (c_{ijkl} \partial_k u_l) + f_i \tag{3}$$

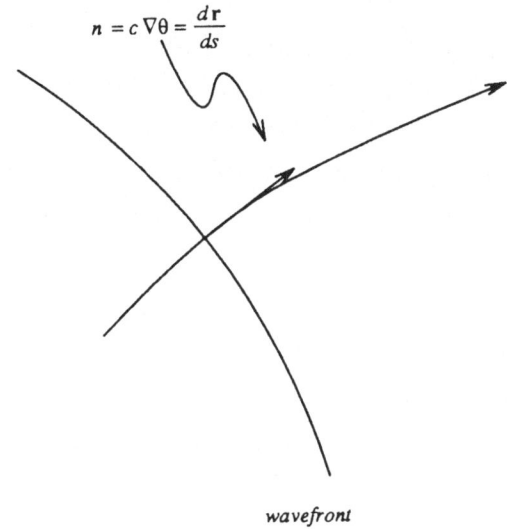

$$n = c \nabla\theta = \frac{d\mathbf{r}}{ds}$$

wavefront

Figure 1

In an isotropic solid, the elasticity tensor c_{ijkl} simplifies to

$$c_{ijkl} = (\kappa - \frac{2}{3}\mu)\delta_{ij}\delta_{kl} + \mu(\delta_{ik}\delta_{jl} + \delta_{il}\delta_{jk})$$

where where μ is the shear modulus and $\kappa - 2\mu/3 = \lambda$, Lamé's parameter. Substitution into (3) gives:

$$\rho\partial_{tt}u_i = (\lambda+\mu)\partial_i\partial_k u_k + \mu\partial_j\partial_j u_i + f_i$$

It is fairly easy to show that the equation of elastic waves now gives rise to two different eikonal equations, one for P and one for S waves. We transform this equation to the frequency domain:

$$-\rho\omega^2 u_i = \partial_i(\lambda\partial_j u_j) + \partial_j[\mu(\partial_i u_j + \partial_j u_i)] \qquad (4)$$

we substitute a trial solution $\mathbf{u}(\mathbf{r},t)=\mathbf{A}(\mathbf{r})\delta[t-\theta(\mathbf{r})]$:

$$\mathbf{u}(\mathbf{r},\omega) = \mathbf{A}(\mathbf{r})e^{i\omega\theta(\mathbf{r})}$$

If we substitute this into (4) we find to first order (ω^2):

$$-\rho A_i = -(\lambda+\mu)\partial_i\theta\partial_j\theta A_j - \mu(\partial_j\theta)^2 A_i$$

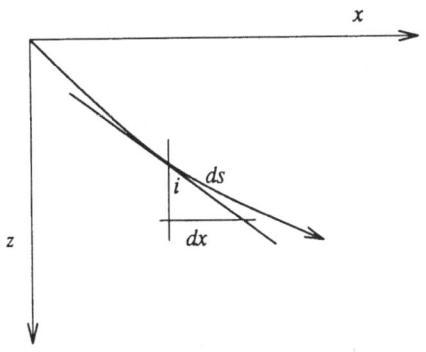

Figure 2

or, in vector notation:

$$-\rho A + (\lambda+\mu)\nabla\theta(\nabla\theta\cdot A) + \mu|\nabla\theta|^2 A = 0$$

This equation contains 3 terms, two of which are directed along A, the other along $\nabla\theta$. Obviously, the equation can be satisfied only if all nonzero terms are parallel, thus either:

$$A = constant \times \nabla\theta \;\; \rightarrow \;\; |\nabla\theta|^2 = \frac{\rho}{\lambda+2\mu} = 1/\alpha^2$$

or

$$A\cdot\nabla\theta = 0 \;\; \rightarrow \;\; |\nabla\theta|^2 = \frac{\rho}{\mu} = 1/\beta^2$$

α is the velocity of P waves, which have their particle motion parallel to the ray and are of compressional type. S waves, which are shear waves with motion in a plane perpendicular to the ray direction, travel with velocity β. We see that both P- and S-waves satisfy their own *eikonal equation* for the location of the wavefront θ. Writing c for either α or β we have:

$$|\nabla\theta|^2 = 1/c^2 \tag{5}$$

The eikonal equation implies that $c\nabla\theta$ is a unit vector. It is a vector perpendicular to the wavefront, and therefore by definition parallel to the ray. Although (5) gives us the location of the wavefront, it is more useful to have an equation that describes the geometry of the rays. Let $d\mathbf{r}$ be a tangent along the ray, with length ds. Then we can write the same unit vector as $d\mathbf{r}/ds$ (Figure 1), or:

$$\nabla\theta = \frac{1}{c}\frac{d\mathbf{r}}{ds}$$

Using the fact that $\mathbf{n}\cdot\nabla\theta = d\theta/ds = 1/c$, and $d(\nabla\theta)/ds = \nabla(d\theta/ds)$:

$$\nabla(\frac{1}{c}) = \frac{d}{ds}\left[\frac{1}{c}\frac{d\mathbf{r}}{ds}\right] \tag{6}$$

In a medium with horizontal stratification we have $c(x,y,z) = c(z)$, so that $\nabla(1/c)$ points in the z-direction. Defining the ray parameter vector by $p_i \equiv c^{-1}dx_i/ds$ we find for a ray in the $x-z$ plane:

$$constant\; p_x \equiv p = \frac{1}{c}\frac{dx}{ds} = \frac{sini}{c}$$

which shows that Snell's law is valid for waves satisfying the ray theory with $\omega \gg 1$ (see figure 2). If we follow a segment of a ray, it will travers a horizontal distance given by

$$X(p) = \int dx = \int tg\; i\; dz = \int \frac{\sin i}{(1-\sin^2 i)^{1/2}}dz = \int \frac{cp}{(1-c^2p^2)^{1/2}}dz$$

in a time:

$$T(p) = \int dt = \int \frac{ds}{c} = \int \frac{dz}{c(1-c^2p^2)^{1/2}}$$

In an Earth with spherical symmetry we find in a similar way:

$$T(p) = \int \frac{\eta^2 dr}{r(\eta^2-p^2)^{1/2}}$$

$$\Delta(p) = \int \frac{pdr}{r(\eta^2-p^2)^{1/2}}$$

where $\Delta(p)$ is the angle in radians, $\eta = r/c$, and $p = \eta\sin\theta$ (θ is the angle of the ray with the radius).

These expressions allow us to find the ray geometry and the theoretical travel times in laterally heterogeneous Earth models. Since ray tracing in symmetric Earth models is so much faster than 3D ray tracing (and speed is essential when one deals with large data sets!), such models form in general the starting point for imaging. The final image then forms the *perturbation* with respect to the starting model (sometimes called the *background model*).

For 3D Earth model, the travel time must be expressed in a more general form:

$$T_i = \int_{S_i} \frac{ds}{c(\mathbf{r})} \qquad i=1,...,N \qquad (7)$$

To derive the Earth model $c(\mathbf{r})$ from measurements T_i using (7) is a very complicated problem, since the unknown $c(\mathbf{r})$ is also implicitly present in the ray path S_i. We therefore use Fermat's Principle to reduce (7) to a *linearized* equations for (small) velocity perturbations with respect to the starting model. Let T_i^0 be the prediction from the starting model:

$$T_i^0 = \int_{S_i^0} \frac{ds}{c_0}$$

Where S_i^0 is the ray trajectory in the starting model. We define the delay time as

$$\delta T_i = T_i - T_i^0 = \int_{S_i} \frac{ds}{c} - \int_{S_i^0} \frac{ds}{c_0} \approx \int_{S_i^0} (\frac{1}{c} - \frac{1}{c_0}) ds$$

or

$$\delta T_i = -\int_{S_i^0} \frac{\delta c(\mathbf{r})}{c_0(\mathbf{r})^2} ds \qquad (8)$$

where $\delta c = c - c_0$. Note that we have used Fermat's principle to substitute the ray path as calculated for the starting model for the (unknown) ray path in the true Earth: we only make a second order error in calculating the time delay.

We shall now wish to solve δc from a large, but finite, number of measurements, each of which gives us a linearized equation such as (8). To do so on a computer, we divide the Earth into a number (M) of cells, and define:

$$h_i(\mathbf{r}) \; = \; 1 \qquad \text{if } \mathbf{r} \; in \; cell \; i$$
$$= \; 0 \qquad elsewhere \qquad (9)$$

The functions h_i span a subspace of the space of all possible Earth models $c(\mathbf{r})$. If the cells do not overlap, the basis functions are orthogonal:

$$\int_V h_i(\mathbf{r}) h_j(\mathbf{r}) d^3\mathbf{r} = \delta_{ij} v_i$$

where v_i is the volume of cell i. An alternative is to expand the Earth's velocity field into a finite number of fully normalized spherical harmonics $Y_l^m(\theta,\phi)$, which form an orthogonal basis as well (Dziewonski, 1984; Morelli and Dziewonski, 1987):

$$h_i(\mathbf{r}) = f_k(r) Y_l^m(\theta,\phi)$$

where $i = \{klm\}$, a renumbering, and where $f_k(r)$ forms a set of depth functions that is orthonormal over the depth region of interest.

Either choice of $h_i(\mathbf{r})$ leads to:

$$\delta c(\mathbf{r}) = \sum_{k=1}^{M} \gamma_k h_k(\mathbf{r})$$

We wish to determine $\delta c(\mathbf{r})$, or γ_k, from the measurements:

$$\delta T_i = \sum_{k=1}^{M} -\int_{S_i^0} \frac{1}{c_0(\mathbf{r})^2} \gamma_k h_k(\mathbf{r}) ds = \sum_k A_{ik} \gamma_k$$

where

$$A_{ik} = - \int_{S_i^o} \frac{h_k(\mathbf{r})}{c_0(\mathbf{r})^2} ds$$

or, in matrix form:

$$\mathbf{A}\,\boldsymbol{\gamma} = \delta\mathbf{T} \qquad\qquad (10)$$

The tomographic problem has now been reduced to a discrete system. At this stage, it is trivial to add additional discrete unknowns to the system, such as event origin time and hypocentral coordinate corrections. In the following, we shall use a *cell*-parametrization. This leads to *sparse* matrices $\mathbf{A}$ (i.e., matrices with a large majority of elements equal to 0), which are advantageous when we wish to solve for detailed models with many parameters, as we shall see in the next section. For a correct treatment of data errors, it is important to scale (10) such that the data have unit a priori variance. In the following we shall assume that this has been done.

THE LARGE SCALE LINEAR INVERSE PROBLEM

Even on very large computers, (10) poses unsurmountable problems with standard numerical techniques for the solution of such linear equations when the number of unknowns is large. The reason is that no computer memory can accommodate all of the matrix elements (even if we take into account that most of them are 0, and need not be stored explicitly). To solve (10) we must make use of iterative techniques. Such techniques use only forward multiplications with the matrix $\mathbf{A}$, which is left undisturbed, and therefore can be accessed row by row. The row elements will fit into almost any computer memory even for very large models since we only need to store the nonzero elements. Strictly speaking, we do not even need to store the whole row, but it leads to more efficiency if we do. In practice, the nonzero matrix elements are stored on disk with a pointer array to tell us its row and column identification, and every row is retrieved into memory whenever we need it. For the following, it is important to realize that not only $\mathbf{A}\boldsymbol{\gamma}$ can be calculated row for row, but that $\mathbf{A}^T \delta\mathbf{T}$ can be handled in the same way, although δT_j is not calculated in the jth step, but slowly growing until all rays passing cell j have been treated.

This section will briefly review the most important iterative techniques available. Much of this section is based on a recent important paper by Van der Sluis and Van der Vorst (1987), and the reader is strongly advised to study that paper for more detail.

To illustrate the iterative technique, we shall first decribe the *Algebraic Reconstruction Technique* or ART, which is still widely employed even though it has severe drawbacks as we shall see. The ART procedure works as follows: we take the first ray from the data set, it will cross a limited number of cells in the model. Of course there are many model adjustments $\boldsymbol{\gamma}$ that will result in a perfect fit of the delay time for this ray. We select the smallest adjustment $(Min \, |\boldsymbol{\gamma}|^2)$. This will leave the cells that are not visited by the ray untouched, and the smallest adjustment is obtained by choosing $\boldsymbol{\gamma}$ parallel to the matrix row that represents this ray. We then take the second ray (row), recalculate the delay for the adjusted model, and proceed similarly as for the first ray. This process is repeated many times. When described in mathematical form the qth iterate is calculated as follows:

$$\gamma_j^{(q+1)} = \gamma_j^{(q)} + \frac{A_{ij} r_i^{(q)}}{\sum_k A_{ik}^2} \quad , \ i = q \ mod \ M$$

where $r_i^{(q)}$ is the residual delay time after q iterations: $r_i^{(q)} = \delta T_i - \sum A_{ij}\gamma_j^{(q)}$. This procedure will always result in a perfect fit of the last delay time $(r_i^{(q+1)} = 0)$. Thus, if we have two identical rays but with different delays because of measurement errors, the solution will oscillate between these two possibilities. Consequently, ART can only converge to a solution if the equations are consistent. This they never are because of data errors, and a better way to proceed is therefore to have a look at all adjustments that the data would impose, before actually changing the model. Techniques that use this delayed adjustment are called *Simultaneous Iterative Reconstruction Techniques*, or SIRT.

The simplest way is to average over all individual ray adjustments for a particular cell. If M_j is the number of ray segments visiting cell j then we may set:

$$\gamma_j^{(q+1)} = \gamma_j^{(q)} + \frac{1}{M_j} \sum_i \frac{A_{ij} r_i^{(q)}}{\sum_k A_{ik}^2}$$

which was first suggested by Dines and Lyttle (1979). The method was shown to be rather slowly convergent by Nolet (1985). A faster convergence is obtained by taking a more sophisticated weighting in the average. If we replace M_j by $C_j = \sum_i |A_{ij}|$, a column sum giving the total ray path length in cell j, and write $R_i = \sum_k |A_{ik}|$ for the row sum (the length of ray i) we obtain the SIRT method used by Hager et al. (1985):

$$\gamma_j^{(q+1)} = \gamma_j^{(q)} + \frac{1}{C_j} \sum_i \frac{A_{ij} r_i^{(q)}}{R_i} \tag{11}$$

Since ray path length is a positive quantity, the absolute value signs are only significant if the system includes source parameters as unknowns.

Some cells may not be crossed by any rays at all, in which case $C_i = 0$. In ART, such cells are not adjusted at all, in SIRT we must explicitly exclude these cells from the iteration (11) to avoid a 0/0 division. In doing so, we impose a *minimum norm* criterion on these unresolved cells. This illustrates a nice property of iterative solvers: they result in solutions even when the system of equations is underdetermined. As we saw earlier, the solution in ART is in this case ambiguous - it all depends on what ray is treated last.

In an extensive study of iterative algorithms for tomographic applications, Van der Sluis and Van der Vorst (1987) show that (11) will converge to the least squares solution of the weighted system $\mathbf{R}^{-\frac{1}{2}} \mathbf{A} \gamma = \mathbf{R}^{-\frac{1}{2}} \delta \mathbf{T}$, where $\mathbf{R} = diag(R_i)$. Thus, SIRT affects the statistics of our solution in the sense that delay times from short rays are weighted more heavily than for long rays. Since there is no a priori reason why the accuracies would differ, this is an unwanted phenomenon. For underdetermined systems, they also show that SIRT selects among all least squares solutions the one for which $|\mathbf{C}^{\frac{1}{2}} \gamma|$ is minimized. This scaled minimum norm criterion has a very attractive property, as a small exercise shows.

Suppose the delays are such that they can be explained by a constant velocity perturbation: $\gamma_i = \gamma$ and that we have perfectly accurate measurements $\delta T_i = \sum_j A_{ij} \gamma_j = \gamma R_i$, then we have $r_i^{(0)} = \delta T_i = \gamma R_i$ since $\gamma_i(0) = 0$. It is simple to check that in this case (11) gives us $\gamma_i^{(1)} = \gamma$: we find the solution with minumum heterogeneity. Unfortunately, the $\mathbf{R}^{-\frac{1}{2}}$ scaling of the least squares fit criterion will blow up the errors for short ray paths in the more realistic case of contaminated data. This effect could be corrected for by multiplying the whole system (10) with $\mathbf{R}$ before iterating, but this will in its turn give us a new matrix $(\mathbf{RA})$ with a subsequent change in C. The analysis given for the unscaled SIRT is then invalid, and we will not find the solution with minimum heterogeneity using SIRT. Within the framework of SIRT it is not possible to minimize error propagation and algorithm-induced heterogeneity at the same time.

This limitation in scaling does not exist for more complicated row-active methods, known as *projection methods* because they project the whole system (10) onto a space of small dimensions, which is spanned by a set of basis vectors that are orthogonal and of which the first vector is $\mathbf{A}^T \delta \mathbf{T}$. The most promising of these algorithms is the LSQR algorithm of Paige and Saunders (1982), which was also shown by Nolet (1985, 1987a) and Van der Sluis and Van der Vorst (1987) to have a rate of convergence that is superior to SIRT. Since LSQR converges to the minimum norm solution $|\gamma|$, a scaling with C must explicitly be introduced if wanted. Some tests by Spakman and Nolet (1987) indicate that a full scaling with the cell ray length may have adverse effects on the propagation of data errors into the solution, and that a more general scaling with the cell volume (the expected ray length for a uniform ray distribution) may be preferable. A short subroutine for LSQR is given by Nolet (1987b).

All iterative methods allow for the introduction of more sophisticated minimum norm criteria if we allow for a bias in the solution. From the statistical theory we know that the allowance of bias may have very beneficial effects on the variance of the solution. Moreover, we must realize that the iterations of SIRT or LSQR will never be continued to anything near a converged solution, not only because this would cost too much computer time, but also because the early iterations tend to give smooth solutions with low variance, whereas models grow erratic because of the propagation of data errors after the first few iterations (Van der Sluis and Van der Vorst, 1987; Spakman and Nolet, 1987). As an example of a slightly more complicated minimum norm criterion, let us minimize the *lateral heterogeneity*, by minimizing the difference between the adjustment in a cell with that of its neighbouring cells: The condition:

$$Min \sum_i \frac{1}{2N_i} \sum_{neighb} (\gamma_i - \gamma_j)^2$$

where N_i is the number of neighbours of cell i, results in M extra equations:

$$\gamma_k - \frac{1}{N_k} \sum_{neighb} \gamma_j = 0 \quad (k=1,...,M)$$

or $S\gamma=0$. If we add these equations to (10), we find:

$$\begin{bmatrix} A \\ S \end{bmatrix} \gamma = \begin{bmatrix} \delta T \\ 0 \end{bmatrix}$$

To investigate the resolution of large systems such as (10), Spakman and Nolet (1987) advocate the use of sensitivity tests. One calculates a synthetic data vector δT^{syn} by devising a synthetic model γ^{syn} with some regularity (spikes or a harmonic velocity variation throughout the model, for example) and calculating the product $\delta T^{syn}=A\gamma^{syn}$. The synthetic data are then inverted with an iterative technique, and the resulting solution γ is compared to γ^{syn}. Such comparisons enable one to identify badly resolved areas in the solution.

SURFACE WAVES

So far we have only considered the inversion of travel times of seismic rays. Of course, there is much more information contained in the seismic wavetrain, and, given the present shortage of seismic stations and their uneven distribution, it would be foolish if we did not attempt to use that information as well.

Much of the wavetrain following the P phase is actually consisting of ray arrivals, but of a more complicated nature: some of them have been converted from P to S, and back to P sometimes, or from S to P. Some have followed complicated ray paths, bouncing from sharp discontinuities in the Earth such as the Moho, the core-mantle boundary and some of the Upper Mantle discontinuities. Combinations of multiple reflections and wave type conversions exist, and all these waves interfere with each other since they arrive in a rather short time window. Take for instance the sequence of surface reflected S waves following the first arriving S wave: SS, SSS, SSSS etc. When one calculates arrival times of the multiples following SSSS, one finds almost equal travel times. In fact, for the vertically polarized S waves, these interfering waves make up most of what we call the *higher mode Rayleigh waves*. Similar mode waves exist for the transversely polarized S wave *(Love waves)*, and in general seismologists use the term *surface waves*.

A complete theory for surface waves includes also the effect of wave conversions and multiple reflections on even minor discontinuities and velocity gradients, and all effects of *diffracted waves*, which we ignored so far. In fact, what we call the *fundamental* mode Rayleigh wave consists mostly of a surface-diffracted S wave. To develop the theory is beyond the scope of a review paper such as this, but we refer the reader to Bullen and Bolt (1986) or Aki and Richards (1980) for details. Here we merely summarize a few basic properties of surface waves that may be used for imaging purposes:

- Each surface wave mode n travels with its own, frequency dependent, phase velocity $C_n(\omega)$ which relates to the wavenumber $k_n(\omega)$ and circle frequency ω as $C_n = \omega/k_n$. The phase of the mode, after travelling a distance x along the surface, is given by $k_n(\omega)x$.

- Neighbouring frequencies of a surface wave mode interfere, such that the maximum energy of the mode travels with the group velocity $U_n(\omega) = (dk_n(\omega)/d\omega)^{-1}$.

- For weak lateral heterogeneities, we may assume that the surface waves travels along the great circle between source and receiver, and that the wavenumber is locally perturbed to $k_n + \delta k_n$, whereby the perturbation δk_n relates in a quasi linear way to the local velocity perturbation.

Dziewonski and Hales (1972) , Panza (1976) and Nolet and Panza (1976) give observational methods to determine phase and group velocities for fundamental and for higher modes of surface waves, respectively. The inverse problem is very similar to that for delay times of P- and S-waves, since the perturbation in the phase velocity acquired over path i is the average of the local phase perturbations $\delta C(\omega,\theta,\phi)$ over the path of length L_i:

$$\delta C_{n,i}(\omega) = \frac{1}{L_i} \int_{S_i} \delta C_n(\omega,\theta,\phi)\, ds$$

and a similar equation for the group velocity U_n.

Using perturbation theory to relate δC_n, or, equivalently δk_n, to the local variation in the S velocity $\delta\beta$:

$$\delta C_n(\omega,\theta,\phi) = \int_0^\infty \left[\frac{\partial C_n}{\partial \beta} \right]_{\omega,z} \delta\beta(\theta,\phi,z)\, dz$$

gives us again a linear equation for velocity perturbations:

$$\delta C_{n,i}(\omega) = \frac{1}{L_i} \int_{S_i} \int_0^\infty \left[\frac{\partial C_n}{\partial \beta} \right]_{\omega,z} \delta\beta(\theta,\phi,z)\, dz\, ds$$

In contrast to the case of delay time tomography, the integrals are now over 2 space variables and not over a line. Thus, we do expect a reduced sensitivity over depth. In general this is true. However, surface wave data are superior in imaging the velocity structure in low velocity layers, such as exist in the Upper Mantle at relatively shallow depth. The phase velocity method works fine with fundamental mode measurements, but is difficult to apply to higher modes, since in general one needs an array of stations to measure higher mode phase velocities explicitly. These difficulties have led to alternative ways to analyze the surface wave part of the seismogram, which we shall develop in the next section.

IMAGING USING WAVEFORMS FROM SURFACE WAVES

The oscillatory nature of the seismograms (the surface wave modes travel like $\exp[ik_n(\omega)x]$), makes it difficult to linearize the relationship between the *time signal* itself and the velocity perturbations $\delta\beta$ for all but very small values for the product $k_n(\omega)x$. Such efforts are generally limited to the lowest frequencies (Woodhouse and Dziewonski, 1984, attacked the problem by repeated linear inversions and limited their data to frequencies below 8 mHz) with consequently a rather severe loss in resolving power because of the long wavelengths involved. Our solution, therefore, is to cast the problem in a form such that it can be attacked by methods of *nonlinear optimization* (Gill et al., 1981). In order to do so, we define a measure for the goodness of fit between one or more observed seismograms, and the corresponding time series predicted by an Earth model and the earthquake source parameters. We shall denote this measure by the objective function $F(\mathbf{p})$, where $\mathbf{p}$ is an M-dimensional vector of all parameters:

$$F(\mathbf{p}) = \frac{1}{2} \sum_{i=1}^{N} \int_{0}^{T_i} |R\,\Psi_i(\mathbf{p},t) - R\,s_i(t)|^2 dt + \frac{1}{2}\gamma \mathbf{p}^T \mathbf{C} \mathbf{p} \tag{12}$$

where N is the total number of time series available, T_i is a (sufficiently large) time span, R is a filtering and windowing operator that describes the different steps of signal processing prior to inversion, Ψ_i is an operator that generates, from a model $\mathbf{p}$, the synthetic seismogram corresponding to the ith time series, and $s_i(t)$ is the observed seismogram. The last term in (12) has the purpose of stabilizing the inverse problem. Without it we may find a very large solution $|\mathbf{p}|$ due to a very weak dependence of $F(\mathbf{p})$ on certain elements of $\mathbf{p}$. $\mathbf{C}$ scales the model parameters and γ allows us to vary the trade off which will exist between the norm of the solution $|\mathbf{p}|$ and of the objective function $F(\mathbf{p})$. It will greatly simplify the mathematics if we define the parameters $\mathbf{p}$ dimensionless, for instance by scaling them form the outset with their a priori uncertainties. In that case $\mathbf{C}$ reduces to the unit matrix $\mathbf{I}$ and we find for the gradient of F with respect to the model parameters:

$$\mathbf{g}(\mathbf{p}) \equiv \nabla F(\mathbf{p}) = \sum_{i=1}^{N} \int_{0}^{T_i} R\,\nabla\Psi_i[R\,\Psi_i(\mathbf{p},t) - R\,s_i(t)]dt + \gamma \mathbf{I} \mathbf{p}$$

where we implicitly assumed that R is independent of the model. For the Hessian matrix, the matrix of second derviatives $\partial^2 F(\mathbf{p})/\partial p_i \partial p_j$ we find:

$$\mathbf{H}(\mathbf{p}) \equiv \nabla \mathbf{g}(\mathbf{p}) = \sum_{i=1}^{N} \int_{0}^{T_i} \left[R\,\nabla\Psi_i(R\,\nabla\Psi_i)^T + R\,\nabla\nabla\Psi_i[R\,\Psi_i - R\,s_i] \right] dt + \gamma \mathbf{I}$$

Nonlinear optimization algorithms work in an iterative way. A starting model $\mathbf{p}$ is updated with a correction $\Delta\mathbf{p}$ to give a new model $\mathbf{p}+\Delta\mathbf{p}$. This new model is then taken as a starting model in the next iteration. Suppose that at some stage we have arrived at model $\mathbf{p}$. With a simple Taylor expansion we may now find an approximation for the step $\Delta\mathbf{p}$ that should bring us close to the minimum $\mathbf{g}(\mathbf{p})=0$:

$$F(\mathbf{p}+\Delta\mathbf{p}) \approx F(\mathbf{p}) + \mathbf{g}(\mathbf{p})^T \Delta\mathbf{p} + \frac{1}{2}\Delta\mathbf{p}^T \mathbf{H}(\mathbf{p})\Delta\mathbf{p}$$

differentiating with respect to $\Delta\mathbf{p}$ gives:

$$\mathbf{g}(\mathbf{p}+\Delta\mathbf{p}) \approx \mathbf{g}(\mathbf{p}) + \mathbf{H}(\mathbf{p})\Delta\mathbf{p}$$

So that $\Delta\mathbf{p}$ can be found by setting $\mathbf{g}(\mathbf{p}+\Delta\mathbf{p})=0$ and solving:

$$\mathbf{H}(\mathbf{p})\Delta\mathbf{p} = -\mathbf{g}(\mathbf{p})$$

In general, the exact calculation of $\mathbf{H}(\mathbf{p})$ is an unattainable goal, and we have to resort to approximate methods. To this end, we search for the minimum of $F(\mathbf{p})$ in a number of directions $\mathbf{d}^i$, starting with the steepest descent direction:

$$\mathbf{d}^0 = -\mathbf{g}(\mathbf{p}^0)$$

We calculate $F(\mathbf{p})$ along this direction in a number of points until we have past a minimum, and then determine the location of the minimum by three-point interpolation. In general it is not advisable to continue in the steepest descent direction after the first step (Gill et al., 1981). An improvement is to use conjugate directions:

$$\mathbf{d}^{i+1} = -\mathbf{g}^{i+1} + \beta_i \mathbf{d}^i$$

where

$$\beta_i = \frac{|g^{i+1}|^2}{|g^i|^2} \quad , \quad \beta_0 = 0$$

and g denotes the gradient of F for the i-th iteration model p^i. To calculate g, one may use finite differences, which gives much flexibility in programming, especially when R is complicated. If computing speed is important (and for large M it almost always is) it may be advisable to use approximations to g (Snieder, 1987), or to do the inversion for simple R in the frequency domain (Nolet, 1987c).

We have to evaluate the forward problem many times during one iteration step. The direct problem, given a laterally homogeneous background model, has to be calculated in a fast way. For that purpose we use a summation of a finite number of modes at each frequency component (at certain steps $\Delta\omega$) to construct a synthetic spectrum, and use an FFT to construct the time signal from that:

$$\Psi_i(\omega) = \sum_n \Psi_{ni}(\omega) = \sum_n A_{ni}(\omega) e^{i\chi_{ni}(\omega) - \alpha_{ni}(\omega)}$$

where $A_{ni}(\omega)$ is calculated from the source parameters. The phase χ and the anelastic damping α can be split into their values for the (laterally homogeneous) background model, and the perturbations to that:

$$\chi_{ni}(\omega) = \int_{S_i} \left[k_n(\omega) + \delta k_n(\omega,\theta,\phi) \right] ds$$

$$\alpha_{ni}(\omega) = \frac{1}{2} \int_{S_i} \left[k_n(\omega) + \delta k_n(\omega,\theta,\phi) \right] \left[Q_n^{-1}(\omega) + \delta Q_n^{-1}(\omega,\theta,\phi) \right] ds$$

where S_i is the surface ray path and $\delta k_n(\omega,\theta,\phi)$ and $\delta Q_n^{-1}(\omega,\theta,\phi)$ are the local perturbations in the wavenumber and the quality-factor. This approximation is only valid for wavelengths much larger than the length scale of the heterogeneities present and for weak horizontal velocity gradients. The total perturbation in the wavenumber is an average of local perturbations, the local contribution to this perturbation at a point (θ,ϕ) along a profile governed by a horizontal and vertical basisfunction as follows:

$$\delta k_n(\omega,\theta,\phi) = \int \left[\frac{\partial k_n}{\partial \beta} \right]_\omega \delta\beta(\theta,\phi,z) dz$$

Nolet (1987c) gives complete expressions for A_{ni} as well as $\partial k_n/\partial\beta$ for Love- and Rayleigh waves.

AN APPLICATION: THE STRUCTURE BENEATH THE NARS ARRAY

In 1983 our Department of Theoretical Geophysics installed 14 digital, broad band seismographs in Western Europe between Denmark and Spain, approximately along a great circle for earthquakes in northern Japan and the west Pacific ocean. This Network of Autonomously Recording Stations (NARS) has operated in this configuration until 1987 giving data from about 200 strong earthquakes (Dost 1984, 1987; Nolet et al., 1986). For an application of the waveform fitting technique outlined above, we selected two earthquakes on either side of the great circle line over the array (Figure 3, Table 1)

Table 1: Earthquake parameters

Date	Time	Epicentre	Depth	Region
May 26, 1983	2:59:58.8	40.5N 139.1E	16	W Coast Honshu
Dec 22, 1983	4:11:29.3	11.8N 13.5W	11	NW Africa

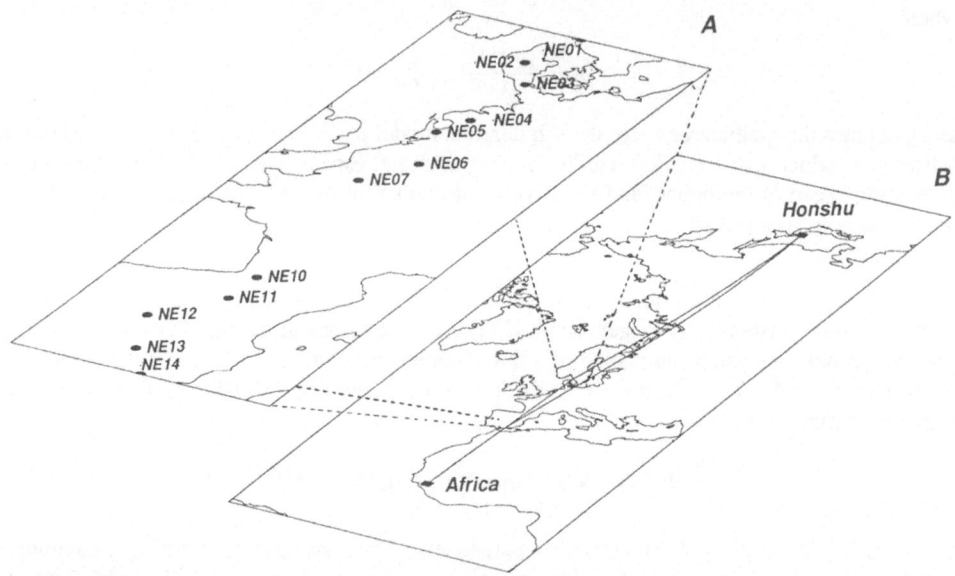

Figure 3 The stations of the NARS array used in this study (A). The NW Africa event and the W Honshu event are approximately located on the great circle through the Nars array (B).

Since the stations are approximately located on a line, we shall image a 2D S velocity structure along the array great circle. To this end we map the spherical coordinates (θ,ϕ) to one parameter x designating distance along the great circle arc, and parametrize:

$$\delta\beta(x,z) = \sum p_{jk} h_j(z) g_k(x)$$

with

$\quad p_{jk}$ local contribution to velocity perturbation
$\quad g_k(x)$ horizontal basis function
$\quad h_j(z)$ vertical basis function

Generally the functions g and h have a boxlike shape giving perturbations with respect to only that particular region or a triangular shape to allow a linear interpolation of the perturbations on adjacent nodes (Figure 4). The original data were recorded at 8 samples/sec. We performed a decimation to a sampling interval of 10 sec and applied lowpass filtering at 25% of the Nyquist frequency, resulting in a corner frequency of 25 mHz. Due to the flat taper used the maximum frequency present was about 35 mHz. In this way the data volume was conveniently reduced. An example is given in Figure 5.

Computation time and computer capacity forced us to use a fairly efficient inversion scheme. In the first trial runs, we used a few selected records only and mapped the more general features of the lateral heterogeneity, notably the long wavepaths between the events and the array boundaries and the average velocity deviation over the array itself. At a later stage we inserted more and more data, and used a more detailed expansion into the basis functions (vertically and horizontally) to obtain more regional information.

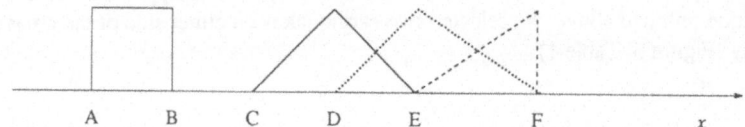

Figure 4 Schematic illustrations of the form of the basis functions $g_k(x)$. Boxcar functions allow a constant perturbation of velocity over a province between A and B. These were used for the propagation path outside the array boundaries. Triangular basis functions were used for the structure below the array itself and allow for linear interpolation (here between node D and F). In a similar way, the depth functions $h_j(z)$ were defined.

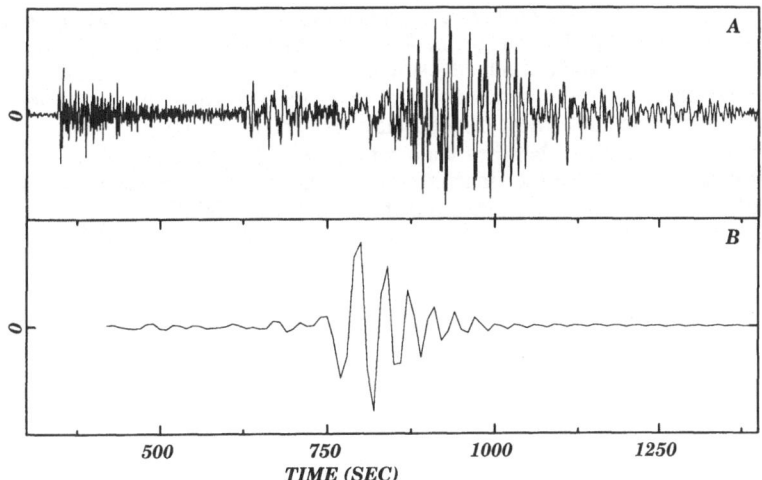

Figure 5 Broad band vertical component Nars registration of the NW Africa event (May 26, 1983) in Granada (A). The same record after low-pass filtering, decimation and windowing for group velocities between 3.0 and 7.0 km/s (B).

We started with fitting the waveform of the NW Africa event, using only the vertical record from the southernmost NARS station NE14. We employed only one horizontal basis function, extending from the epicentre to the station. In this first inversion we used a high model smoothing factor γ to allow only small perturbations with respect to the starting model. This factor was later reduced. After obtaining a good fit we decided to include the northernmost NARS station NE01 (Göteborg) in the experiment, now using two horizontal basisfunctions, covering Africa and the region under the NARS array. Some experimenting at this stage with variations in the source mechanism showed us that this was of little influence to the final variance reduction that could be obtained. This may be different for data with a complete azimuth coverage.

After this we included the NE02 and NE11 records from the second event (W. Honshu) in the data set and extended the horizontal basis functions with another region, covering the Eurasian structure between NE01 and Japan. We now found that the background model employed so far was unsatisfactory: it showed very large time delays in the synthetics of the fundamental mode with respect to the observations, even though the time of the S and SS pulses was approximately right. The wavepath of this event crosses the Verkhoyansk suture, the Anabar shield, the Kara and Barentz shelves in Asia and the Baltic shield in Scandinavia. Thus, the region crossed consists for a large part of older lithosphere, making it very likely that a low velocity region is absent or only weakly present. Some experimenting finally led us to select a rather simple background model, with a constant S velocity of 4.5 km/s between depths of 60 and 300 km, which is listed in Table 2.

Table 2: Background Model

Depth (km)	V_s (km/s)
29	4.32
40	4.40
60	4.50
300	4.50
400	4.77
400	4.93
500	5.22
600	5.52

A look at Figure 6 reveals the misfit of the synthetics generated by this starting model. We see that the observed fundamental Rayleigh mode for the Africa event is early in the first station NE14, but is apparently slowed down by the structure under Europe: when it arrives in NE06 near the Belgian-French border, the synthetics and observations align at the beginning of the fundamental mode wavetrain. Most of the delay appears to be obtained between NE12 and NE07, indicating low velocities between the central part of the array. This is also visible in the fundamental mode signals from the Honshu event, where the observed fundamental is in phase with the synthetic when the wavefront first hits the array (station NE02 in northern Denmark) but is then delayed, again in the central part, somewhere between NE05 and NE10.

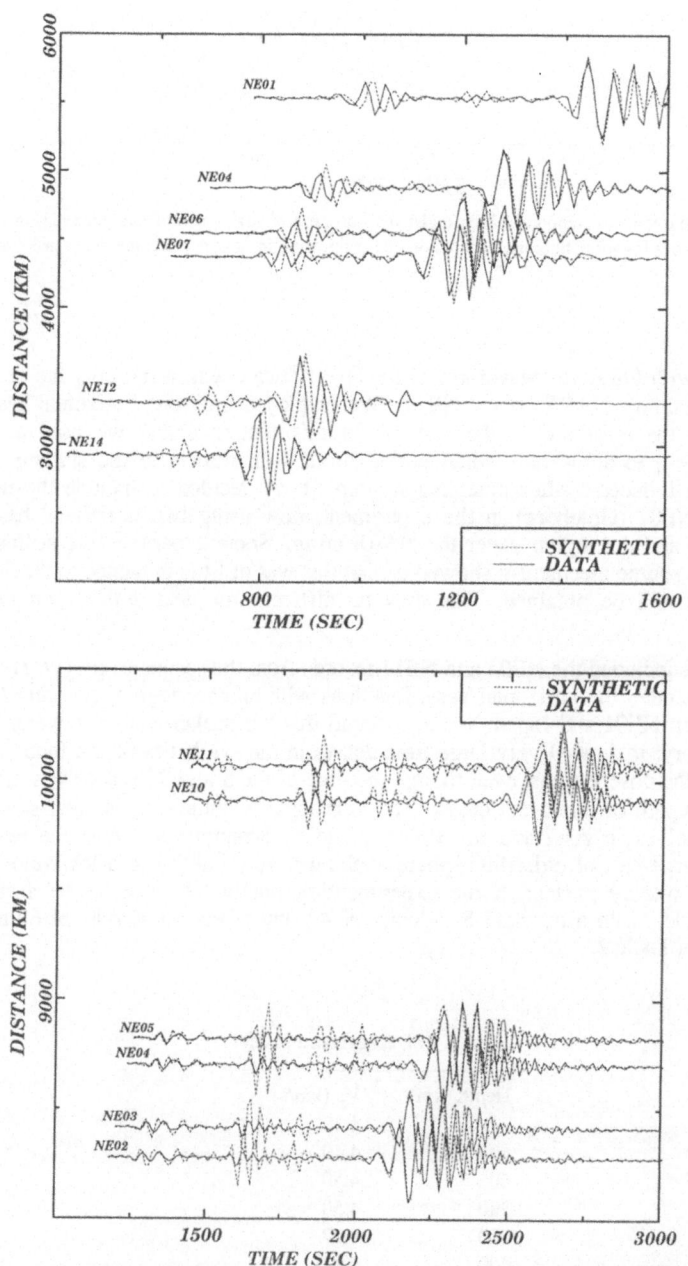

Figure 6 Initial fit of synthetic seismograms (dashed) with the data (solid) using the background model of Table 1. Top: NW Africa event, bottom: W Honshu event.

40

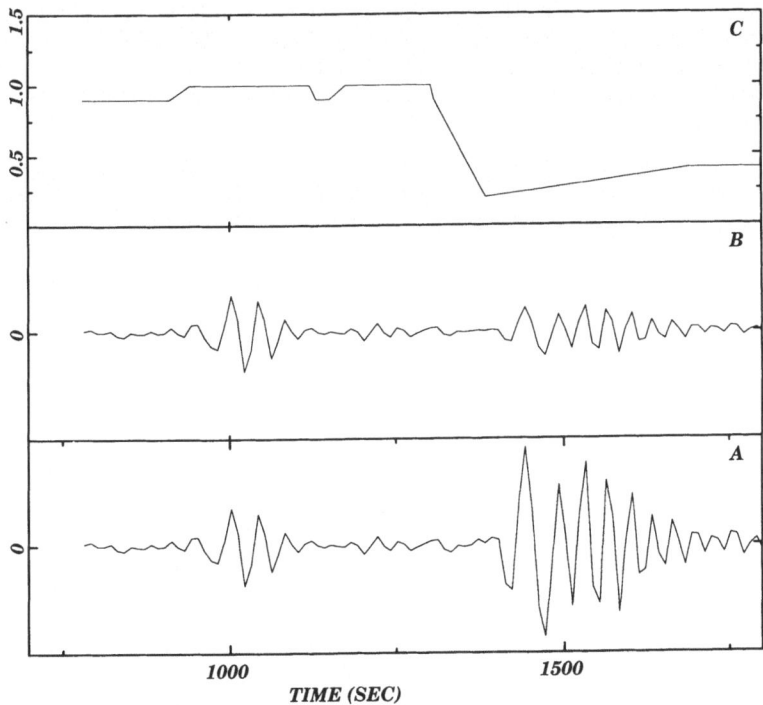

Figure 7 An illustration of the effect of variable weights in the operator *R* using the Goteborg record (NE01) of the Africa event. The original record is given in A. Multiplication of this signal with the window (C) results in a signal (B) in which the contribution of the S phase is comparable to the fundamental mode.

Although the fundamental mode from the African event aligns with the predicted wave in the northern part of the array, the S phase (near T=1000 s) arrives some 8 seconds before the predicted time in NE01. In contrast with this, both the S and SS arrivals from the Honshu event are in phase with the synthetics over most of the array.

There are some important inconsistencies with the amplitudes. For example, the S wave which is predicted in NE14 for the African event is not present in the observations. Also, the SS and SSS amplitudes for the Honshu event (arrivals near T=1600s and 2000s in NE11) are predicted with very much larger amplitudes than actually observed. Interestingly, these waves have turning points between depths of 600 and 1000 km, indicating that the data contain some definite information about the Earth at those depths. In fact, we discovered later a misprint in one of the background model input cards at a depth of 670 km. Even though this is not serious (the inversion merely tries to correct for it), we shall limit the results to be reported in this paper to the region between 0 and 600 km.

We now subdivided the region under the NARS array into three major regional zones, Spain, France (including the Paris Basin) and NW Europe and allowed for a linear velocity gradient through the use of (5) triangular basis functions. Together with the constant velocity perturbations over Africa and Eurasia we thus used 7 independent $g_k(x)$. With 8 independent depth functions $H_j(z)$, we obtained 56 parameters to be varied in the nonlinear inversion.

In the subsequent inversion runs, we found that we could very well fit the fundamental mode, but convergence slowed down when this was accomplished, even though a severe misfit was still present for the S and SS amplitudes and arrival times. A strong increase in convergence was obtained by modifying the windowing operator *R* such as to weigh different parts of the seimogram with different scaling factors. This is illustrated in Figure 7. The effect of this weighting was beneficial. In the first 14 (unweighted) iterations, the value of $F(p)$ decreased 19%. The next 25 iterations showed an appreciable slowing down of convergence (a decrease of 1%). After introduction of the new *R*, it took 17 iterations to bring $F(p)$ down another 11% -

41

yielding a total reduction of 31%. These percentage values say little by themselves, of course, as they are highly dependent on the initial value of $F(p)$. Figure 8 shows the data fit for the actual seismograms. For the African event, the S phases in the stations close to the event ($\Delta \approx 28°$) are still badly fitted in amplitude, as is the amplitude of the low frequency Rayleigh wave in NE11 and NE10.

The resulting model is given in Figure 9b. It shows a low velocity body with a velocity contrast of about 0.08 km/s between the Pyrennees and the Netherlands. The maxima of the triangular shaped basis functions $g_k(x)$ are indicated by the squares in Figure 9c. These locations

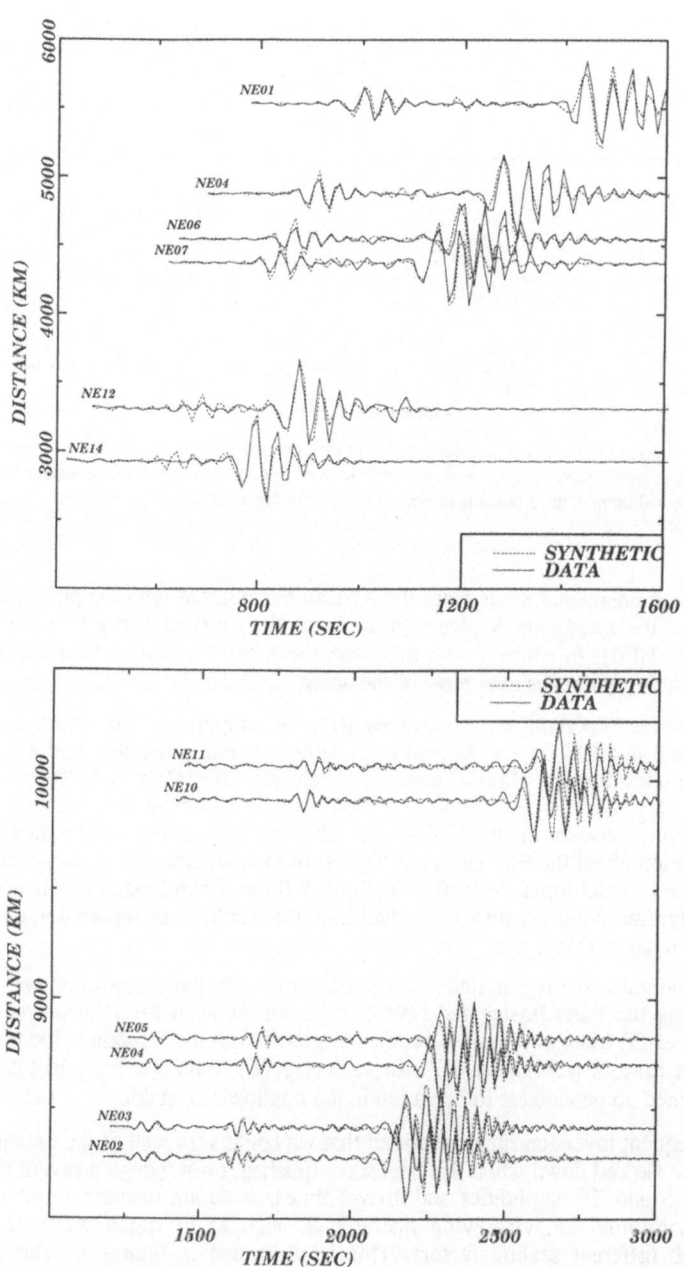

Figure 8 Final fit with background model given in figure 9b, using a parametrization according to the surface geology. Top: NW Africa event, bottom: W Honshu event.

were selected with some preconceived ideas about the subsurface geology. It could be that the selection of the basis is of considerable influence in the final result. Another factor, which needs investigation, is the fact that the half-triangles that allow for a linear interpolation in the border regions (Spain, Denmark) are more limited in extent and have therefore less influence. Could it be that Figure 9b reflects this effect instead of a really limitation in the extent of the low velocity layer under wester Europe? In order to test this we have repeated the inversion with a completely different horizontal parameterization.

The parameterization adopted in our final inversion is illustrated in Figure 10. We have selected 5 functions giving increasing detail. Triangle maxima are distibuted evenly over the full array span. Of course the first function, a boxcar over all of the array, will carry the highest weight in the inversion since all seismograms are influenced by it. We therefore expect that the algorithm will prefer a constant perturbation over the whole array if possible. The next perturbation then gives a gradient, and so on - not unlike an expansion in harmonic functions when we do a Fourier analysis.

The inversion with the new basis resulted in very much the same shape of the low velocity layer, with exception of the northern boundary, which was shifted somewhat more towards Denmark. The absolute value of the velocity perturbations increased, apparently because the hierarchy in the new $g_k(x)$ now allowed for an improved convergence speed. The resulting model, shown in Figure 9a, has a low velocity layer with a maximum velocity deviation of 0.12 km/s but an improved data fit (Figure 11).

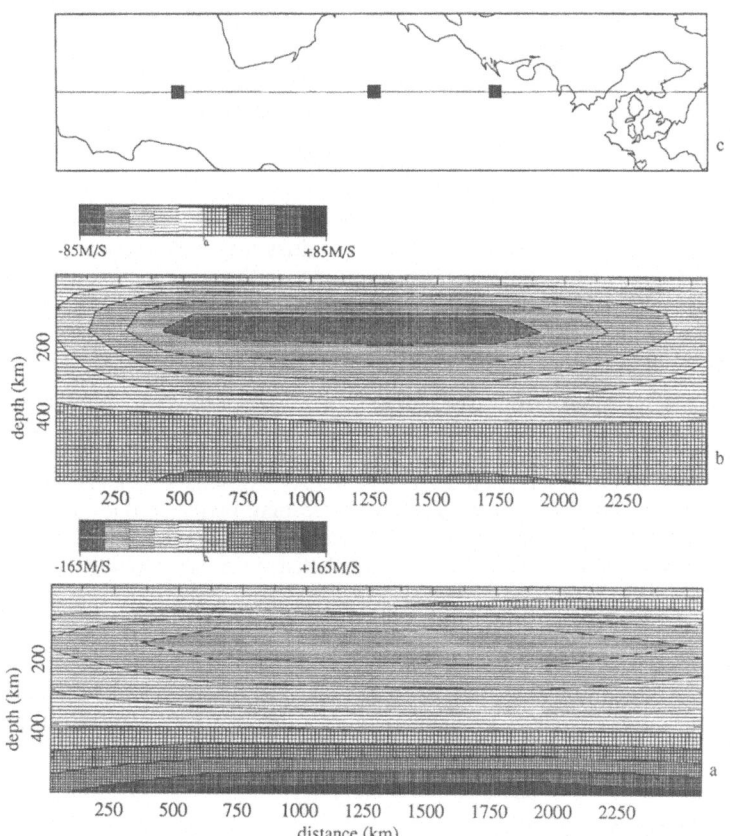

Figure 9 Final results in shear wave velocity perturbations with respect to the background model:

a) β velocity perturbations in the Upper Mantle across Western Europe. Horizontal parametrization independent from the local geology.

b) β velocity perturbations with geology dependent parametrization. The maxima of $g_k(x)$ are given by the square symbols in the top figure.

c) Geographic location of the cross section. Solid squares mark the grid-points of geology dependent basisfunctions.

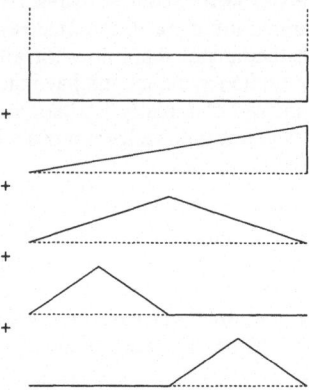

Figure 10 Basis functions $g_k(x)$ in a geology-independent parametrization spanning the length of the NARS array.

During the inversions we have experimented both with gradients calculated with 2-point differences and with 3-point difference gradients. The use of a 2-point difference formula is attractive as it saves half of the function evaluations. However, although a 2-point formula works fine in the initial iterations, once we approach the minimum we do need a higher accuracy in the gradient, which is offered by the 3-point formula. Some gain in efficiency was obtained by excluding those parameters from later iterations that had shown little potential for improvement in the initial stages. For this we used estimates of the curvature in the parameter directions.

The low velocity layer between 80-275 km depth has a minimum velocity of 4.37 km/s. Thus the velocity is somewhat higher over most of the model than the 'average' LVZ velocity in the WEPL2 model (Dost, 1987) between 60 and 260 km depth, which is 4.38 km/s. The LVZ is most pronounced under France, Belgium and The Netherlands, more southwards it disappears nearly. A steep gradient between 275-350 km corresponds exactly with the gradient in the WEPL2 model (Dost, 1987) at that depth. Between the two discontinuities from 400-670 km depth the velocity gradient under NW Europe is steeper than anywhere else in the cross section and reaches values observed by Dost (1987). We conclude that the perturbed shear velocity model obtained shows a broad agreement with the WEPL2 model from Dost (1987) except for the pronounced LID zone, which could not be modelled with our restricted vertical parametrization.

Generally the same features in the LVZ have been observed by Panza et al. (1980). In their results - based on fundamental mode measurements only - the shear wave velocity varied in the central part between 4.20 and 4.30 km/s (France) up to 4.40-4.50 at the edges (Spain, The Netherlands). The relative higher velocities in the LID zone of 4.35-4.65 km/s, which was found by these authors, is not modelled here. The difference may partly be due to the difference in crustal thickness, since a thickness in excess of the 29 km adopted here implies higher lid velocities. The low β velocity under Belgium predicted by Souriau (1980) is not present in Panza's results as it is not in ours.

Paulssen (1987) used modelling of P and S waves using NARS data from Eastern Mediterranean events and found much lower values for the low velocity region under central Europe. 4.20 - 4.38 km/s extending from 120 - 300 km. The higher velocity found for the cross section that is studied in this paper presumably reflects the relatively undisturbed nature of the Upper Mantle under western Europe near the continental margin.

HYBRID WAVEFORM INVERSION

The CPU times consumed by the nonlinear waveform inversions were several hours on a minicomputer for the inversions with all records included. Such CPU times prohibit a full 3D inversion of the whole Earth when many more recorded wave signals are inverted simultaneously. This, however, does not seem to be necessary, since most of the wavepaths will be widely different and could in principle be inverted independently. These considerations led us to propose a hybrid method that will be described in this section.

Essentially, in hybrid waveform inversion we apply nonlinear waveform inversion only to one record at a time, or only to those records that have traveled along the same minor arc (different components from the same event, for instance). For record j we then define the average velocity perturbation along the path as:

$$\overline{\delta\beta(z)}^{(j)} = \int_{S_j} [\beta(\theta,\phi,z) - \beta_0(z)^{(j)}]ds$$

Develop

$$\overline{\delta\beta(z)}^{(j)} = \sum_{i=1}^{M} p_i^{(j)} h_i(z) \tag{13}$$

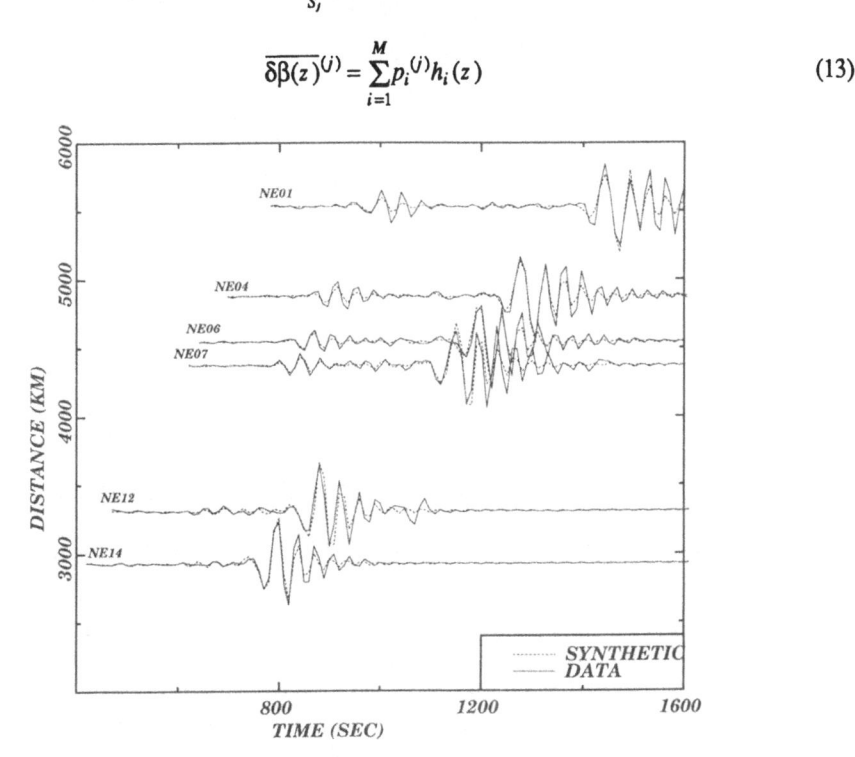

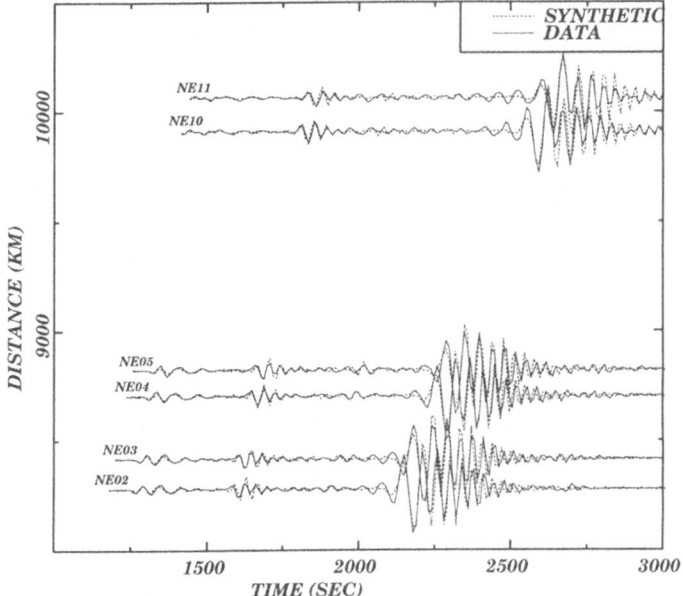

Figure 11 Final fit with background model given in figure 9a, using a parametrization without *a priori* geological information. Top: NW Africa event, bottom: W Honshu event.

We then find for the average wavenumber perturbation $\overline{\delta k_n(\omega)}^{(j)}$ for mode n along this path:

$$\overline{\delta k_n(\omega)}^{(j)} = \int_{S_j} [k_n(\omega,\theta,\phi) - k_{n0}(\omega)] ds$$

$$= \iint_{S_j} \left[\frac{\partial k_n}{\partial \beta} \right]_\omega [\beta(\theta,\phi,z) - \beta_0^{(j)}] dz ds$$

$$= \int \left[\frac{\partial k_n}{\partial \beta} \right]_\omega \overline{\delta\beta(z)}^{(j)} dz$$

$$= \sum_{i=1}^{M} p_i^{(j)} \int \left[\frac{\partial k_n}{\partial \beta} \right]_\omega h_i(z) dz$$

For each path j, $p_i^{(j)}$ may now be determined with nonlinear waveform fitting, as in the last section, taking care that the system is overdetermined, such that $p_i^{(j)}$ is unique. If we design the basis functions $h_i(z)$ such that they are orthonormal - for instance by selecting nonoverlapping layers:

$$\int h_i(z) h_j(z) dz = \delta_{ij}$$

then the waveform inversion of this one record results in M equations when we multiply (13) by $h_j(z)$ and integrate:

$$\iint_{S_j} [\beta(\theta,\phi,z) - \beta_0(z)^{(j)}] h_i(z) dz = p_i^{(j)} \quad i=1,...,M \tag{14}$$

Inverting many events will result in a large number of such linear equations that can be solved with any of the methods outlined in section 4.

Hybrid waveform inversion has several other advantages. We may add to (14) other linear information of different kind, such as delay times, in order to increase the resolution. Note also that we may adapt the background model to the general character of each wavepath (oceanic, continental) separately, thus increasing the accuracy of the perturbation theory involved in the nonlinear inversion. Hybrid waveform inversion can be applied with any perturbation theory that results in *path integrals*. Woodhouse and Wong (1986), Snieder (1987) and Snieder and Romanowicz (1987) have shown how amplitude perturbations of individual modes relate to such path integrals over the second derivative of the velocity normal to the wavepath, whereas path integrals for anisotropic perturbations are given by Romanowicz and Snieder (1987). Snieder and Nolet (1987) finally, showed how scattered surface wave signals are linearly related to the velocity perturbations as long as these are weak enough, thus opening the possibility to include information on sharp boundaries in (14) as well. Tests of the hybrid waveform inversion method have been planned and will be published elsewhere.

REFERENCES

Aki, K. and P.G. Richards, 1980. *Quantitative Seismology*, W.H. Freeman and Co., San Francisco, 932 pp.

Bullen, K. and B.A. Bolt, 1987, *Introduction to the theory of seismology*.

Dost, B., A. van Wettum and G. Nolet, 1984. The NARS array, Geol. Mijnb., 63, 381-386.

Dost, B., 1987, *The NARS array*, PhD thesis, Utrecht University, 117pp.

Dziewonski, A.M., 1984. Mapping the lower mantle: Determination of lateral heterogeneity in *P* velocity up to degree and order 6, J.Geophys.Res., 89, 5929-5952.

Dziewonski, A.M. and D.L. Anderson, 1981. Preliminary reference Earth model, Phys. Earth Planet. Inter., 25, 267-356.

Dziewonski, A.M. and A.L. Hales, 1972, Numerical Analysis of dispersed seismic waves, *Methods in Computational Physics*, 11, 39-85, Acad. Press, New York.

Gilbert, J.F. and A.M. Dziewonski, 1975. An application of normal mode theory to the retrieval of structural parameters and source mechanisms from seismic spectra, Phil. Trans. Roy. Soc. Lond., 278, 187.

Gill, P.E., W. Murray and M.H. Wright, *Practical Optimization*, Acad. Press, London, 401pp.

Hager, B.H., Clayton, R.W., Richards, M.A., Comer, R.P., Dziewonski, A.M., 1985. Lower mantle heterogeneity, dynamic topography and the geoid. Nature, **313**, 541-545.

Nolet, G., 1985. Solving or resolving inadequate and noisy tomographic systems. J. Comp. Phys., **61**, 463-482.

Nolet, G. (ed.), 1987a. *Seismic Tomography, with Applications in Global Seismology and Exploration Geophysics*, Reidel, Dordrecht, 386pp.

Nolet, G., 1987b. Seismic wave propagation and seismic tomography, *in: Seismic Tomography*, ed. G. Nolet, Reidel, Dordrecht.

Nolet, G., 1987c. Waveform tomography, *in: Seismic Tomography*, ed. G. Nolet, Reidel, Dordrecht.

Nolet, G. and G.F. Panza, 1976. Array analysis ofseismic surface waves: limits and possibilities, Pure Appl. Geophys., **114**, 776-790.

Nolet, G., B. Dost and H. Paulssen, 1986. Intermediate wavelength seismology and the NARS experiment, Ann. Geophys, **4**,, 305-314 (with erratum in Ann. Geophys., **4**, 593, 1987).

Nolet, G., J. van Trier and R. Huisman, 1986. A formalism for nonlinear inversion of seismic surface waves, Geophys. Res. Lett., **13**, 26-29.

Morelli, A. and A.M. Dziewonski, 1987, The harmonic expansion approach to the retrieval of deep Earth structure, in: *Seismic Tomography*, ed. G. Nolet, Reidel, Dordrecht, 251-274.

Paige, C.C., Saunders, M.A., 1982. LSQR: An algorithm for sparse linear equations and sparse least squares. ACM Trans. Math. Softw. **8**, 43-71.

Panza, G.F., 1976, Phase velocity determination of fundamental Love and Rayleigh waves, Pure Appl. Geophys., **114**, 753-764.

Panza G.P., St. Mueller and G. Calcagnile, 1980. The gross features of the lithosphere-asthenosphere system in Europe from seismic surface waves and body waves, Pageoph, **118**, 1209-1213.

Paulssen, H., 1987. Lateral heterogeneity of Europe's upper mantle as inferred by modelling of broad band body waves, Nars progress rep., **4**, 35-62.

Romanowicz, B. and R. Snieder, 1987. A new formalism for the effect of lateral heterogeneity on normal modes and surface waves: - II: General anisotropic perturbations, subm. to Geophys. J. R. astr. Soc.

Snieder, R., 1986. 3-D linearized scattering of surface waves and a formalism for surface wave holography, Geophys. J. R. astr. Soc., **84**, 581-605.

Snieder, R., 1987. *Surface wave scattering theory*, PhD thesis, University of Utrecht, 175pp.

Snieder, R. and G. Nolet, 1987, Linearized scattering of surface waves on a spherical Earth, J. Geophys., **61**, 55-63.

Snieder, R. and Romanowicz, B., 1987. A new formalism for the effect of lateral heterogeneity on normal modes and surface waves - I: isotropic perturbations, perturbations of interfaces and gravitational perturbations, submitted to Geophys. J. R. astr. Soc.

Spakman, W., and G. Nolet, 1987, Imaging algorithms, accuracy and resolution in delay time tomography, in: *Mathematical Geophysics, a survey of recent developments in seismology and geodynamics*, eds. N.J. Vlaar et al., Reidel, Dordrecht.

Souriau, A., 1981. Le manteau superiéur sous la France, Bull. Soc. géol. France, **23**, 65-81.

Van der Sluis, A., and H.A. Van der Vorst, 1987, Numerical solution of large, sparse linear algebraic systems arising from tomographic problems, in: *Seismic Tomography*, ed. G. Nolet, Reidel, Dordrecht, 49-83.

Wielandt, E., 1987, On the validity of the ray approximation for interpreting delay times, in: *Seismic Tomography*, ed. G. Nolet, Reidel, Dordrecht, 85-98.

Woodhouse, J.H. and A.M. Dziewonski, 1984. Mapping the upper mantle: three-dimensional modelling of Earth structure by inversion of seismic waveforms, J. Geophys. Res., **89**, 5953-5986.

Woodhouse, J.H. and Wong, Y.K., 1986. Amplitude, phase and path anomalies of mantle waves. Geophys. J. R. astr. Soc., **87**, 753-774.

WAVEFORM SYNTHESIS BY RAY THEORETICAL METHODS

Raúl Madariaga

Laboratoire de Sismologie
Université Paris 7 et Institut de Physique
de Globe, 4 Place Jussieu, Tour 14
75252 Paris Cedex 05, France

1. INTRODUCTION

The purpose of these notes is to present in a relatively simple form several techniques to construct synthetic seismograms in the high frequency regime. The simplest and most widely used of the high frequency methods is classical geometrical ray theory, which is the basis of most practical methods for the modeling of seismograms and the inversion of travel times in seismology and applied geophysics. A comprehensive discussion of several aspects of ray theory may be found in the books of Cerveny et al. (1977) and Bleistein (1984), and in the notes by Burridge (1976). Many programs that perform ray tracing have been written in order to calculate travel times and synthetic seismograms based on ray theory. One of the most difficult problems in generating synthetics was the calculation of geometrical spreading. A major advance in this field was made by Popov and Psencik (1978) who proposed a new technique for the calculation of geometrical spreading. This method is usually called dynamic ray tracing. Later work demonstrated that dynamic ray tracing was in fact a subset of what is called paraxial ray theory in optics, i.e., the study of rays propagating in the vicinity of another ray. This method is new to seismology but it has been used in optics for several decades (see, for example, Deschamps, 1972). In fact, as demonstrated by Farra and Madariaga (1987), paraxial ray theory, dynamic ray tracing, Gaussian beams and many other problems in ray theory can all be derived from ray perturbation theory as described in all Classical Mechanics books (e.g., Landau and Lifschitz, 1981; Goldstein, 1981). Recognition of this similarity between ray theoretical methods and classical analytical mechanics dates back to the last century and was the major contribution of Hamilton to modern physics. With the identification of paraxial ray theory as perturbation theory a wealth of powerful theoretical methods becomes available. For instance, the propagation of paraxial rays across curved interfaces reduces to a simple canonical transformation problem as shown by Farra and Madariaga (1987). A few simple applications of Hamiltonian methods to seismic ray theory will be presented in these notes.

Classical ray theory presents a number of practical problems due to the presence of singularities of the ray field (caustics and focal points), and numerical instabilities due to small scale perturbations in the velocity model. In the last ten years or so, a number of methods based on the spectral decomposition of the wavefield at the source have been proposed in order to alleviate some of these problems. Among these methods the best known are WKB for vertically stratified velocity models (Chapman, 1978), Maslov or asymptotic Fourier transforms (Chapman and Drummond, 1972) and Gaussian beams (Popov, 1982 and Cerveny et al., 1982). In the later part of this paper we will discuss these various techniques in the simple case of a homogeneous medium and we will show that they all have a common theoretical background in ray theory.

The basic methods that will be used in these notes are ray theory and ray perturbation theory. Ray theory is used to trace rays and determine travel times for a given set of initial conditions, for instance a point source. Ray perturbation theory is used to evaluate ray amplitudes and to iteratively solve two-point ray tracing problems. The most important perturbations in this context are perturbations of initial and final values of position and slowness. Perturbation theory provides a method to calculate the trajectories of paraxial rays that propagate in the vicinity of a reference ray. The study of the divergence and convergence of paraxial rays provides a method for the calculation of ray amplitudes and travel time extrapolation for different conditions at the source: plane waves, Snell-waves, point sources, Gaussian beams and other wavefront configurations that are needed in waveform synthesis and inversion. Paraxial rays provide also a natural way to interpolate rays so as to solve the two point ray tracing problem: to find a ray that passes through a given source and receiver. Thus we find a common background to such apparently disparate techniques as Gaussian beam summation, ray bending, and the WKB method.

2. CLASSICAL RAY THEORY

Ray theory is based on an ansatz or hypothesis about the form of the solution of elastic field. As proposed by Babich (1956) and Karal and Keller (1959) we look for elastic waves of the form:

$$u(x,\omega) = A(x,x_o,\omega)\sqrt{\frac{\rho_o c_o}{\rho c J(x,x_o)}}\, e^{i\omega\theta(x,x_o)} \tag{1}$$

where $u(x,\omega)$ is the Fourier transformed displacement at point x in the elastic medium and ω is the circular frequency, ρ and c are the density and wave velocity. This expression is valid both for P and S waves. For the former $c = \alpha$, the P-wave velocity, while for the latter $c = \beta$, the shear wave velocity. $\theta(x,x_o)$ is diversely known as the eikonal, phase or travel time function. In (1) we have explicitly introduced the position of the source x_o as well as the density ρ_o and velocity c_o at this source. The parameter $J(x,x_o)$ appearing under the square root is the ray Jacobian or geometrical spreading of the wavefronts; it will be defined later in the paper. In many applications J may be negative or complex so that the proper branch of the square root of J in (1) should be chosen. The vector amplitude $A(x,x_o,\omega)$ is a complex function of x. From a strict theoretical point of view x_o should not appear in (1), but since we know that solutions propagate along rays we introduce the source and geometrical spreading from the beginning. A more detailed formal development of (1) may be found in Cerveny (1985) or in the references cited in that paper.

In expression (1) there is no approximation since A is a general function of x and ω. In order to obtain the ray theoretical approximation we expand the vector amplitude into a series of inverse powers of ω:

$$A(x,x_o\omega) = s(\omega) \sum_{i=0}^{\infty} A_i(x,x_o)\omega^{-i} \tag{2}$$

and retain only the first few terms. In practice, however, only the lowest order term in the series is actually used. In this case, vector A_o contains the polarization of the wave, the radiation pattern of the source and, if appropriate, a product of reflection and transmission coefficients. For P-waves the polarization is along the ray direction, while S-waves are polarized on a plane that is tangent to the wavefront. In (2), $s(\omega)$ is the source wavelet which contains information about the development of rupture for natural earthquakes, or the source time function for explosive sources. Using only the first term in the series (2), (1) reduces to

$$u(x,\omega) = A_o(x,x_o)\sqrt{\frac{\rho_o c_o}{\rho c J(x,x_o)}}\, s(\omega)e^{i\omega\theta(x,x_o)}. \tag{3}$$

In some applications of ray theory, such as Gaussian beams, or in the presence of attenuation θ, J and A_o may all be complex.

Equation (3) is an approximation to the wave equation valid only at high frequencies when the higher order terms in (2) may be ignored. The ray theoretical approximation assumes that A_o, J and θ are slowly varying functions of space; the only rapidly varying term in (3) being the exponential. This form of the solution simplifies the calculation of seismograms in a substantial way. It is in fact simple to do the inverse Fourier transform of (1) in order to obtain the time-domain version of (3). Since $u(x,t)$ is a real function, the Inverse Fourier Transform has the following form:

$$f(t) = \frac{1}{\pi} \operatorname{Re}\left\{\int_o^{\infty} f(\omega)e^{-i\omega t}d\omega\right\} \tag{4}$$

where Re denotes the Real part of the complex function in brackets. The inverse of (3) is straightforward:

$$u(x,t) = \frac{1}{\pi}\, s(t)\, * \operatorname{Re}\left\{A_o\sqrt{\frac{\rho_o c_o}{\rho c J}}\,\frac{1}{i(t-\theta-i\Delta t)}\right\}, \tag{5}$$

here Δt is a very small real quantity that is used to pull a possible pole at $t = \theta$ out of the Real t axis. Once the real part is evaluated Δt can be made to tend to zero, in which case (5) may be written in terms of generalized functions or distributions. In the computer implementation of (5) it is preferable to keep a small positive Δt in order to stabilize the numerical evaluation of this expression near the pole at $t = \theta$. As proposed by Madariaga and Papadimitriou (1985), a convenient value for Δt is the time step used in the discrete evaluation of u. Comparing with (3), the inclusion of this small imaginary part is equivalent to multiplying the frequency domain expression (3) by $f(\omega) = \exp(-\omega\Delta t)$ for $\omega > 0$. The time domain transform of this function is:

$$f(t) = \frac{\Delta t}{t^2 + \Delta t^2}. \tag{6}$$

Thus, adding the small imaginary part $i\Delta t$ to θ is equivalent to a convolution of the time domain displacement u with the function (6). If Δt is equal to the time step, the effect of (6) is practically negligible below one half the Nyquist frequency of the signal.

When the travel time $\theta(x,x_o)$ is real we can evaluate (5) in a more familiar form in terms of the source time function $s(t)$ and its Hilbert transform. Letting $\Delta t \to 0$, we get

$$
\begin{aligned}
u(x,t) = \mathrm{Re} &\left\{ A_o \sqrt{\frac{\rho_o c_o}{\rho c J(x,x_o)}} \right\} s[t - \theta(x,x_o)] \\
+ \mathrm{Im} &\left\{ A_o \sqrt{\frac{\rho_o c_o}{\rho c J(x,x_o)}} \right\} s^\dagger[t - \theta(x,x_o)]
\end{aligned}
\tag{7}
$$

where $s^\dagger(t)$ is the Hilbert transform of $s(t)$:

$$
s^\dagger(t) = \frac{1}{\pi} \, \mathrm{P.V.} \int_{-\infty}^{\infty} \frac{s(\tau)}{\tau - t} \, d\tau = \frac{1}{\pi} \, \mathrm{Im} \left\{ \int_o^\infty s(\omega) e^{-i\omega t} d\omega \right\}.
\tag{8}
$$

P.V. denotes the principal value of the integral. The last Fourier integral provides the most practical way of computing the Hilbert transform of $s(t)$. It is easy to see that in most computer calculations it is preferable to use (5) to calculate time-domain seismograms.

Let us return now to the ray theoretical expression (3). The main property of ray theoretical seismograms is the clear separation between the kinematics of rays and wavefronts represented here by θ, and the amplitudes and waveforms controlled by J, A_O and $s(t)$. At high frequencies the quantities outside the exponential vary slowly with position, while the exponential term varies very fast because ω is large. This simplicity of ray solutions comes from the neglect of the interaction of the waveform with the heterogeneities of the propagation medium. there is no scattering along the ray trajectory: rays are bent and deviated by the structure but energy is conserved along ray tubes. As shown by (7), in classical ray theory (θ is real) the only effect of propagation upon waveforms is an eventual Hilbert transformation (phase shift) of the signal. Unfortunately, the limits of validity of this approximation are difficult to evaluate and, except in simple cases, there is no general method for determining the validity of ray theory. A recent discussion of the limits of applicability of ray theory may be found in Ben Menahem and Beydoun (1986).

3. RAY TRACING

In order to calculate the different terms that make up the ray theoretical ansatz (3) we substitute it in the elastodynamic equations:

$$
\rho(x)\omega^2 u(x,\omega) = \mathrm{Div} \, \sigma(x,\omega)
\tag{9}
$$

where $\rho(x)$ is the density and $\sigma(x,\omega)$ is the stress tensor which is related to strain by the Lame parameters $\lambda(x)$ and $\mu(x)$:

$$
\sigma(x,\omega) = \lambda(x)\mathrm{Div} \, u(x,\omega) \, I + 2\mu(x)\varepsilon(x,\omega)
\tag{10}
$$

where I is the identity matrix and ε is the strain tensor:

$$
\varepsilon = \frac{1}{2} [\mathrm{Grad} \, u + (\mathrm{Grad} \, u)^T]
\tag{11}
$$

and the superscript T denotes transposition.

After collecting terms of the same order in ω one finds two sets of independent solutions (see, for example, Cerveny et al., 1977). From the highest order terms in ω one finds for P waves:

$$
(\nabla\theta)^2 = \alpha^{-2} \qquad A_o \times (\nabla\theta) = 0
\tag{12}
$$

and for S waves:

$$(\nabla\theta)^2 = \beta^{-2} \qquad A_o \cdot (\nabla\theta) = 0. \qquad (13)$$

From the next order term in ω, we get the equations for the Jacobian J, which will be discussed later in the text.

In order to interpret these equations we define as in Figure 1, a wavefront as the surface $\theta(x)$ = constant. Then the vector

$$p = \nabla\theta$$

is perpendicular to the wavefront and its length is the slowness of the P-wave, α^{-1}, or of the S-wave, β^{-1}. p is simply the slowness vector. The right hand side equations in (12) and (13) define the polarization of vector A_o. This is parallel to p for P-waves, and perpendicular to it for S-waves.

The left hand side of equations (12) and (13) may be rewritten in the standard form:

$$(\nabla\theta)^2 = u^2 \qquad (14)$$

where $u = c^{-1}$ stands for the slowness α^{-1} of P-waves or β^{-1} of S-waves, and c for the corresponding wave velocity. This is a first-order non-linear equation for the travel time θ, that is usually called the eikonal equation, from eikon, image in greek. The standard method to solve it is the method of characteristics as developed, for instance, by Courant and Hilbert (1966). In ray theory the characteristics are called rays. Let us define the wavefronts as surfaces of equal travel time in x space: $\theta(x,x_o) = t$ = constant. The characteristics of the eikonal equations, or rays, are defined as the trajectories that are orthogonal to the wavefronts. The set of rays and wavefronts depend on the initial conditions for θ. Given an initial wavefront $t_o = \theta_o(x,x_o)$, the rays and successive wavefronts may be calculated by ray tracing.

In order to obtain the ray tracing equations we introduce the ray coordinate s, a parameter that measures position along the ray. There are many choices for this parameter, for instance it may be the curvilinear distance along the ray, the travel time θ itself, or other discussed by Cerveny (1985). In the following we will use curvilinear distance s as the ray parameter. The ray tracing equations may be easily rewritten for any of the other ray parameters using the Hamiltonian formulation to be discussed later in the paper. From Figure 1, we remark that

$$p = \nabla\theta = u(x) \frac{dx}{ds} \qquad (15)$$

so that the slowness is parallel to the local ray tangent dx/ds. This is the first ray tracing equation, the other one may be obtained taking the gradient of the eikonal equation (14). We write the ray tracing system in the following way:

$$\frac{dx}{ds} = u^{-1}p = cp$$
$$\qquad (16)$$
$$\frac{dp}{ds} = \nabla u.$$

The latter equation is closely related to ray curvature. It shows that rays deviate from a straight trajectory because of the gradient of the slowness. The ray curvature is actually given by:

$$\kappa = u^{-1}n.\nabla u = -c^{-1}n.\nabla c \qquad (17)$$

where n is the unit normal to the ray.

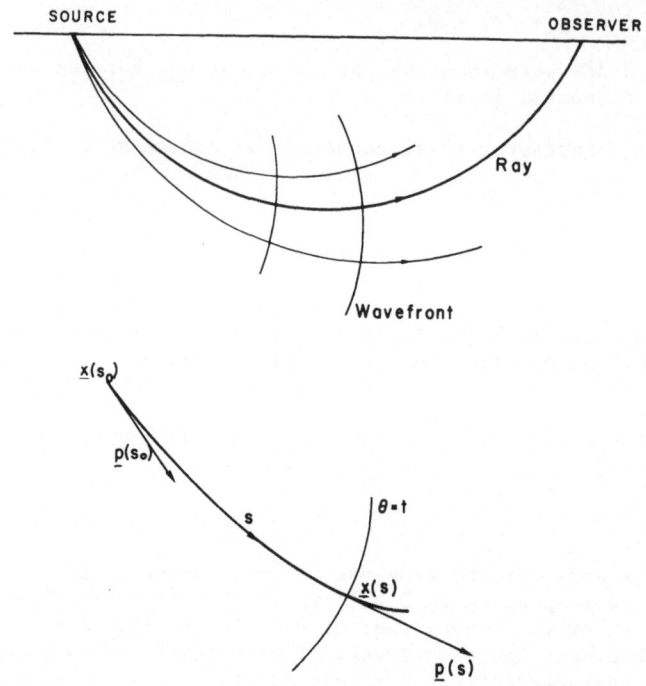

Fig. 1. Geometry of Rays and Wavefronts, p denotes the slowness vector.

Solution of (16) requires the specification of initial or boundary
conditions. The simplest problem is to specify the initial position
$x(s_0)$, and slowness vector $p(s_0)$ for each ray on some initial surface.
The direction of the initial slowness vector is arbitrary but its norm has
to satisfy the eikonal equation (14). For a point source, for instance,
$x(s_0)$ is the same for all the rays, while $p(s_0)$ changes from ray to ray.
Once the initial conditions are specified the ray tracing system (16) can
be integrated numerically, for instance, by the Runge Kutta method. There
exist also a few models of slowness or velocity distribution for which
equations (16) may be integrated analytically. Let us remark that the
system of six ray tracing equations (16) has to be solved together with
the eikonal equation (14) so that in fact only 5 of the equations are
independent. In practice, when (16) is being solved numerically, the
eikonal equation (14) can be used as a consistency check. Other ways to
reduce the system based on coordinate transformations will be discussed in
section 5. Once the rays have been traced, the travel time $\theta(x,x_0)$ may be
calculated by direct integration of

$$\frac{d\theta}{ds} = u \tag{18}$$

along each ray.

Solution of the initial value ray tracing problem is relatively
straightforward. In most seismological applications, however, the usual
problem is to trace a ray that passes through two points x_0 and x_1. In
this case, one has to find the initial value of the slowness p_0 for the
ray that satisfies the two boundary conditions. This problem may be
solved by iterative methods using the paraxial ray tracing techniques to
be discussed later in this paper.

Given appropriate initial conditions, the set of rays and wavefronts
is uniquely determined in those regions of space that are illuminated by

54

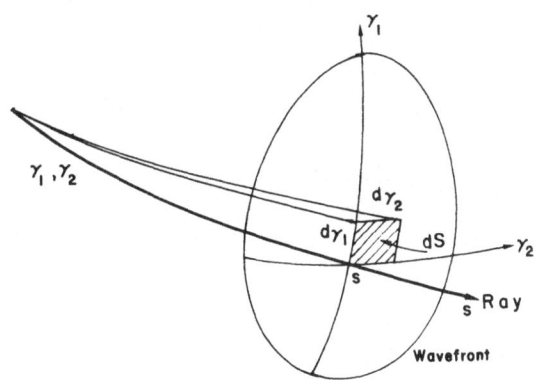

Fig. 2. Definition of the Ray coordinates (γ_1, γ_2) and the elementary surface area on a wavefront dS.

the initial data. Because the ray tracing system (16) is nonlinear the ray field may present singularities. In order to understand these problems and to determine geometrical spreading we remark that the set of rays and wavefronts form a curvilinear coordinate system. As shown in Figure 2 we introduce orthogonal curvilinear coordinates γ_1 and γ_2 on the wavefronts in addition to the ray coordinate s. Each pair (γ_1, γ_2) defines a ray. The curvilinear coordinate set (s, γ_1, γ_2) defined in this form is usually called the ray coordinate system. Any point P in the region illuminated by rays may be defined by its ray coordinates. In this coordinate system the volume element is

$$dV = dxdydz = J(x, x_o)dsd\gamma_1 d\gamma_2 \qquad (19)$$

where J is the Jacobian of the transformation from cartesian to ray coordinates. Since ds is a curvilinear abscissa along the ray the cross section dS (see Figure 2) of a beam of rays defined by the four rays with coordinates γ_1, γ_2, $\gamma_1 + d\gamma_1$ and $\gamma_2 + d\gamma_2$ is given by

$$dS = J(x, x_o)d\gamma_1 d\gamma_2 \qquad (20)$$

so that in fact J is a measure of the variation of the cross section of this beam. J is usually called geometrical spreading, because it measures the spreading of the wavefront around the ray (γ_1, γ_2).

We can now explain the presence of $J^{-1/2}$ in the expression for ray theoretical seismograms (3). Elastic energy flow across a wavefront element of cross section dS, see Figure 2, is:

$$dE = 1/2 \, \rho c |\dot{u}|^2 dS = 1/2 \, \rho c |\dot{u}|^2 J d\gamma_1 d\gamma_2.$$

Since in the ray approximation energy flows along a beam of rays without lateral scattering, the energy flux

$$F = dE/d\gamma_1 d\gamma_2 = \rho c J |\dot{u}|^2$$

is conserved. Therefore the amplitudes of the ray theoretical velocity and displacement field are necessarily of the form:

$$|u| = \Phi \sqrt{\frac{\rho_o c_o}{\rho c J}}$$

where ρ_o and c_o have been introduced for convenience and Φ is a source excitation factor to be computed from the solution of a certain canonical problem.

As mentioned earlier, the transformation to ray coordinates may be singular. Near these singularities the usual expressions of ray theory as given by (3) fail and other methods, like WKB or Gaussian beam summation, have to be used. The most common singularity is a caustic which appears when $J \to 0$. An example of simple caustic formation in a two-dimensional reflexion seismogram is shown in Figure 3. The medium is homogeneous but the reflector is curved, rays deflected by the interface cross each other forming a cusp. We observe that the set of rays is tangent to two curves or caustics. Inside the area delimited by the two caustics three rays arrive at each observer, while only one ray reaches observers outside the caustics. The geometry of caustics may be described by catastrophe theory as shown, for instance, by Nye (1985).

Finally, let us return to the expression (18) for the computation of travel times along a ray. Actually, travel times may be calculated not just by integration along rays but by integration along any curve that may be convenient for the problem at hand. In fact since $p = \nabla\theta$, for any curve Γ joining two points x_1 and x_2 in the region illuminated by rays we get

$$\theta_2 = \theta_1 + \int_\Gamma p.dx \tag{21}$$

where θ_1 and θ_2 are the wavefronts passing through points x_1 and x_2, respectively. The curvilinear integral (21) is path independent in the regions where the ray field is regular. As long as no caustics are crossed the path may be chosen arbitrarily. This curvilinear integral will be used later to calculate wavefront approximations for paraxial rays and beams.

4. VARIATIONAL FORMULATION

The ray tracing problem has been posed so far in its differential form. The problem may also be posed in a variational form which may be used to develop alternative methods of solution of these equations, to introduce perturbation theory, to calculate wavefronts, etc. The starting point for this formulation is Fermat's principle which may be stated in the following form: among all trajectories joining two fixed points x_0 and x_1, a ray is the trajectory for which the travel time is stationary. We write this condition in the form:

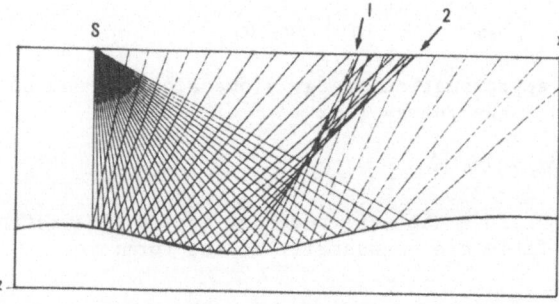

Fig. 3. Formation of a caustic in a reflection seismogram. The change in curvature of the reflector produces two caustics (1 and 2) in the reflected rays issued from a point source S.

$$\delta\theta(x_o, x_1) = \delta \int_o^1 u(x) \left| \frac{dx}{ds} \right| ds = 0 \tag{22}$$

where s is as before the curvilinear distance along the ray. Let us note that since the ray tracing problems are highly nonlinear several rays may satisfy the variational condition (22). The rays that render (22) stationary may be found by standard techniques of the calculus of variations. The solution satisfies the following system of Euler equations:

$$\frac{d}{ds} \left(u \frac{dx}{ds} \right) - \nabla u = 0 \tag{23}$$

which, taking into account (15) is seen to be equivalent to the ray tracing equations (16). This result demonstrates the equivalence between the variational formulation based on Fermat's principle and the standard ray tracing system derived from the eikonal equation.

Let us note that the variational principle (22) looks for an extremal trajectory without any constraint upon s. The total distance s from x_o to x_1 is allowed to change so that the time is extremal. Fermat's principle corresponds to the so-called Maupertius principle of minimum reduced action in analytical mechanics. A more powerful variational principle is that of Hamilton that states that the functional (22) is to be minimized under the constraint that the value of s at the initial point x_o and at the final point x_1 remain constant. The perturbations to the functional that we consider are virtual so that they do not satisfy the condition that $|dx/ds| = 1$. Thus s has to be considered as a truly independent parameter. Under this condition we rewrite $|dx|$ in the form:

$$|dx| = \sqrt{ds^2 - (|dx|^2 - ds^2)}$$

where $|dx|^2 - ds^2$ is the variation of the squared length vector. For small variation we can approximate:

$$|dx| \simeq \frac{1}{2} ds(1 + |\dot{x}|^2)$$

where $\dot{x} = dx/ds$. We can write Hamilton's principle in the standard form:

$$\delta\theta = 0 = \delta \int_o^1 L ds \tag{24}$$

where the Lagrangian function is:

$$L(x, \dot{x}) = \frac{1}{2} u(x)(1 + |\dot{x}|^2).$$

On the true ray the variation is equal to zero and $|\dot{x}| = 1$, so that the travel time function $\theta = \int uds$ obtained from Fermat's principle (22) and for Hamilton's principle (24) are the same. Otherwise, these two principles are quite different.

We can now introduce the Hamiltonian formulation of the ray tracing problem. For that purpose we remark that the ray tracing equations define a ray by a couple of variables x(s), p(s). x(s) describes the ray in configuration space, the physical space where rays are being traced. Rays are more naturally described in a six dimensional phase space, where the variables are x and p. Rays are trajectories in this space, where they have a number of well-known properties. For instance, two distinct ray trajectories never cross each other, because two rays that pass through the same point in phase space are identical. In order to obtain the Hamiltonian we introduce the Legendre transformation:

$$H(x, p) = p\dot{x} - L(x, p)$$

where the derivative $\dot{x}$ is replaced by the generalized momentum p using the classical relationship $p = \partial L / \partial \dot{x} = u(x)\dot{x}$. Comparing with the definition of slowness in (15) we see that the generalized momentum associated with the coordinate x is the local slowness vector. Thus we find the Hamiltonian:

$$H(x,p) = \frac{1}{2} u^{-1}(x)[p^2 - u^2(x)] \tag{25}$$

The term in brackets is just the eikonal equation (14) so that the Hamiltonian is constant and equal to zero for all ray trajectories. Finally, the ray equations (16) may be found in a straightforward manner from Hamilton's canonical equations (Landau and Lifschits, 1981 or Goldstein, 1981). These equations state that the trajectories that satisfy the variational principle (24) are given by the following set of canonical equations:

$$\frac{dq}{ds} = \frac{\partial H}{\partial p} \qquad \frac{dp}{ds} = -\frac{\partial H}{\partial q} \tag{26}$$

where, following conventional notations in analytical mechanics, q stands for the vector x, p for the vector p and partial derivatives are to be interpreted as gradients in phase space. It may be easily verified inserting the Hamiltonian (25) in the canonical equations (26) that these equations are just a shorthand notation for the ray tracing equations (16).

5. REDUCED RAY TRACING SYSTEMS

We remarked above that the standard ray tracing system (16) contains only five independent equations because p has to satisfy the eikonal equation $p^2 = u^2$. In addition to this, the ray parameter s is not really independent and may also be eliminated taking any of the q or p coordinates as the independent variable. The new ray tracing system will contain only four equations. The best way to implement this reduction of the ray tracing equations is to use the Hamiltonian formulation. In fact, if we redefine the Hamiltonian in any convenient way that satisfies the eikonal equation we can generate new ray tracing equations using the canonical equations (26). We will present two examples of reduction for the two most frequently used coordinate systems: cartesian coordinates in this section, and ray centered coordinates in the following.

In most ray tracing problems in the Earth it is natural to consider ray tracing as a function of depth. This is always done in the case of vertically varying media (see, for example, Bullen and Bolt, 1986) where the ray tracing equations reduce to only 2 equations because the ray is contained in a plane through the source and the observer. In this section we will consider that the vertical coordinate z is used as an independent parameter. In this case we define a reduced Hamiltonian:

$$H_r(q,p,z) = -p_z = -\sqrt{u^2(q,z) - p^2} \tag{27}$$

where q stands for the two-dimensional position vector $q = (x,y)$ and p for the two-dimensional conjugate momentum $p = (p_x, p_y)$. The reduced Hamiltonian (27) was obtained solving for p_z from the original Hamiltonian (25). Thus, the complete slowness vector (p_x, p_y, p_z) still satisfies the eikonal equation. The reduced Hamiltonian (27) contains the same information as (25), but it depends on only four variables. This apparent simplicity is offset by the fact that H_r depends on the independent variable z. Inserting (27) in the canonical ray tracing equations (26) we find explicitly:

$$\frac{dx}{dz} = \frac{p_x}{p_z}$$

$$\frac{dy}{dz} = \frac{p_y}{p_z}$$

$$\frac{dp_x}{dz} = \frac{\partial u/\partial z}{p_z} \tag{28}$$

$$\frac{dp_y}{dz} = \frac{\partial u/\partial z}{p_z}$$

where p_z is given by (27). This form of the ray tracing system is entirely equivalent to (15). One may at first sight think that it is easier to solve (28) in order to trace rays in cartesian coordinates. This is not necessarily so because solutions to (28) may be multiple-valued. This is typically the case in seismology where rays penetrate to a maximum depth and then return to the surface. At the maximum depth, d/dz is not defined, or equivalently $p_z = 0$, so that special care must be taken to integrate (28) across this point.

6. RAY CENTERED COORDINATES

In the course of the study of rays in a multimirrored resonator Popov (1969), see also Babich and Buldyrev (1972), introduced an orthogonal curvilinear coordinate system centered around a reference curve. Later, Popov and Psencik (1978), Psencik (1979), Cerveny and Hron (1980) among many others, found that this coordinate system centered around a reference ray was very convenient to calculate geometrical spreading. The procedure to calculate geometrical spreading by this method is usually called dynamical ray tracing. The ray centered coordinate system was also found to be useful for the development of paraxial ray theory and one of its main applications: the calculation of Gaussian beams (see, for example, Cerveny, Klimes and Psencik, 1984; Madariaga, 1984; Klimes, 1984; Cerveny, 1985). Recently, Farra and Madariaga (1987) showed that this coordinate system is also very convenient for the calculation of ray perturbations and the modeling of slightly heterogeneous media. The ray centered coordinate system will be discussed in some detail in the following.

Referring to Figure 4, we consider a curve parameterized by the curvilinear abscissa s. Around this curve we generate an orthogonal coordinate system (s,q) where q is the position vector on a plane orthogonal to the ray. This coordinate system is regular in the vicinity of any curve with finite curvature. The system becomes double valued once

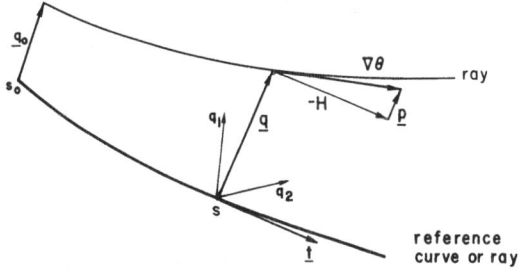

Fig. 4.   Geometry of the curve centered coordinate system (s,q_1,q_2). Vector q is in the plane (q_1,q_2) perpendicular to the local ray tangent t.

the distance to the reference curve is greater or equal to the radius of curvature of the reference curve. The system is chosen so that it is cartesian in the plane (q_1, q_2) normal to the curve. In this system of coordinates, the slowness vector p is given by:

$$p = \nabla\theta = h_s \frac{\partial\theta}{\partial s} t + \frac{\partial\theta}{\partial q_1} e_1 + \frac{\partial\theta}{\partial q_2} e_2 \qquad (29)$$

where the e_i are the unit vectors in the plane perpendicular to the curve, t is the tangent to the curve, and h_s is the scale factor of the curvilinear coordinate s. The scale factor h_s takes into account the curvature κ of the reference curve, and is given by:

$$h_s(s,q) = 1 - \kappa(s)n \cdot q$$

where n is the normal to the curve.

Let us introduce the components of the slowness vector in the plane perpendicular to the reference curve. From (29) we get:

$$P_1 = \frac{\partial\theta}{\partial q_1} \qquad P_2 = \frac{\partial\theta}{\partial q_2} \qquad (30)$$

In the nomenclature of Hamiltonian theory, these two components of the slowness vector are the momenta conjugate to the position vectors q_1 and q_2. A four-dimensional vector (q_1, q_2, p_1, p_2) completely defines the trajectory of a ray in the four-dimensional phase space of position and slowness. Following Popov and Psencik (1978, Eq. 3.14) we define the reduced Hamiltonian in this phase space as:

$$H_s(q,p,s) = - h_s\sqrt{u^2(s,q) - p^2} \qquad (31)$$

where as in the case of cartesian coordinates of the previous section, H_s is the negative component of the slowness along the curve, i.e., $p_s = -H_s$, as may be easily verified by reference to (29). Inserting this Hamiltonian in the canonical ray tracing equations (26) we find the ray tracing system in curve centered coordinates:

$$\frac{dq}{ds} = h_s \frac{p}{\sqrt{u^2 - p^2}}$$

$$\qquad (32)$$

$$\frac{dp}{ds} = h_s \frac{\nabla_T u}{\sqrt{u^2 - p^2}} + \nabla_T h_s \sqrt{u^2 - p^2}$$

where ∇_T denotes the gradient on the (q_1, q_2) plane. Cerveny and Psencik (1979) derived an equivalent expression by a more complex method. As long as the trajectory q(s), p(s) does not deviate beyond the evolute of the reference curve, the system (32) is regular and rays may be traced by numerical integration. As will be shown below the ray tracing system (32) is always linearized in order to perform paraxial ray tracing. I am not aware of any examples of ray tracing using the full nonlinear system (32) although it may be convenient in the ray bending and continuity methods.

The ray tracing system (32) was derived for any reference curve. One particular curve of interest is a ray that has been previously traced by some numerical method. In this case we define the coordinate system around this reference ray as shown in Figure 4. Rays have some particular properties that simplify the calculation of h_s. The curvature of a ray is given by (17). This may be used to simplify the expression for the scale factor. Since q.t = 0,

$$h_s = 1 - u^{-1}q \cdot \nabla u,$$

an expression that was derived by Popov and Psencik (1978).

7. PARAXIAL RAY THEORY

The solution of two-point ray tracing, the calculation of Geometrical spreading, Gaussian beams, etc. become much simpler using the so-called paraxial ray theory, which is just a particular application of perturbation theory to the ray tracing equations. We define paraxial rays (see Figure 5) as those rays that propagate in the vicinity of another ray that is taken as a reference. Suppose that we have succeeded in tracing a ray in a medium of slowness $U(x)$. We denote by $[x_0(s),p_0(s)]$ the trajectory as a function of s of this ray in phase space. In generalized coordinates this trajectory is written $y_0(s) = [q_0(s),p_0(s)]$, where q_0 represents the generalized position coordinates, p_0 the generalized slownesses, and y_0 is the canonical vector, a vector in phase space. Because of the generality of Hamilton's theory we will derive the paraxial ray equations in generalized coordinates. Results for particular coordinates systems may be determined by straightforward operations. A paraxial ray is described by a perturbation of the ray trajectory in phase space:

$$q(s) = q_0(s) + \delta q(s) \qquad p(s) = p_0(s) + \delta p(s). \tag{33}$$

Tracing paraxial rays consists in finding the canonical perturbation vector $[\delta q, \delta p]$ in phase space. These perturbations in the trajectory are due to small changes in the initial conditions of the ray at the initial point of the ray s_0. Let

$$\delta q(s_0) = \delta q_0 \text{ and } \delta p(s_0) = \delta p_0 \tag{34}$$

be the perturbations in initial conditions. Inserting the perturbation vectors (33), into the canonical ray tracing equations (20) and developing to first-order we find the following linear system for the calculation of paraxial rays:

$$\begin{aligned}
\frac{d\delta q}{ds} &= \frac{\partial^2 H}{\partial p \partial q} \delta q + \frac{\partial^2 H}{\partial p^2} \delta p \\
\frac{d\delta p}{ds} &= -\frac{\partial^2 H}{\partial q^2} \delta q - \frac{\partial^2 H}{\partial p \partial q} \delta p
\end{aligned} \tag{35}$$

where all the derivatives of the Hamiltonian are calculated on the reference ray.

In order to find the paraxial ray trajectories the linear system (35) has to be solved numerically together with the ray tracing system. Let us note however that since the system is linear all the paraxial rays in the neighborhood of a certain reference ray may be computed by simple linear operations. In order to see this we write the solution of (35) in the form of a propagator matrix (see, for instance, Gilbert and Backus, 1966):

$$\begin{bmatrix} \delta q \\ \delta p \end{bmatrix} = \Pi(s,s_0) \begin{bmatrix} \delta q_0 \\ \delta p_0 \end{bmatrix} \tag{36}$$

where $\Pi(s,s_0)$ is the paraxial ray propagator from s to s_0 of the paraxial rays. This propagator has a number of very useful properties, the most important for us is that it may be easily inverted. Given then the

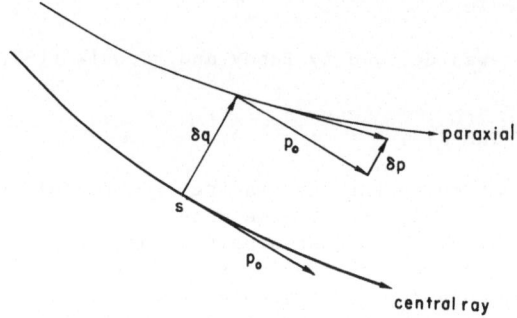

Fig. 5. Geometry of a paraxial ray propagating in the vicinity of a
         central ray. δq and δp are the perturbations in position and
         slowness in ray centered coordinates, respectively.

initial perturbations of position and slowness at s_o, and knowing the
propagator, the later position and slowness of the paraxial ray are
entirely determined by (36). The paraxial approximation remains valid as
long as the perturbation vector is small. The validity of this
approximation is unfortunately very difficult to establish in the general
case of a heterogeneous reference medium.

7.1. Beams

The elements of the propagator Π have a relatively simple physical
interpretation if we introduce the concept of a beam. We define a beam as
a one parameter family of paraxial rays such that the perturbed initial
conditions are related by:

$$[\delta q_o] = \varepsilon [\delta p_o] \tag{37}$$

where ε is a complex scalar that defines the shape of the beam. The
scalar ε may be replaced by a constant complex matrix, but we will not
discuss this possibility since we will not use it here.

Let us introduce the following partition of the propagator matrix:

$$\Pi(s,s_o) = \begin{bmatrix} Q_1 & Q_2 \\ P_1 & P_2 \end{bmatrix} \tag{38}$$

where, depending on the number of dimensions, Q_i and P_i are submatrices or
scalars. Inserting (37) and (38) into equation (36) we find the following
solution for individual paraxial rays in a beam of parameter ε:

$$\delta q = (\varepsilon Q_1 + Q_2)\delta p_o = (Q_1 + \varepsilon^{-1}Q_2)\delta q_o$$
$$\delta p = (\varepsilon P_1 + P_2)\delta p_o = (P_1 + \varepsilon^{-1}P_2)\delta q_o \tag{39}$$

so that, finally, we may write $\delta p = M(s,s_o)\delta q$, where matrix M is

$$M(s,s_o) = (\varepsilon P_1 + P_2)(\varepsilon Q_1 + Q_2)^{-1}. \tag{40}$$

This matrix is also the tensor of second order derivatives of the travel
time function θ as will be demonstrated below. From (37) we observe that
the initial value of this matrix is simply $M(s_o,s_o) = \varepsilon I$, where I is the
identity matrix.

The shape of a beam is controlled by ε. Two extreme values of ε give simple fundamental ray beams. For $\varepsilon = 0$ we get a point source since all the rays in the beam leave from the same point in space q_0 with slightly different values of $p(s_0)$. The set of paraxial rays forms a solid angle at the source. This corresponds to the point source (see Figure 6a). On the other hand, when $\varepsilon = \infty$ $\delta p_0 = 0$ in (37), so that all the paraxial rays share the same initial slowness vector, but leave from slightly different positions $q(s_0) = q_0 + \delta q(s_0)$ in space. The geometry of the beam with $\varepsilon = \infty$ depends on the coordinate system under consideration (Madariaga, 1984). For cartesian coordinates the paraxial rays form what is sometimes called a Snell-wave, i.e., as shown in Figure 6b, a wave such that all the rays make a constant angle with respect the (x,y) coordinate plane. In ray-centered coordinates, the $\varepsilon = \infty$ beam corresponds, as shown in Figure 6c, to an initially plane wave. As we mentioned before, complex values of ε are legitimate provided that the corresponding travel times satisfy causality conditions. This occurs for $\text{Im}\varepsilon < 0$, in which case we get Gaussian beams as will be shown below.

Now that we have traced a beam and its paraxial rays we have to calculate their travel times for the paraxials. This is conveniently done using the curvilinear integral (21). We first write:

$$\theta[q(s_0),q(s_1)] = \theta_0[s_0,s_1] + \delta\theta_0 + \delta\theta_1 \tag{41}$$

where θ_0 is the travel time along the central ray from $q_0(s_0)$ to $q_0(s_0)$ and $\delta\theta_0$ and $\delta\theta_1$ are the travel time perturbations at the initial and final point, respectively. From (21) the end point perturbations are simply:

$$\delta\theta = \int_0^{\delta q}(p_0 + \delta p)\cdot d\delta q. \tag{42}$$

Replacing the relation $\delta p = M\delta q$ in (42) and integrating we get:

$$\delta\theta = p_0\cdot\delta q + \frac{1}{2}\delta q^T M\delta q. \tag{43}$$

This is the second order Taylor expansion of the travel time θ around the central ray. From (43) we observe that $M = \nabla_T\nabla_T\theta$, so that M is the tensor of second order derivatives of the travel time with respect to the (q_1,q_2) variables. The second order terms are needed in the generation of Gaussian beams and in paraxial approximations to the calculation of synthetic seismograms. In most of the following discussion we will consider point sources so that $\delta\theta_0$ in (41) will be zero.

7.2. Geometrical Spreading of a Beam

Finally, in order to actually compute the ray amplitudes we need to calculate geometrical spreading J. Following Figure 7, we consider an elementary cross section of the beam at s_0 and from each point on this cross section we trace a paraxial ray. As the beam propagates, the paraxial rays may contract or dilate, so that the beam cross section at the end point s is a measure of the geometrical spreading. Consider as in Figure 7 two distinct paraxial rays denoted by their paraxial position vectors δq_1 and δq_2. At any point s along the beam, the cross section of the beam described by these paraxials and the central beam is:

$$dS(s) = (\delta q_1 \times \delta q_2)\cdot t \tag{44}$$

where t is tangent to the ray. Let $dS(s_0)$ be the initial beam cross section at position s_0 along the central ray (see Figure 5). We can follow the change in cross section as a function of s, tracing paraxial rays by means of the paraxial ray equations (39). At the curvilinear abscissa s the paraxial position vectors δq_1 and δq_2 are given by:

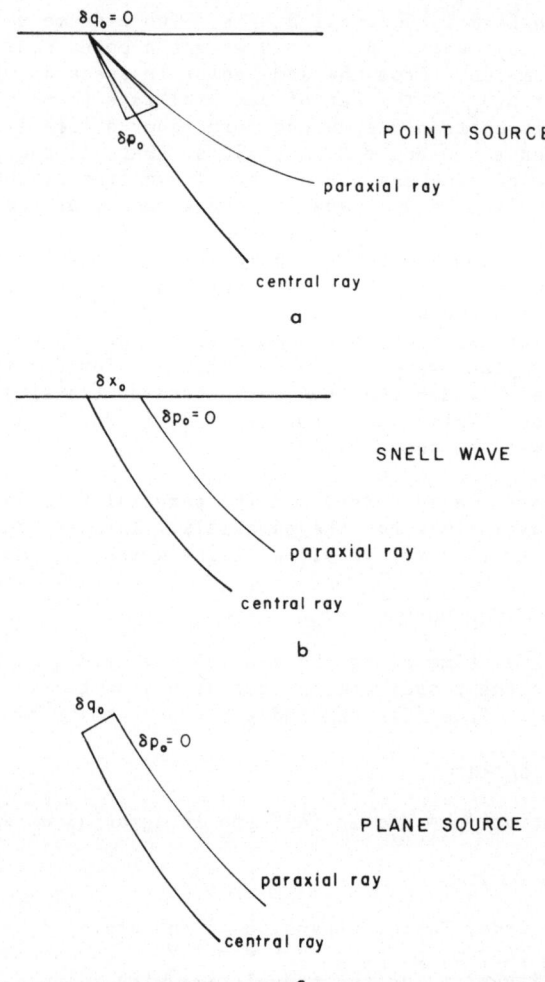

Fig. 6.   The three basic geometries for the paraxial rays. On top, a, the
point source. At the center the "Snell wave" where the paraxial
leaves the reference plane with the same take off direction as
the central ray. At the bottom, c, the plane source of Cerveny
et al. (1982). Here the rays form a plane wave at the source.

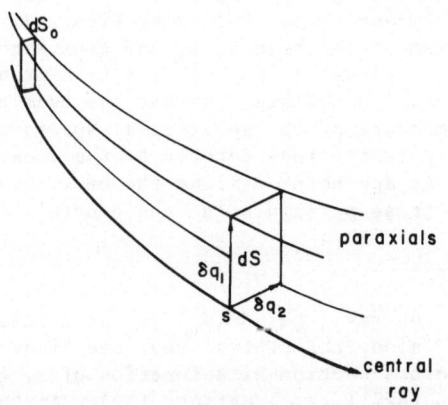

Fig. 7.   Geometry of a beam for the calculation of geometrical spreading.

$$\delta q_1(s) = (Q_1 + \varepsilon^{-1}Q_2)\delta q_1(s_o) \qquad \delta q_2(s) = (Q_1 + \varepsilon^{-1}Q_2)\delta q_2(s_o) \qquad (45)$$

where as usual ε is the beam parameter. Replacing these vectors in (44) we can find the cross section dS at abscissa s:

$$dS(s) = Det(Q_1 + \varepsilon^{-1}Q_2)\cos\phi |\delta q_1(s_o)||\delta q_2(s_o)| \qquad (46)$$

where ϕ is the angle between the normal to the plane defined by the vectors δq_1 and δq_2 and the tangent to the ray. Comparing (47) with the corresponding expression for the cross section at the initial coordinate s_o, we get:

$$dS(s) = Det(Q_1 + \varepsilon^{-1}Q_2) \frac{\cos\phi}{\cos\phi_o} dS(s_o). \qquad (47)$$

Finally we use the very well-known result that the geometrical spreading J is just the ratio of the cross section of the beam at s to the initial cross section:

$$J(s,s_o) = Det(Q_1 + \varepsilon^{-1}Q_2) \frac{\cos\phi}{\cos\phi_o}. \qquad (48)$$

This is a general expression for J that is independent of the coordinate system used to trace the rays.

Let us remark that the minors $Det(Q_1)$ and $Det(Q_2)$ of the propagator matrix Π have a clear physical meaning. $Det(Q_1)$ measures the geometrical spreading of a plane or Snell wave because in this case $\varepsilon = \infty$. For a point source, $\varepsilon \to 0$, and J defined as in (48) becomes singular. For point sources we redefine J so that:

$$J = \lim_{\varepsilon \to 0} \frac{J}{\varepsilon} = Det(Q_2) \frac{\cos\phi}{\cos\phi_o}. \qquad (49)$$

This expression permits to calculate geometrical spreading for point sources from the propagator for paraxial rays. This method of calculating J is more stable than the more traditional one that consists in tracing several rays around the central ray, and calculating J by the ratio of cross sections.

Let us conclude this section by remarking that our derivation of the paraxial ray equations, the travel time perturbation and geometrical spreading are independent of the coordinate system used to trace the rays. It is equally valid for the original ray tracing system (16) as well as for the reduced ray tracing systems (28) in cartesian coordinates, or (32) in ray centered coordinates.

8.   RAY CENTERED COORDINATES AND DYNAMIC RAY TRACING

The expressions derived in the previous section for paraxial rays are much simpler in ray centered coordinates, which were the coordinates used by Popov and Psencik (1978) in their derivation of dynamic ray tracing. In ray centered coordinates the central ray is simply given by $q_o = p_o = 0$, since the reference ray is the origin of the coordinates q and p. Thus in this coordinate system the paraxial ray coordinates (33) have the simple expressions $q(s) = \delta q(s)$ and $p(s) = \delta p(s)$. Ray tracing of paraxial rays may actually be performed exactly using the nonlinear system (32). In (32) there is no approximation in the region where the ray centered coordinate system is regular, i.e., inside a region limited by the smallest radius of curvature of the ray. In the paraxial approximation, where δq and δp are considered to be first-order perturbations, we use the system (35) which is obtained by linearization of (32). In ray centered coordinates $\partial H / \partial q \partial p = 0$, so that (35) simplifies to:

$$\frac{d\delta q}{ds} = u^{-1} I \delta p$$

$$\frac{d\delta p}{ds} = - u^2 V \delta q \tag{50}$$

where V is the matrix of second-order derivatives of the velocity field with respect to the ray centered coordinates q_i:

$$V = [V_{ij}] \text{ with } V_{ij} = \frac{\partial^2 v}{\partial q_i \partial q_j} . \tag{51}$$

This expression was originally derived by Popov and Psencik (1978) from the Ricatti equation for the wavefront curvature. Our derivation shows that (50) is just a particular application of the general paraxial ray approximation (35) to ray centered coordinates.

For a heterogeneous distribution of velocity, the numerical solution of equation (50) can be obtained simultaneously with the solution of the ray tracing equations (16). The only difficulty is in the calculation of the matrix of second-order derivatives of the velocity in (51). This requires that velocities be interpolated with continuous second-order derivatives. Local or global cubic splines are the preferred method for interpolating the velocity field with this condition.

The dynamic ray tracing or paraxial system in ray-centered coordinates may be integrated and the solution expressed as before in terms of a propagator matrix:

$$\Pi = \begin{bmatrix} Q_1 & Q_2 \\ P_1 & P_2 \end{bmatrix}$$

where the submatrices Q_1, P_1 are the propagators for "plane wave" initial conditions, while Q_2, P_2 are the propagators for a point source. Most of the other results obtained in the previous section apply with some slight modification to ray centered coordinates. The most important result is that the travel time of a paraxial ray is given by:

$$\theta(s, \delta q) = \theta(s, 0) + \frac{1}{2} \delta q^T M \delta q \tag{52}$$

where $\theta(s, 0)$ is the travel time along the central ray. Compared to the general expression (44) the linear term in δq has disappeared because in ray centered coordinates δq is by definition perpendicular to p_o, which is parallel to the ray tangent. Looking at (52) we realize that the matrix M has a clear physical meaning: it is the Gaussian curvature matrix of the wavefront $\theta(s, q)$ = constant. Matrix M may be diagonalized in order to find the two principal radii of curvature of the wavefront. In two-dimensions, M is scalar and is simply the curvature of the wavefront. The form (52) is a local paraboloidal approximation to the wavefront θ = constant, for this reason Cerveny, Popov and Psencik (1982) called it the parabolic approximation. In fact in order to derive the paraxial approximation they incorporated the parabolic approximation directly into the ray anzatz (1) and then derived the equations for δq and δp from the wave equation. In our opinion, the method used here is simpler and has a clearer physical meaning.

Finally, geometrical spreading J is given by either (47) or (48) with $\cos\phi = \cos\phi_o = 1$, since for ray centered coordinates the normal to the coordinate plane q is always parallel to the local tangent to the ray t. Thus once the paraxial ray tracing propagator Π has been calculated, travel times, geometrical spreading, etc. are easily calculated in ray centered coordinates.

9. SEISMIC SOURCES

The last element that we need to calculate ray theoretical seismograms using the frequency domain expression (3) or its time domain counterpart (5) is the vector amplitude A and the source function $s(\omega)$. We write the excitation of P- and S-waves in the following form (Madariaga, 1982). For P-waves:

$$A_o = t \frac{1}{4\pi\rho_o \alpha_o^3} \qquad s(t) = M_{RR}(t) \tag{53}$$

For S-waves:

$$A_o = q \frac{1}{4\pi\rho_o \beta_o^3} \qquad s(t) = M_{Rq}(t). \tag{54}$$

Where t is the vector tangent to the ray at the observation point, and $\dot{M}(t)$ is the moment rate tensor. M_{RR} is the radial component of the moment tensor in local spherical coordinates around the source. Similarly M_{Rq} is the tangent component of the moment tensor in spherical coordinates projected on a plane perpendicular to the take off direction of the ray. Let t_o be the tangent to the ray at the source, then the projection of M on the plane perpendicular to t_o is:

$$M \cdot t_o = M_{RR} t_o + M_{Rq} q_o$$

where q_o the unit vector in the direction of $t_o \times (M \cdot t_o)$. This unit vector points in the direction of the shear component of $M \cdot t_o$. With this notation for the components of the moment tensor, the source time functions may be rewritten as:

$$\begin{aligned} M_{RR} &= t_o \cdot M_o \cdot t \\ M_{Rq} &= t_o \cdot M_o \cdot q_o. \end{aligned} \tag{55}$$

Vector q is the polarization vector for the S-wave, it may be obtained by propagation of the unit vector q_o. These expressions permit to calculate ray amplitudes in any elastic medium where the ray tracing problem has been solved. The calculation of synthetic seismograms of far field body waves for the study of source processes are one example of the use of ray synthetics in practical applications. Calculation by ray theory is limited to rays that penetrate into the lower mantle, otherwise we would have to deal with triplications and caustics due to the upper mantle discontinuities. For this reason classical ray theory is limited to the calculation of synthetics in the range from $30°$ to $90°$. At shorter distances WKB (Chapman, 1978) or Gaussian beam summation (Madariaga and Papadimitriou, 1985) provide a practical method to calculate synthetics. This kind of synthetic seismogram is widely used in order to study fault mechanisms and the distribution of asperities on the faults.

10. SPECTRAL METHODS

The methods we have described so far are all based on the direct use of the ray theoretical expressions (3 and 5). These expressions become singular or unstable under many practical circumstances. For instance, as shown on Figure 8a, near caustics when neighboring rays cross each other.

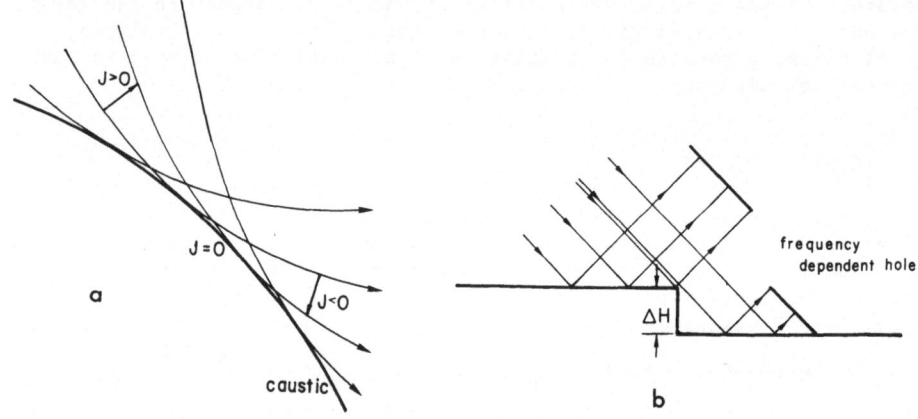

Fig. 8. Two examples of problems with ray theory. On top a caustic, at
the bottom a small discontinuity in a reflector.

In the vicinity of the caustic the geometrical spreading function J tends
to zero and changes sign once the rays have crossed the caustic. When
J → 0, ray theory predicts infinite amplitude although we know that in
reality amplitudes are finite around the caustic. This failure of ray
theory comes from its inability to deal with finite frequency phenomena.
Another problem, also illustrated in Figure 8b is the extreme sensitivity
of ray theory to small local perturbations in the slowness field. This
problem is due again to the assumption that frequency is infinite. At
finite frequencies, small discontinuities should only affect waves with a
wavelength that is similar or shorter than the dominant wavelength of the
diffracting object. At low frequencies, when the wavelength is longer
than the characteristic size of the heterogeneity, the effect of the
perturbation is to generate weak Rayleigh scattering, but the travel times
and geometrical spreading of the rays should not be severely affected.
Spectral methods provide a partial solution to this problem without
loosing the physical appeal and simplicity of ray theory.

In a spectral method, the field is not calculated directly from (3)
but from a sum of beams of the form (3). The best known of these methods
is WKB for vertically heterogeneous media. In this method, originally
proposed in cartesian coordinates, the source is developed into a sum of
Snell-waves (see Figure 6b), then each Snell-wave is propagated
independently and a seismogram is calculated summing all the propagated
Snell-waves. This process is illustrated in Figure 9. Since a seismogram
is calculated evaluating a sum of beams, it is much easier to control the
frequency contents of the final synthetic. Also, the problem with
caustics is partially suppressed because the Snell-waves usually behave
regularly near caustics in configuration space. Unfortunately, caustics
may also appear in the individual Snell-waves and may render the
calculation unstable. Appropriate combination of both classical ray
theory and WKB seem to be the best answer to problems with the stability
of the ray field.

Another method that has appeared relatively recently in the
literature is the Gaussian beam summation method. Just as with WKB, the
source is expanded into a series of Gaussian beams each of which is
propagated independently. A synthetic is calculated by the summation or
stacking of individually propagated Gaussian beams. The method is very
similar to WKB, but presents the additional advantage that Gaussian beams
do not have caustics and may be calculated everywhere. In the following
we will briefly discuss these two methods.

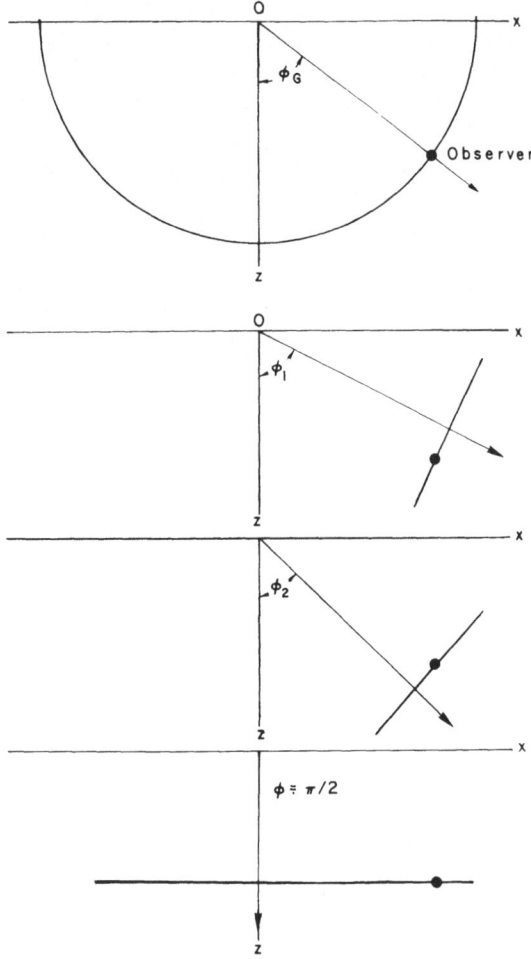

Fig. 9. Construction of a cylindrical wave by the summation of plane waves in the Weyl integral representation of the two-dimensional Green function.

11. THE WKB METHOD

The WKB method introduced by Chapman (1978) is probably the most widely used method for the calculation of high frequency synthetic seismograms. Derived originally only for vertically heterogeneous media, it was later extended to media with arbitrary heterogeneity (Chapman and Drummond, 1982). Its main limitation is that of ray theory: the heterogeneity of the medium should be smooth in comparison to the wavelength under consideration. For simplicity of the exposition we will develop the WKB method in a two-dimensional medium, the three-dimensional case is treated by Chapman (1978).

The starting point for the WKB method is Weyl's integral in two-dimensions or the Sommerfeld integral in three-dimensions. In order to simplify the presentation we consider a line explosion in a homogeneous medium as a model source. P-wave radiation from this source is written in the form of a Green function:

$$G(x,t) = \frac{1}{2\pi} A_o (t^2 - x^2/c^2)^{-1/2} H \left(t - \frac{|x|}{c} \right) \tag{56}$$

where $c = u^{-1}$ is the constant velocity of the medium. x is the position of the observer with respect to the source, and the vector amplitude A_0 is parallel to the radial vector x, its scalar value depends on the source under consideration, and will be discussed later. The two-dimensional Green function (56) has the classical inverse square-root singularity at the wavefront. Its time Fourier transform is:

$$G(x,\omega) = \frac{i}{4} A_o H_o^{(1)} \left(\frac{\omega |x|}{c} \right) \tag{57}$$

where i is the imaginary unit. The Fourier transform in (57) is defined as in (4). The Green functions (56) and (57) are exact. In order to compare these solutions with ray theory we have to calculate their high frequency approximation. Using the asymptotic expansion of the Hankel function, we get:

$$G(x,\omega) = \frac{1}{4\pi} A_o \left[\frac{2c}{|x|} \right]^{1/2} \lambda(\omega) e^{i\omega\theta_2} \tag{58}$$

where the source function

$$\lambda(\omega) = (\pi/\omega)^{1/2} e^{i\pi/4} \tag{59}$$

is the Fourier transform of the inverse square-root pulse:

$$\lambda(t) = t^{-1/2} H(t). \tag{60}$$

The travel time is

$$\theta_2 = x/c \tag{61}$$

and comparing with (3) we observe that the geometrical spreading

$$J_2 = x. \tag{62}$$

In (61) and (62) a subindex 2 has been added to θ and J, in order to indicate that this is for two-dimensional propagation. With these definitions we observe that (58) is in the form of the ray theoretical approximation (3) if in the latter we assume that the medium is homogeneous. The source function $s(\omega)$ in (2) has been replaced by $\lambda(\omega)$. We can calculate the time domain inverse of (58) using (7). Since all the amplitude terms are real we get:

$$G(x,t) = \frac{1}{4\pi} A_o \left[\frac{2c}{J_2} \right]^{1/2} \lambda(t - \theta_2) \tag{63}$$

with θ_2 and J_2 given by (61) and (62).

This completes the demonstration that ray theory in a homogeneous medium is identical to high frequency asymptotics. In quantum mechanics these high frequency methods are usually designated WKB approximations (see also Bleistein, 1984).

The spectral decomposition of (57) into plane waves is very well-known and will be the starting point for our development of the WKB method. By means of Weyl's integral we can write (57) for z > 0 in the form:

$$G(x,\omega) = \frac{i}{4\pi} \int_{-\infty}^{\infty} A_o e^{i\omega(px + pz)} \frac{dp}{q} \tag{64}$$

for $\omega > 0$. The integral representation (64) is in fact the Fourier transform with respect to x of the Green function (57). Equation (64) may be rewritten in order to show that it is in fact a superposition of plane waves. Let us define position and slowness as usual:

$$x = (x,z) \text{ and } p = (p,q). \tag{65}$$

Because of the eikonal equation, $|p| = c^{-1}$, q is not independent of p and may be written:

$$q = (c^{-2} - p^2)^{-1/2} \text{ with } Im(q) \geqslant 0 \tag{66}$$

for z > 0.

Finally, following the definitions in Figure 10, we can change the integration variable in (3) to the take off angle ϕ_0 defined by

$$p = \frac{\sin\phi_0}{c} \qquad q = \frac{\cos\phi_0}{c} \tag{67}$$

where the angle ϕ_0 should be taken along the contour L in the complex ϕ_0 plane shown in Figure 11. This contour is the mapping of the Real p axis into the complex ϕ_0 plane. We then rewrite (64) in the form:

$$G(x,\omega) = \frac{i}{4\pi} \int_L A_0 e^{i\omega\theta_2} d\phi_0 \tag{68}$$

where $\theta_2 = p.x$ is the travel time from the origin to x of a plane wave that leaves the origin in the direction p. (68) is plainly a sum of plane waves of amplitude $i/4\pi A_0$ and phase θ_2.

The plane wave decomposition provides an alternative way to calculate synthetic seismograms. A problem with it is the presence of inhomogeneous waves. The two vertical branches of L shown in Figure 11, represent inhomogeneous plane waves that propagate horizontally along the z = 0 line and decrease exponentially with depth. These waves contribute to the near field of the Green function (56) and may be neglected as long as the observer is not too close to the x axis. In order to see this we remark that in the far field when $|x| \gg 0$, the main contribution to the integral (68) comes from the vicinity of the stationary phase point:

$$\frac{\partial\theta_2(\phi_0)}{\partial\phi_0} = c^{-1}(x \cos\phi_0 - y \sin\phi_0) = 0. \tag{69}$$

Let us call ϕ_G the value of the angle ϕ_0 at the stationary point, solving (69) we find $\phi_G = atan(x/y)$. ϕ_G is just the take off angle of the ray that joins the source to the observation point x. Using the stationary phase approximation in the spectral integral (68), we obtain the ray theoretical solution (58). Thus a simple guess is that most of the

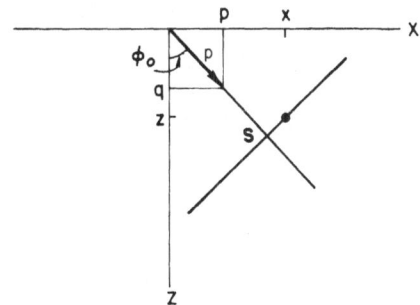

Fig. 10. Geometry of an elementary plane wave in the Weyl integral.

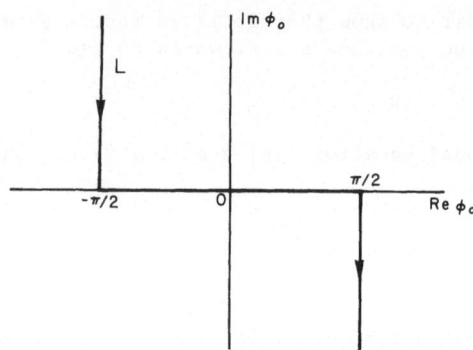

Fig. 11. Contour of integration in the complex ϕ_o-plan of the WKB or plane wave spectral decomposition of the two-dimensional Green function.

contribution to a seismogram calculated using (68) will come from take off angles ϕ_o close to ϕ_G. For this reason we restrict our integral to the real axis $-\pi/2 < \phi_o < \pi/2$ and calculate the time domain transform of (68). Let us consider an individual plane wave in (68):

$$g(x,\omega) = \frac{i}{4\pi} A_o e^{i\omega\theta_2(\phi_o)}. \tag{70}$$

This plane wave may be transformed into the time domain using (5):

$$g(x,t) = \frac{1}{4\pi^2} A_o \text{Re}\left[\frac{1}{t-\theta_2}\right]. \tag{71}$$

We observe that the plane waves that are used to generate a synthetic seismogram by the sum (68), present a non causal behavior. In fact the time signal in (71) is the Hilbert transform of a delta function.

Using the Fourier transform (71) in the sum (68) we get the WKB approximation to the Green function:

$$G(x,t) = \frac{1}{4\pi^2} \text{Re}\left[\int_L \frac{1}{t-\theta_2(\phi_o)} d\phi_o\right]. \tag{72}$$

This way of calculating the synthetics, doing first the time Fourier transform and later the sum over the take-off angle was introduced by Chapman (1978) and is the basis of the success of the WKB and the Gaussian beam summation method. As long as the inverse Fourier transform (71) of the individual plane wave can be computed exactly, the plane wave sum can be easily computed. Let us finally remark that evaluating the integral (72) is not straightforward because of the singularity that occurs when t = θ. This problem may be avoided by several different approximations described by Chapman (1978). In the following and by analogy with Gaussian beams we will use the method we proposed in equation (5), i.e., we add a small Imaginary part iΔt to the plane wave phase θ. The resulting Green function, is smoothed but the smoothing is perfectly compatible with the discretization path Δt. More serious is the problem with the cut off phases. These phases that arrive before the geometrical onset of the Green function are due to the limitation of the integral (72) to a segment of the real axis. This problem comes from the fact that all plane waves contribute with the same amplitude to the WKB synthetic calculated using (72). Thus the arrest of the integration at a certain angle ϕ_1 produces a rather large effect on the integral. One of the possible ways to reduce the effect of these cut off phases is to filter the integral near the ends of the integration range, this may be done by simple tapering of the amplitudes A_0 or by more sophisticated techniques. Another possibility is to replace the plane waves by Gaussian beams as will be shown in the next section.

The generation of the Green function (56) by the sum of plane waves is quite curious. The individual waves that are summed in (72) have their peak amplitude near $t = \theta_2$. Thus the contribution of each plane wave in the sum will be concentrated around θ_2. But θ_2 varies as a function of ϕ_0 in such a way that it has maximum at $\phi_0 = \phi_G$, the stationary phase point already calculated above. Thus the maximum contribution of each plane wave occurs before the geometrical arrival time!

The WKB method can be easily extended to vertically heterogeneous media for which it was originally proposed by Chapman. It may also be applied in laterally heterogeneous media where Chapman and Drummond (1982) prefer to call it the Maslov method, because in laterally heterogeneous media the plane wave decomposition (64) is only asymptotically valid.

11. GAUSSIAN BEAMS

As we mentioned above, one of the main problems with the calculation of the plane wave sum (72) is the spurious cut off phases that appear when the integral is limited to a finite segment of ϕ_0. One successful method for reducing the influence of these cut off phases is to use Gaussian beams. Before introducing Gaussian beams we must rewrite the plane wave (70) as a beam. Referring to Figure 10 we note that this expression may be rewritten in terms of the distance s along the ray leaving the origin in the direction ϕ_0. This ray plays the role of the central ray in the paraxial theory presented in section 8. We can rewrite θ_2 in the very simple form:

$$\theta_2 = \frac{s}{c} \tag{73}$$

which is the travel time of the plane wave that passes through x. s is the distance along the central ray and plays the role of the independent parameter s in ray theory. The set of rays associated with this plane wave are parallel to the central axis. In the vicinity of the central ray we define the paraxial zone. In this zone the ray centered coordinates of a paraxial ray is simply $\delta q = \delta q_0$, $\delta p = \delta p_0 = 0$ which states the trivial fact that paraxial rays are parallel to the central ray. With respect to the paraxial theory developed in section 7 we observe that the plane waves that are used in the WKB integral form what we called a plane beam. Since the medium is homogeneous the plane beam remains plane as it propagates away from the source.

In Gaussian beam summation, we replace the plane beams of the WKB method by Gaussian beams. In order to construct Gaussian beams we slightly perturb the plane beam introducing a small complex part in p_0. Referring to (37) in section 7 we take:

$$\delta p_0 = \varepsilon^{-1} \delta q_0 = i \delta \delta q_0. \tag{74}$$

The small parameter δ produces a fanning of the plane wave. For real ε the plane wavefront deforms into a parabolic front and the paraxial rays tend to move away from the central ray. Let us demonstrate this: (74) gives the initial conditions for tracing of the paraxial ray that leaves the source plane at a distance δq_0 from the origin (see Figure 12). We can then trace this paraxial ray with the help of the paraxial propagator (37). In a homogeneous medium this propagator takes the simple form:

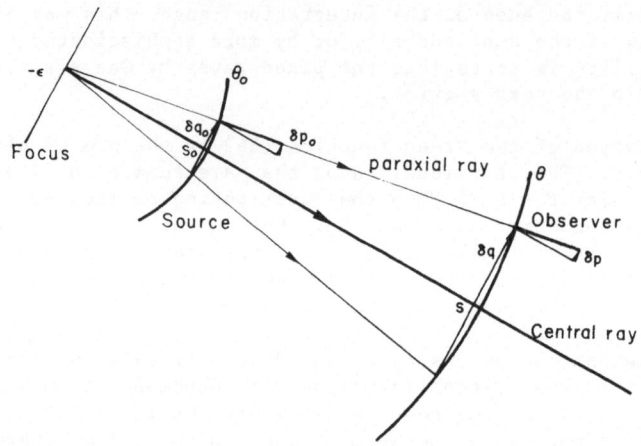

Fig. 12. Geometry of a parabolic beam. The focus does not coincide with the source.

$$\Pi(s,s_o) = \begin{bmatrix} 1 & c(s-s_o) \\ 0 & 1 \end{bmatrix} \tag{75}$$

so that the paraxial ray is given by (39):

$$\delta q = [1 + \varepsilon^{-1}c(s-s_o)]\delta q_o \qquad \delta p = \varepsilon^{-1}\delta q_o = \delta p_o. \tag{76}$$

Thus the paraxial ray is a straight line as one expects for a homogeneous medium, but the direction of the ray differs from that of the central ray. Finally we can compute the travel time θ along the perturbed beam from (43):

$$\theta_2 = \frac{s}{c} + \frac{1}{2} M\delta q^2 \tag{77}$$

where M the curvature of the wavefront is given by

$$M = \frac{\varepsilon^{-1}}{[1 + \varepsilon^{-1}c(s-s_o)]} = [\varepsilon + c(s-s_o)]^{-1}.$$

For $\varepsilon > 0$, M^{-1} grows with distance from the origin and is zero for a point $s - s_o = -\varepsilon/c$. Thus as shown in Figure 12 it appears as if the perturbed beam was coming from a source situated behind the origin at a distance $-\varepsilon/c$. The paraxial approximation for the travel time replaces the spherical wavefront emanating from the focus at $x = -\varepsilon/c$ by a parabolic wavefront. In this sense the paraxial ray theory is a parabolic approximation to the wave equation.

We can now introduce Gaussian beams. They are directly obtained from the paraxial approximation (76) replacing ε by a large imaginary value. The paraxial rays are complex and the travel time θ_2 is also complex. If we collect the results obtained so far we find that a Gaussian beam is given by:

$$g(x,\omega) = \frac{i}{4\pi} J^{-1/2} e^{i\omega\theta_R} e^{-\omega\theta_I} \tag{78}$$

where θ_R and θ_I denote the real and imaginary part of the travel time (77). For Im $\varepsilon < 0$, we can verify that $\theta_I > 0$ and therefore (78) presents a Gaussian amplitude decrease in the direction perpendicular to the central ray:

$$e^{-\omega\theta}I = e^{-\omega 1/2\,\text{Im}(M)\delta q^2}$$

so that the Imaginary part of the curvature of the wavefront determines the attenuation of the beam away from the central ray.

Finally, in order to calculate (78) we need J, but this is given by (48) for a beam of arbitrary shape:

$$J = 1 + \varepsilon^{-1}c(s - s_o). \tag{79}$$

Thus replacing (79) and (77) in (78) and taking $s_o = 0$ we have a simple practical way for computing a Gaussian beam centered around a ray passing through the source situated at the origin of coordinates. The calculation of these Gaussian beams is not much more difficult than the calculation of a plane wave, the only additional difficulty is estimating the distance δq from the observation point to the central ray. This is not difficult to evaluate in the present problem but may be a serious problem in heterogeneous media. In that case it would be preferable to use cartesian coordinates for the calculation of Gaussian beams. This has been discussed by Madariaga (1984).

We can now replace every plane wave in the WKB integral (72) by its corresponding Gaussian beam. One could in principle take any suitable value of ε for each beam. In practice however, ε should be constant for all rays when evaluating the Gaussian beam sum. We get then

$$G(x,\omega) = \frac{i}{4\pi} \int A_o J^{-1/2} e^{i\omega\theta} d\phi_o \tag{80}$$

where J and θ are the complex geometrical spreading and travel time determined above. Finally using (5) we find the time domain Gaussian beam sum:

$$G(x,t) = \frac{1}{4\pi} \text{Re}\left[\int A_o J^{-1/2} \frac{1}{t - \theta} d\phi_o \right]. \tag{81}$$

Gaussian beams synthetics are evaluated discretizing this integral. If one uses the broadening parameter Δt introduced in (5) this discretization presents no stability problems. However, in more complex situations one has to ensure an appropriate density of rays in order to avoid interference problems in calculating it.

A large number of examples of calculation of theoretical seismograms using classical ray theory or Gaussian beam summation have appeared in the literature. Among the numerous publications where the method has been used to generate realistic seismograms we can cite: Cerveny (1985a), Cormier and Spudich (1984), Nowack and Aki (1985), Madariaga and Papadimitriou (1985), etc. These authors show that the Gaussian beam summation method can be used to solve many of the problems encountered when using straightforward ray theory, without completely losing the simplicity and physical appeal of rays. Comparison with finite difference calculations by George et al. (1987) showed that the method works for caustics giving not only the right amplitudes in the illuminated zone but also on the shadow regions. The same authors showed that edge diffraction and head waves are difficult if not impossible to model with Gaussian beam summation unless special techniques are used. Recently, White et al. (1987) made a careful study of the conditions of validity of the sum of Gaussian beams for several simple structural models. We refer the interested reader to these papers for further information. We conclude presenting with an example of a realistic structure with five discrete reflectors presented in Figure 13. At the top of the figure is a diagram

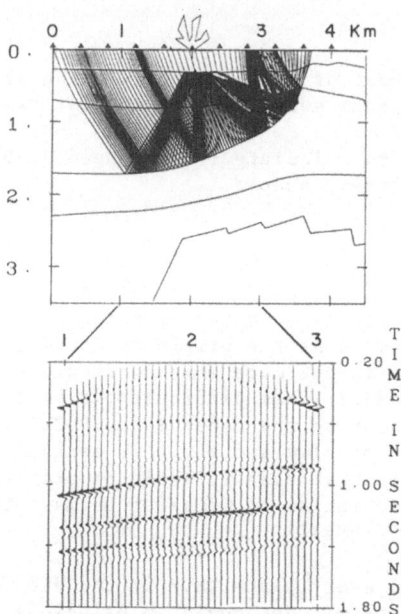

Fig. 13. Example of the calculation of a seismic reflection cross section by the Gaussian beam summation method. The model is a multilayered structure. All contributing reflections and multiples have been considered in the computation.

of the model together with an example of the ray tracing that is needed to obtain good precision in the Gaussian beam summation. For the construction of the cross section at the bottom we have to calculate a series of reflected rays from each interface. The presence of small kinks and imperfections in the interfaces creates numerous caustics which would render impossible to calculate these profiles with classical ray theory. Gaussian beams on the other hand can smooth these imperfections yielding continuous reflections in the seismograms presented at the bottom. The calculation was carried out in a minicomputer and included several thousand rays.

ACKNOWLEDGEMENTS

This research was supported by CNRS through the programs ATP Geophysique Appliquée and ATP Propagation d'ondes dans des milieux hétérogènes: application au génie parasismique. We also acknowledge the support of a consortium of French Oil Companies (ELF, CFP, IFP and CGG). Numerous discussions with V. Farra, Th. George and J. Virieux are gratefully acknowledged. IPG Contribution No. xxx.

REFERENCES

Babich, V. M., 1956, Ray method for the computation of the intensity of wavefronts, Doklady Akad. Nauk USSR, 110:355-357.
Babich, V. M., and Buldyrev, V.S., 1972, "Asymptotic Methods in Short Wave Diffraction Problems", Nauka, Moscow.
Babich, V. M., Molotov, I. A., and Popov, M. M., 1985, Gaussian beams solutions concentrated around a line and their use, Preprint, Institute of Radiotechnics and Electronics (in Russian).

Ben Menahem, A., and Beydoun, W., 1985, Range of validity of seismic ray and beam methods in inhomogeneous media: 1 General theory, Geophys. J. R. Astron. Soc., 82:207-234.

Bleistein, N., 1984, "Mathematical Methods for Wave Phenomena", Academic Press, New York.

Bullen, K., and Bolt, B., "An Introduction to the Theory of Seismology", Cambridge University Press, Cambridge, UK.

Burridge, R., 1976, "Some Mathematical Topics in Seismology", Lecture Notes, Courant Institute of Mathematical Sciences, New York University.

Cerveny, V., 1985, The application of ray tracing to the numerical modeling of seismic wavefields in complex structures, in: "Handbook of Geophysical Exploration", Section 1 (Seismic Exploration), 15A:1-119.

Cerveny, V., 1985a, Gaussian beam synthetic seismograms, J. Geophys., 58:44-72.

Cerveny, V., and Hron, F., 1980, The ray series method and dynamic ray tracing system for three-dimensional inhomogeneous media, Bull. Seismol. Soc. Am., 70:47-77.

Cerveny, V., Klimes, L., and Psencik, I., 1984, Paraxial ray approximations in the computation of seismic wavefields inhomogeneous media, Geophys. J. R. Astron. Soc., 79:89-104.

Cerveny, V., Molotov, I. A., and Psencik, I., 1977, "Ray Methods in Seismology", Karlova Univerzita, Prague.

Cerveny, V., Popov, M. M., and Psencik, I., 1982, Computation of wave fields in inhomogeneous media - Gaussian beam approach, Geophys. J. R. Astron. Soc., 70:109-128.

Cerveny, V., and Psencik, I., 1979, Ray amplitudes of seismic body waves in laterally inhomogeneous media, Geophys. J. R. Astr. Soc., 57:97-109.

Cerveny, V., and Psencik, I., 1983, Gaussian beams and paraxial ray approximations in three-dimensional elastic inhomogeneous media, J. Geophys., 53:1-15.

Cerveny, V., and Psencik, I., 1984, Gaussian beams in elastic two-dimensional laterally varying layered structures, Geophys. J. R. Astron. Soc., 78:65-91.

Chapman, C. H., 1978, A new method for computing synthetic seismograms, Geophys. J. R. AStron. Soc., 54:481-513.

Chapman, C. H., and Drummond, R., 1982, Body-wave seismograms in inhomogeneous media using Maslov asymptotic theory, Bull. Seismol. Soc. Am., 72:S277-S317.

Cisternas, A., Jobert, G., and Compte, P., 1984, Classical theory of Gaussian beams (in Spanish), Rev. de Geofisica, 40:27-32.

Cormier, V. F., and Spudich, P., 1984, Amplification of ground motion and waveform complexity in fault zones: examples from the San Andreas and Calaveras faults, Geophys. J. R. Astron. Soc., 79:135-152.

Courant, R., and Hilbert, D., 1966, "Methods of Mathematical Physics", Interscience, New York.

Deschamps, G. A., 1972, Ray techniques in electromagnetics, Proceedings of the IEEE., 60:1022-1035.

Farra, V., and Madariaga, R., 1987, Seismic waveform modeling in heterogeneous media by ray perturbation theory, J. Geophys. Res., B92:2697-2712.

George, Th., Virieux, J., and Madariaga, R., 1987, Seismic wave synthesis by Gaussian beam summation: a comparison with finite differences, Geophysics, 52:1065-1073.

Gilbert, F., and Backus, G. E., 1966, Propagator matrices in elastic wave and vibration problems, Geophysics, 31:326-333.

Goldstein, H., 1981, "Classical Mechanics", Addison-Wesley Publ. Co., Reading, Mass.

Karal, F. C. Jr., and Keller, J. B., 1959, Elastic wave propagation in homogeneous and heterogeneous media, J. Acoust. Soc. Am., 31:694-705.

Klimes, L., 1984, The relation between Gaussian beams and Maslov asymptotic theory, Studia Geophys. and Geod., 28:237-247.

Landau, L., and Lifchitz, E., 1981, "Mécanique", Mir, ed., Moscow, USSR.

Madariaga, R., 1983, Earthquake source theory: a review, in: "Earthquakes, Observation Theory and Interpretation", E. Boschi and H. Kanamori, eds., North Holland, Amsterdam.

Madariaga, R., 1984, Gaussian beam synthetic seismograms in a vertically varying medium, Geophys. J. R. Astron. Soc., 79:589-612.

Madariaga, R., and Papadimitriou, P., 1985, Gaussian beam modeling of upper mantle phases, Ann. Geophysicae, 6:799-812.

Nowack, R., and Aki, K., 1985, The two-dimensional Gaussian beam synthetic method: testing and application, J. Geophys. Res., 89:7797-7819.

Nye, J. F., 1985, Caustics in Seismology, Geophys. J. R. Astron. Soc., 83:477-485.

Popov, M. M., 1969, Normal modes of a multimirror resonator, Vestnik Leningrad University, 22:42-54.

Popov, M. M., 1982, A new method for the calculation of wave fields using Gaussian beams, Wave Motion, 4:85-97.

Popov, M. M., and Psencik, I., Computation of ray amplitudes in inhomogeneous media with curved interfaces, Studia Geophys. Geod., 22:248-258.

Psencik, I., Ray amplitudes of compressional, shear and converted body waves in three-dimensional laterally inhomogeneous media with curved interfaces, J. Geophys., 45:381-390.

White, B. S., Norris, A., Bayliss, A., and Burridge, R., 1987, Some remarks on the Gaussian beam summation method, Geophys. J. R. Atron. Soc., 89:579-636.

ATTENUATION MEASUREMENTS BY MULTIMODE SYNTHETIC SEISMOGRAMS

Giuliano F. Panza

Istituto di Geodesia e Geofisica
Università di Trieste and
International School for Advanced Studies
34100 Trieste, Italy

INTRODUCTION

"Were it not for the intrinsic attenuation of sound in the earth's interior, the energy of earthquakes of past would still reverberate through the interior of the earth today." In this way Knopoff in his classical paper entitled Q (Knopoff, 1964) nicely describes the importance of the anelastic behavior of the Earth.

The attenuation of seismic energy in the Earth, being a direct measure of anelasticity, is also an important source of information regarding the composition, state and temperature of the earth interior. Unfortunately, amplitude information, reliable for attenuation studies, cannot easily be obtained since instrumental effects, local geology, phase conversion, scattering (Snieder, 1986; Snieder and Nolet, 1987) and source characteristics tend to obscure the amplitude variation due to true energy dissipation.

Amplitudes of body waves are particularly difficult to interpret because of phase conversion at interfaces, scattering and complicated spreading losses.

For this reason much research, in recent years, has been directed toward understanding the anelastic properties of the Earth, as reflected in the attenuation of seismic surface waves (e.g., Hermann and Mitchell, 1975; Yacoub and Mitchell, 1977; Burton, 1977; Canas and Mitchell, 1978; Lee and Solomon, 1978; Hwang and Mitchell, 1986). These studies neglect more or less the effects of lateral inhomogeneities, i.e., the rearrangement of surface wave energy-depth profile when propagating in continuously inhomogeneous media or when transferring from one block of the lithosphere to another, as well as focusing and defocusing of the wave energy in the presence of lateral inhomogeneities.

Any rigorous theoretical support for the estimation of these effects is hardly possible now, since the theoretical formalism allowing to calculate wave fields in arbitrary laterally inhomogeneous media has not been yet worked out. However, the study of multimode wave attenuation (Cheng and Mitchell, 1981; Kijko and Mitchell, 1983) in combination with the estimate of the possible bias due to the effects of lateral variations

79

as proposed by Levshin (1985), may allow the determination of the anelastic properties of the lithosphere with satisfactory accuracy, also in tectonically active regions, usually characterized by lateral variations.

In this paper, the single source - single station multimode method (Cheng and Mitchell, 1981) will be extended to the time domain. The analysis in the time domain, making use of the phase information, usually neglected when comparing amplitude spectra, allows a better constraint on the possible fault plane solutions than the analysis in the frequency domain. The method makes use of standard long period seismograms, but it seems also particularly suitable to analyze data gathered by broad-band instruments (e.g., Panza et al., 1986). As a general rule there is a trade-off between the desirability for long paths and short paths. Since the effect of attenuation on amplitudes is usually small, especially for high-Q regions, longer paths lead to more accurate measurements. Longer paths are, however, also likely to include more lateral heterogeneities. Here applications will be shown for the study of the anelastic properties of the crust along relatively short paths. By utilizing short paths, effects of lateral variations as well as the effects of refraction, reflection and multipathing can be minimized. Furthermore, utilizing the results of Levshin (1985), it is possible to define approximatively the period ranges where the effect of lateral variations is more or less the same on the amplitude of each mode. This may definitely help in satisfying the crucial assumption that the mechanism which causes seismic waves to attenuate affects fundamental and higher modes in the same way.

DEPARTURE FROM PERFECT ELASTICITY DUE TO TIME EFFECTS

When a perfectly elastic isotropic body is in equilibrium under an assigned stress distribution, p_{ij}, the strain distribution, e_{ij}, is uniquely determined by

$$p_{ij} = (k - 2/3\mu)\theta_{ij} + 2\mu e_{ij} \tag{1}$$

where k and μ are the bulk and the shear modulus respectively, and θ_{ij} is the cubic dilatation.

Introducing the deviatoric stress tensor

$$P_{ij} = p_{ij} - 1/3 \ p_{11}\delta_{ij} \tag{2}$$

and deviatoric strain tensor

$$E_{ij} = e_{ij} - 1/3 \ e_{11}\delta_{ij} \tag{3}$$

we can write

$$P_{11} = 3k\theta \tag{4}$$

$$P_{ij} = 2\mu E_{ij} \tag{5}$$

which are equivalent to (1). If the applied stress is symmetrical, then relation (4) describes adequately the behavior of the material, while any departures from symmetry is described by (5). The main imperfections of elasticity that are observed arise only under non-symmetrical stresses, therefore they can be described by modification of relation (5) keeping relation (4) always the same. For a continuous viscous fluid with zero bulk loss, relation (5) is replaced by

$$P_{ij} = 2\nu \ d/dt \ E_{ij} \tag{6}$$

while relation (4) continues to hold. In relation (6) ν is the shear viscosity of the fluid. If $\nu = 0$, (6) reduces to $P_{ij} = 0$, which gives the stress-strain relation for a perfect fluid

$$P_{ij} = k\theta\delta_{ij}. \tag{7}$$

Summing (5) and (6) one may write

$$P_{ij} = 2\mu E_{ij} + 2\nu \, d/dt \, E_{ij}. \tag{8}$$

Relation (8) corresponds to Kelvin-Voigt or firmoviscous behavior. It has been used to indicate the attenuation in seismic wave transmission but is deficient in not having instantaneous elastic response. This can be achieved qualitatively by adding to (8) a term in dP_{ij}/dt obtaining:

$$P_{ij} + \tau d/dt \, P_{ij} = 2\mu E_{ij} + 2\nu \, d/dt \, E_{ij}. \tag{9}$$

If in relation (9) we put $\mu = 0$, we obtain the Maxwell or elasticoviscous model

$$P_{ij} + \tau d/dt \, P_{ij} = 2\nu \, d/dt \, E_{ij}. \tag{10}$$

CONSTITUTIVE LAWS FOR ANELASTICITY

The fundamental relations (4) and (5) may be generalized to contain a time variable, i.e., it is possible to write relations of the type

$$p(\mathbf{x},t) = Ke(\mathbf{x},\tau) \qquad -\infty < \tau < t. \tag{11}$$

Let the strain change by a step $e(x,\tau) = AH(\tau)$, then

$$p(\mathbf{x},t) = AKH(\tau) \qquad -\infty < \tau < t$$
$$= A\psi(t) \tag{12}$$

where $\psi(t)$ is the creep or relaxation function, equal to zero for $t < 0$ and continuous and non-decreasing for $t > 0$. Suppose now that the strain is changed by a series of n step increments $H(t_j)$ at times $t = t_j$, $j = 1,2,\ldots$ The corresponding stress is

$$p(\mathbf{x},t) = \sum_{j=1}^{n} H(t_j)\psi(t - t_j) \tag{13}$$

with $t_i < t_j$ for $i < j$.

Equation (13) can be written as the Stieltjes convolution integral

$$p(\mathbf{x},t) = \int_o^t \psi(t - \tau)de(\tau) = \int_o^t \psi(t - \tau)\dot{e}(\tau)d\tau. \tag{14}$$

Integration by parts gives, for $t > 0$ and $\psi(0-) = 0$

$$p(\mathbf{x},t) = m_o e(t) + \int_o^t \dot{\psi}(t - \tau)e(\tau)d\tau \tag{15}$$

where m_o is the instantaneous elastic modulus governing the instantaneous response, while the second term represents the relaxation or creep. Equation (15) can thus be used to write the generalized visco-elastic stress-strain relations

$$P_{11}(x,t) = 3k_o e_{11}(x,t) + 3 \int_o^t \dot{R}_k(t - \tau)e_{11}(x,\tau)d\tau \tag{16}$$

$$P_{ij}(x,t) = 2\mu_o E_{ij}(x,t) + 2 \int_o^t \dot{R}_\mu(t - \tau)E_{ij}(x,\tau)d\tau \tag{17}$$

where R_k, R_μ are the dilatational and shear relaxation functions and k_o, μ_o are the equivalent of the Lame parameters for the elastic response. The case of perfect elasticity is obtained simply by putting $\dot{R}_k$ and $\dot{R}_\mu$ equal to zero.

Taking the one sided Fourier transform, relations (16) and (17) become

$$P_{11}(x,\omega) = 3e_{11}(x,\omega)[k_o + k_1(\omega)] \tag{18}$$

$$P_{ij}(x,\omega) = 2E_{ij}(x,\omega)[\mu_o + \mu_1(\omega)] \tag{19}$$

where

$$k_1(\omega) = \int_o^\infty \dot{R}_k(t)e^{i\omega t}dt \tag{20}$$

$$\mu_1(\omega) = \int_o^\infty \dot{R}_\mu(t)e^{i\omega t}dt \tag{21}$$

are the imaginary parts of the complex elastic moduli

$$\begin{aligned} \bar{k} &= k_o + k_1(\omega) \\ \bar{\mu} &= \mu_o + \mu_1(\omega) \end{aligned} \tag{22}$$

Recalling (2) and (3), the complete stress-strain relation for a visco-elastic body is given by

$$P_{ij}(x,\omega) = \bar{\lambda}e_{11}(x,\omega)\delta_{ij} + 2\bar{\mu}e_{ij}(x,\omega) \tag{23}$$

where

$$\bar{\lambda} = \bar{k} - 2/3\ \bar{\mu} = \lambda_o + \lambda_1(\omega). \tag{24}$$

Relation (23) indicates that in the regime of linear departures from elastic behavior, the stress-strain relation at frequency ω is the same as that for an elastic medium, but with complex moduli.

Relations (18), (19) and (23) can be considered as linear filter equations. The filter, represented by the visco-elastic complex moduli, must be causal, i.e., in the time domain, the filter output $p_{ij}(x,t)$ must not start earlier than the filter input $e_{ij}(x,t)$. This requirement imposes relations between the imaginary part of the elastic moduli which are called dispersion or Kramers-Krönig relations, which can be written, for instance in the case of $\bar{\mu}$,

$$Re[\mu_1(\omega)] = 1/\pi P \int_{-\infty}^\infty Im[\mu_1(\omega')]/(\omega' - \omega)d\omega' \tag{25}$$

where P denotes the Cauchy principal value.

THE LOSS FACTOR

A convenient measure of the rate of energy dissipation is the loss factor $Q^{-1}(\omega)$, which may be defined as

$$Q^{-1}(\omega) = -\Delta E(\omega)/2\pi E_o(\omega)) \tag{26}$$

where $\Delta E(\omega)$ is the energy loss in a cycle at frequency $f = \omega/2\pi$ and E_o is the sum of the strain and kinetic energy calculated with just the instantaneous elastic moduli.

For purely dilatational disturbances

$$Q^{-1}{}_k(\omega) = (\tfrac{1}{2}i\omega k_1(\omega)|e_{11}|^2)/(\tfrac{1}{2}\omega k_o|e_{11}|^2) = - \text{Im } k_1(\omega)/k_o \qquad (27)$$

and for purely deviatoric waves

$$Q_\mu^{-1}(\omega) = (\tfrac{1}{2}i\omega\mu_1(\omega)|E_{ij}|^2/(\tfrac{1}{2}\omega\mu_o|E_{ij}|^2) = - \text{Im}[\mu_1(\omega)]/\mu_o. \qquad (28)$$

Combining (25) and (28) it is possible to write

$$\text{Re}[\mu_1(\omega)] = -2\mu_o/\pi P \int_o^\infty \omega' Q_\mu^{-1}(\omega')/(\omega'^2 - \omega^2)d\omega'. \qquad (29)$$

Relation (29) indicates that we cannot have dissipative effects, i.e., $\text{Im}[\mu_1(\omega)] \neq 0$ and $Q_\mu^{-1}(\omega) \neq 0$, without some frequency dependent modification of the elastic moduli.

The distribution of the loss factor with depth in the earth, $Q_\mu^{-1}(z)$, is still imperfectly known, because of the difficulties in isolating all the factors which affect the amplitude of a recorded seismic wave. However, most models show a moderate loss factor in the crust ($Q_\mu^{-1} \sim 0.004$) with an increase in the uppermost mantle ($Q_\mu^{-1} \sim 0.01$) and then a decrease to crustal values, or lower, in the mantle below 1000 km. Over the frequency band 0.0001 - 10 Hz the intrinsic loss factor $Q_\mu^{-1}(z)$ appears to be essentially constant, but in order for there to be a physically realizable loss mechanism, $Q_\mu^{-1}(z)$ must depend on frequency outside this band. A number of different forms have been suggested (Azimi et al., 1968; Liu, Anderson and Kanamori, 1976; Jeffreys, 1958) but provided $Q_\mu^{-1}(z)$ is not too large ($Q_\mu^{-1}(z) < 0.01$) these lead to the approximate relation

$$\text{Re}[\mu_1(\omega)] = 2\mu_o \ln(\omega a)/(\pi Q_\mu) \qquad (30)$$

in terms of some time constant a.

A similar development may be made for the complex bulk modulus in terms of $Q_k^{-1}(\omega)$.

In the hypothesis of Q^{-1} small and independent of frequency from (30) we can deduce the dispersion relation for shear-waves in anelastic media (Panza, 1985)

$$B_1(\omega) = B_1(\omega_o)/\{1 + [(2/\pi)B_1(\omega_o)B_2(\omega_o)\ln(\omega_o/\omega)]\} \qquad (31)$$

where $B_1(\omega)$ is the S-wave phase velocity and $B_2(\omega_o)$ is the S-wave phase attenuation related to Q_μ^{-1} by:

$$Q_\mu^{-1} = 2B_1(\omega_o)B_2(\omega_o). \qquad (32)$$

For compressional waves the situation is a little more complicated since the anelastic effects in pure dilatation and shear are both involved; however if $Q_k^{-1} \ll Q_\mu^{-1}$ it is possible to derive the approximated relation suggested by Anderson et al. (1965)

$$Q^{-1}{}_A = 4/3(\beta_o/\alpha_o)^2 Q_\mu^{-1} \qquad (33)$$

e.g., see Kennett (1985), and then the dispersion relation for compressional waves in anelastic media can be written as:

$$A_1(\omega) = A_1(\omega_o)/\{1 + [(2/\pi)A_1(\omega_o)B_2(\omega_o)\ln(\omega_o/\omega)]\}$$ (34)

with obvious meaning of the symbols.

In equations (31) and (34) it is necessary to fix a reference frequency ω_o; usually $\omega_o = 2\pi$ radians.

The assumption of a constant Q^{-1} to obtain (31) and (34) may be not correct since Q^{-1} appears to increase with decreasing period, at least at periods shorter than 2 or 3 s (Mitchell, 1980). However, it is not likely, using multimode data and presently available methods, that a frequency dependence of Q_μ^{-1} could be detected at periods between about 3 and 50 s. This could be possible in the near future when broad-band digital instruments will supply high quality data also at short periods (e.g., Panza et al., 1986).

ANELASTICITY AND SCATTERING

In addition to the dissipation of elastic energy by anelastic processes, the apparent amplitude of a seismic wave can be diminished by scattering which redistributes the elastic energy. The choice of elastic moduli, following the constitutive relations (1) or (23), defines a reference medium whose properties smooth over local irregularities of the material. The fluctuations of the true material about the reference will lead to scattering of the seismic energy out of the primary wave, which will be cumulative along the propagation path, and the apparent velocity of transmission of the scattered energy will vary from that in the reference. Since the material may locally be faster or slower than the assigned wave speed, the effect of scattering is to give a pulse shape which is broadened and diminished in amplitude relative to that in the reference medium, with an emergent onset before the reference travel time. At a frequency ω we may describe the effect of the scattering by a loss factor $_sQ^{-1}(\omega)$ and the changing character of the scattering process leads to a strong frequency dependence. As the wavelength diminishes, the effect of local irregularities becomes more pronounced and so $_sQ^{-1}$ tends to increase until the wavelength is of the same order as the size of the scattering region.

This scattering mechanism becomes important in areas of heterogeneity and its influence seems largely to be confined to the lithosphere.

For each wave type the overall rate of seismic attenuation Q^{-1}, which is the quantity which would be derived from observations, will be the sum of the loss factors from intrinsic anelasticity and scattering. Thus for S-waves

$$Q^{-1}_\beta(\omega) = Q^{-1}_\mu(\omega) + {_sQ^{-1}}_\beta(\omega).$$ (35)

For P-waves

$$Q^{-1}_\alpha(\omega) = Q^{-1}_A(\omega) + {_sQ^{-1}_\alpha}(\omega);$$ (36)

since the scattering component here is arising from a totally distinct mechanism than the dissipation there is no reason to suppose that $_sQ^{-1}_\alpha$, $_sQ^{-1}_\beta$ are related in a similar way to (33).

Recent observational results (Aki, 1981) suggest that the contributions Q^{-1}_μ and $_sQ^{-1}_\beta$ are separable via their different frequency behavior (Figure 1). The intrinsic absorption Q^{-1}_μ is quite small and nearly frequency independent over the seismic band, and superimposed on

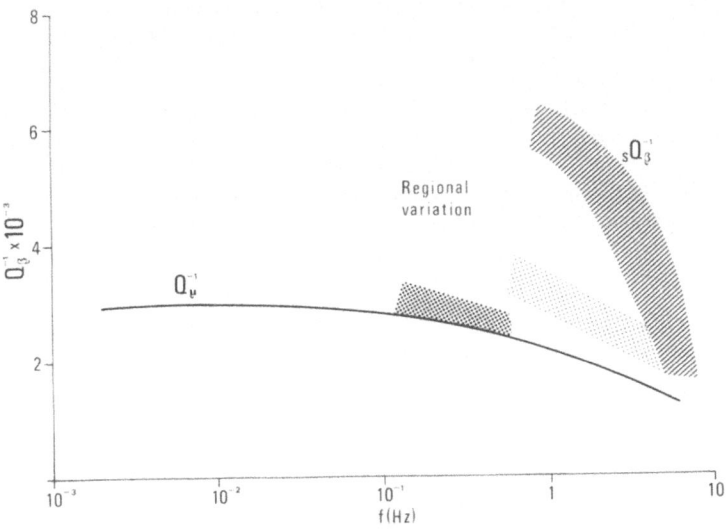

Fig. 1. Frequency separation of intrinsic loss factor Q_μ^{-1} from scattering contribution $_sQ^{-1}{}_\beta$ (after Kennett, 1985).

this, with characteristics which vary from region to region, is the more rapidly varying scattering loss. Other studies also indicate such an increase in the loss factors Q^{-1} as the frequency rises towards a few Hz. The results of Figure 1 could be fitted with some postulated dependence of Q^{-1} on frequency. However, the wave speed dispersion estimated by (29) from such relations would be very misleading. It is only for the anelastic portion Q_μ^{-1}, $Q^{-1}{}_A$ that we have dispersive wave speed terms. The scattering contribution $_sQ^{-1}{}_\beta$, $_sQ_\alpha^{-1}$ does not have the same restriction to a local "memory" effect, and there is no consequent dispersion.

MULTIMODE COMPUTATION IN ANELASTIC MEDIA

Following Schwab and Knopoff (1972) the quantities appearing in (31) and (34) are related to the complex body wave velocities $\bar{\alpha}(\omega)$ and $\bar{\beta}(\omega)$, describing the properties of anelastic media, by

$$\bar{\alpha}^{-1}(\omega) = A^{-1}{}_1(\omega) - iA_2(\omega) \tag{37}$$

$$\bar{\beta}^{-1}(\omega) = B^{-1}{}_1(\omega) - iB_2(\omega). \tag{38}$$

A similar relation holds for surface waves

$$\bar{c}^{-1}(\omega) = C^{-1}{}_1(\omega) - iC_2(\omega) \tag{39}$$

where $C_1(\omega)$ is the attenuated phase velocity and $C_2(\omega)$ is the phase attenuation.

With these generalizations, the algorithms developed for perfectly elastic media can readily be applied to anelastic media. Extensive descriptions of the algorithms and examples of computation of synthetic signals as multimode superposition are given by Panza (1985), Panza et al. (1986) and Panza and Suhadolc (1987).

The modal summation is definitely one of the most appropriate approaches for attenuation studies. In fact it is extremely efficient when inverting for the source depth and mechanism (Panza et al., 1986), a critical step in the single source-single station multimode method (Cheng and Mitchell, 1981). Furthermore, for epicentral distances of about $10°$, the generation of signals comparable with records from standard long period seismographs (seismometer period about 15 s, galvanometer period about 100 s) can be made summing up only about 20 modes, thus requiring a very limited amount of time. Typical values for an IBM 370/168 computer are: about 200 s for the computations in the frequency domain and about 20 s for the computation of the first time series. This figure decreases by a factor of 10 for all subsequent seismograms computed for sources with any orientation but same focal depth.

In the following, some examples of computation of the quality factor for travelling waves, $Q^{-1}_x(\omega)$, will be described. According to Schwab and Knopoff (1972)

$$Q^{-1}_x(\omega) = 2C_1(\omega)C_2(\omega).$$ (40)

Note that in Eq. (40) $Q^{-1}_x(\omega)$ varies with frequency even if we have considered models for the intrinsic attenuation independent of frequency.

$Q^{-1}_x(\omega)$ represents the energy lost by a mode after travelling one wavelength and is related to the energy lost per cycle $Q^{-1}_t(\omega)$ by

$$Q^{-1}_t(\omega) = u(\omega)C^{-1}_1(\omega)Q^{-1}_x(\omega)$$ (41)

where u is the group velocity of the mode.

In Figure 3 we have plotted the quality factor, i.e., the inverse of the loss factor, for the first 20 modes of the structural models shown in Figure 2. These models can be taken as representative of some portions of the Mediterranean region, namely Po Valley (Figure 2a)., Apennines (Figure 2b), Adriatic (Figure 2c) and Mediterranean ridge (Figure 2d). The main feature of the structural model PAD2A, representing the average properties of the Po Valley, is a rather thick sedimentary cover - about 6 km - characterized by very low velocities but very high Q. The model APNC, representing the average properties of the Apennines, is characterized by a sequence of layers of total thickness of about 15 km, simulating the nappes cover and presents a gentle low velocity layer between 30 km and 40 km of depth. Q values are rather low - 100 in the first 20 km and 300 for greater depths. The model ADRI04, representing the average properties of the Adriatic plate, is characterized by a water layer of 50 m on top of a sedimentary sequence about 6 km thick. The Q is equal to 500 over the whole lithosphere. Finally the structure SOA9A, representing the average properties of the Mediterranean ridge, is characterized by a water layer 1 km thick overlying a sedimentary sequence of 4 km. The crustal thickness is of continental type and Q values are very low - 100 in the first 20 km and 250 in the remaining lithosphere.

In the four structural models considered, in each layer, it has been assumed $Q_A\alpha(z) = 2Q_\beta(z)$ and this, taking into account equation (33), implies some energy loss also in compression. Indexes α and β refer to P- and S-waves respectively.

The values of Q_x shown in Figure 3a, pertinent to the model PAD2A, as expected, are very large since the intrinsic quality factor $Q_\beta(z)$ is constantly equal to 1000. For each mode Q_x is a strongly varying function of frequency even if $Q_\beta(z)$ is frequency independent. The envelopes of minima around a value of 600 are due to the presence of the thick low velocity sedimentary cover. The situation is different in Figure 3b, where the effect of the layering in Q(z) is quite clearly evidenced by the

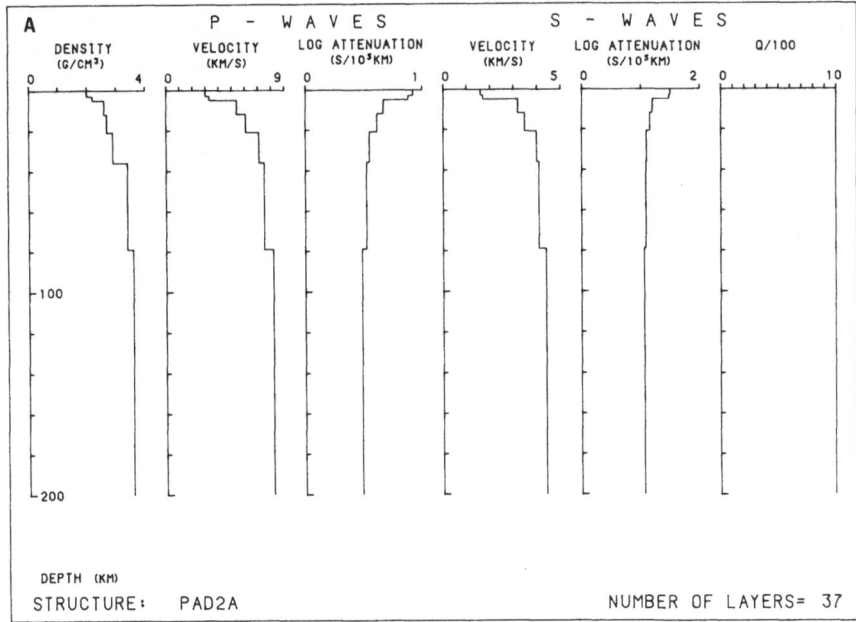

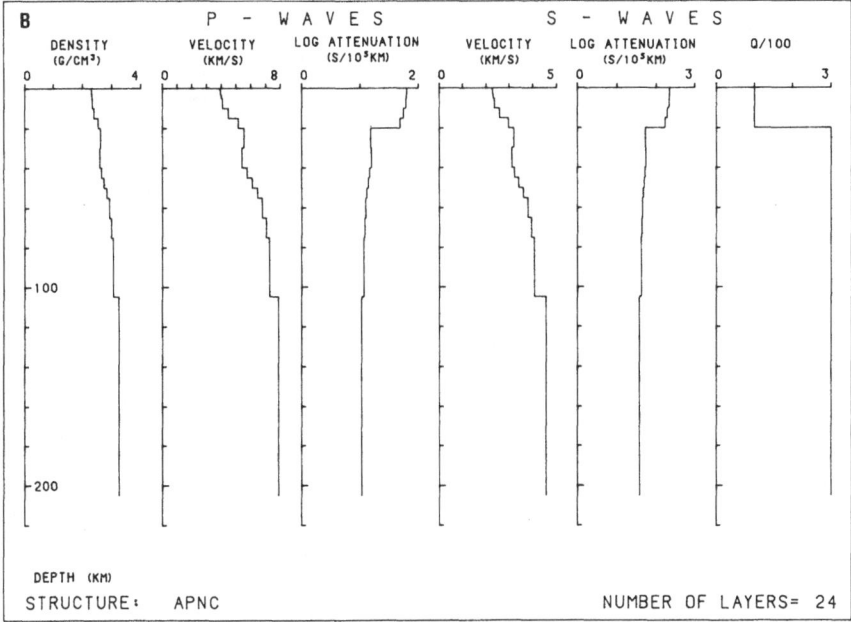

Fig. 2a. Elastic and anelastic properties of the structural model PAD2A, representative of the Po Valley.

b. Elastic and anelastic properties of the structural model APNC, representative of the Apennines.

envelopes of Q tending asymptotically to the values of 100 and 300. Figure 3c is an example of the effect that an asthenospheric low velocity layer may have on Q_x spectra even when $Q(z)$ is constant. Finally the most remarkable feature of Figure 3d is the behavior of Q_x for the fundamental mode; for frequencies larger than about 0.6 Hz it tends to infinity because the fundamental mode at these frequencies samples only the water layer which is characterized by $Q_{A}\alpha = \infty$.

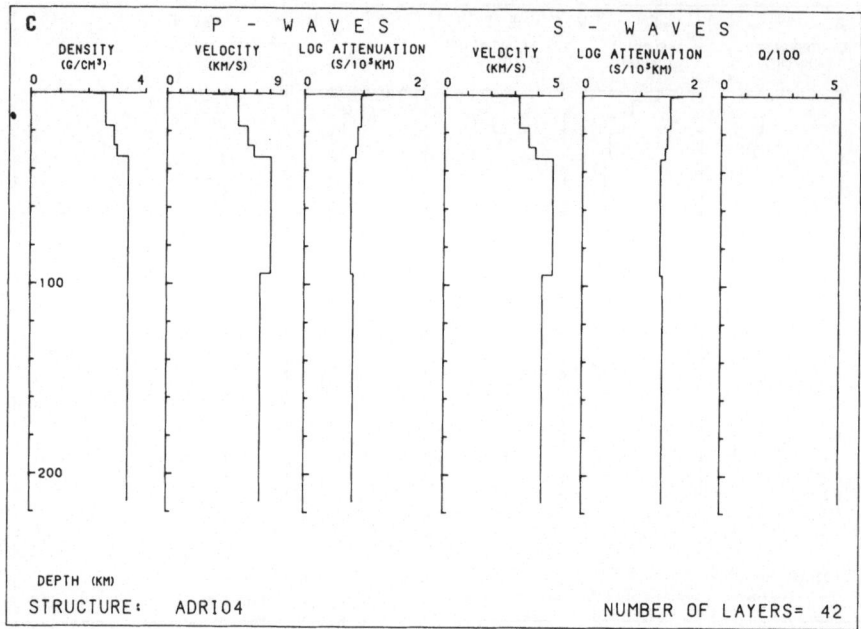

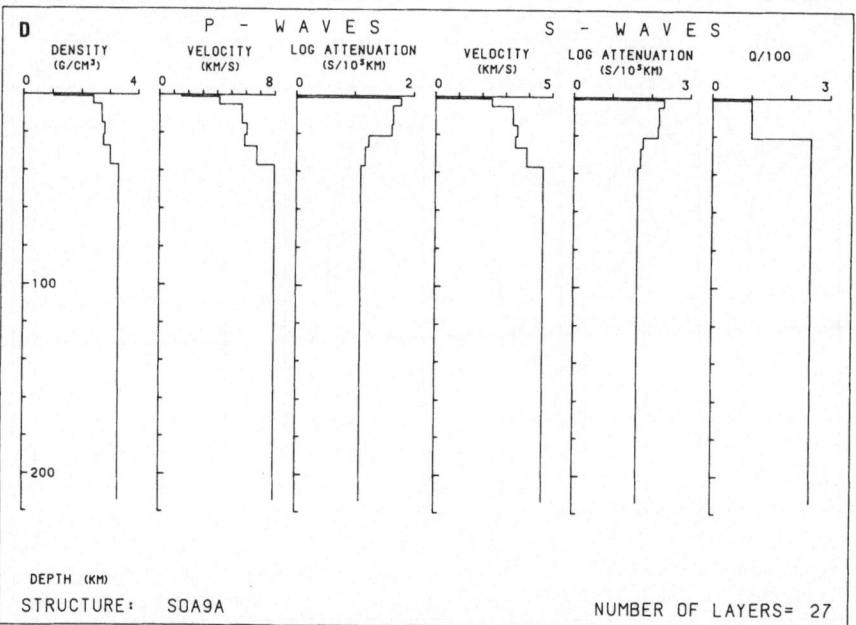

Fig. 2c. Elastic and anelastic properties of the structural model ADRIO4, representative of the Adriatic plate.

  d. Elastic and anelastic properties of the structural model SOA9A, representative of the Mediterranean ridge.

A common feature to all Q_x spectra is the presence, over quite broad frequency bands, of a population of modes with Q_x ranging from the lowest to the highest possible value. This indicates that the preferable procedures to obtain good estimates of the intrinsic Q(z) are those like the multimode attenuation studies proposed by Cheng and Mitchell (1981). In fact, when individual modes are not separable, a straightforward analysis of seismic records would only result in measuring the envelope of

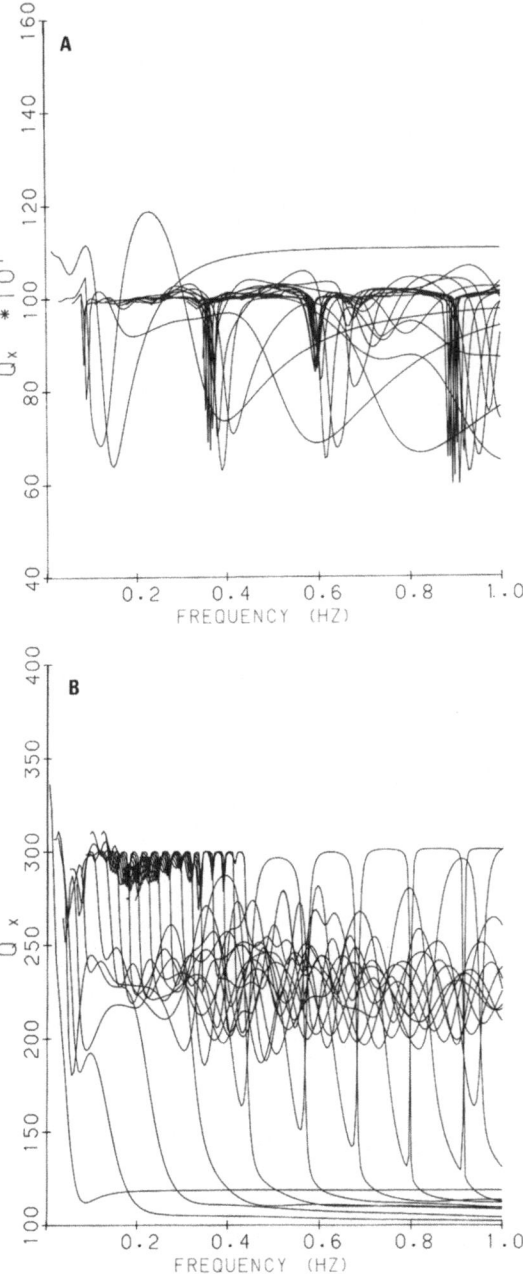

Fig. 3a. Spectra of quality factor, Q_x, for the first 20 modes of P-SV
waves for the model PAD2A.
b. Spectra of quality factor, Q_x, for the first 20 modes of P-SV
waves for the model APNC.

maximum values of Q_x which cannot be assigned to any specific part of the
structure for a subsequent physical interpretation. This means that in
order to make physically interpretable measurements of Q_x it is necessary
to apply quite accurate time or group velocity windows to the records
before any further processing. The need of applying accurate group
velocity windows over quite large frequency bands makes the availability

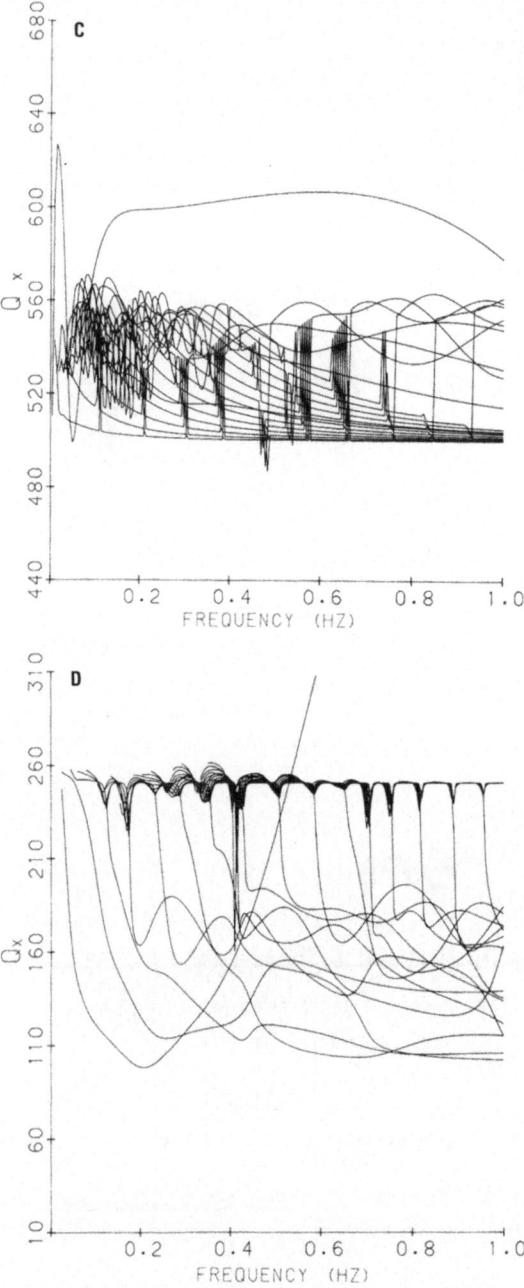

Fig. 3c. Spectra of quality factor, Q_x, for the first 20 modes of P-SV waves for the model ADRI04.

d. Spectra of quality factor, Q_x, for the first 20 modes of P-SV waves for the model SOA9A.

of well calibrated broad-band instruments extremely valuable for the achievement of the very difficult task of determining physically meaningful Q_x values.

THE SINGLE SOURCE-SINGLE STATION MULTIMODE METHOD

The method consists of generating theoretical seismograms formed by fundamental and first 20 higher modes surface waves and comparing them to corresponding observed signals. Since we will consider only vertical components of standard WWSSN long period seismograms we will limit our attention to Rayleigh waves and P-SV higher modes.

Theoretical seismograms were calculated using the formulation of Panza (1985). The calculations require a knowledge of the source depth and fault plane solution as well as the velocity-density model along the path of propagation. The synthetic seismograms are very sensitive to variations of focal depth and of elastic structural parameters, while anelastic structural parameters and fault plane geometry represent a second-order effect. Potentially serious is the effect of source finiteness, but a proper choice of the size of the earthquake and of the period range of investigations makes this effect negligible.

Therefore the determination of the internal friction profile, $Q_\beta(z)$, for the lithosphere can be made in two stages:

(1) choice of elastic structural parameters and range of focal depths,
(2) perturbation of anelastic structural parameters and of fault plane solution parameters.

The initial values of the elastic structural parameters can be chosen either on the basis of already existing information or directly from the record, which will be used for the determination of $Q_\beta(z)$, by means of multiple-filter techniques (Levshin et al., 1972). The initial value of the focal depth can be taken from the available seismological bulletins (e.g., ISC or NEIS), from which it is also possible to extract some information about the focal mechanism, if the event was not studied earlier. With these input data it is possible to construct a multimode synthetic seismogram to be compared with the experimental record.

By trial-and-error it is possible to define the depth range of the source, for instance from the period of the maximum of the fundamental mode (Panza et al., 1973, 1975a,b), as well as a set of elastic structural parameters, which give satisfactory agreement between theory and observations. At present a trial-and-error method is the only applicable when comparing time series of high complexity like the ones obtainable summing some tens of modes. To minimize the integral of the square of the difference between the experimental and the synthetic seismograms is wrong. In fact the contribution to the integral from the fundamental mode part dominates over everything else obscuring the contribution from the higher modes. To avoid this, weighting functions must be introduced whose definition, at present, is missing a satisfactory theoretical basis.

Once the first order parameters have been chosen, keeping them fixed it is possible to see the effects of variations in anelastic structural parameters and in fault plane geometry. Since the method is based on the analysis of relative amplitudes of fundamental and higher modes and not on absolute values, here the relevant fault plane parameters are dip and rake. On the base of the resolving power of the data it is reasonable to consider for the lithosphere two-valued $Q_\beta(z)$ models (Cheng and Mitchell, 1981; Kijko and Mitchell, 1983), and steps in dip and rake of $20°$ (for the definition of the parameters describing the source geometry see, for instance, Panza, 1985). The ranges of $Q_\beta(z)$ and of the fault plane geometry are then determined by trial-and-error. The availability of reliable fault plane solutions, determined in other ways, may help in reducing the uncertainties.

Before concluding this section some remarks on sources of error are necessary. The possible sources of error may be separated into two groups: those which are associated with uncertainties in the source specifications and those which may be due to propagation effects between the source and the receiver.

Errors due to source effects include uncertainties in focal depth, strike, dip and rake as well as the effects of finiteness of the fault in space and time. Errors due to propagation effects include those due to an incorrect velocity model and lateral variations of elastic and anelastic properties along the path of propagation.

The effects of uncertainties in the angles of strike, dip and rake are relatively small while the effect of focal depth is quite severe (e.g., Panza et al., 1973, 1975a,b; Kijko and Mitchell, 1983). If standard long period instruments are used, the effects of source finiteness are relevant only when using earthquake sources which produce significant fault movement. On the other hand if broad-band seismographs are used the source finiteness can be matter of investigation, even if it is reasonable to expect some trade-off between the effects of source size and those of anelasticity.

In general, the source time function assumed for earthquakes is a step. Here, the relevant effect of departures from a step function source would produce differences in the spectral shape but would not affect the relative amplitude of the fundamental and higher modes.

The effects of changes in the velocity model can be controlled satisfying the group velocity dispersion of the fundamental mode. In fact, in the period range of interest, this quantity has a very weak dependence on the source apparent initial phase (e.g., Panza et al., 1973). What we may expect are changes in individual phases but not in the general shape of the wave train represented by the signal envelope. In fact the effects of changes in the velocity model upon spectral amplitudes have been found to be insignificant compared to those produced by changes in the $Q_\beta(z)$ model, as long as the velocity model is reasonably close to the true structure (Mitchell, 1980). In any case our experience has shown that even using different models, consistent with the dispersion data, the results inferred about $Q_\beta(z)$ do not vary significantly.

The effects of lateral changes in elastic properties may definitely play a relevant role. A quite important control on these effects can be obtained from the study of Levshin (1985), which allows a semi-quantitative estimate of the effect of lateral inhomogeneities on mode amplitude analyzing quantities of the type

$$R(\omega) = (\sqrt{uI_1})_{M1} / (\sqrt{uI_1})_{M2} \tag{42}$$

where u is the group velocity, I_1 is the energy integral (e.g., Panza, 1985) and M1 and M2 indicate the medium where respectively the receiver and the source are located.

In general, $R(\omega)$ is strongly frequency dependent and the measure of the effects of lateral variations is given by how different R is from unity. In other words R can be taken as a rough estimate of the, frequency-dependent, percentage variation in amplitude due to lateral inhomogeneities.

The numerical examples given by Levshin (1985) clearly indicate how large can be the bias introduced by lateral heterogeneities, when single mode amplitude measurements are performed. The situation is much better when the multimode method is applied. In this case, in fact, the absolute amplitude value is not relevant, and meaningful results can be obtained as long as the comparison of amplitudes is limited to the frequency ranges over which the values of $R(\omega)$, for the different modes, are close to each other. Thus, for the first few modes, a preliminary analysis of $R(\omega)$ computed for structural models representative of the considered region may help in reducing significantly the effect of lateral inhomogeneities on amplitude measurements.

To conclude this section mention must be made of the effects of lateral changes in anelastic properties. These effects at present are the most difficult to assess. On the basis of the existing literature and from the consideration that, in general, the anelastic behavior is a second-order property of Earth's materials, we do not expect that anelastic lateral variations may significantly affect the records we are processing. This effect may be of some interest if broad-band data are used. In this case the only reasonable solution, at present seems to be the capability of formally treating the propagation of multimode signals in laterally varying anelastic media. A recent step in this direction has been made by Gregersen et al. (1988).

DATA PROCESSING FOR THE DETERMINATION OF Q MODELS

In this section the complete data processing process will be described, making use of real examples. Ideally, the method should be applied to situations where the elastic properties along the path are known as well as the location, depth and fault plane solution of the source. For intermediate sized events, i.e., for events for which the point-source approximation is valid in the period range of interest, in general only the source location is available with sufficient accuracy and a set of starting values must be chosen for the other parameters.

The earthquakes considered are listed in Table 1.

For the first event the record of the long-period vertical component of the station of Bari (BAI 40.878°N, 17,204°E) has been used, while for the second use has been made of the long period vertical component record of the station of Trieste (TRI 45.709°N, 13.759°E). Both events have been digitized, corrected for the instrument response and processed by means of the multiple-filter technique called FTAN (Levshin et al., 1972). The resulting plots of spectral amplitudes contoured as a function of period and group velocity are shown in Figure 4a and 4b. In the left side of these figures the time series is plotted, after removal of the instrument transfer function.

From the group velocity dispersion of the fundamental mode it is possible to infer the distribution versus depth of elastic parameters. Structural models consistent with the two dispersion curves and used in the subsequent computations are shown in Figure 2c and 2d. The parameter to be fixed next is the hypocentral depth. This can easily be done

Table 1. Events Used

Date m/d/y	Origin Time h : m : s	Latitude °N	Longitude °E	M
04.07.74	14 22 47.1	34.7	24.7	4.7
06.19.75	10 11 13.0	41.6	15.7	4.9

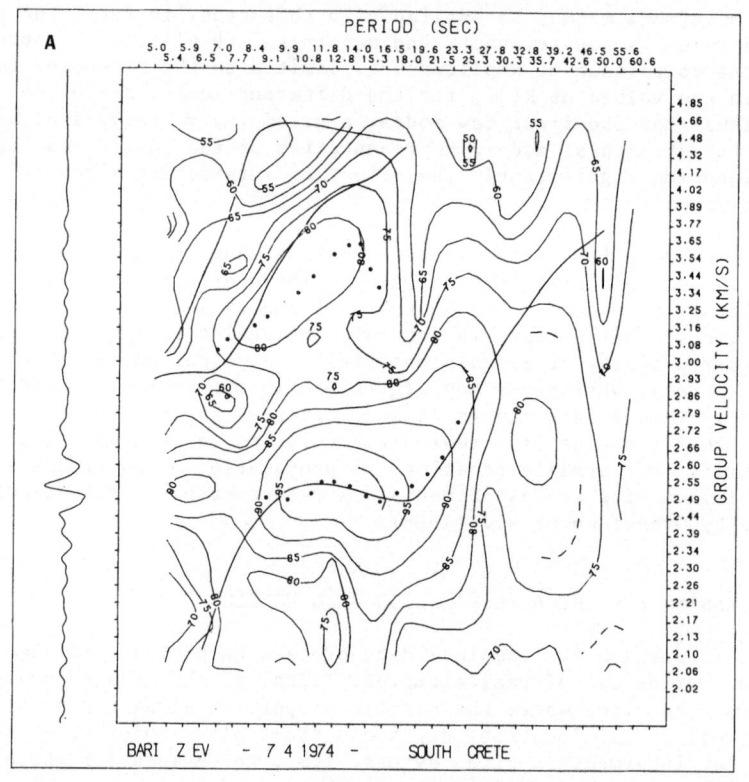

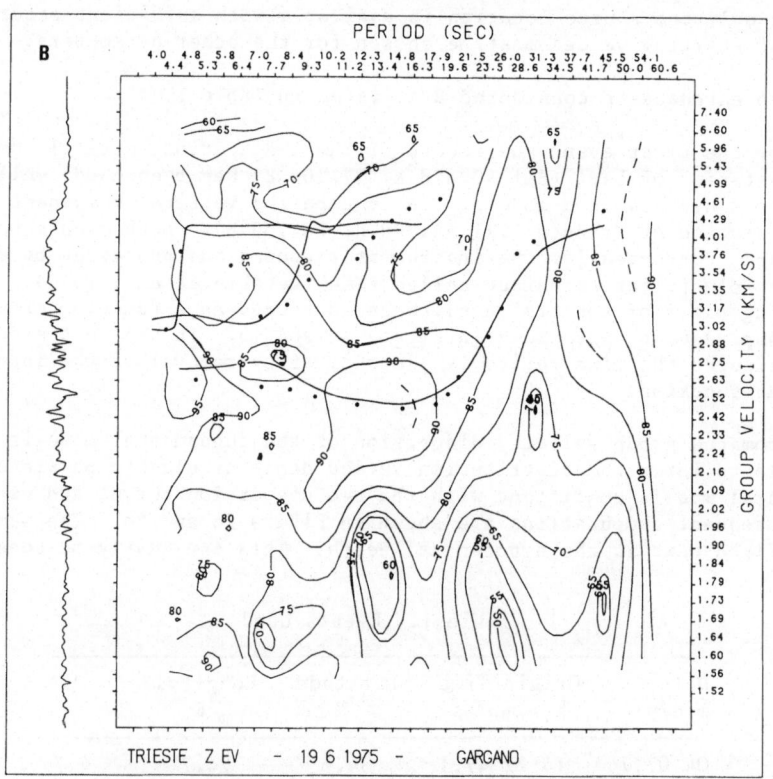

Fig. 4a and b. See next page for legend.

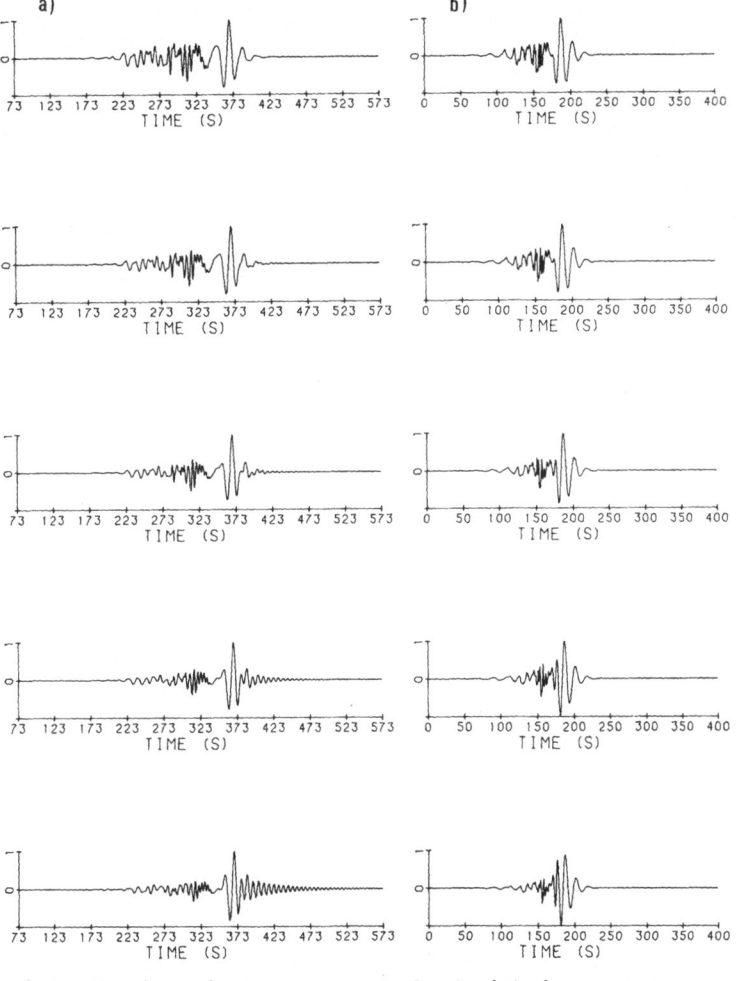

Fig. 5.   a) Synthetic seismograms, convolved with instrument response of
          station BAI, computed for different values of focal depth, h, for
          the study of event 04/07/74: from top to bottom h = 30,25,20,15,
          10 km respectively. Epicentral distance equal to 922 km. Input
          structure is given in Figure 1d, while the input fault plane
          solution is given in Table 2; b) synthetic seismograms, convolved
          with instrument response of station TRI, computed at different
          focal depths for the study of event 06/19/75: from top to bottom
          h = 28,26,22,18,14 km respectively. Epicentral distance equal to
          478 km. Input structure is given in Figure 1c, while the input
          fault plane solution is given in Table 2.

Fig. 4a. Spectral amplitudes contoured as a function of period and group
          velocity for the first event of Table 1. Epicentral distance
          equal to 922 km. The time series is plotted on the left side of
          the figure. Dots indicate the experimental dispersion values,
          while the continuous line gives the group velocities
          corresponding to the model of Figure 2d.

      b. Spectral amplitudes contoured as a function of period and group
          velocity for the second event of Table 1. Epicentral distance
          equal to 478 km. The time series is plotted on the left side of
          the figure. Dots indicate the experimental dispersion values
          while the continuous line gives the group velocities
          corresponding to the model of Figure 2c.

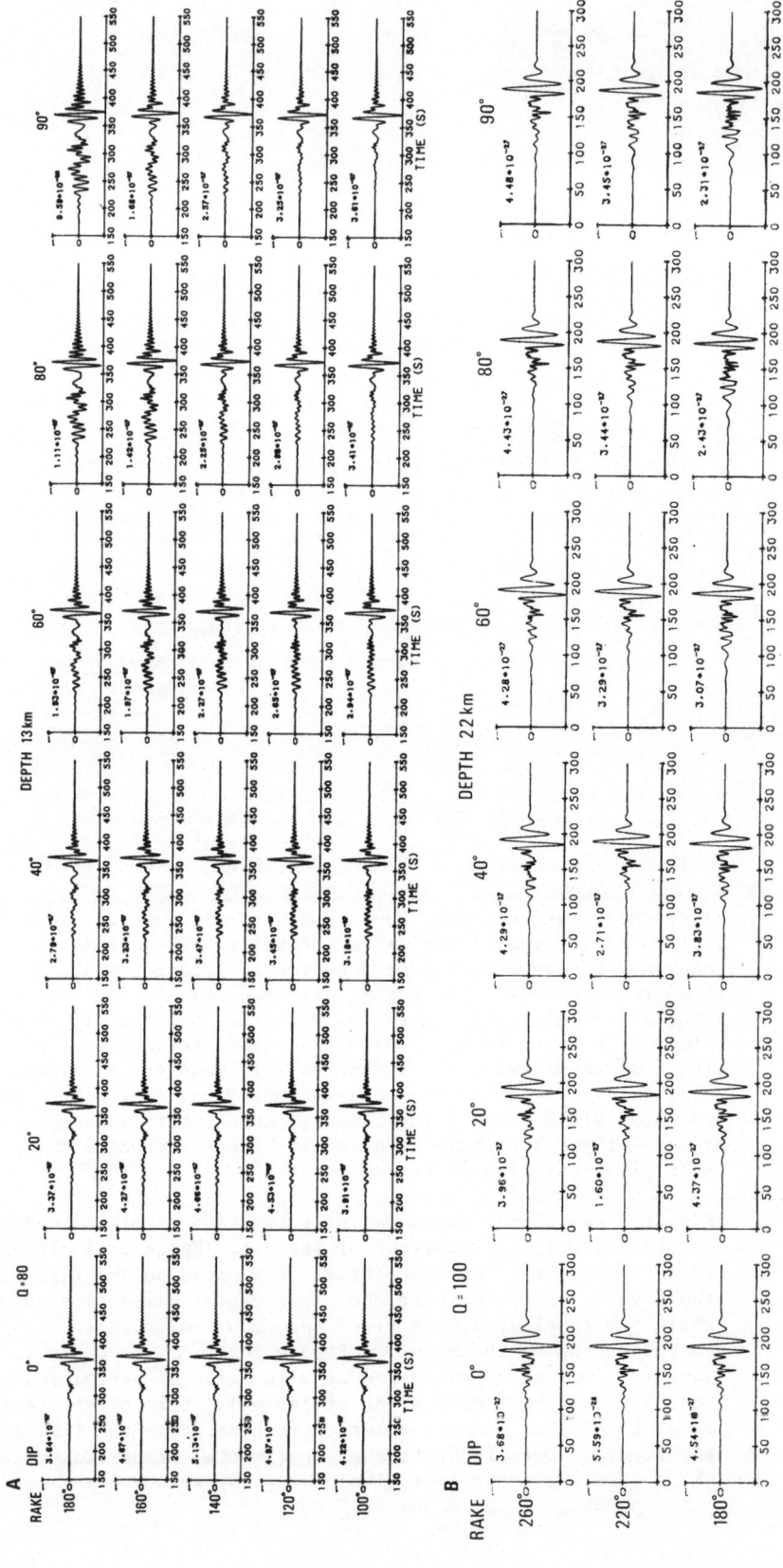

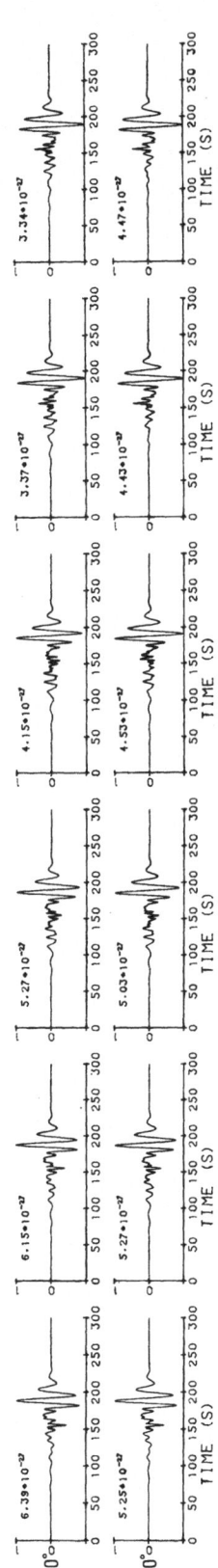

Fig. 6a. Examples of matrices of synthetic seismograms, convolved with the instrument response of the station BAI, for the event 04/07/74. Input elastic model, at 1 Hz, given in Figure 2d, $Q_\beta(z)$ = 80, h = 13 km.

b. Examples of matrices of synthetic seismograms, convolved with the instrument response of the station TRI, for the event 06/19/75. Input elastic model, at 1 Hz, given in Figure 2c, $Q_\beta(z)$ = 100, h = 22 km.

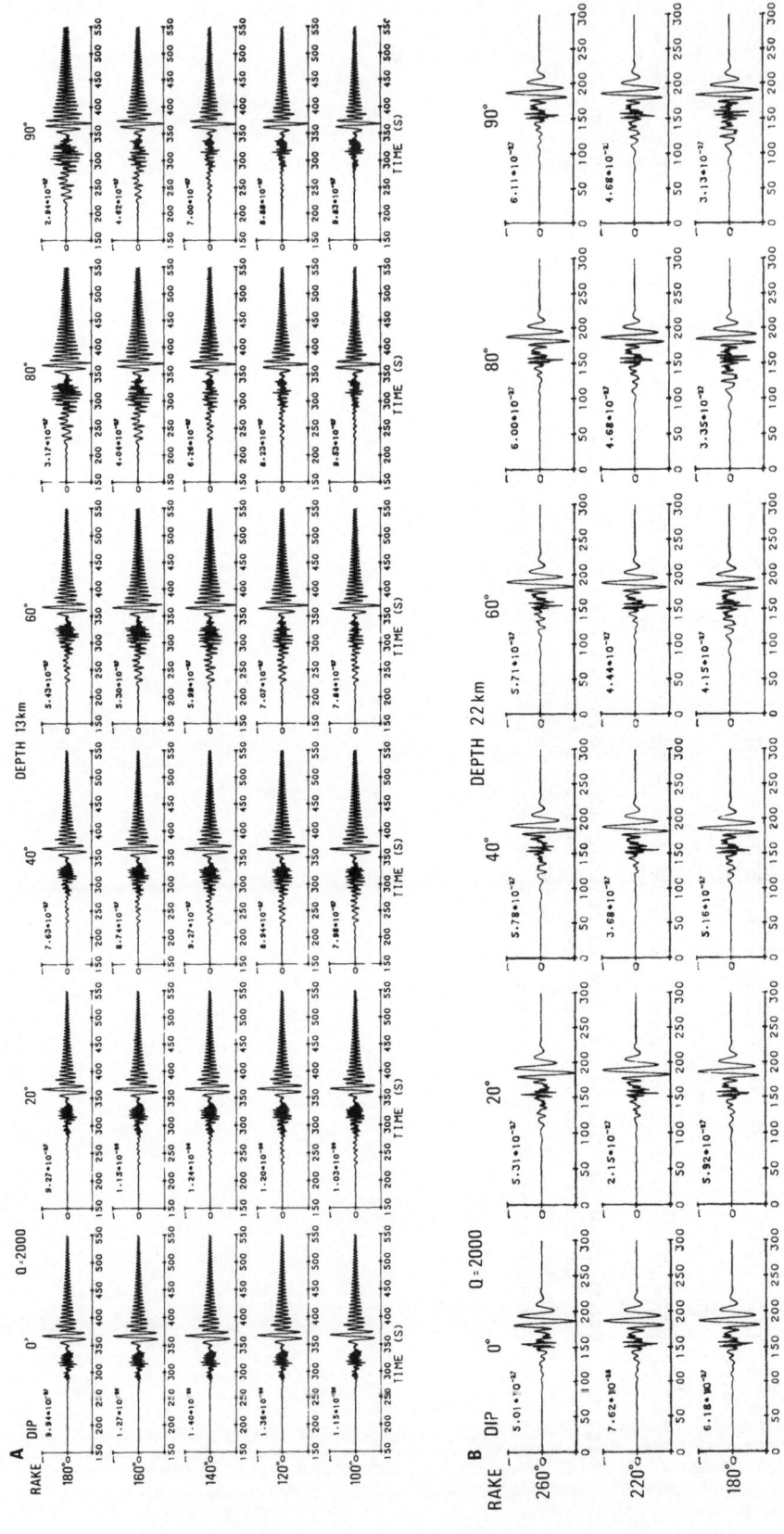

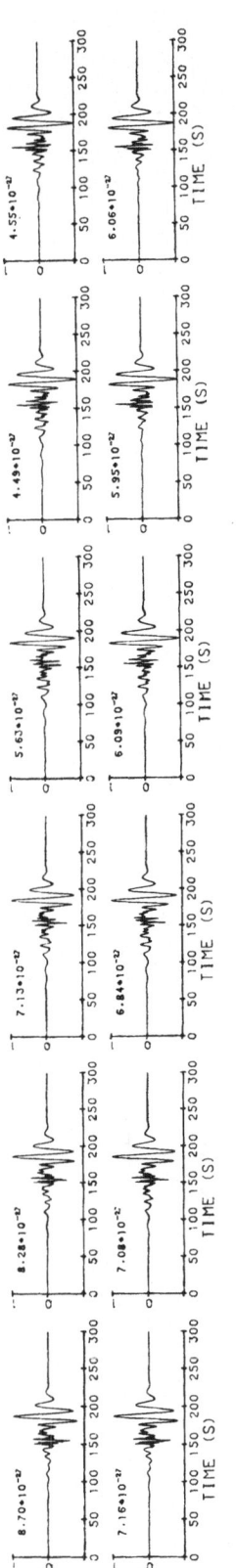

Fig. 7.   a. As in Figure 6a but $Q_\beta(z) = 2000$.   b. As in Figure 6b but $Q_\beta(z) = 2000$.

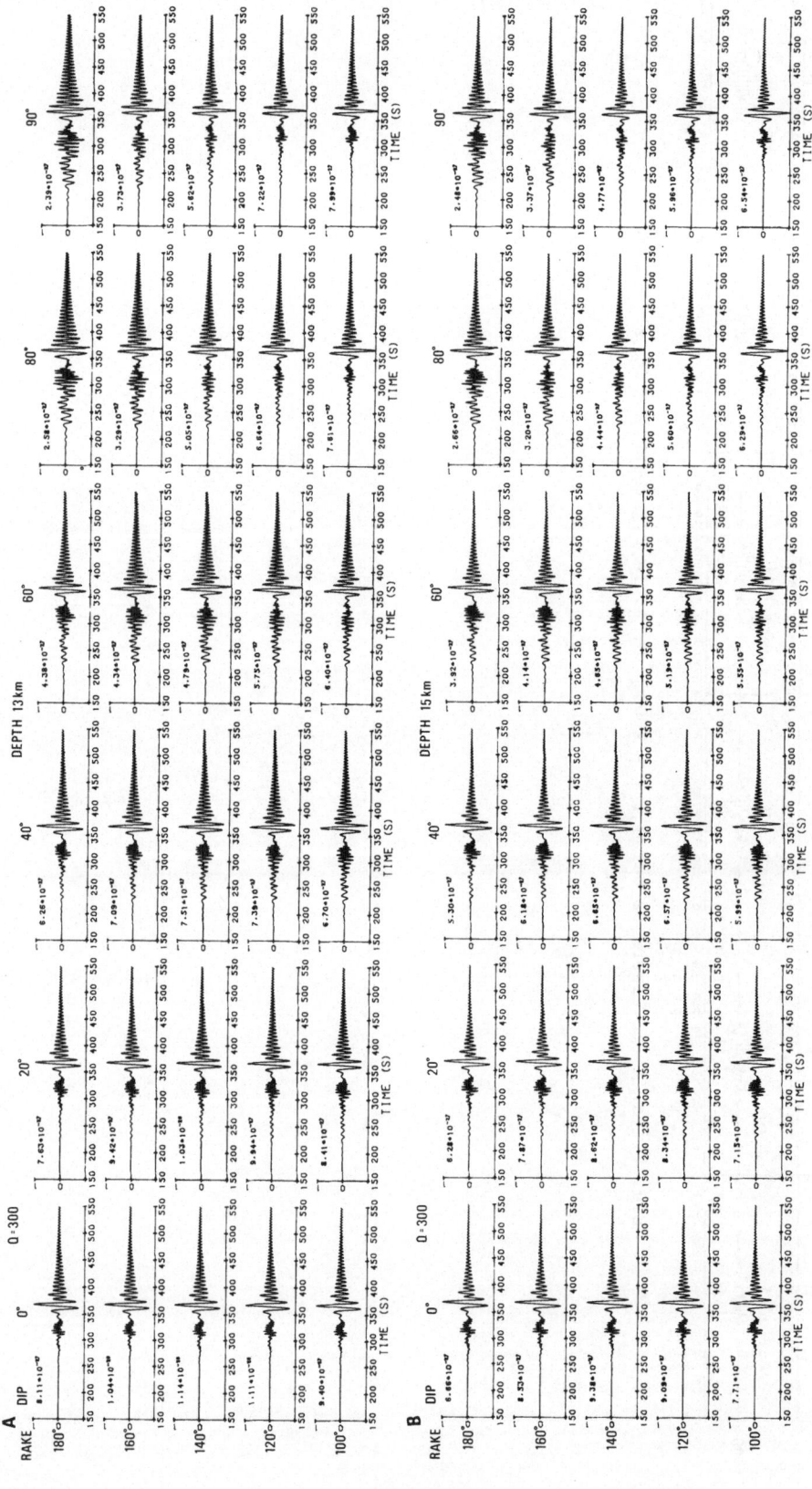

Fig. 8. a. As in Figure 6a but $Q_\beta(z) = 300$. b. As in Figure 8a but h = 15 km.

Table 2. Fault Plane Solutions

Event m/d/y	Plane A strike(θ) dip(δ) rake(λ) (degrees)			Plane B strike(θ) dip(δ) rake(λ) (degrees)		
04.07.74	278	39	174	14	86	50
06.19.75	306	32	208	185	72	232

choosing a set of values properly spaced (few kilometers) around the depth given by bulletins. For the different focal depths, using as source mechanisms the one given by D'Ingeo et al. (1980) for the event of 04/07/74 and the one obtained from P-waves polarities for the event of 19/06/75 (see Table 2), synthetic seismograms are computed (see Figure 5a and 5b).

From the relative amplitudes of the fundamental and higher modes it is possible to determine the focal depth of the equivalent point source with a better precision than the one given by bulletins (Tsai and Aki, 1970; Panza et al., 1973, 1975a,b). From a comparison of Figure 5 with the observed time series it can be deduced that the focal depth, h, varies in the ranges shown in Table 3.

At this stage it is possible to start the procedure for determining $Q_\beta(z)$. It consists in generating matrices of synthetic seismograms where each element corresponds to a value of rake and dip. These computations are performed in correspondence of the extremes of the focal depth ranges. Dip angles are variable between $0°$ and $90°$, while rake angles are varied around a value consistent with predetermined fault plane solutions. These computations are performed for two extreme values of $Q_\beta(z)$: about 100 and about 2000. Examples referring to the events considered here are given in Figure 6 and Figure 7.

In general the comparison with the experimental data allows us to exclude one of the two models. Thus in our case we can exclude the model with low Q for the event 04/07/74 (event n. 1) and the one with high Q for the event 06/19/75 (event n. 2). In this way, by iteration, the range of possible variations of $Q_\beta(z)$ is reduced. The result of this trial and error procedure indicates that for the path sampled by the event n. 1 $Q_\beta(z)$ is lower than 300 (see Figure 8a,b), while for the one sampled by event n. 2 the quality factor is around 500 (see Figure 9a,b).

At this point an attempt can be made to get a finer $Q_\beta(z)$ structure. On the basis of the resolving power of the available data, the best we can do is to consider a two-valued $Q_\beta(z)$, with the surface of separation between the two values placed at variable depth. In such a way we obtained for the path sampled by the first event $Q_\beta(z)$ about 80 in the

Table 3. Focal Depths Determined from Preliminary Waveform Fitting

Event m/d/y	Depth (h) Range (km)
04.07.74	10 - 15
06.19.75	22 - 28

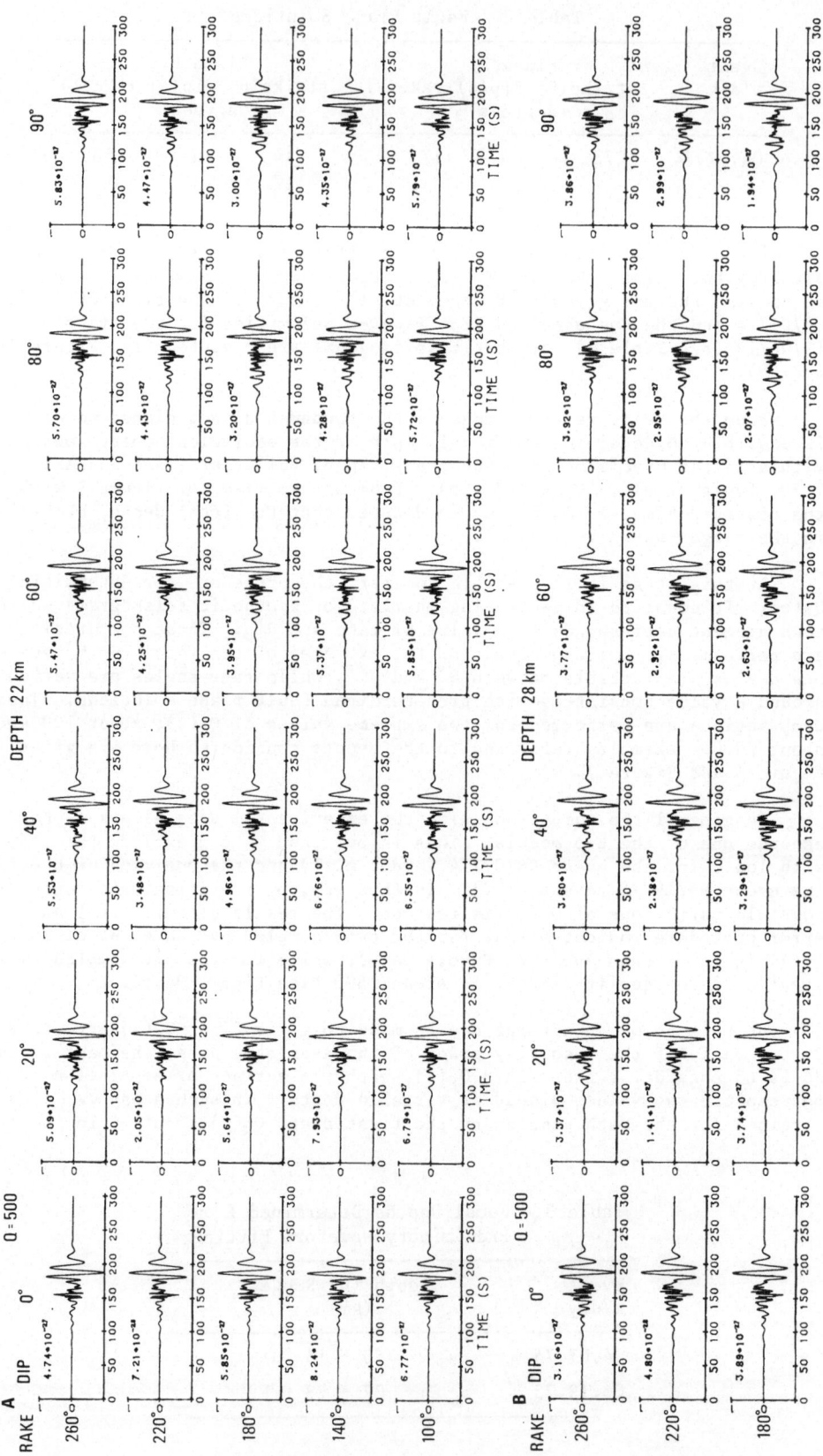

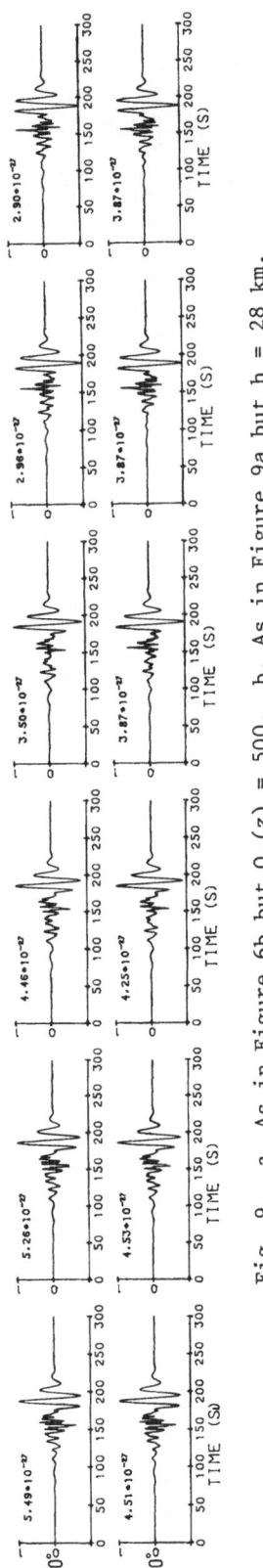

Fig. 9. a. As in Figure 6b but $Q_\beta(z) = 500$. b. As in Figure 9a but h = 28 km.

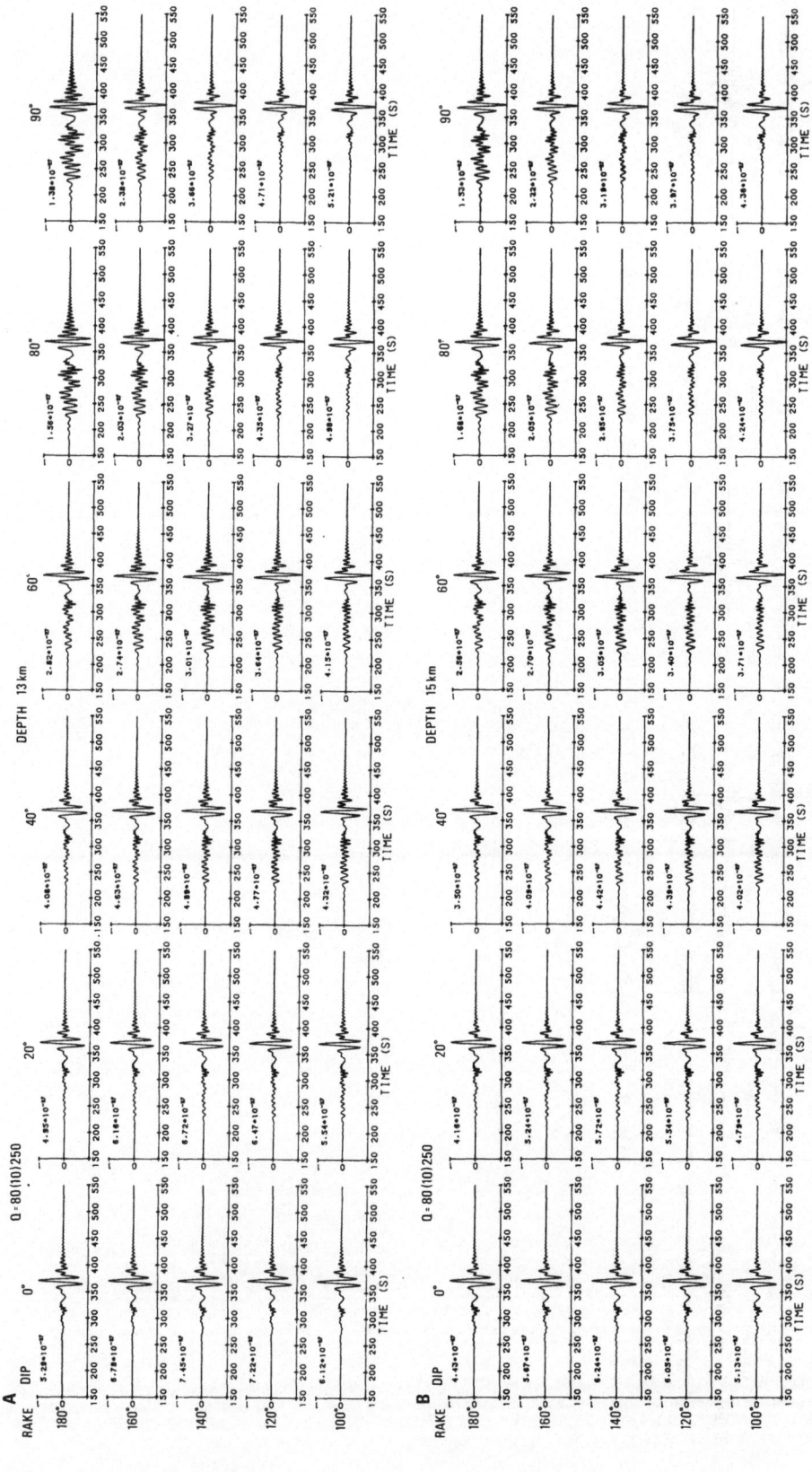

Fig. 10. a. As in Figure 6a, but now $Q_\beta(z)$ is a two valued function: $Q_\beta(z) = 80$ if $0 \leqslant z \leqslant 10$ km and $Q_\beta(z) = 250$ if $z \geqslant 10$ km. b. As in Figure 10a, but h = 15 km.

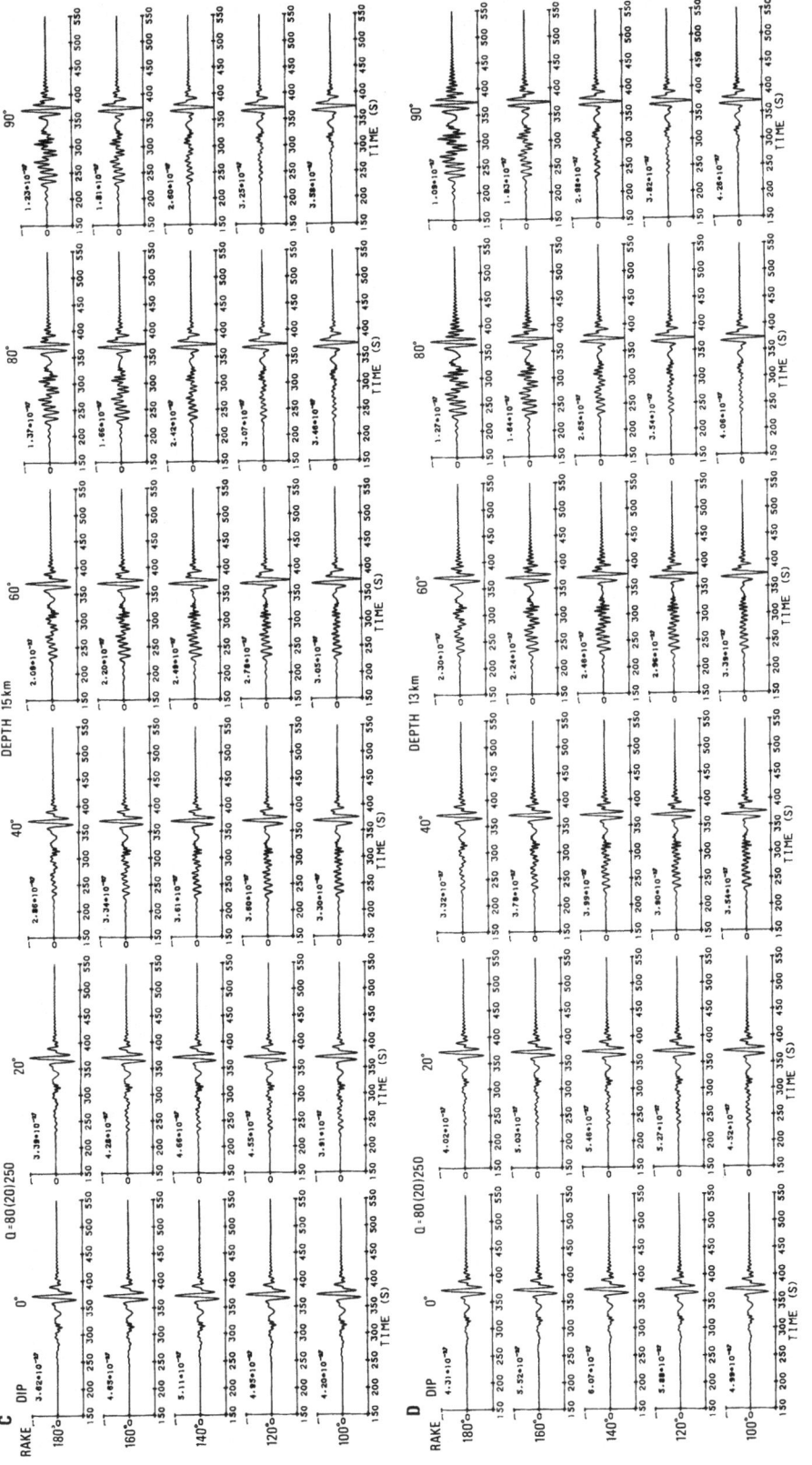

Fig. 10. c. As in Figure 10b but $Q_\beta(z) = 80$ if $0 \leq z \leq 20$ km and $Q_\beta(z) = 250$ if $z \geq 20$ km. d. As in Figure 10c but with h = 13 km.

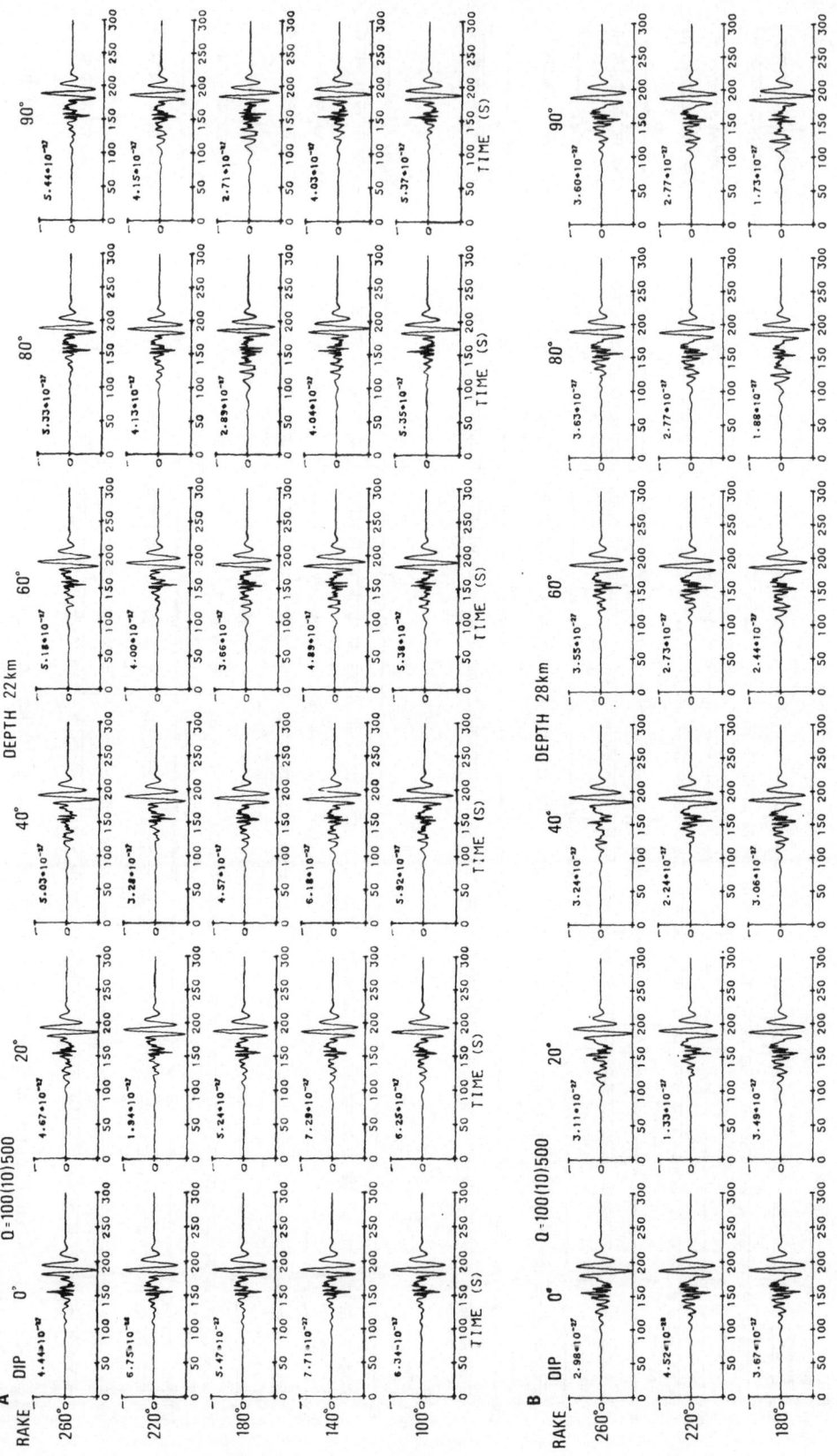

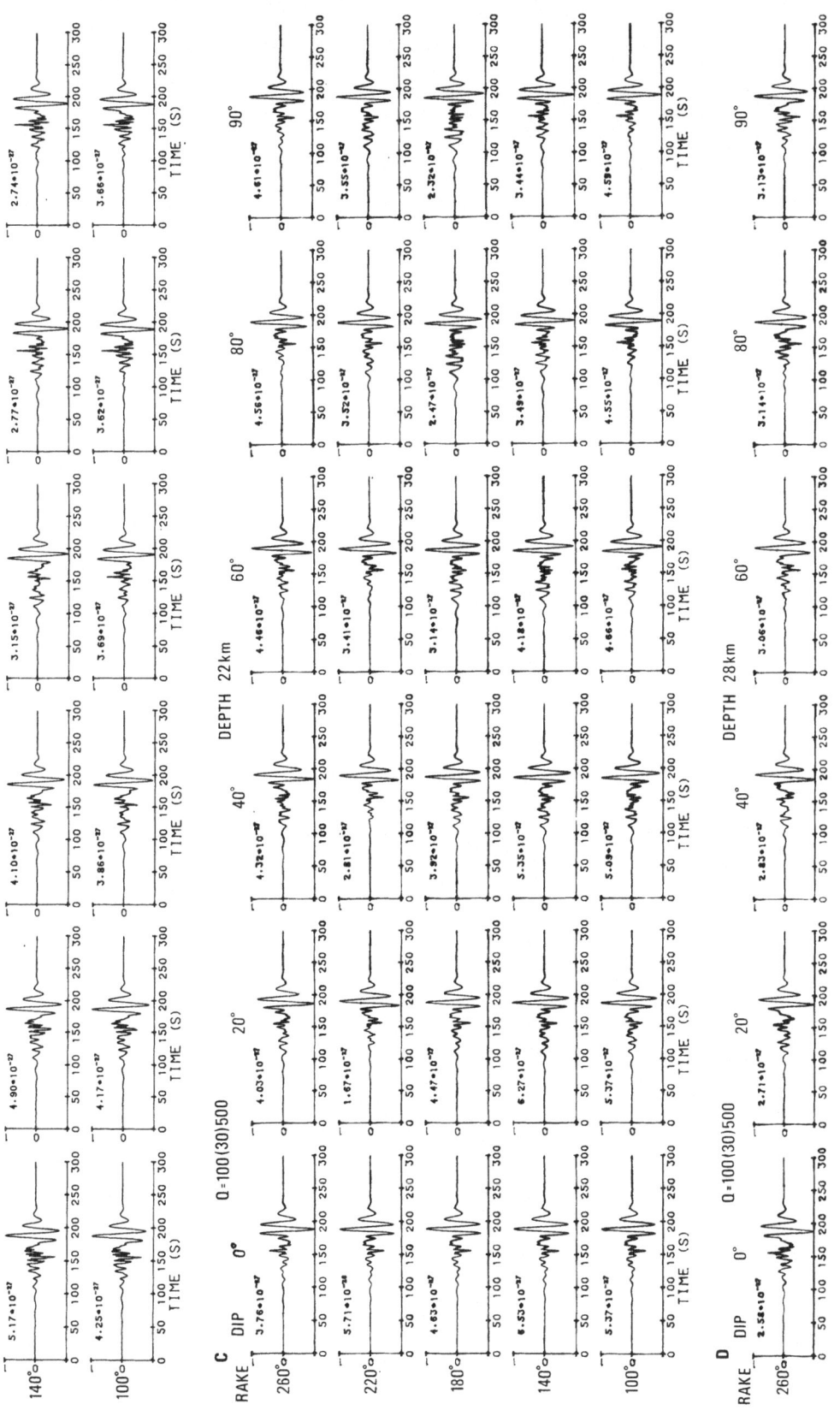

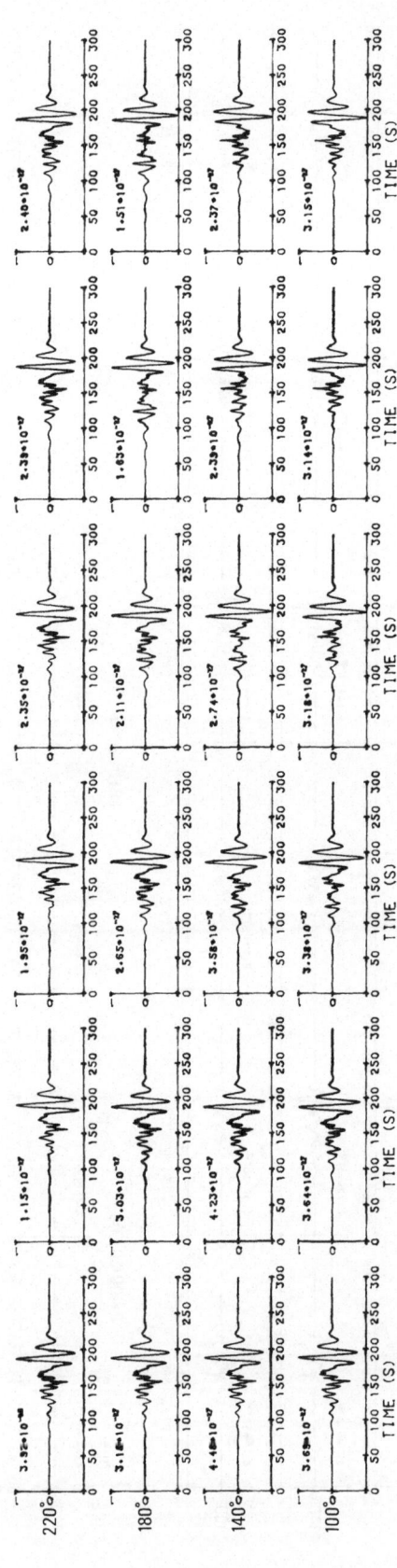

Fig. 11. a. As in Figure 6b but now $Q_\beta(z)$ is a two valued function: $Q_\beta(z) = 100$ if $0 \leqslant z \leqslant 10$ km and $Q_\beta(z) = 500$ if $z \geqslant 10$ km.

b. As in Figure 11a but h = 28 km.

c. As in Figure 11a but $Q_\beta(z) = 100$ if $0 \leqslant z \leqslant 30$ km and $Q_\beta(z) = 500$ if $z \geqslant 30$ km.

d. As in Figure 11c but with h = 28 km.

first 10 to 20 km of the crust and $Q_\beta(z)$ about 250 underneath (see Figure 10a,b,c,d). The result is quite different in the other case where a layering in Q_β does not seem to be resolved (see Figure 11a,b,c,d).

Let us now consider Figure 12a,b where the experimental traces are compared with some of the elements of the space of solutions. The choice of these elements is based on a trial-and-error approach and therefore suffers some subjectivity. In both cases the agreement between theory and observations is quite satisfactory. As already mentioned in the section describing the method, any numerical calculation of the similarity between experimental and theoretical data is at present meaningless, since a proper theory for the definition of weighting functions is missing.

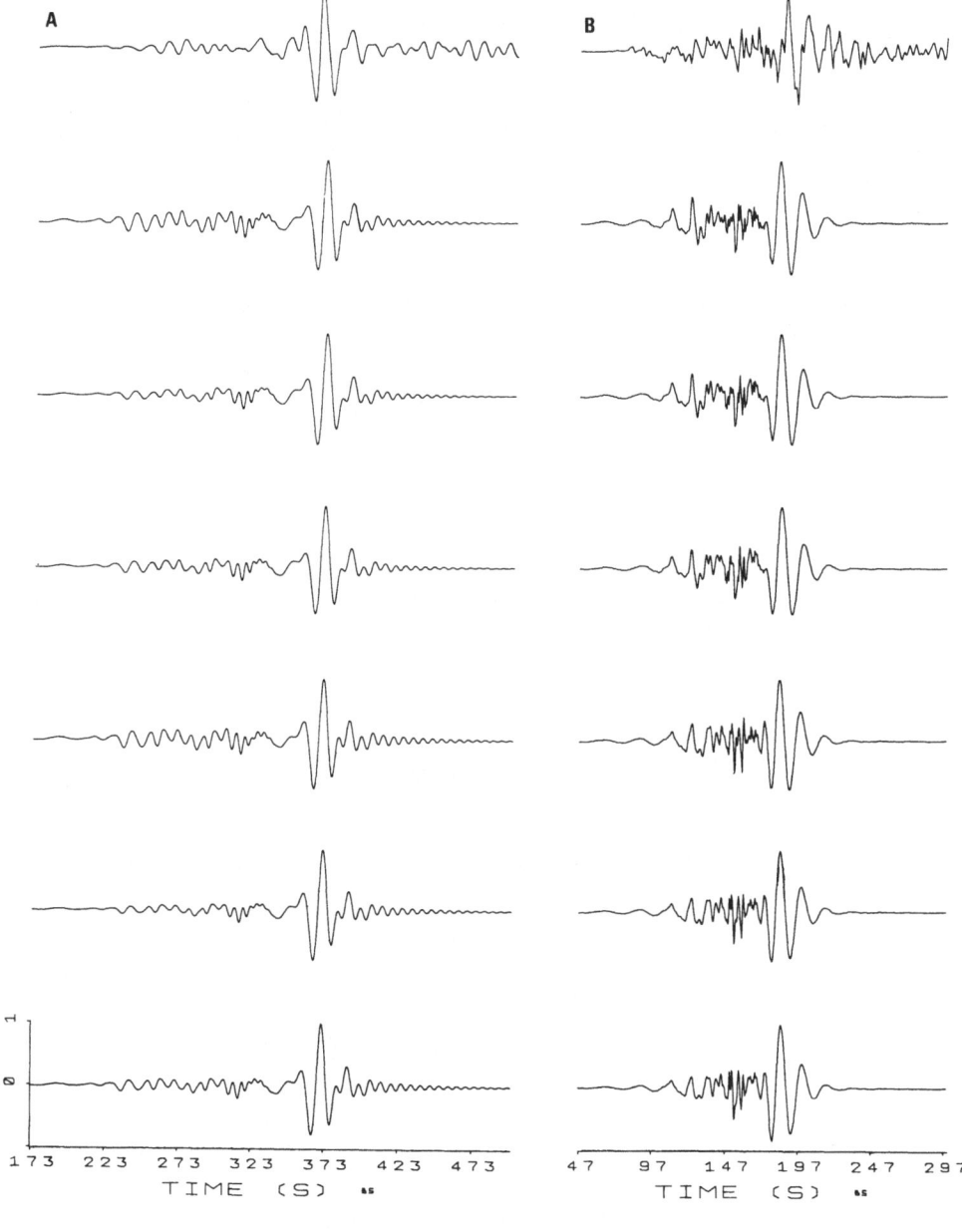

Fig. 12. See next page for legend.

In the record of the event n. 2, in the time interval 50-57 s, is present a long period perturbation which, on the base of inspection of the rest of the daily record, can be attributed to deficiencies of the instrument (Figure 13). Working in the time domain allows the easy removal of most of this undesired signal. In fact, differences can be computed between the theoretical traces of Figure 12b and the experimental one obtaining signals like the ones shown in Figure 14. In this figure the lower trace corresponds to the minimum difference between theoretical and experimental signal; therefore it is natural to smooth it, as indicated by dashed line in Figure 14, and to subtract the smoothed curve from the experimental record. The resulting time series is the top trace of Figure 12b.

To conclude this section let us compare the focal parameters, including source depth, which can be extracted from bulletins and those consistent with the computation of synthetic seismograms. As can be seen from Table 4, where the agencies' focal depths are reported in the first two columns and the result of the present study in the last, the variations are quite significant. Therefore, once broad-band digital data become available, it is advisable to determine the source depth also by means of waveform modelling, mainly for studies where the focal depth has a key role, as, for instance, in the identification of "seismogenetic layers".

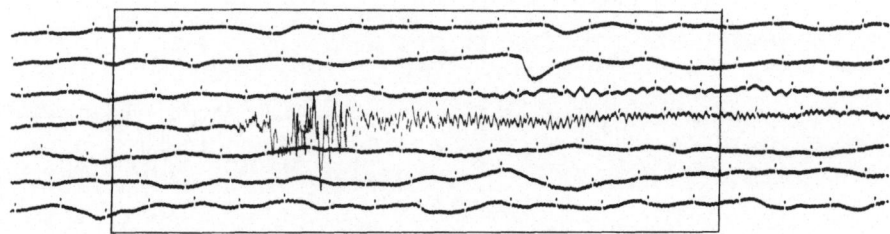

Fig. 13.   Reproduction of a portion of the record of station TRI showing
           long-period perturbations which very likely are also
           contaminating the signal corresponding to event 06/19/75. Ticks
           are minute marks.

Fig. 12.   Comparison between the observed (upper trace) and some
           theoretical time series chosen by trial-and-error:
           a) event 04/07/74: synthetic traces correspond, from top to
           bottom, to the following source parameters:
           $h = 13$ km, $\theta = 315°$, $\delta = 20°$ and $\lambda = 120°$
           $h = 13$ km, $\theta = 315°$, $\delta = 20°$ and $\lambda = 160°$
           $h = 13$ km, $\theta = 315°$, $\delta = 40°$ and $\lambda = 160°$
           $h = 15$ km, $\theta = 315°$, $\delta = 20°$ and $\lambda = 120°$
           $h = 15$ km, $\theta = 315°$, $\delta = 20°$ and $\lambda = 160°$
           $h = 15$ km, $\theta = 315°$, $\delta = 40°$ and $\lambda = 160°$
           b) event 06/19/75: synthetic traces correspond, from top to
           bottom, to the following source parameters:
           $h = 22$ km, $\theta = 320°$, $\delta = 60°$ and $\lambda = 180°$
           $h = 22$ km, $\theta = 320°$, $\delta = 70°$ and $\lambda = 180°$
           $h = 22$ km, $\theta = 320°$, $\delta = 80°$ and $\lambda = 180°$
           $h = 28$ km, $\theta = 320°$, $\delta = 60°$ and $\lambda = 180°$
           $h = 28$ km, $\theta = 320°$, $\delta = 70°$ and $\lambda = 180°$
           $h = 28$ km, $\theta = 320°$, $\delta = 80°$ and $\lambda = 180°$
           For both events, the zero of time axis coincides with the origin
           time.

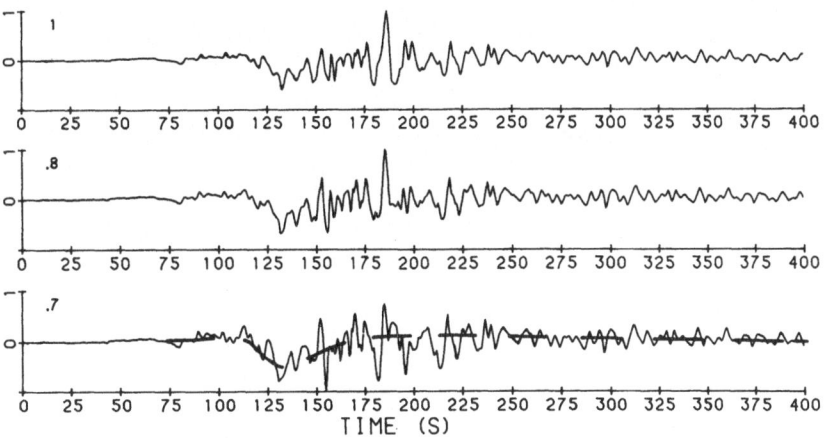

Fig. 14. Difference between observed and some theoretical time series, corresponding to event 06/19/75. The lower trace corresponds to minimum difference between theoretical and experimental data.

The choice of appropriate dip and rake values is made by direct comparison of the experimental data with the elements of the matrices of synthetic seismograms. In this way a range of possible solutions can be defined corresponding to a given $Q_\beta(z)$. Figure 15 compares the values of dip and rake determined with the standard P-waves polarity method with those obtained from waveform modelling. Taking into account the intrinsic uncertainty of the P-wave polarity method, particularly high for medium size earthquakes, the agreement is quite remarkable.

ANALYSIS OF THE FACTOR R

Making use of equation (42) it is possible to assess some of the effects of lateral variations in velocity structures on attenuation measurements made using the multimode method. For this purpose we have computed the value of R in (42) for several combinations of the models of Figure 2. Some of the results are shown in Figure 16a-d.

In each figure the ratio between the value of R for the n-th higher mode, R(N), to the one for the fundamental mode, R(F), is given. In Figure 16a the results referring to the coupling of the structures PAD2A and APNC are illustrated; from the figure it is evident that the comparison of the amplitudes of the fundamental mode with the ones of the higher modes may give meaningful results only in the frequency ranges around 0.07 - 0.1 Hz, 0.13 - 0.17 Hz and 0.19 - 0.25 Hz. The situation is definitely more satisfactory if the coupling of the structures APNC and ADR04 is considered (Figure 16b). In this case, in fact, the comparison may give satisfactory results over most of the frequency band considered.

Table 4. Comparison between Focal Depths given by Agencies and those Determined by Waveform Fitting

Date m/d/y	NEIS	Focal Depth (km) ISC	This Study
07.04.74	29	38 ± 6	13 - 15
06.19.75	16	18 ± 7	22 - 28

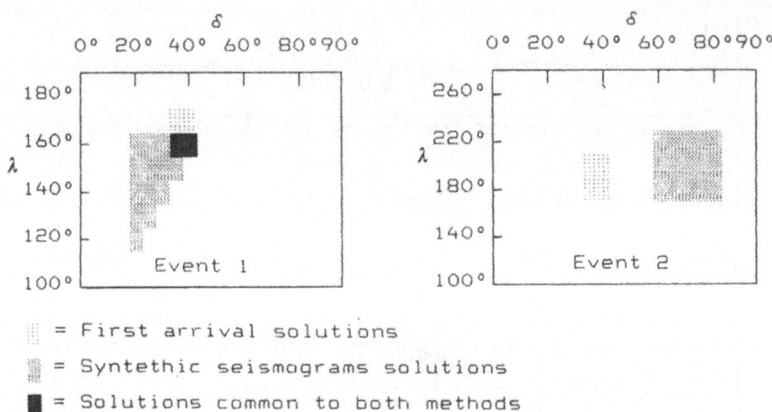

= First arrival solutions

= Syntethic seismograms solutions

= Solutions common to both methods

Fig. 15. Dip and rake angles as determined from P-waves polarities and from waveform fitting.

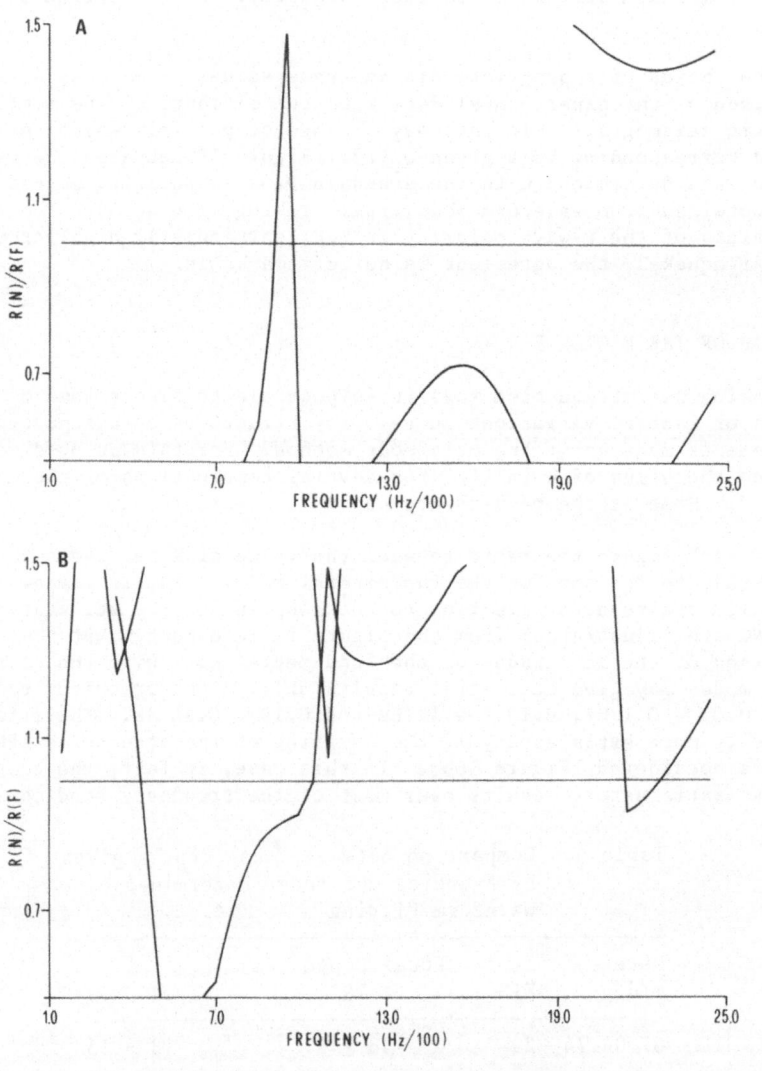

Fig. 16. See next page for legend.

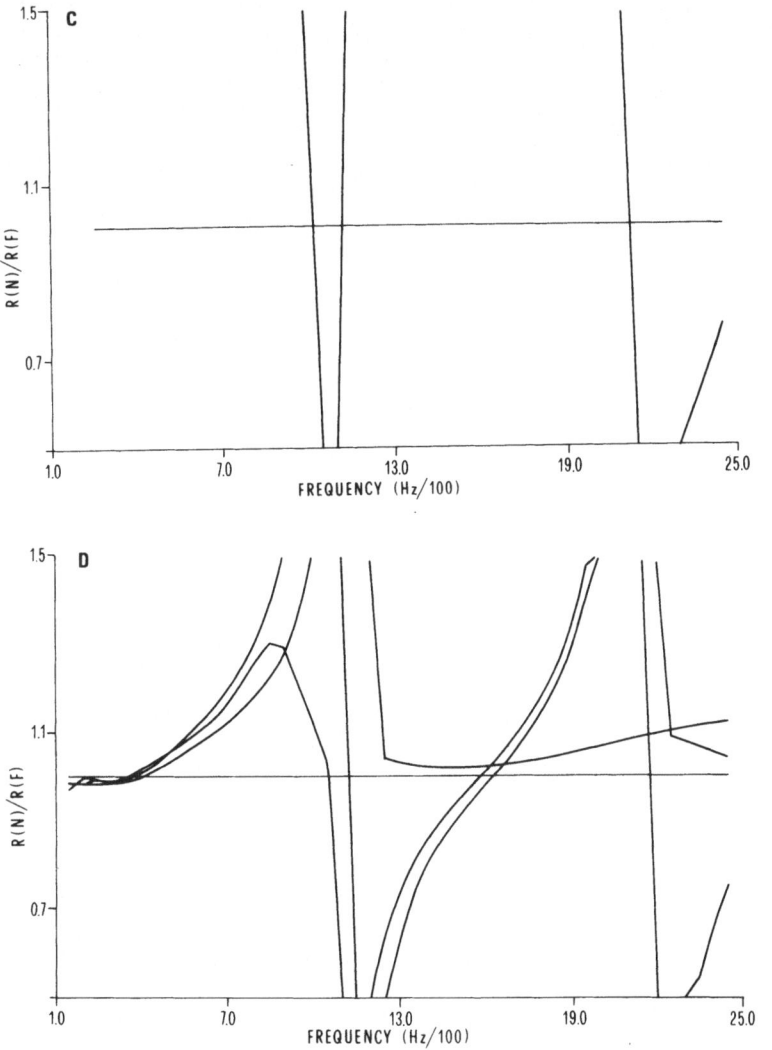

Fig. 16.   Plot of R(N)/R(F) for the coupling between the structures:
a) PAD2A - APNC, b) APNC - ADR04, c) PAD2A - ADR04, d) two
variants of a possible structural model for the Adriatic
plate.

On the other hand the coupling between the structures PAD2A and ADR04
(Figure 16c) would give absolutely unsatisfactory results. To conclude
this section let us consider Figure 16d where the coupling between two
quite similar structures is given, the variation being limited to the
first 17 km, where the S-wave velocities differ by 0.1 km/s. In this case
satisfactory results could be achieved over the entire frequency band
considered.

CONCLUSIONS

The extension to the time-domain of the single source-single station
multimode method for the determination of the anelastic properties of the
Earth's lithosphere allows the simultaneous estimation of source
parameters and of $Q_\beta(z)$. For path's length not exceeding $10°$, using

standard LP data, it is possible to resolve only the gross layering in anelastic properties of the lithosphere. A better resolving power can be expected using broad-band digital data. A proper exploitation of these data requires the computation, with satisfactory efficiency, of multimode synthetic seismograms for laterally heterogeneous models as well as the definition of proper criterion functions to be used in the comparison between observed and synthetic signals.

ACKNOWLEDGEMENTS

I am indebted to Professor B. Mitchell for his comments and for the accurate revision of the manuscript. Special thanks go to Dr A. Craglietto who supplied most of the material used to draw the figures. The help of Mrs Irene Galante for the patient and accurate typing of the manuscript is gratefully acknowledged.

This research was supported by Consiglio Nazionale delle Ricerche (CNR) (contributions n. 85.00933.05 and n. 86.00666.05) and Ministero Publica Istruzione (MPI) (40% and 60%) funds.

REFERENCES

Anderson, D. L., Ben Menahem, A., and Archambeau, C. B., 1965, Attenuation of seismic energy in the upper mantle, J. Geophys. Res., 70:1441-1448.

Aki, K., 1981, Attenuation and scattering of short period seismic waves in the lithosphere, in: "Identification of Seismic Sources", E. S. Husebye, ed., Noordhof, Leiden.

Azimi, Sh., Kalinin, A. V., Kalinin, V. V., and Pivorov, B. L., 1968, Impulse and transient characteristics of media with linear and quadratic absorption laws, Izv. Physics of Solid Earth, 2:88-93.

Burton, P. W., 1977, Inversion of high frequency $Q_\gamma^{-1}(f)$, Geophys. J. R. Astron. Soc., 36:167-189.

Canas, J. A., and Mitchell, B. J., 1978, Lateral variation of surface-wave anelastic attenuation across the Pacific, Bull. Seism. Soc. Am., 68:1637-1650.

Cheng, C. C., and Mitchell, B. J., 1981, Crustal Q structure in the United States from multimode surface waves, Bull. Seism. Soc. Am., 71:161-181.

D'Ingeo, F., Calcagnile, G., and Panza, G. F., 1980, On the fault plane solutions in the central-eastern Mediterranean Region, Boll. Geof. Teor. Appl., 22:13-22.

Gregersen, S., Panza, G. F., and Vaccari, F., 1988, Developments toward computations of synthetic seismograms in laterally inhomogeneous media, Phys. Earth. Planet. Int., 51:55-58.

Herrman, R. B., and Mitchell, B. J., 1975, Statistical analysis and interpretation of surface wave anelastic attenuation data for the stable interior of North America, Bull. Seism. Soc. Am., 65:1115-1128.

Hwang, H. J., and Mitchell, B. J., 1986, Interstation surface wave analysis by frequency-domain Wiener deconvolution and modal isolation, Bull. Seism. Soc. Am., 76:847-864.

Jeffreys, H., 1958, A modification of Lomnitz's law of creep in rocks, Geophys. J. R. Astron. Soc., 1:92-95.

Kennett, B. L., 1985, "Seismic Wave Propagation in Stratified Media", Cambridge University Press, Cambridge.

Kijko, A., and Mitchell, B. J., 1983, Multimode Rayleigh wave attenuation and Q_β in the crust of the Barents shelf, J. Geophys. Res., 88:3315-3328.

Knopoff, L., 1964, Q, Rev. Geophys. Space Phys., 2:626-660.

Lee, W. B., and Solomon, S. C., 1978, Simultaneous inversion of surface wave phase velocity and attenuation: love waves in Western North America, J. Geophys. Res., 83:3389-3400.

Levshin, A. L., 1985, Effects of lateral inhomogeneities on surface wave amplitude measurements, Ann. Geophys., 3,4:511-518.

Levshin, A. L., Pisarenko, V. F., and Pogrebinsky, G. A., 1972, On a frequency-time analysis of oscillations, Ann. Geophys., 2:211-218.

Liu, H. P., Anderson, D. L., and Kanamori, H., 1976, Velocity dispersion due to anelasticity: implications for seismology and mantle composition, Geophys. J. R. Astron. Soc., 47:41-58.

Mitchell, B. J., 1980, Frequency dependence of shear wave internal friction in the continental crust of Eastern North America, J. Geophys. Res., 85:5212-5218.

Panza, G. F., Schwab, F., and Knopoff, L., 1973, Multimode surface wave response for selected focal mechanisms, I: Dip-slip sources on a vertical fault plane, Geophys. J. R. Astron. Soc., 34:265-278.

Panza, G. F., Schwab, F., and Knopoff, L., 1975a, Multimode surface wave response for selected focal mechanisms, II: Dip-slip sources, Geophys. J. R. Astron. Soc., 42:931-943.

Panza, G. F., Schwab, F., and Knopoff, L., 1975b, Multimode surface waves response for selected focal mechanisms, III: Strike-slip sources, Geophys. J. R. Astron. Soc., 42:945-955.

Panza, G. F. 1985, Synthetic seismograms: the Rayleigh waves modal summation, J. Geophys., 58:125-145.

Panza, G. F., Suhadolc, P., and Chiaruttini, C., 1986, Exploitation of broad-band networks through broad-band synthetic seismograms, Ann. Geophys., 4:315-328.

Panza, G. F., and Suhadolc, P., 1987, Complete strong motion synthetics, in: "Computational Techniques", 4, B. A. Bolt, ed., Academic Press, New York.

Schwab, F. A., and Knopoff, L., 1972, Fast surface wave and free mode computations, in: "Methods in Computational Physics", 11, B. A. Bolt, ed., Academic Press, New York.

Snieder, R., 1986, 3D linearized scattering of surface waves and a formalism for surface wave holography, Geophys. J. R. Astron. Soc., 84:581-605.

Snieder, R., and Nolet, G., 1987, Linearized scattering of surface waves on a spherical Earth, J. Geophys., 61:55-63.

Tsai, Y. B., Aki, K., 1970, Precise focal depth determination from amplitude spectra of surface waves, J. Geophys. Res., 75:5729-5743.

Yacoub, N. K., and Mitchell, B. J., 1977, Attenuation of Rayleigh wave amplitudes across Eurasia, Bull. Seism. Soc. Am., 67:751-769.

SEISMIC SOURCE STUDIES FROM WAVEFORM

MODELLING OF STRONG MOTION DATA

Peter Suhadolc

Istituto di Geodesia e Geofisica
Università di Trieste
Trieste, Italy

INTRODUCTION

A substantial evidence has accumulated in these last years showing that most natural earthquakes are complex multiple events. They are, generally, caused not by a single rupture on a smooth fault plane, but rather by a fracture encountering asperities (concentrated high stress drop patches) or barriers (unbreakable patches).

This is true not only for large ($M \geqslant 7$) earthquakes, but also for smaller ones. The point-source approximation may be valid for $M < 7$ earthquakes at teleseismic distances, due to the lack of resolution of far-field observations, but is certainly not appropriate if these events are observed at wavelengths and at distances comparable to or smaller than their rupture length. At these "small" distances and "high" frequencies ($f > 0.1$ Hz) the seismogram contains information on the details of the earthquake rupturing process, and the event cannot be therefore modelled as instantaneous and point-like.

At the wavelengths proper of strong motion data the fault zone is not infinitesimal any more. The projections of faults on the surface of the earth show changes in directions, bifurcations and echeloning. One can assume the three-dimensional geometry is correspondingly complex. However, the rupturing process depends not only on fault geometry (e.g., King, 1986), but also on rock properties, i.e., the strength distribution along the fault plane, which is likely to be the dominant effect. As the distribution of earthquakes in space and time shows self-similarity properties (Kagan and Knopoff, 1980), one can assume that these features are present also in the smaller scale range of rock fracturing processes.

The source modelling has developed from a point-source concept (Burridge and Knopoff, 1964) to kinematical models (e.g., Haskell, 1964, 1966) and finally to dynamic or crack models based on linear (e.g., Das and Aki, 1977a, 1977b; Madariaga, 1977, 1983; Knopoff and Chatterjee, 1982) and nonlinear (Newman and Knopoff, 1982) fracture mechanics. The nonlinearity of the problem stems from the fact that the motion of the crack tip depends on the stress redistribution due to the radiation from the just-fractured portions of the crack, and this implies the solution of linear partial differential equations with macroscopically moving boundaries.

With the expanding possibilities of computing synthetic seismograms, waveform modelling has gained more and more recognition also in source studies. It closely followed the limits and developments of synthetic seismogram programs. The first applications have been, in fact, in the low-frequency range and with teleseismic surface waves (Kanamori, 1970a, 1970b) or normal oscillation (e.g., Gilbert and Dziewonski, 1975; Dziewonski et al., 1981) data. Langston and Helmberger (1975) proposed a method to model the source using waveform data of body-wave phases, which extended the modelling to much shorter periods. These studies, however, were aimed at obtaining the focal parameters of the event, assumed to be a point-source.

Only in recent years waveform modelling of synthetic and real ground motion (see Spudich and Archuleta, 1987) allowed the modelling of the behavior of the earthquake source for frequencies up to 1 Hz. The attention has also shifted on the time and space complexity of earthquake sources by modelling strong-motion data (e.g., Hartzell and Helmberger, 1982; Olson and Apsel, 1982; Archuleta, 1984; Bernard and Madariaga, 1984; Hartzell and Heaton, 1986; Takeo, 1987).

Usually only strong ground motion from earthquakes of magnitude about 5 and greater is of much interest. Furthermore, with only a few exceptions, potentially damaging effects are confined to within several tens of kilometers of the causative fault, even in great earthquakes. Outside of this range, although motion is still perceptible, it is typically associated only with nonstructural damage or damage to particularly weak, poorly engineered structures.

Thus strong-motion seismology is concerned with high-frequency seismic waves from large earthquakes, the type of motion which is the least understood phenomenon in earthquake seismology, also because strong motion in the vicinity of large earthquakes is not recorded very often. Nevertheless, significant progress has been made in the past decade in understanding how high-frequency waves are generated during the rupture of an earthquake fault (e.g., Madariaga, 1976, 1977, 1983; Archuleta and Day, 1980; Das, 1981, 1985, 1986; Das and Kostrov, 1983, 1985).

Recently an increasing number of seismologists has been involved in both collecting and interpreting strong-motion data (see Erdik and Toksöz, 1987; Bolt, 1987). Various deterministic and stochastic earthquake source models have been applied in their interpretation. Theoretical results on fracture mechanics as well as laboratory results on rock failure have led to a better understanding of the physics of fault ruptures. Various attempts have also been made to relate the earthquake rupture process to the observable behavior of earthquake faults. Such a relation is vitally important in evaluating the ground motion expected for earthquake faults mapped by geologists.

Strong motion, in fact, depends crucially on the details of the rupture process, which cannot be seen clearly through the long-period window. It is thus necessary to extend the window to higher frequencies, a difficult task, since high-frequency seismic waves are subject to complex propagation effects, which involve strong scattering and attenuation.

Usually, the frequency content of the strong motion is broad-band. That is, the acceleration amplitude spectra are roughly constant over a range of frequencies from 5 Hz down to 0.5 Hz or less. At lower frequencies the amplitude decreases with decreasing frequencies. The point where this occurs, called corner frequency, f_o, depends primarily on the duration of the strong shaking or equivalently on the size of the source, with the longer motions, from larger earthquakes, delivering more energy at lower frequencies.

At higher frequencies, above 5 to 10 Hz, experience has shown that there is a tendency for the motion to be strongly attenuated. In this case, the frequency at which attenuation begins, defined as f_{max} by Hanks (1982), appears to depend primarily on the distance from the source of the motion and on the nature of the local geology. Papageorgiou and Aki (1983) on the other hand propose that f_{max} originates from the nonelasticity of the fault, and in particular that it varies inversely with the size of the cohesive zone behind the propagating crack tip. As might be expected, records obtained close to the source on competent rock show the most high-frequency motion. Even in this case, the motion attenuates significantly with increasing frequency above about 15 to 20 Hz. There is some evidence that measurable motion may be present at frequencies of 30 Hz or higher, but this motion is of interest only in special cases.

An exception to this general description occurs when the surficial deposits at the recording site are very soft. Under these circumstances, the soil deposits can "amplify" the motions at some frequencies and reduce it at others, for example, the accelerograms obtained in Mexico City during the May 11, 1962 and the September 19, 1985 earthquakes. Path effects along the source-to-site distance, 260 km, have helped in reducing the high-frequency motion and increasing the duration of shaking. However, the "amplification" of motion at periods near 2.5 s is believed to be caused primarily by the response from the soft soils of the old lake bed on which Mexico City is located (see, for example, Jennings, 1983). This "amplification" (e.g., Shearer and Orcutt, 1987) is due not only to the free-surface effect and the impedance contrast between the soft sediments and the underlying medium, but is also largely due to the influence of the soft sediments on the polarization of the surface ground motion. In fact, sedimentary waves (Chiaruttini et al., 1985; Panza et al., 1986) have dominant horizontal motion in large frequency intervals.

The availability of digital accelerograms, combined with the increasing demand from structural engineers for more quantitative methods of analysis and interpretation, makes the development of efficient theoretical tools for the construction of synthetic records very important.

The construction of synthetic signals of long duration having a broad-band frequency content can hardly be done using the ray approach (see, for example, Chapman, 1985; Červený, 1985). At present, a very suitable tool for this purpose seems to be the modal summation (Panza, 1985; Panza and Suhadolc, 1987), which has already been applied at lower frequencies (Liao et al., 1978).

In this paper some examples of strong motion modelling and interpretation using synthetic seismograms constructed with the normal mode summation method are presented.

SOME THEORETICAL CONSIDERATIONS

By summing the modes of oscillation of a given structure approximated with flat parallel layers and using realistic models for the seismic source it is easy to construct complete synthetic seismograms. The use of Rayleigh modes gives the vertical and radial components, while the use of Love modes gives the transverse one.

For construction of complete synthetic accelerograms using mode summation, it is convenient to consider models of the earth with elastic and anelastic properties specified to depths on the order of 100 km which is a representative thickness of the lithosphere. The handling of structural models extending to these depths in an efficient way, makes it possible to synthesize early P-wave arrivals without the necessity of introducing any unrealistic high-velocity half-space with the consequent generation of spurious S-wave arrivals as in the case of the locked mode approximation (Harvey, 1981). Moreover, the computation is not limited to layered models, because it is possible to simulate also gradients by means of a sequence of thin layers.

For distances of the order of the fault length, which are commonly encountered in strong-motion studies, the Ben Menahem (1961) finiteness factor cannot be applied. The fault of finite length can be modelled in this case as a series of point sources on a defined grid placed along the fault line with an appropriate spacing Δ s. In order not to enhance artificially certain frequencies, Δ s must be such that the apparent time separation between individual sources, as seen at the receiver, is smaller than the Nyquist period. Also, Δ s depends on the rupture velocity, since the time interval between the "ruptures" of two nearby points has to be smaller or equal to the time sampling interval of the seismograms. Another "equivalent" condition is that the smallest considered wavelength has to be sampled by at least three grid points. The same considerations apply if the grid is two- or three-dimensional.

Let us consider a two-dimensional vertical fault plane. The fault is 3 km long and 1 km wide, is between 5 km and 6 km deep and has been modelled with a rectangular grid with Δ s = 100 m. The geometry is taken from Spudich and Frazer (1984), and the observer is 70° off strike at an epicentral distance of 10 km. The hypocenter is placed 5.6 km deep at 0.5 km from one end of the fault. The structural model used in the computations is FRIUL7A (see Figure 9). Since we use as the upper phase velocity cut-off the value 0.98 β_n (where β_n is the S-wave velocity in the underlying half-space), the generated synthetic signals will contain all phases whose phase velocity is less than 4.55 km/s.

Two examples have been considered, both having a constant rupture velocity v_r = 0.9β_n, the rupture spreading circularly from the hypocenter. In the first example, all points except the central ones have been given equal weights in order to model a background smooth fault rupture with embedded asperities. The locations and weights of the asperities, adjusted after Spudich and Frazer (1984), are shown in Figure 1. The corresponding accelerogram is seen in Figure 2a. The result shown in Figure 2b, on the other hand, has been obtained by summing just the 10 seismograms relative to the asperities of Figure 1, without considering other grid points simulating the background fault rupture. As expected, this last trace is very similar to the previous one, even if there are some differences, especially in the amplitudes of the initial peaks. A reference accelerogram for an instantaneous point-source located at the hypocenter is shown in Figure 2c. This example makes us confident, that we can model a complex earthquake rupture in the near field by considering just the contribution of some point-source modelling asperities and from which most of the high-frequency radiation is coming.

The latest developments regarding the generation of high-frequency (f > 0.1 Hz) synthetic seismograms using the normal mode summation technique have been extensively described in Panza (1985) and Panza and Suhadolc (1987). Panza (1985) gives the details of eigenvalues and eigenfunctions computations for P-SV waves, and discusses the double-couple point-source excitation. Panza and Suhadolc (1987) describe, among

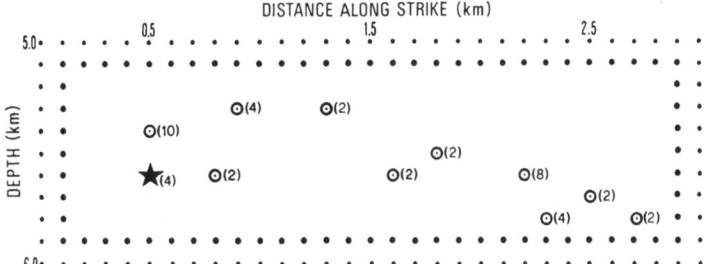

Fig. 1. Grid modelling of a two-dimensional source. This example is adjusted after Spudich and Frazer (1984). Only the edge lines and some interior points are shown. The small points are given a weight of 0.25, and the big dot points are given a weight of 0.50 when cosine tapering is applied. The star represents the point of rupture initiation - the hypocenter; the open circles are points with nonunitary weights. The numbers in parenthesis near the interior points are the weights. The rupture spreads out uniformly at a velocity of 0.9β.

other things, the mode follower algorithm and the implications of the introduction of the body-wave dispersion. They also give several theoretical examples of synthetic signals arising from both point sources and extended fault sources. The power and limits of this approach are extensively discussed in Panza et al. (1986).

This complete synthetic seismogram technique has been applied to different seismological problems, ranging from the influence of source parameters on the shape of the isoseismals (Suhadolc et al., 1988a), the determination of the elastic and anelastic properties of the crust (Craglietto et al., 1987), to the study of the influence on the peak ground acceleration of the crustal model and source parameters (Suhadolc and Chiaruttini, 1987). Examples of waveform modelling of strong motion displacement data for frequencies lower than 1 Hz, using the modal

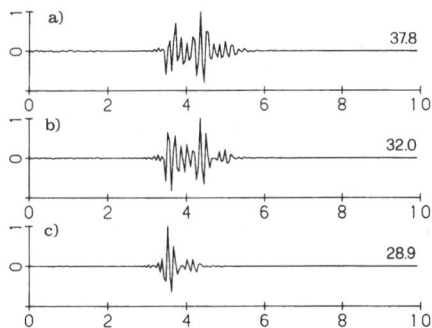

Fig. 2. Examples of vertical component accelerations due to the two-dimensional fault of Figure 1 modelled by a grid of instantaneous point sources with $|M_o| = 1$ dyn cm. The receiver is placed $70°$ off strike at an epicentral distance of 10 km. (a) Nonequal weights applied: nonunitary weights are shown in Figure 1, the other point sources being given unitary weights with cosine tapering applied at the edges. (b) Only the central 10 sources with nonunitary weights are summed. (c) Only the point source signal corresponding to the hypocenter marked by a star in Figure 1 is shown. The peak acceleration is in units of 10^{-23} cm/s^2.

summation approach, have been already given by Suhadolc and Panza (1985) and Panza and Suhadolc (1987).

The aim of this paper is to show how the waveform modelling of observed high-frequency accelerogram traces might be used to retrieve information on the corresponding rupturing process. We will focus our attention on two events: the November 23, 1980 Irpinia (southern Italy) earthquake and the September 11, 1976 Friuli (north-eastern Italy) aftershock.

IRPINIA, ITALY, 1980 EVENT

The earthquake source behavior is relatively well understood by waveform matching synthetic and observed ground motions for frequencies up to 1 Hz. If we move to higher frequencies, smaller scale details of the earthquake source process and the structure surrounding the source volume become essential for a deterministic prediction of the strong ground motion. Since these details are usually not known at present, a statistical approach has been taken up to now to predict ground motion above 1Hz (see, for example, Boore and Joyner, 1978; Boatwright, 1982; Koyama, 1985).

In the following we discuss a direct method to unravel the rupturing process of the November 23, 1980 Irpinia earthquake (M_s = 6.9) in southern Italy (see Bernard and Zollo (1988) and Westaway and Jackson (1987) for a detailed discussion of this event). The starting data (Figure 3) are represented by the Ente Nazionale Energie Alternative - Ente Nazionale Energia Elettrica (ENEA-ENEL) strong ground motion recordings (Berardi et

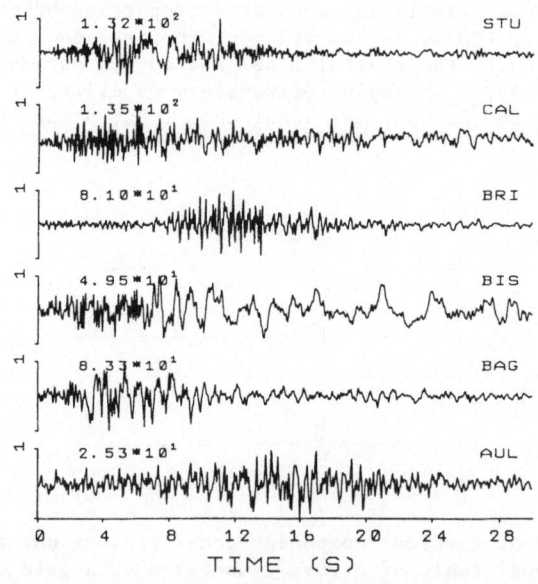

Fig. 3. Observed accelerations, after Gaussian filtering with a cut-off frequency at 10 Hz, due to the 1980 Irpinia, Italy, earthquake. From top to bottom, the signals refer to the stations: Sturno, Calitri, Brienza, Bisaccia, Bagnoli, Auletta. For station locations see Figure 4. The maximum peak ground acceleration in units of cm/s^2 is shown in the upper left part of each record. The zero of the time axis does not coincide with the earthquake origin time.

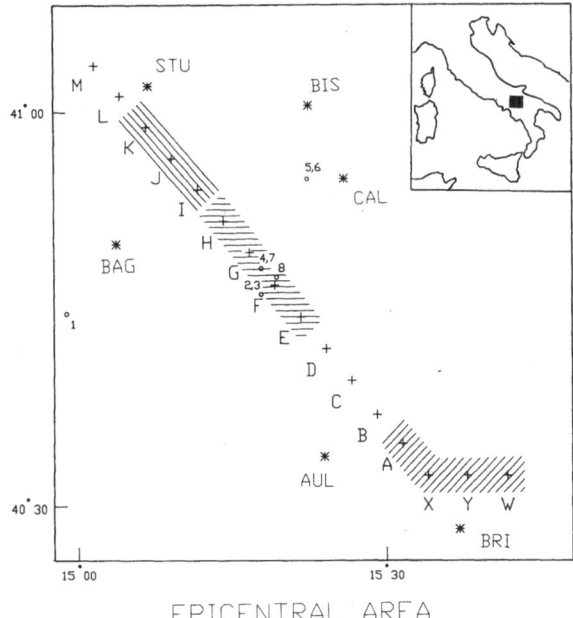

Fig. 4. Map showing the positions of the epicenters (see text), used in
the construction of the synthetic signals of Figure 7, and the
considered strong-motion stations. The three hatchings denote
the locations of the three rupture events described in the
obtained source model (see Figure 6).

al., 1981). Since we are at source-receiver distances comparable with the
fault length and the event is a complex one, we will use only the vertical
components recorded at the six accelerometric stations closest to the
epicentral area (Figure 4). We will also restrict our study to the first
30 s of the recordings. The cut-off frequency of computation - 10 Hz -
largely exceeds the limit - a few Hz - below which the present knowledge
of the elastic and anelastic properties of the crust allows a
deterministic modelling of wave propagation. In the comparison between
synthetic observed accelerograms we will therefore give more relevance to
envelope fitting rather than to single peaks matching.

The basic crustal model used to compute synthetics is the "A" model
proposed by Del Pezzo et al. (1983). The model has been modified: at the
top, where we have introduced a positive gradient in the P- and S-wave
velocities of the first 8 km to obtain a more realistic structural model
reproducing the loading effect on sediments, and at the bottom, where more
layers have been added in order to model the lithosphere (Panza et al.,
1980). Due to the difficulty in measuring the intrinsic attenuation of
the Earth in a physically meaningful way (Panza et al., 1986), a constant
Q model (Knopoff, 1964) has been assumed in the layers. The Q in the
crustal layers have been assigned according to the average values given by
Knopoff (1964) for layers with similar S-wave velocities. For the
subcrustal region information on the average Q has been obtained from
Craglietto et al. (1987). This structure is reproduced in Figure 5.

Comparing the duration of the observed ground motion with the
examples of "complete" synthetic signals obtained for comparable distances
(Panza and Suhadolc, 1987) one can see that the single-point source is by
no means a realistic representation of a large earthquake's source process
in the near field. Moreover, looking at the recordings we note that the

123

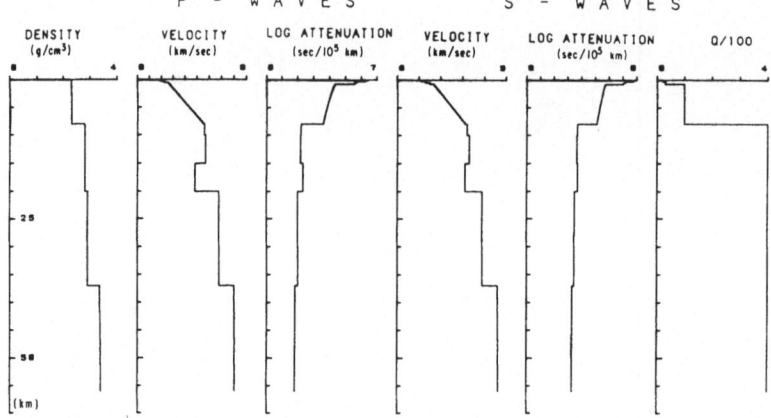

Fig. 5. Structure IRPGRA1 used in the Irpinia modelling. To minimize the effect of anelasticity we have chosen the relation $Q_\alpha = 2.5\ Q_\beta$ between the quality factors for P- and S-waves. In the figure only Q_β is plotted.

accelerations at Auletta are particularly low if compared to the ones obtained at Brienza, a station located almost in the same direction but at a larger distance. This fact cannot be explained in terms of different geological setting, but, as we will see, contains very useful information to constrain the orientation of the point sources used to model the strong motion records.

The fault strike is taken to be parallel to the axis of the Apennine chain, which is oriented N140°E; the fault length, as deduced from the spatial distribution of aftershocks (Deschamps and King, 1984), is taken equal to 70 km. Initially, thirteen evenly spaced locations have been selected along the fault; subsequently three more had to be added in near the Brienza station (Figure 4). The spacing of these locations - hereafter called epicenters - is 5.7 km. Point sources of different hypocentral depth are placed in correspondence of each epicenter.

It has been shown by Suhadolc et al. (1988b) that in order to model the observed recordings, sources with finite durations and a mechanism having a rake of 240° on a fault dipping 70° to the NE had to be taken into account. For the sake of simplicity the duration of 0.6 s has been kept constant for all the point sources. Moreover, Suhadolc et al. (1988b) found that rupture velocities smaller than the S-wave velocity in the pertinent layers constrain the location of the shock activating the rupturing process in epicenter G.

In order to obtain the final synthetic signals it has been necessary to consider twelve point sources, properly distributed in space and time (Figure 6). Such a sequence of point sources may be considered a schematic representation of asperities of different size. The rupturing process seems to have propagated in a bilateral fashion, mainly along the path drawn as dotted line in Figure 6, with a subsonic velocity. The fracturing process, deduced from this modelling, nucleates with a point source of relatively small seismic moment, even if earlier nucleations with an even smaller size cannot be excluded. The downward propagating fracture gives rise to the maximum energy release, located just above a methamorphic layer with high P- and S-wave velocities (Mueller, 1977). The fracture propagating upwards from the nucleation point seems to be continuous both in space and time. The south-western branch of the

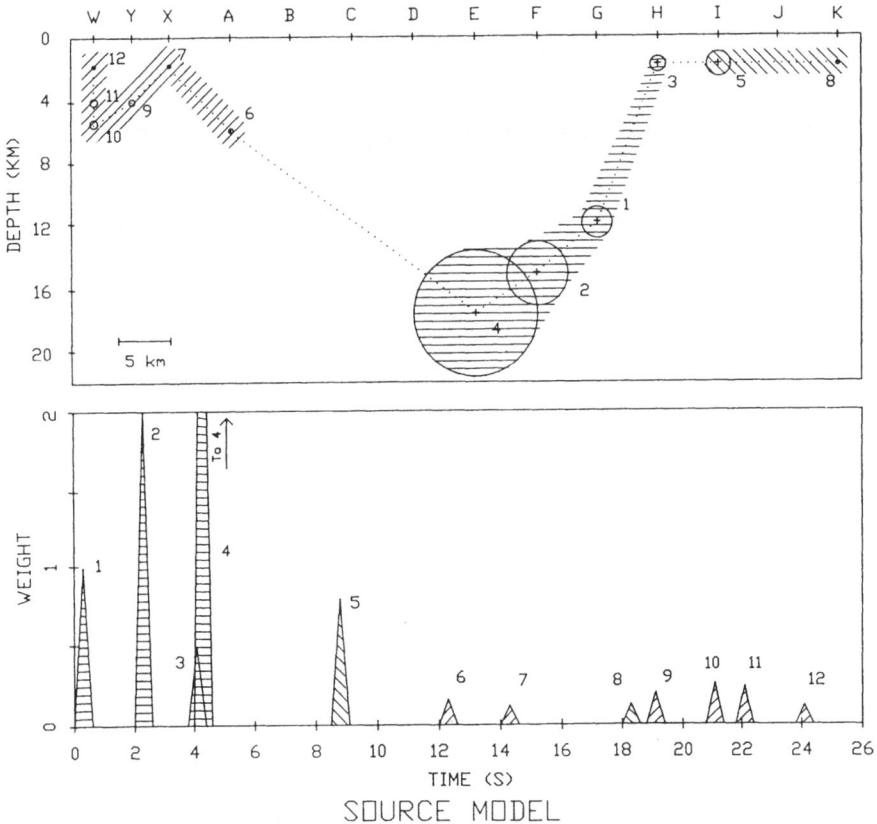

SOURCE MODEL

Fig. 6. Distribution in space (top) and time (bottom) of point sources
used to construct the synthetic signals of Figure 7. The numbers
of the point sources are in order of their excitation. The
different hatchings denote the three rupture events in which the
overall rupturing process can be divided.

fracture, on the other hand, is not so well linked with the main central
fracture process and could be well taken to represent an almost
"independent" subevent. The later parts of the strong-motion records are
therefore modelled by means of some shallow point sources, with varying
seismic moment, simulating the stopping of the rupture within the
sedimentary cover. The seismic moments of these shallow sources are
always smaller than the ones of the point sources simulating the beginning
of the rupturing process, but their effect is very relevant in all the
records.

The final synthetic signals are shown in Figure 7, which is made up
of six parts, each corresponding to one of the six stations considered. In
each part the synthetic seismograms due to the twelve considered point
sources, as seen at the corresponding station, are shown on the top. In
the lower part of the figure two records are shown: above, the sum of the
twelve synthetic contributions, below, the observed accelerograms Gaussian
filtered with a cut-off frequency of 10 Hz. The time shift between these
last time series is not the one which gives the best accord between the
synthetic and the observed accelerations.

Five point sources have been used to model the initial part of the
recordings. Without affecting the theoretical signals at the other
considered stations, the remarkable peak acceleration observed at Sturno

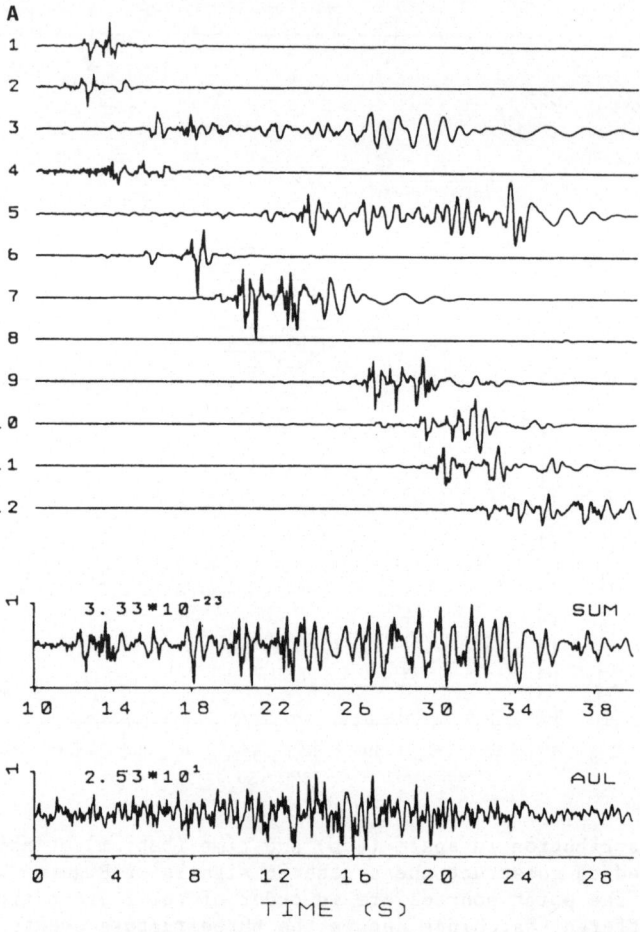

Fig. 7. (a) Top: synthetic accelerations relative to the twelve point
sources, whose locations, weights and time shifts are reported in
Figure 6, as seen at the station of Auletta (see Figure 4). The
fault strike, dip and rake of all sources are respectively 310°,
70° and 240°, except for point source number 5, its values being
40°, 70° and 270°. The amplitude of each trace corresponds to a
seismic moment $|M_o|$ = 1 dyn cm. Center: synthetic accelerations
sum of the twelve acceleration records shown above. The
amplitude, in units of cm/s^2, is shown in the upper left part of
each record. Bottom: the accelerations, Gaussian filtered with a
cut-off frequency of 10 Hz, observed during the Irpinia 1980
earthquake at the station of Auletta. The peak acceleration, in
units of cm/s^2, is shown in the upper part of the record.

(Figure 7f) at about 11 s has been modelled with a 2 km deep point source
located at epicenter K. To model the long-period oscillation, which
appears in the experimental time series recorded at Sturno between 5 and
8 s after the starting of the record, it has been necessary to add a
signal corresponding to a normal-fault point-source, 2 km deep in
epicenter I. This is the only point source with a focal mechanism
different from all the others. The fault strikes in a direction
perpendicular to the Apenninic chain, and the rake is 270°. This
reorientation is necessary in order to modify the waveform of the Sturno
record without affecting those obtained so far at Bagnoli (Figure 7b),
Bisaccia (Figure 7c) and Calitri (Figure 7e) stations. The existence near

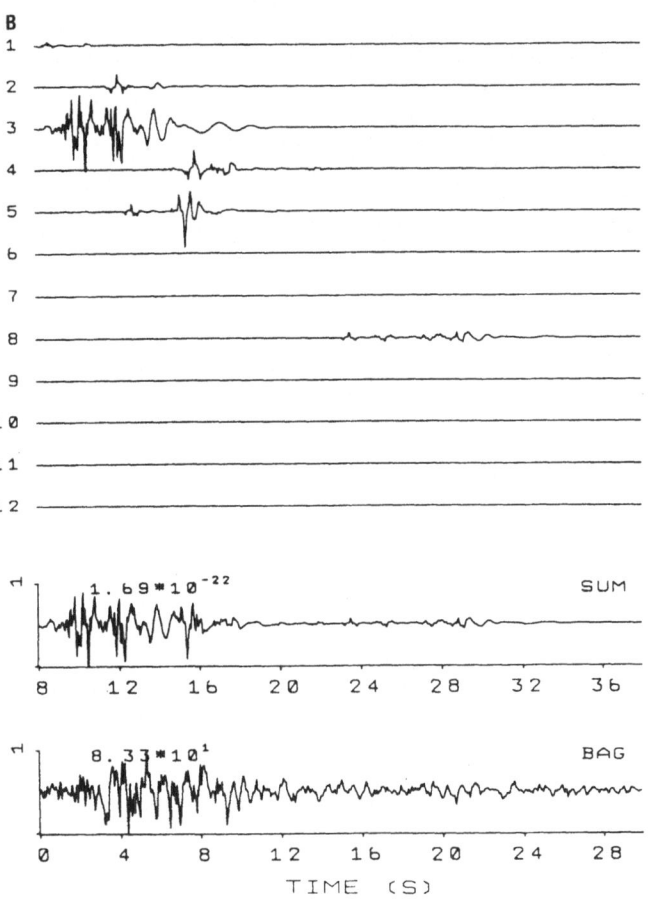

Fig. 7. (b) Same as Figure 7a, but for the station of Bagnoli.

epicenter I of a fault orthogonal to the main one has been also suggested by Crosson et al. (1986) to explain the vertical movements observed in the region after the earthquake. Other solutions (e.g., Bernard and Zollo, 1988; Westaway and Jackson, 1987) might be of course possible, but up to now we have not investigated the parameter space in detail.

The energy maximum, concentrated between 8 and 14 s after the starting of the observed Brienza (Figure 7d) record, cannot be due to point sources located in correspondence of epicenters A to M, as can be deduced from an analysis of the possible P- and S-wave travel times. It must be therefore associated with one or more point sources becoming active many seconds after the first one and close to the station itself. Placing these point sources in correspondence of epicenters X, Y, W (Figure 4) and properly spacing them in time and depth, produced at Brienza a peak acceleration three times larger than at Auletta.

In general, some signals with higher (> 3 Hz) frequency content should be added to the obtained synthetic signals, especially at Brienza, Calitri and Sturno stations. This lack of high frequencies is due to the fact that the source duration of all point sources has been kept the same.

The relative amplitudes of the theoretical accelerograms, however, agree quite well with those of the experimental records. The Auletta station synthetics (Figure 7a), as a consequence of the assumed rake

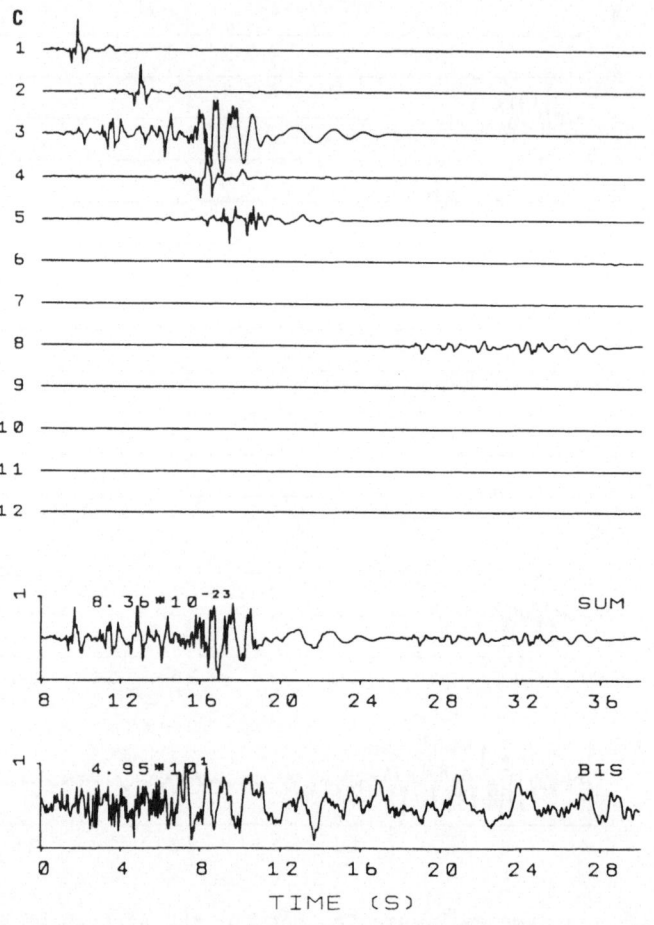

Fig. 7. (c) Same as Figure 7a, but for the station of Bisaccia.

value, is also nicely reproduced both in the peak value and in the waveforms. Only the Bagnoli theoretical signal (Figure 7b), in spite of the fact that it resembles the recorded waveform, has a relative peak value, which is larger than in the observed case. A possible explanation could be found in terms of different geological setting around that station.

Suhadolc et al. (1988b) have linearly scaled the observed peak acceleration recorded at the Sturno station to the weighted sum, at the same instant of time, of the accelerations from the contributing point sources. They found a total seismic moment, M_O, released by the twelve point sources of about 2.4×10^{25} dyn cm, the seismic moment of the largest point source (number 4 in Figure 6) being 10^{25} dyn cm. Even if one takes into account the moment associated to the subevent recorded around 40 s after the triggering (about one fifth of the total released moment, according to Bernard and Zollo, 1988), the found value would be still smaller than the one obtained with body waves by Deschamps and King (1983) and would be about one order of magnitude smaller than those obtained with long-period surface waves (Kanamori and Given, 1982). This shows the great difference between the average seismic moment estimated from long-period surface waves, where the approximation of considering the whole rupture process as due to a "point source" is valid, and the corresponding value obtained from short-period data, which give

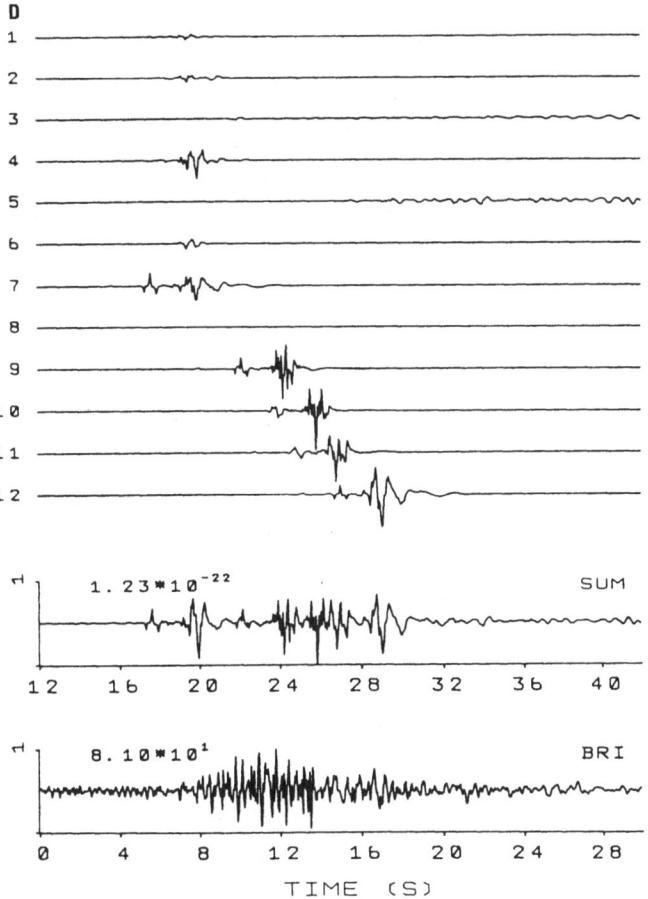

Fig. 7. (d) Same as Figure 7a, but for the station of Brienza.

information on local fault heterogeneities. The fact that in the case of very large earthquakes, which can be seen as multiple events, the seismic moment release, which can be determined in the point source approximation, is larger than the sum of each individual asperity, has been found also from long-period P-wave data by Beck and Ruff (1987).

To estimate the average stress drop over the whole fault area, the long-period seismic moment (3×10^{26} dyn cm) should be used. Considering the area (see the top part of Figure 6) roughly comprised around the dotted line, we get a surface of about 650 km. This yields an average stress drop, according to the formula $\Delta\sigma = (7/16) M_o/r$ (Brune, 1971), where r is the radius of the equivalent circular fault, of about 40 bar. If we consider the whole area shown in Figure 6, down to depths of about 15 km, the corresponding stress drop turns out to be around 20 bar.

To estimate the local stress drop associated with single asperities, it seems reasonable to associate to each point source an area equal to the square of the spacing between the point sources. This choice should give the asperity area resolvable with this approach. As seen before, one should obviously use "local" seismic moments determined according to the method described in Suhadolc et al. (1988b). In this way local stress drops of the order of several hundred bars are obtained, which are in the

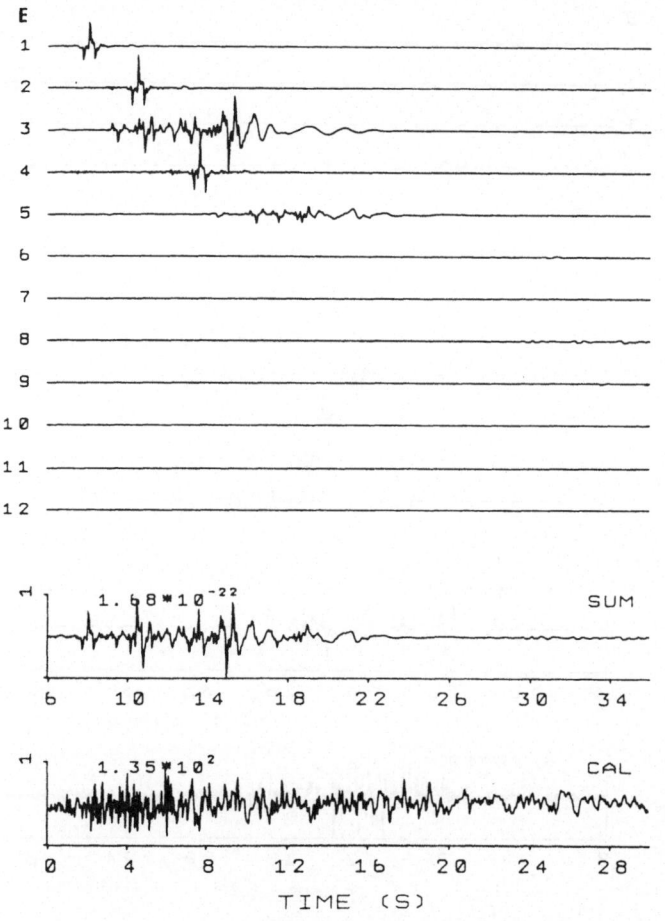

Fig. 7. (e) Same as Figure 7a, but for the station of Calitri.

same magnitude range as local stress drop estimates obtained with different methods (e.g., Papageorgiou and Aki, 1983).

The waveform fitting can be obviously improved considering a higher number of point sources, but to do this an inversion scheme has to be devised (e.g., Takeo, 1987; Harabaglia et al., 1987). A close waveform matching at these frequencies is also not significant, at least until a better understanding of the high-frequency behavior of earthquake rupturing will be achieved, and until the structural parameters will be known on the scale of the involved wavelengths. This requires the development of codes for the treatment of detailed two- and three-dimensional laterally heterogeneous structures, presently in progress (Gregersen et al., 1987).

FRIULI, ITALY, 1976 AFTERSHOCK

On September 11, 1976, at 16.35, an M_L = 5.6 aftershock of the May 6, 1976 event occurred in Friuli (46.1800°N, 13.1123°E, according to ISC). The fault-plane solution of this event is not well constrained (Lyon-Caen, 1980), and it is only possible to infer that it is consistent with the one of the main shock: a thrust event on a very shallow NW dipping plane, as

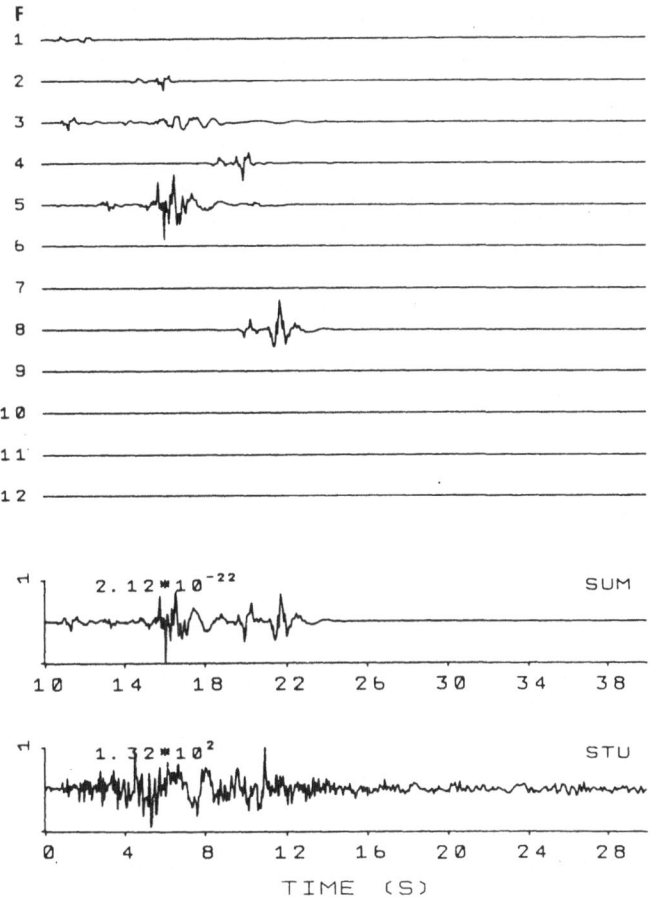

Fig. 7. (f) Same as Figure 7a, but for the station of Sturno.

confirmed in a recent study by Slejko and Renner (1984). The depth of the
September event according to these authors is around 3.7 km.

    This earthquake has been recorded by various accelerographic stations
(CNEN-ENEL, 1977). The vertical component recorded at the station of Buia
(46.1312°N, 13.0522°E) has been chosen in order to make a first analysis
of the possible source solutions responsible for this accelerogram trace.
This strong motion record is shown in the upper part of Figure 8. In the
middle part the same record is shown, Gaussian filtered with a cut-off
frequency of 10 Hz, while in the lower part the corresponding amplitude
spectrum is given.

    We present examples of exact computations for a continental structure
containing both low-velocity layers in the crust and a sedimentary cover
(see Figure 9). This structure is meant to simulate the elastic and
anelastic properties of the southern pre-Alps close to the May 6, 1976,
Friuli earthquake, and is obtained from the interpretation of deep seismic
sounding experiments (see, for example, Italian Explosion Seismology
Group, 1981; Angenheister et al., 1972) and surface wave dispersion
measurements (Calcagnile and Panza, 1981). The structural properties are
specified to depths of 50 km, where the S-wave velocity reaches 4.65 km/s.
The intrinsic attenuation of the layers has been determined with the same
procedure described for the Irpinia case.

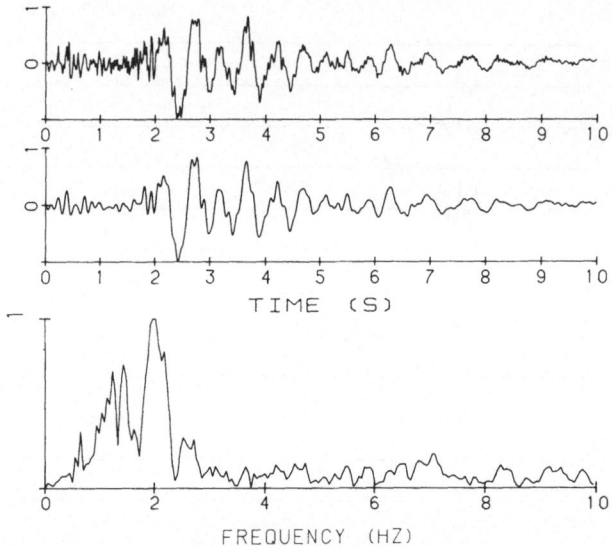

Fig. 8.   Top: accelerogram (CNEN-ENEL, 1977) relative to the Friuli,
September 11, 1976, 16.35, M = 5.6 aftershock recorded at the
station Buia.   The peak ground acceleration (normalized to one in
the figure) is 91 cm/s^2.   Center: the same accelerogram Gaussian
filtered at 10 Hz.   Bottom: the amplitude spectrum corresponding
to the filtered accelerogram.

        Several tries have been made starting from the source parameters
given in the literature to model the observed waveforms.   Good results
have been obtained by considering a very shallow source, as can be seen
from the upper part of Figure 10, where the synthetic signal due to a
point source located at a depth of 0.5 km at a distance of 17 km and with
the strike, dip and rake being respectively 270°, 24° and 75°, is shown.

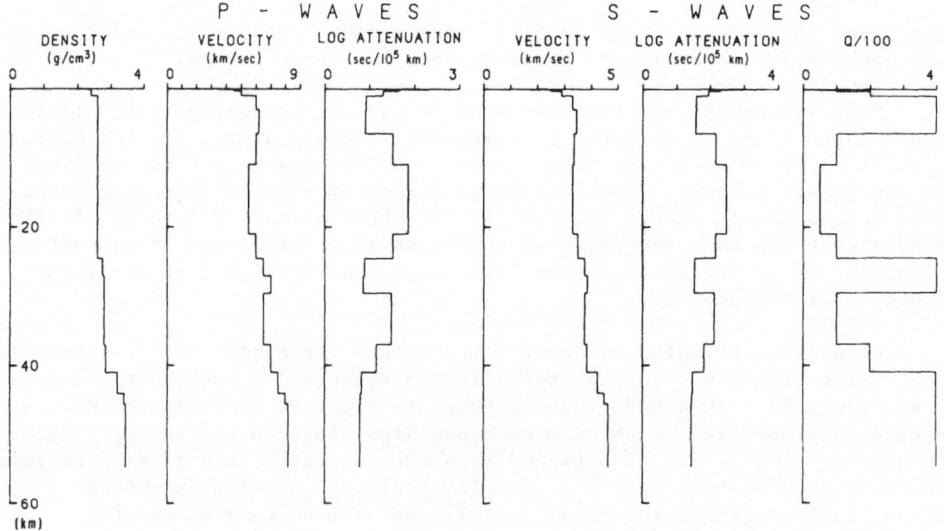

Fig. 9.   Distribution versus depth (km) for 44 layers of elastic and
anelastic properties for the structural model FRIUL7A.   It has
been assumed that Q_α = 2.5 Q_β.

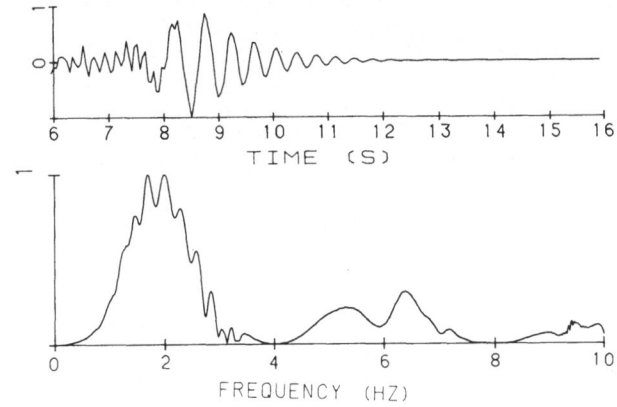

Fig. 10. Above: the synthetic accelerogram due to one point source (see
          text) corresponding to the observed trace of Figure 8 (center).
          The zero-to-peak amplitude (for a 1 dyn cm seismic moment) is
          5.5×10^{-22} cm/s^2. Below: the amplitude spectra corresponding
          to the trace shown above.

In the lower part of Figure 10 the corresponding amplitude spectrum is
shown. It is evident, that the amplitude spectrum of the observed signal
is more complex, containing several minima, the lowest one at around 1.5
Hz. This effect could be due to the interference of waves coming from
different sources. In fact, the agreement between experimental and
theoretical data is greatly enhanced, if more than one point source is
considered. The result of summing three point sources all having the same
focal depth and mechanism as the single point source of Figure 10, but
with different weights and time shifts (0.4, 1.0, 0.8 and 0 s, 0.6 s, and
1.5 s respectively) is shown in Figure 11. Now the accord between the
observed and synthetic data is much better, both in the time and frequency
domain. The seismic moment of the three sources turns out to be about 0.5
x 10^{23} dyn cm, 1.3×10^{23} dyn cm and 1×10^{23} dyn cm, respectively.

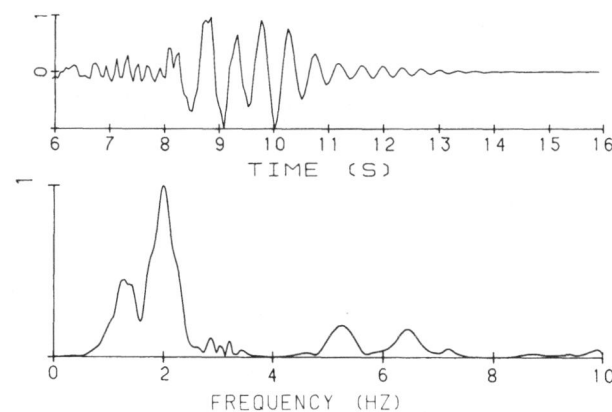

Fig. 11. Above: the synthetic accelerogram (sum of three point sources,
          see text) corresponding to the observed trace of Figure 8
          (center). The zero-to-peak amplitude (for the three sources
          having seismic moments 0.4, 1.0 and 0.8 dyn cm) is 5.8×10^{-22}
          cm/s^2. Below: the amplitude spectra corresponding to the trace
          shown above.

The conclusions which can be drawn from this modelling are the following. The accelerogram seems to be either due to a very shallow complex source or the rupture(s) did initiate at greater depths but propagated almost to the surface. The second hypothesis seems to be more sound, if we think of the relatively big magnitude - 5.6 - of this event. At the surface the rupture might have generated several stopping phases, giving rise to waves propagating in the sedimentary layers, which form the dominant part of the signal.

The solution proposed above is certainly not unique and we are presently investigating other possible sources, that produce waveforms similar to the observed one (Mao and Suhadolc, 1987). However, the implications we can draw remain valid: very shallow asperities or stopping phases may excite waves in sedimentary layers, that become the dominant part of the observed motion in the case of stations located on sediments.

CONCLUSIONS

It has been shown that strong-motion data provide a very appropriate data set for seismic source studies. Even if at present frequencies higher than a few Hz cannot be modelled deterministically, the high frequency content of strong-motion records permits to "see" better the details of the rupturing process along the fault.

Since S-waves and surface waves are responsible for the dominant part of an accelerogram trace, the modal summation approach is extremely appropriate to compute the desired synthetic signals. Routinely it is possible to generate, with satisfactory efficiency, "complete" synthetic strong-motions to frequencies as high as 10 Hz, and to consider Earth models made up of 70 layers and more.

To constrain better the rupturing models derived from strong-motion data one should have, on the experimental side, possibly digital data with a good azimuthal coverage, at various distances, up to 100 km, and with absolute timing; on the theoretical side the capability of treating lateral heterogeneities becomes increasingly important. Once this is achieved, it will be possible to attack the basic problem of understanding the high-frequency radiation associated with earthquake rupturing.

ACKNOWLEDGEMENTS

I am greatly indepted to Prof. G. F. Panza for critically reading the manuscript and for his valuable suggestions. Some of the figures were drawn by Mr G. Cavicchi.

This research was financially supported by ENEA contract n. 16094 and by MPI 40% and 60% funds.

REFERENCES

Angenheister, G., Bogel, H., Gebrande, P., Giese, P., Schmidt-Tome, P., and Zeil, W., 1972, Geol. Runds., 61:349-395.
Archuleta, R., 1984, A faulting model for the 1979 Imperial Valley, California, earthquake, J. Geophys. Res., 89:4559-4585.
Archuleta, R., and Day, S. M., 1980, Dynamic rupture in a layered medium: the 1966 Parkfield earthquake, Bull. Seism. Soc. Am., 70:671-689.
Beck, S. L., and Ruff, L. J., 1987, Asperity interaction and multiple event ruptures along the Kurile Islands and Colombia-Ecuador

subduction zones, 19 General Assembly IUGG, Vancouver, <u>Abstracts</u>, 1:343.

Ben Menahem, A., 1961, Radiation of seismic surface waves from finite moving sources, <u>Bull. Seism. Soc. Am.</u>, 51:401-435.

Berardi, R., Berenzi, A., and Capozza, F., 1981, Campania-Lucania earthquake on 23 November 1980 accelerometric recordings of the main quake and relating processing, <u>in</u>: " Contributo alla caratterizzazione del territorio italiano", ENEA-ENEL, Roma, pp 1-103.

Bernard, P., and Madariaga, R., 1984, A new asymptotic method for the modelling of near-field accelerograms, <u>Bull. Seism. Soc. Am.</u>, 74:539-558.

Bernard, P., and Zollo, A., 1988, The Irpinia (Italy) 1980 earthquake: detailed analysis of a complex normal fault, submitted to <u>J. Geophys. Res.</u>

Boatwright, J., 1982, A dynamic model for far-field acceleration, <u>Bull. Seism. Soc. Am.</u>, 72:1049-1068.

Bolt, B. A., 1987, "Seismic Strong Motion Synthetics", Computational Techniques 4, Academic Press, Orlando.

Boore, D. M., and Joyner, W. B., 1978, The influence of rupture incoherence on seismic directivity, <u>Bull. Seism. Soc. Am.</u>, 68:283-300.

Brune, J. N., 1971, Tectonic stress and spectra of seismic shear waves from earthquakes (correction), <u>J. Geophys. Res.</u>, 76:5002.

Burridge, R., and Knopoff, L., 1964, Body force equivalents for seismic dislocations, <u>Bull. Seism. Soc. Am.</u>, 54:1875-1888.

Calcagnile, G., and Panza, G. F., 1981, The main characteristics of the lithosphere-asthenosphere system in Italy and surrounding regions, <u>PAGEOPH</u>, 119:865-879.

Cerveny, V., 1985, Gaussian beam synthetic seismograms, <u>J. Geophys.</u>, 58:44-72.

Chapman, C. H., 1985, Ray theory and its extension - WKBJ and Maslov seismograms, <u>J. Geophys.</u>, 58:27-43.

Chiaruttini, C., Costa, G., and Panza, G. F., 1985, Wave propagation in multilayered media: the effect of wave guides in oceanic and continental earth models, <u>J. Geophys.</u>, 58:189-196.

CNEN-ENEL, 1977, Uncorrected accelerograms, Accelerograms from the Friuli, Italy, earthquake of May 6, 1976 and aftershocks; part 3, Rome, Italy, November 1977.

Craglietto, A., Panza, G. F., Mitchell, B., and Costa, G., 1987, Anelastic properties of the crust in the Mediterranean area, 19 General Assembly IUGG, Vancouver, <u>Abstracts</u>, 1:70.

Crosson, R., Martini, M., Scarpa, R., and Key, S. C., 1986, The southern Italy earthquake of 23 November 1980: an unusual pattern of faulting, <u>Bull. Seism. Soc. Am.</u>, 76:381-384.

Das, S., 1981, Three-dimensional rupture propagation and implications for the earthquake source mechanism, <u>Geophys. J. R. Astron. Soc.</u>, 67:375-393.

Das, S., 1985, Application of dynamic shear crack models to the study of the earthquake faulting process, <u>Int. J. Fracture</u>, 27:263-276.

Das, S., 1986, Comparison of the radiated fields generated by the fracture of a circular crack and a circular asperity, <u>Geophys. J. R. Astron. Soc.</u>, 85:601-615.

Das, S., and Aki, K., 1977a, Fault plane with barriers: a versatile earthquake model, <u>J. Geophys. Res.</u>, 82:5658-5670.

Das, S., and Aki, K., 1977b, A numerical study of two-dimensional spontaneous rupture propagation, <u>Geophys. J. R. Astron. Soc.</u>, 50:643-668.

Das, S., and Kostrov, B. V., 1983, Breaking of a single asperity: rupture process and seismic radiatiion, <u>J. Geophys. Res.</u>, 88:4277-4288.

Das, S., and Kostrov, B. V., 1985, An elliptical asperity in shear:

fracture process and seismic radiation, Geophys. J. R. Astron. Soc., 80:725-742.

Del Pezzo, E., Iannaccone, G., Martini, M., and Scarpa, R., 1983, The 23 November 1980 southern Italy earthquake, Bull. Seism. Soc. Am., 73:187-200.

Deschamps, A., and King, G. C. P., 1983, The Campania-Lucania (southern Italy) earthquake of 23 November 1980, Earth Plan. Sci. Lett., 62:296-304.

Deschamps, A., and King, G. C. P., 1984, Aftershocks of the Campania-Lucania (Italy) earthquake of 23 November 1980, Bull. Seism. Soc. Am., 74:2483-2517.

Dziewonski, A. M., Chou, T.-A., and Woodhouse, J. H., 1981, Determination of earthquake source parameters from waveform data for studies of global and regional seismicity, J. Geophys. Res., 86:2825-2852.

Erdik, M. O., and Toksöz, M. N., 1987, "Strong Motion Seismology", NATO ASI Series C 204, Reidel, Dordrecht.

Gilbert, F., and Dziewonski, A. M., 1975, An application of normal mode theory to the retrieval of structural parameters and source mechanisms from seismic spectra, Philos. Trans. R. Soc., London, Ser. A, 278:187-269.

Gregersen, S., Vaccari, F., and Panza, G. F., 1987, 2-D and 3-D synthetic seismograms based on surface wave modes, 19 General Assembly IUGG, Vancouver, Abstracts, 1:311.

Hanks, T. C., 1982, f_{max}, Bull. Seism. Soc. Am., 72:1867-1879.

Harabaglia, P., Suhadolc, P., and Panza, G. F., 1988, Il terremoto irpino del 23 Novembre 1980: modelli di rottura dall' inversion di dati accelerometrici, Atti 6° Convegno GNGTS, CNR, Rom, in press.

Hartzell, S. H., and Helmberger, D. V., 1982, Strong-motion modelling of the Imperial Valley earthquake of 1979, Bull. Seism. Soc. Am., 72:571-596.

Hartzell, S. H., and Heaton, T. H., 1986, Rupture history of the 1984 Morgan Hill, California, earthquake from the inversion of strong-motion records, Bull. Seism. Soc. Am., 76:649-674.

Harvey, D. J., 1981, Seismograms synthesis using normal mode superposition: the locked mode approximation, Geophys. J. R. Astron. Soc., 66:37-69.

Haskell, N. A., 1964, Total energy and energy spectral density of elastic wave radiation from propagating faults, Bull. Seism. Soc. Am., 54:1811-1841.

Haskell, N. A., 1966, Total energy and energy spectral density of elastic wave radiation from propagating faults. Part II: a statistical source model, Bull. Seism. Soc. Am., 56:125-140.

Italian Explosion Seismology Group, 1981, Crust and upper mantle structures in the southern Alps from deep seismic sounding, Boll. Geof. Teor. Appl., 23:267-330.

Jennings, P. C., 1983, Engineering seismology, in: "Earthquakes: Observation, Theory and Interpretation", H. Kanamori and E. Boschi, eds., North Holland Publishing Co., Amsterdam, pp 138-173.

Kagan, Y. Y., and Knopoff, L. 1980, Spatial distribution of earthquakes: the two-point correlation function, Geophys. J. R. Astron. Soc., 62:303-320.

Kanamori, H., 1970a, Synthesis of long-period surface waves and its applications to earthquake source studies - Kurile Islands earthquake of October 13, 1963, J. Geophys. Res., 75:5011-5028.

Kanamori, H., 1970b, The Alaska earthquake of 1964: radiation of long-period surface waves and source mechanism, J. Geophys. Res., 75:5029-5040.

Kanamori, H., and Given, J. W., 1982, Use of long-period surface waves for rapid determination of earthquake source parameters, 2. Preliminary determination of source mechanisms of large earthquakes (MS $\geqslant$ 6.5) in 1980, Phys. Earth Plan. Int., 30:260-268.

King, G. C., 1986, Speculations on the geometry of the initiation and termination processes of earthquake rupture and its relation to morphology and geological structure, PAGEOPH, 124:567-585.

Knopoff, L., 1964, Q, Rev. Geophys., 2:625-660.

Knopoff, L., and Chatterjee, A. K., 1982, Unilateral extension of a two-dimensional shear crack under the influence of cohesive forces, Geophys. J. R. Astron. Soc., 68:7-25.

Koyama, J., 1985, Earthquake source time-function from coherent and incoherent rupture, Tectonophysics, 118:227-242.

Langston, C. A., and Helmberger, D. W., 1975, A procedure for modelling shallow dislocation sources, Geophys. J. R. Astron. Soc., 42:117-130.

Liao, A., Schwab, F., and Mantovani, E., 1978, Computation of complete theoretical seismograms for torsional waves, Bull. Seism. Soc. Am., 68:317-324.

Lyon-Caen, H., Seismes du Frioul (1976): modéles de source a l'aide de seismogrammes synthètiques d'ondes de volume, Ph.D. Thesis, Univ. of Paris VII.

Madariaga, R., 1976, Dynamics of an expanding circular fault, Bull. Seism. Soc. Am., 66:639-666.

Madariaga, R., 1977, High-frequency radiation from crack (stress drop) models of earthquake faulting, Geophys. J. R. Astron. Soc., 51:625-651.

Madariaga, R., 1983, High-frequency radiation from dynamic earthquake fault models, Ann. Geophys., 1:17-23.

Mao, W. J., and Suhadolc, P., 1988, The Friuli seismic area: inversion of travel times for a structural model and strong ground motion waveform modelling, Atti 6° Convegno GNGTS, CNR, Roma, in press.

Mueller, S., 1977, A new model of the continental crust, in: "The Earth's Crust", J. Heacock, ed., AGU, Monogr. 20, pp 289-317.

Newman, W. I., and Knopoff, L., 1982, Crack fusion dynamics: a model for large earthquakes, Geophys. Res. Lett., 9:735-738.

Olson, A. H., and Apsel, R. J., 1982, Finite faults and inverse theory with applications to the 1979 Imperial Valley earthquake, Bull. Seism. Soc. Am., 72:1969-2001.

Panza, G. F., 1985, Synthetic seismograms: the Rayleigh waves modal summation, J. Geophys., 58:125-145.

Panza, G. F., Mueller, St., and Calcagnile, G., 1980, The gross features of the lithosphere - asthenosphere system in Europe from seismic surface waves and body waves, PAGEOPH, 118:1209-1213.

Panza, G. F., Suhadolc, P., and Chiaruttini, C., 1986, Exploitation of broad-band networks through broad-band synthetic seismograms, Ann. Geophys., 4B:315-328.

Papageorgiou, A. S., and Aki, K., 1983, A specific barrier model for the quantitative description of inhomogeneous faulting and the prediction of strong ground motion, I: description of the model, Bull. Seism. Soc. Am., 73:693-722.

Shearer, P. M., and Orcutt, J. A., 1987, Surface and near-surface effects on seismic waves - theory and borehole seismometer results, Bull. Seism. Soc. Am., 77:1168-1196.

Slejko, D., and Renner, G., 1984, in: "Finalità ed esperienze della rete sismometrica del Friuli-Venezia Giulia", Regione autonoma Friuli-Venezia Giulia, Trieste, pp 75-91.

Spudich, P., and Frazer, L. N., 1984, Use of ray theory to calculate high-frequency radiation from earthquake sources having spatially variable rupture velocity and stress drop, Bull. Seism. Soc. Am., 74:2061-2082.

Spudich, P., and Archuleta, R. J., 1987, Techniques for earthquake ground-motion calculation with application to source parametrization of finite faults, in: "Seismic Strong Motion Synthetics Computational

Techniques 4", B. A. Bolt, ed., Academic Press, Orlando, pp 205-265.

Suhadolc, P., and Chiaruttini, C., 1987, A theoretical study on the dependence of the peak ground motion acceleration on source and structure parameters, in: "Strong Ground Motion Seismology", NATO ASI, Series C 204, Reidel, Dordrecht, pp 143-183.

Suhadolc, P., and Panza, G. F., 1985, Some applications of seismosynthesis through the summation of modes of Rayleigh waves, J. Geophys., 58:183-188.

Suhadolc, P., Cernobori, L., Pazzi, G., and Panza, G. F., 1988a, Synthetic isoseismals: application to Italian earthquakes, Proc. CSEM Summer School on "Seismic Hazard in Mediterranean Regions", Strasbourg, July 21 - August 1, 1986.

Suhadolc, P., Vaccari, F., and Panza, G. F., 1988b, Strong motion modelling of the rupturing process of the November 1980 Irpinia, Italy, earthquake, Proc. CSEM Summer School on "Seismic Hazard in Mediterranean Regions", Strasbourg, July 21 - August 1, 1986.

Takeo, N., 1987, An inversion method to analyze the rupturing processes of earthquakes using near-field seismograms, Bull. Seism. Soc. Am., 77:490-513.

Westaway, R., and Jackson, J., 1987, The earthquake of 1980 November 23 in Campania-Basilicata (southern Italy), Geophys. J. R. Astron. Soc., 90:375-443.

WAVE PROPAGATION IN SUBDUCTED LITHOSPHERIC SLABS

E. R. Engdahl, J. E. Vidale* and V. F. Cormier**

National Earthquake Information Center
US Geological Survey, Denver Federal Center
PO Box 25046, Mail Stop 967
Denver, Colorado 80225, USA
*Earth Sciences Board of Study
University of California
Santa Cruz, California 95064, USA
**Department of Geology and Geophysics
University of Connecticut, Box U-45
Storrs, Connecticut 06268, USA

INTRODUCTION

The strongest lateral velocity gradients in the Earth are found in subduction zones. In these regions, the temperatures associated with colder, downgoing slabs can lead to compositional variations and to seismic velocities up to 10 percent higher than the surrounding mantle. These major sources of heterogeneity are known to have significant effects on the travel times and ray trajectories of waves travelling within them (e.g., Sleep, 1973; Davies and Julian, 1972). However, until only recently, the amplitude, extent, and location of the slab velocity anomaly have been examined primarily by observations of travel time anomalies. The deployment of broad-band instrumentation and advances in analysis methods now permit more extensive use of waveform data to model slab effects, including those that are non-geometrical.

Studies of wave propagation in subducted slabs may help to resolve fundamental questions about the structure of subduction zones. These include the position of earthquakes within the slab, slab thickness, vertical extent of slab penetration, lateral slab extent, and slab dip. A typical question of current interest is whether subducted lithospheric slabs extend below the 670 km discontinuity and the implications for mantle convection. The details of slab dip and shape at and beyond its aseismic extension provide constraints on both the temperature and viscosity of the surrounding mantle. These constraints would be valuable to efforts to model mantle flow by either a one- or two-layered system. Amplitude and waveform distortions produced by slab penetration may help refine results based so far entirely on travel-time anomaly patterns from deep earthquakes.

Studies that seismically probe slabs generally have one of three geometries: the station is above the slab to be probed, and the source is far away; the source is in the slab, and the station is above the slab; or the source is within the slab, and the station is far away. Many earlier studies (e.g., Fukao, Kanjo and Nakamura, 1978; and, more recently, Owens and Crosson, 1986) fall in the first category. In the second category, relatively short-period digital and analog seismograms from regional networks have been utilized to study updip and lateral propagation of waves from intermediate-depth earthquakes. Important information about structural features is provided by phase conversion and refraction that generally takes place when seismic waves encounter discontinuities between the earthquake source and the recording station. Chiu, Isacks, and Cardwell (1985) analyzed seismograms of intermediate-depth earthquakes (h > 70 km) recorded in the distance range up to 350 km by a seismograph network in the Vanatu (formerly New Hebrides) island arc. They identified three seismic phases associated with the slab of subducted lithosphere. They are interpreted as: (1) a P-wave refracting into the inner portion of the slab; (2) a P-wave travelling mainly through the outer (upper) portion of the slab; and (3) a lower frequency P to S converted wave in which the conversion occurs across the upper surface of the slab (see also Matsuzawa et al., 1986). Modelling of the two P arrivals from intermediate-depth earthquakes suggests that these events are not located in the strong and cold high-velocity elastic core of the slab, but within an outer zone of high thermal and velocity gradients.

Ansell and Gubbins (1986) studied waveforms from events in the Tonga-Kermadec subduction zone along paths parallel to the strike of the slab to the New Zealand network of seismometers. The seismograms show two distinct phases: an early, emergent, first phase with energy in the high-frequency band 2 - 10 Hz and a distinct later phase, containing lower frequency energy, arriving at about the time predicted by standard travel-time tables. Ray tracing shows that the first phase can be attributed to propagation through the high-velocity slab. The absence of low frequency energy in the first phase is due to the narrowness of the high-velocity slab which transmits only short-wavelength waves. The second phase, which contains low frequencies, is identified as a P-wave travelling primarily beneath the subducted slab in the normal upper mantle. Separation of frequencies between the phases is a non-geometrical effect, suggesting the slab is quite thin.

Slab propagation effects have been more widely studied for the case of a source within the slab and stations at teleseismic distances, undoubtedly because of the greater number and variety of source to receiver paths. Noticeable distortions of amplitudes and waveforms of teleseismic signals have been observed from earthquakes and explosions in the central Aleutians (Engdahl and Kind, 1986; and Vidale and Helmberger, 1986) and in the Kuriles (Silver and Chan, 1986; Beck and Lay, 1986; Cormier, 1986; Vidale and Helmberger, 1986; and Vidale, 1987). Some of these studies used new techniques that permit the analysis of wave propagation effects in slabs not accounted for by geometrical wave theory. Here, we will briefly review these new techniques and their application to two subduction zones, the central Aleutian Islands and the Kurile Islands.

SYNTHETIC SEISMOGRAMS

Techniques

It is important to be able to calculate waveform distortions due to slab structure, both to test hypotheses about slab structure and to separate effects of slab geometry from effects of other lateral velocity variations in the Earth. Several techniques have been developed to simulate the teleseismic signals of earthquakes propagating through slabs. These techniques can be classified as being either numeric or asymptotic.

Fully numerical solutions of the elastic wave equation are necessary
to accurately account for frequency dependent effects of internal slab
reverberations and P to S conversions at slab boundaries. These
boundaries may be either discontinuities or narrow zones of strong
gradient in density and elastic velocity. Examples of the numerical
techniques currently being applied to the slab problem include the finite
difference method (Vidale and Helmberger, 1986) and the pseudo-spectral
method (Witte, 1987). In either approach, the effects of arbitrarily
narrow slabs, slab boundaries, and solid-solid transition zones within the
slab can be calculated to any desired accuracy given a sufficiently fine
network of knots over which difference derivatives are calculated. Both
techniques, however, are practically limited to two-dimensional geometry
because of their computational expense.

Asymptotic techniques include ray theoretical approximations that
neglect internal slab reverberations and conversions at slab boundaries
and transition zones. The possible approaches include dynamic ray
tracing, Kirchoff integration, superposition of Gaussian beams and WKBJ
Maslov plane waves. These techniques are limited in application to slab
models in which velocity (V) does not vary rapidly with respect to
wavelength (λ), i.e., $\lambda \ll V/|\nabla V|$. Since they are numerically cheap,
application to three-dimensional geometry is practical. Cormier (1987),
for example, has applied superposition of Gaussian beams to the slab
problem in three-dimensions for the slab models of Creager and Jordan
(1986). The velocity structures predicted from the thermal models of
Sleep (1973) and Minear and Toksoz (1973) can be parameterized by
continuous functions of space in which the $\lambda \ll V/|\nabla V|$ criterion of
asymptotic validity is satisfied throughout the body wave band of
frequencies between 0.03 to 5 Hz.

Although the asymptotic techniques make calculations in three-
dimensional models feasible, the numerical methods are still the only
technique available for the investigation of arbitrarily thin slabs and
the effects of fine scale structure at slab boundaries and transition
zones. Numerical methods are also necessary for checking the validity of
the asymptotic and paraxial approximations used in the asymptotic methods
(e.g., Ben-Menahem and Beydoun, 1985; White et al., 1987).

Results of Synthetic Modelling

Vidale (1987) computed synthetic displacement seismograms appropriate
for a record section in a plane perpendicular to the strike of the slab
using a coupled finite-difference and Kirchoff method. The effects of a
80 km thick high-velocity slab at teleseismic distances may be seen in
Figure 1. For the case of a receiver that is directly below the source (i
= 0°) in the slab, the initial arrival, with a take-off angle near the
vertical dip angle of the slab, has a reduced amplitude and an earlier
arrival time than if the slab had not been present. This is the higher
frequency geometrical arrival that travels in the high-velocity slab.
Amplitude loss occurs through diffraction out of the slab. At the same
receiver, the longer period pulse, about 8 sec behind the first arrival in
the seismograms in Figure 1, results from long-period energy that travels
in the slower surrounding media and must diffract around the slab to reach
the receiver. The second arrival has no analogue in the case where there
is no slab and is not a geometrical arrival. As the take-off angle
increases, the high-frequency arrival loses amplitude and precedes the
second arrival by less time. At just past 20°, the two arrivals merge as
the energy that travels down the fast slab no longer precedes the energy
that travels straight from the source to the receiver, and there is little
in the seismograms to indicate even the presence of a fast slab. These
results demonstrate that a high-velocity slab is an anti-waveguide.

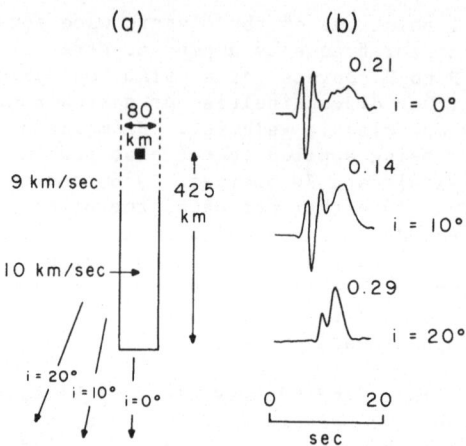

Fig. 1. Synthetic P-waves (from a point source) distorted by propagation
through a high-velocity, isotropic slab. The model is shown in
(a), with the source indicated by a filled square, 425 km above
the base of the slab. The receivers (b), at a distance of 10,000
km, are in the directions indicated by the arrows at 0°, 10°, and
20° take-off angles. The amplitudes relative to a P-wave
propagating in a slabless model are printed above each
trace. From Vidale (1987).

Waves are continuously refracted out of the slab, leading to waveform
distortion and a reduction in amplitude.

Further calculations that predict noticeable distortions of
amplitudes and waveforms of teleseismic signals from subduction zone
earthquakes have been presented by Cormier (1987). Cormier used
computations in Gaussian beam or WKBJ/Maslov seismograms to study the
effects of slab structure on waveforms. The synthetics for a source
within a 60° dipping slab demonstrated a phenomenon of slab diffraction
that acts to lengthen the apparent pulse width of the S-wave. Mild, but
smoothly varying, waveform distortions are predicted for downdip ray
paths. These slab diffraction effects can be expected to persist up to
the slab strike direction, reaching a maximum along strike and then very
abruptly shutting off at azimuths away from the slab dip. Cormier (1986)
also noted that small changes of take-off angle, such as the difference in
angles between P and PcP or SKS and SKKS, can make large differences in
the amplitude ratios of these phases because of the effects of slab
focusing and defocusing.

CENTRAL ALEUTIAN SLAB

The central Aleutian slab provides an excellent opportunity to study
wave propagation in subduction zones. Engdahl and Gubbins (1987) applied
a combined location and velocity inversion technique to travel-time data
from 151 well-recorded central Aleutian earthquakes. For the rectangular
region shown in Figure 2, they found a steep northward-dipping slab
structure with a thickness of 80 - 100 km and a downdip length of about
400 km, well below the deepest seismic activity. The slab is
characterized by seismic velocities as much as 11 percent higher than the
surrounding mantle in its upper portions and 4 - 6 percent higher at
depth. A sharp velocity gradient and lower velocities occur directly
beneath the volcanic arc near the top of the slab.

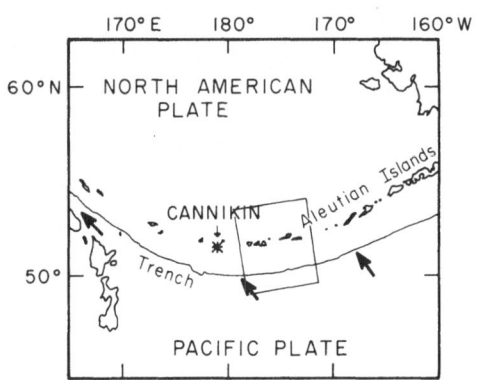

Fig. 2. Central Aleutian subduction zone. Slab model has been determined
by Engdahl and Gubbins (1987) for the rectangular region. The
location of the nuclear explosion CANNIKIN is indicated by the
asterisk.

Engdahl and Kind (1986) used broad-band Graefenberg array (GRF) data
from moderate-size, shallow-depth earthquakes with well-determined source
parameters in the same region as Engdahl and Gubbins (1987) to study the
effects of focal depth and slab structure on waves propagating downdip.
Because all of the earthquakes had similar simple source mechanisms and
source durations, any variations in GRF waveforms should be almost
entirely the result of variations in source depth and structure across
the arc. They observed that the character of vertical-component
summation seismograms at different passbands for earthquakes located more
than about 85 km in distance from the trench are markedly different than
those that are closer to the trench. For the examples of displacement
seismograms shown in Figure 3, the more complex and broader P waveforms
correspond to ray paths to the array that are directly down the slab
structure determined by Engdahl and Gubbins (1987), while the simpler
waveforms have ray paths that are below the lower boundary of the
descending slab. Engdahl has extended the Engdahl and Kind (1986) study
by examining broad-band waveforms from other stations on the downdip side
of the slab. He found an effect similar to that observed on GRF
seismograms, that is, trenchward earthquakes along the active thrust zone
have much narrower P waveforms than more northerly earthquakes along the
same zone. One explanation for the observed complexity and waveform
broadening, suggested by Engdahl and Kind, is multi-pathing in the plate
produced by discontinuities at the plate interfaces or by sharp local
changes in arc structure. A more likely explanation for the waveform
broadening is slab diffraction.

Similar effects on central Aleutian waveforms can be seen in LPZ
(WWSSN) P-wave seismograms from the nuclear explosion CANNIKIN (Figure 2).
Vidale and Helmberger (1986) observed that waves propagating in downdip
slab direction from CANNIKIN were anomalously broad and lower in amplitude
(Figure 4). More recently, Vidale and Garcia-Gonzalez (1987) found that
shallow events (< 50 km depth) in the Aleutians show broader P-wave
arrivals in the downdip direction (Europe) than in other directions.
Vidale and Helmberger (1986) computed synthetic seismograms for the
Engdahl and Gubbins (1987) slab structure (on the left of Figure 5), but
found that the effects were less pronounced than shown by the observed
data (Figure 6, on the left). Consequently, with the method of Vidale
(1987), they tested two modified versions of the Engdahl and Gubbins model
- one twice as narrow and another with only the top edge narrower (Figure
5, middle and right, respectively). The resulting theoretical waveforms

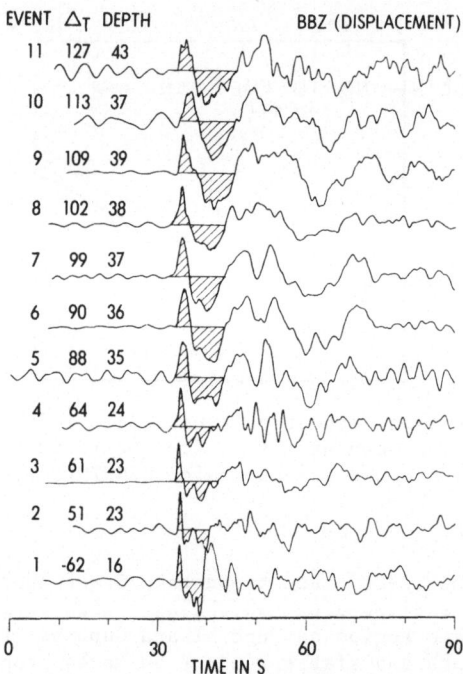

EVENT Δ_T DEPTH   BBZ (DISPLACEMENT)

EVENT	Δ_T	DEPTH
11	127	43
10	113	37
9	109	39
8	102	38
7	99	37
6	90	36
5	88	35
4	64	24
3	61	23
2	51	23
1	-62	16

0   30   60   90
TIME IN S

Fig. 3. Shown are, vertical-component, GRF displacement seismograms for
         11 central Aleutian earthquakes. The records have been aligned
         by P onset and normalized in amplitude. They are arranged from
         the bottom upwards by increasing arc normal distance to the
         trench axis (Δ_T). Depth and (Δ_T) are in kilometers. From Engdahl
         and Kind (1986).

for the modified structure show slab diffraction effects and are in better
agreement with observed data. The thinner slab model is not in
disagreement with the results of Engdahl and Gubbins (1987), as the
resolution of their slab structure was limited by the mesh spacing (50 x
40 km) and by the calculation of the ray paths with a radially symmetric

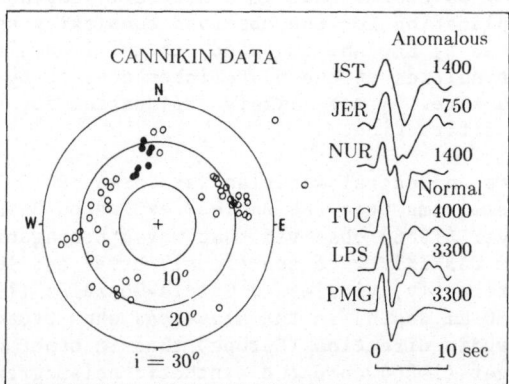

Fig. 4. LPZ (WWSSN) P-waves from the nuclear explosion CANNIKIN plotted
         on the left by take-off angle and azimuth. Anomalously broad and
         lower amplitude waveforms are indicated by the filled circles.
         On the right are examples of anomalous and normal P waveforms
         with observed amplitudes.

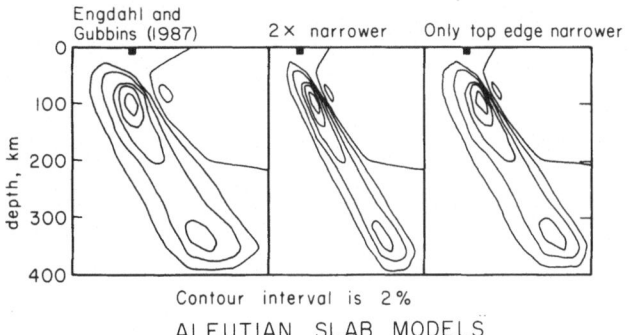

Fig. 5. Two-dimensional Aleutian slab models. On the left is the model taken from Engdahl and Gubbins (1987). The model in the center is a 2X narrower version, and the model on the right is a version with only the top edge narrower. Velocity perturbations are contoured at 2 percent intervals. The position of CANNIKIN is indicated by the solid square.

Earth model. Inversions carried out with a finer grid (20 x 20 km) suggested an even thinner slab, but, without ray tracing, could not resolve slab features at depth. The important result of Vidale and Helmberger's work is that diffraction effects can be highly sensitive to slab width, demonstrating that waveforms can provide important new constraints on slab structure independent of information provided by their travel times.

KURILE SLAB

The question of whether the subducted slab in the Kurile Islands penetrates the 670 km discontinuity is of great concern to Earth scientists. Creager and Jordan (1984, 1986) have proposed, on the basis of P-wave travel time anomalies, that the Kurile slab extends to a depth greater than 1000 km. Although these studies use only travel-time data to

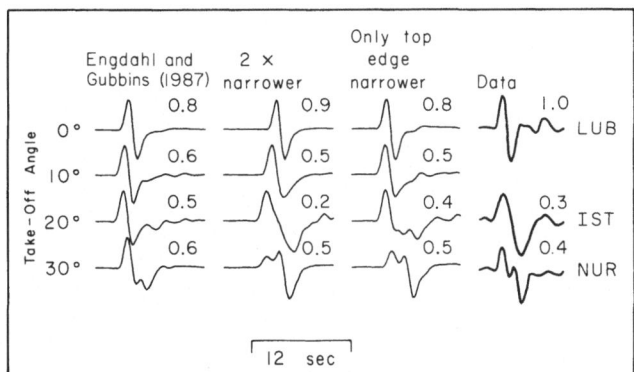

Fig. 6. Synthetic seismograms constructed from the models shown in Figure 5 for P-waves with a surface reflection, and a long-period WWSSN instrument response for take-off angles of 0°, 10°, 20°, and 30°. Examples of observed data are shown at the right (not corresponding to the take-off angles shown). Amplitudes are relative to a P-wave propagating outside of the slab. The source time function (RDP) is taken from Burdick et al. (1981).

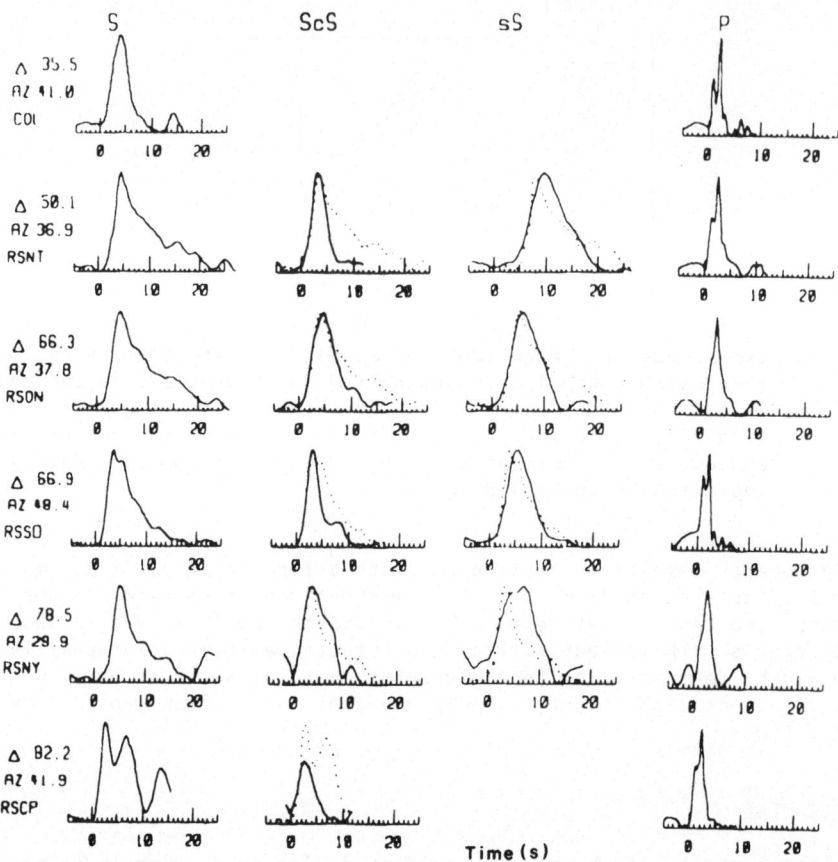

Fig. 7.   Deconvolved S, ScS, sS, on the transverse component, and P, on
          the vertical component, for well-recorded phases of a deep-focus
          Sea of Okhotsk earthquake recorded by stations in North America.
          The P-waves indicate a source duration of 4 s. Superimposed on
          the traces of sS and ScS is the S trace (dashed).   From Silver
          and Chan (1986).

constrain velocity models of subducting slabs, the models have
implications for amplitude and waveform effects of deep slabs. In this
section, we first review some of the relevant waveform data and then
calculate the effects of representative deep slab structures on the
amplitudes and waveforms of teleseismic signals from deep earthquakes.

        Silver and Chan (1986) noted anomalously broad and complex
horizontally polarized shear waves recorded at stations in North America
from intermediate- and deep-focus events in Western Pacific subduction
zones. Figure 7 shows examples of the unusual broadening of the phase S
with respect to both ScS and sS for a deep event beneath the Sea of
Okhotsk. By comparing S-ScS travel times to S-wave duration, they found
that the broadest phases had the largest integrated amplitudes and the
fastest travel times, the opposite of what would be expected from
attenuation. Ray tracing through a modified version of the Creager and
Jordan (1984) slab model reveals multi-pathing along the strike of the
slab where broadening of the S-waves is observed (Figure 8). They
conclude that the observed S-waves are experiencing a high-velocity
feature along the path and that these phases consist of multiple arrivals
caused by refractions from the subducted slab. The existence of such

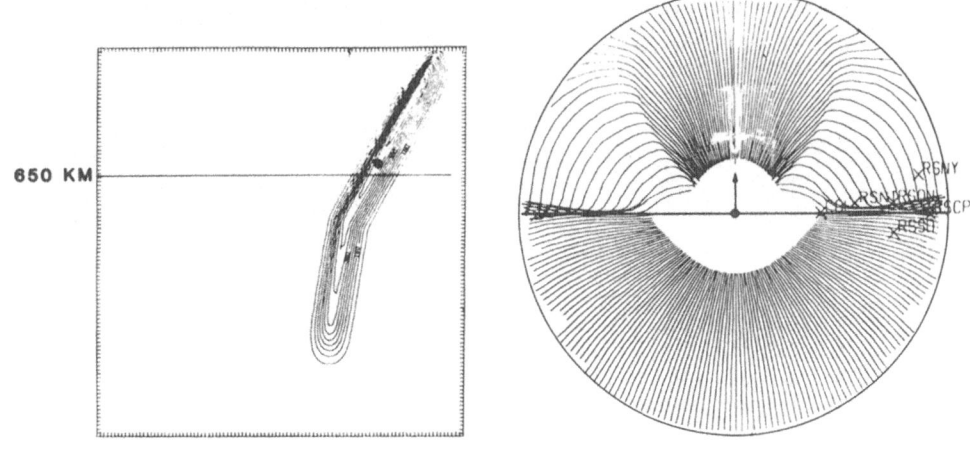

| | a | DEEP FOCUS | b |

Fig. 8. (a) Slab model for the southern Kuriles region (from Creager and Jordan, 1984) modified with a bend at 900 km depth. Velocity contours are in units of meters per second. Solid circle indicates location of event at 600 km depth. (b) Resulting ray diagram. Map is azimuthal equidistant projection of the Earth's surface. Each line represents the joining of a family of rays shot with constant azimuth and variable take-off angle. Crossing azimuth lines represent regions where multi-pathing is expected. The epicenter is in the center of the plot with the slab striking horizontally as indicated by the dark line. Arrow gives downdip direction. Station locations are indicated by X's. From Silver and Chan (1986).

phases for a deep focus event implies that the slab underneath the Sea of Okhotsk continues several hundred kilometers beyond the 670 km discontinuity.

Beck and Lay (1986) also studied S and ScS phases from deep-focus events beneath the Sea of Okhotsk with a greatly expanded data set. They also found direct S pulses with a broader, more complex waveform than the corresponding ScS pulses. However, they also show observations along strike with no waveform anomalies and some stations at all azimuths which show the S-wave broadening. Array observations of broadening in Europe, in the downdip slab direction, may be the result of defocusing by the high-velocity slab extension. This should also produce a distance trend. They conclude on the basis of the entire data set that lateral heterogeneities both near the source (possibly due to slab penetration) and deep in the mantle near the S-wave turning points probably contribute to S and ScS complexity.

Vidale and Helmberger (1966) and Vidale (1987) calculated synthetic seismograms for P- and S-waves using P- and S-wave velocity structures derived from the thermal model of Creager and Jordan (1986) for the slab in the Kurile-Kamchatka subduction zone. Figure 9 shows the Creager and Jordan P-wave model and two variants on that model. Displacement seismograms at teleseismic distances for a source at 515 km depth in these slab models, calculated for take-off angles ranging from $0°$ to $40°$, are plotted in Figure 10. We note that, relative to the case of no anomalous slab, the slab models produce travel-time advances of up to 5 sec, amplitude reduction by up to a factor of 2, and waveform broadening of up

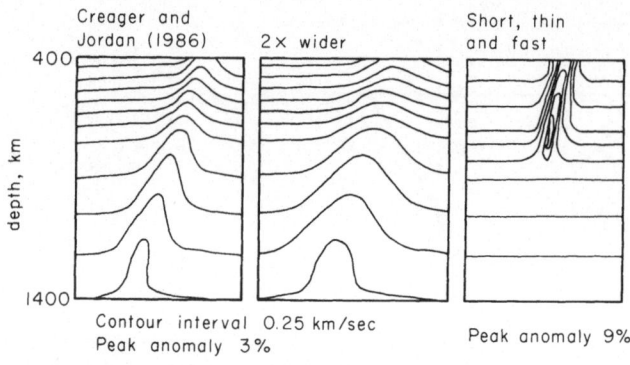

KURILE SLAB MODELS

Fig. 9.   Possible P-wave velocity structures for the Kurile-Kamchatka
          subduction zone. Velocities anomalies are contoured at 0.25
          km/sec intervals. Left model is taken from Creager and Jordan
          (1986). Center model is for a slab twice as wide as the Creager
          and Jordan model. Slab on the right ends at 700 km depth and is
          thinner and faster with a 9 percent velocity anomaly in its
          center.

to 5 sec. However, the amplitude reduction and waveform broadening are
less pronounced for the 2x broader and the short and fast slab models.
Figure 11 shows the corresponding WWSSN P-waves.

    Figure 12 shows a profile of long-period WWSSN records and
measurements of P waveform broadening for a 135 km deep Kurile earthquake.
The inner contour corresponds to broadening of 5 or more seconds and the
outer contour to broadening of 2 or more seconds. The broadening is
measured from the onset to the second zero crossing. Clearly, there is
distortion in the P waveforms at stations for which energy left the source
region along the plane of the slab. P waveform data are also plotted for
a 540 km deep Kurile earthquake in Figure 13. In comparison to the
previous example, there is no obvious effect of the slab. In addition, no

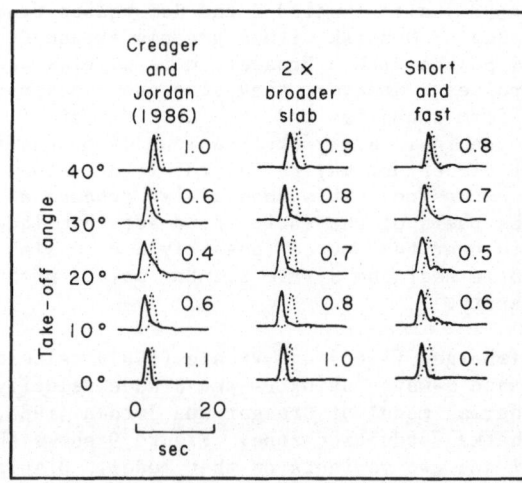

Fig. 10. Synthetic displacement seismograms calculated at teleseismic
          distances for take-off angles ranging from 0° to 40°. Source is
          at 515 km depth in the slab models shown in Figure 9. The
          dotted traces are for the case of no slab.

148

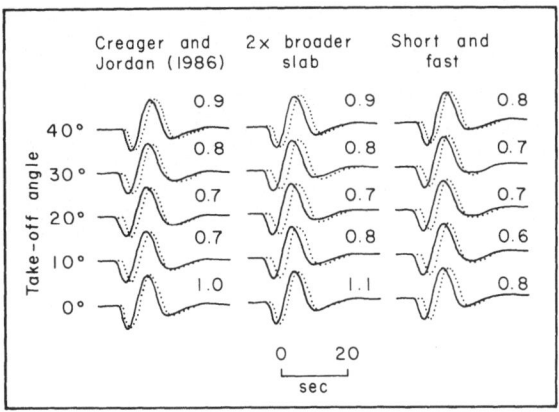

Fig. 11. Same as for Figure 10, with seismograms passed through a LPZ
(WWSSN) filter.

obvious difference is seen between record sections along the strike of the
Kurile slab and record sections perpendicular to the strike. This
suggests that either the slab becomes thicker with depth or that the slab
velocity anomaly is less than has been proposed by Creager and Jordan
(1986) below this event. However, this interpretation is complicated by
the observation that several other events at depths similar to the 135 km
deep event show less distortion in the P waveforms. Until we understand
the variations between events, these results remain uncertain.

If the amount of distortion is noticeable in the P-wave data, then
the distortions in S-waves are likely to be greater. Alternative slab
geometries for S-wave velocity structures are shown in Figure 14. The
"trim" slab is derived from Creager and Jordan's (1986) model. The
"flaring" slab is a simple modification of the trim slab that might be
produced by higher lower mantle viscosity where at 1200 km depth the slab
has a width of 500 km rather than 200 km. The short, thin, and
anomalously fast slab was constructed to produce approximately the same
travel-time anomaly as the trim slab. The related SH seismograms for the

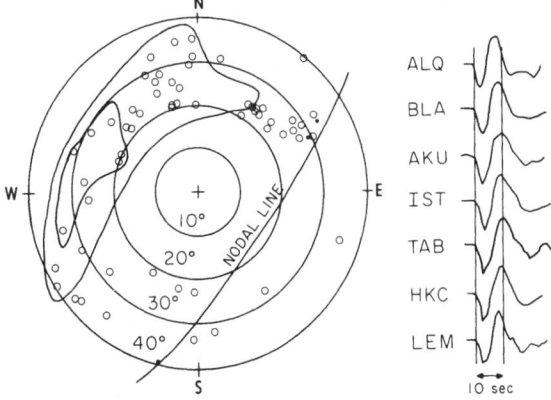

Fig. 12. Measurements of P waveform broadening plotted as a function of
take-off angle and azimuth from a 135 km deep Kurile earthquake.
Inner contour corresponds to 5 or more seconds broadening and
the outer contour to 2 or more seconds. Representative LPZ
(WWSSN) records are shown on the right.

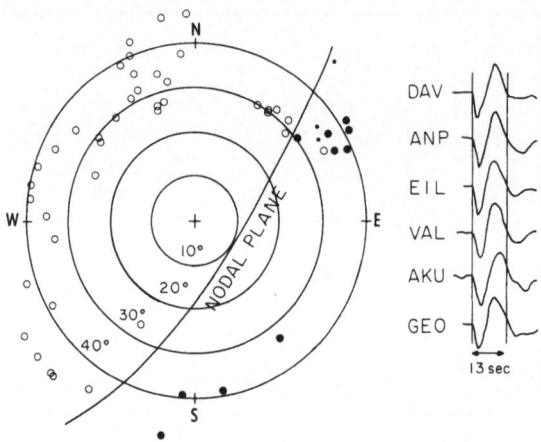

Fig. 13. Same as Figure 12 except for a 540 km deep Kurile earthquake.
Solid circles are compressional first motions. No systematic
broadening is observed.

models in Figure 14 are shown in Figure 15. For the trim slab, the
amplitudes of downdip seismograms are a factor of three lower than for the
case of no slab. The waveforms are broadened up to 20 sec by the non-
geometrical arrival late in the record. The flaring slab also produces a
long-period bump, although of smaller amplitude than the trim slab (factor
of 2.5). The thin fast slab has such a narrow waveguide that the arrival
begins emergently and the upswing is broadened. The amplitude anomaly is
a factor of 2.5 and the broadening is up to 10 sec. Vidale (1987)
concludes that the anomalous effects of the slab are spread over a range
of 20 - 30° in take-off angle in the downdip direction, that the flared
slab produces a wider band of fast arrivals than the trim slab, and also

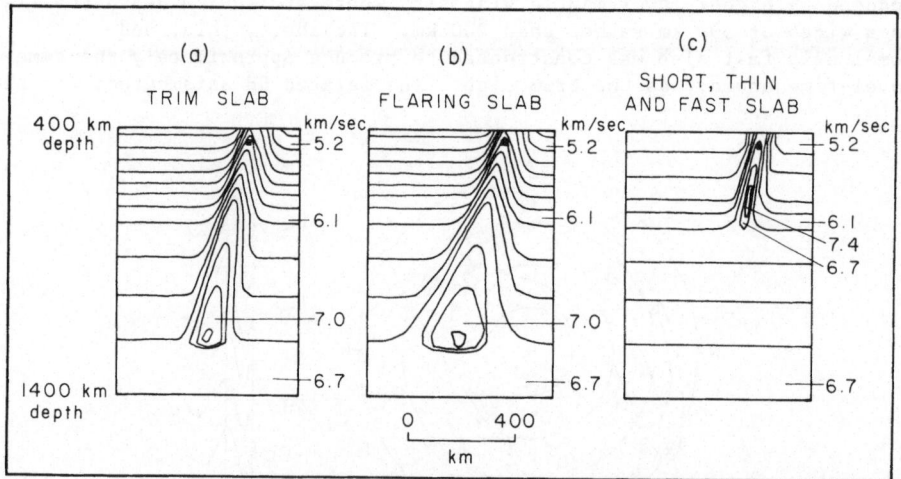

Fig. 14. Possible S-wave velocity structures for subducting slabs. The
trim slab has twice the percentage anomaly as the P-wave slab
model shown in Figure 9. The flaring slab model is for a slab
that broadens out as it sinks deeper into the mantle. The
short, thin and fast slab ends at 700 km depth, has a 20 percent
velocity anomaly in its center, and is only half as wide as the
trim slab. From Vidale (1987).

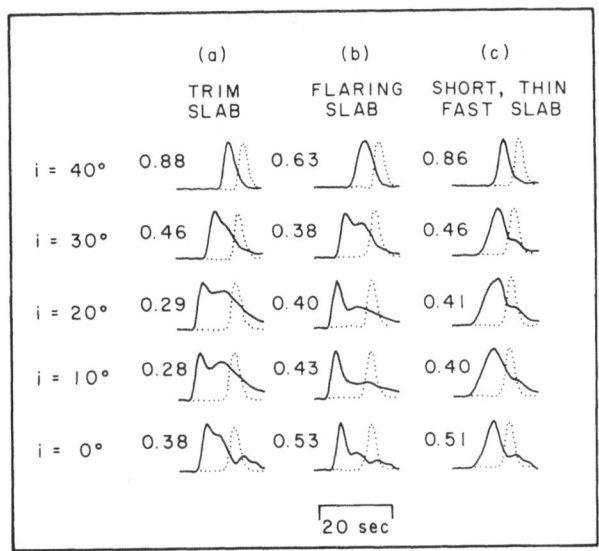

Fig. 15. SH waves distorted by propagation through the models of S-wave velocity structure shown in Figure 14. The SH waves are computed at teleseismic distances for take-off angles ranging from 10° to 40°. The dotted SH waves are calculated for a model with no slab velocity anomaly. Printed above each trace is the amplitude ratio of the SH waves affected by the slab to the SH waves with no slab. The take-off angle of i = 20° corresponds to rays taking off directly downdip. From Vidale (1987).

that the best discriminant between the slab models combines waveform, amplitude, and travel-time data.

Fully three-dimensional modelling, incorporating the frequency dependent effects of slab interaction, has been done by Cormier (1986) for the Kurile slab model. Figure 16 shows the synthesis of SH waves at 50° and variable azimuths from a 700 km deep earthquake located in a velocity structure derived from a model determined by Creager and Jordan (1986) from P travel-time residuals. Stations at azimuths on the dipping side of the slab (270° and 315°) arrive earlier than stations on the side away from the dip direction (45° and 90°). Stations at azimuths on the dipping side have been defocused by the high-velocity slab and are smaller in peak amplitude compared to stations at azimuths on the side away from the dipping direction. Stations on the dipping side exhibit a tail of energy or slab diffracted wave that persists for up to 10 or more seconds after the peak. Slab diffraction was found in the entire azimuthal range on the downdip side, up to the direction parallel to slab strike, and was stronger in a direction oblique to the normal to the strike of the slab. Cormier (1986) concludes that slab diffraction should persist up to the slab strike direction, reaching a maximum along strike, and then abruptly shutting off at azimuths away from the dip. This may be consistent with the range of the scatter in pulse widths seen along strike in the study by Beck and Lay (1986). Confining these comments to variations in azimuthal angle alone, however, is a simplification. The phenomena strongly depends, among other things, on the positions of the earthquake in the slab, the slab thickness, the vertical extent of the slab penetration, the lateral slab extent, and the local take-off angle.

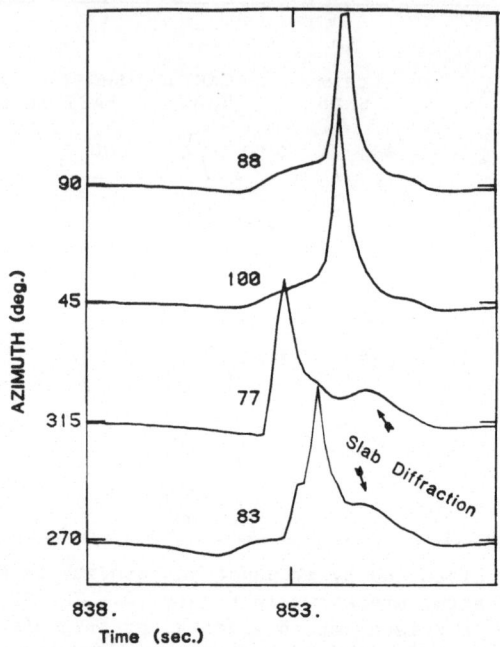

Fig. 16.   Synthetic SH waves at a range of 50 great circle degrees and
variable azimuths from a 700 km deep earthquake located in a 60°
dipping slab that penetrates to 1200 km. The numbers at the
left of each pulse measure peak amplitude.

DISCUSSION

Some general features of wave propagation in subducted slabs have
been highlighted. The anomalous features are of three kinds: (1) faster
material in the slab advances the arrivals that leave the source region in
the plane of the slab; (2) faster material in the slab tends to defocus
energy leaving the source region in the plane of the slab; and (3) the
waveforms of energy that leave the source region in the plane of the slab
can be distorted, with emergent first arrivals and a long-period, late-
diffracted arrival, which can make the full waveform appear as a broadened
pulse. Slab diffraction is akin to a head wave travelling along the
underside of the slab. It arrives late in the waveform because it is
sensitive to the slower velocity structure that surrounds the higher
velocity slab. In general, thinner slabs of a given length produce more
waveform broadening and smaller amplitudes for signals that leave the
source in the plane of the slab. It is also predicted that slab
diffraction will be observed in all azimuths on the downdip side of a
steeply dipping slab, strongest in a direction oblique to the normal to
the strike of the slab, evolving into a prominent secondary phase along
strike, and rapidly decaying on the side opposite to the dip. Slab
diffraction effects have been demonstrated for both P- and S-waves, but
are more pronounced in the case of S-waves. However, attenuation affects
S-waves more than P-waves, so that both types of body waves are needed to
study waveform distortions due to slabs.

There seems to be strong evidence for slab diffraction effects in P-
waves from central Aleutian earthquakes. The observations reported by
Vidale and Garcia-Gonzalez (1987) and Engdahl and Kind (1986), and the
modelling reported by Vidale and Helmberger (1986) suggest that the
central Aleutian slab is relatively thin, probably no more than 40 - 50 km

thick. Further development of the inversion techniques of Engdahl and Gubbins (1987) and more investigation of broad-band data and modelling is needed to confirm this result.

Thus far, the primary source of information on penetration of the Kurile slab beyond 670 km has been travel-time anomalies. However, slab length will still have some trade-off with lateral variations in slab shape and velocity, and the magnitude of velocity perturbation within the slab. A priori constraints of thermal models and phase changes can reduce, but not eliminate these trade-offs. The possibility of using amplitudes and waveforms as additional constraints on the modelling of slab structure has been demonstrated. Vidale and Helmberger (1986) presented evidence for slab diffraction effects along an azimuth perpendicular to the strike of the slab for a 135 km deep Kurile event, but not for an event at 540 km depth. Silver and Chan (1986) showed that a zone of multi-pathed arrivals can exist for S-waves from deep focus Kurile earthquakes observed along azimuths parallel to the strike of the slab. The duration of the predicted multi-pathing, however, is too small to account for the duration of the tail in broad-band waveforms of the S-wave. Cormier (1987), based on modelling studies, suggests that the mechanism of slab diffraction is sufficient to explain the duration of observed S pulses in many azimuths without additional ray theoretical arrivals. Multi-pathing along strike, however, is consistent with the observations of high variability in S waveform widths and complexity for azimuths close to the strike of the slab (e.g., Silver and Chan, 1986; Beck and Lay, 1986). Thus, there is conflicting evidence in amplitude and waveform data on the geometry of the Kurile slab near and below the 670 km discontinuity and the data may require independent lower mantle lateral heterogeneity as well.

CONCLUSIONS

Early work with body waveforms for slab modelling used ray trajectories and densities to estimate ray theoretical amplitudes and to identify multi-pathing, but could not calculate the frequency dependent effects of head waves and diffraction along slab boundaries and radiation having wavelengths on the order of or larger than slab width. Recently, however, the techniques of forward modelling of body waves in two- and three-dimensional structures have been systematically advanced to show these effects to be highly sensitive to slab width and its variations with depth. Application of these new techniques to the central Aleutian and Kurile subduction zones demonstrate that waveforms of P- and S-waves can provide important new constraints on slab structure that are independent of the information provided by their arrival times. In the central Aleutians, it has been possible to demonstrate that the slab is thinner than previously believed, and in the Kuriles, it has been shown that reconciliation of amplitude and waveform distortions of body waves with travel-time data is necessary to understand the geometry of slabs near and below the 670 km discontinuity.

We propose that much can be learned about slab structure by the deployment of high-resolution, broad-band arrays in regions where pronounced waveform changes produced by wave propagation in slabs have been observed or are predicted. Waveform behavior can be highly sensitive to variables such as the position of an earthquake within the slab, slab thickness, slab dip, vertical extent of slab penetration, and lateral slab extent. Hence, the need for high-quality, broad-band, digital data with adequate spatial coverage.

ACKNOWLEDGEMENTS

We thank K. Creager and G. Choy for helpful reviews. Partial support for this research was received through NSF grants EAR-8707578 (JEV) and EAR-8709456 (VFC). Contribution number 44 of the Charles Richter Seismological Laboratory and the Institute of Tectonics of the University of California, Santa Cruz.

REFERENCES

Ansell, J. H., and Gubbins, D., 1986, Anomalous high-frequency wave propagation from the Tonga-Kermadec seismic zone to New Zealand, Geophys. J. R. Astron. Soc., 85:93-106.

Beck, S. L., and Lay, T., 1986, Test of the lower mantle slab penetration hypothesis using broad-band S-waves, Geophys. Res. Lett., 13:1007-1010.

Beydoun, W. B., and Ben-Menahem, A., 1985, Range of validity of seismic ray and beam methods in general inhomogeneous media - I and II, Geophys. J. R. Astron. Soc., 82:207-262.

Burdick, L. J., Cole, D. M., Helmberger, D. V., Lay, T., and Wallace, T., 1981, Effective source functions from local surface measurements, Woodward-Clyde Consultants, report number WWCP-R-82-01.

Chiu, J.-M., Isacks, B. L., and Cardwell, R. K., 1985, Propagation of high-frequency seismic waves inside the subducted lithosphere from intermediate-depth earthquakes recorded in the Vanatu arc, J. Geophys. Res., 90:12,741-12,754.

Cormier, V. F., 1986, Focusing and defocusing of S, SKS, and ScS waves by a descending lithospheric slab (abstract), EOS, Trans. Am. Geophys. Union, 44:1099.

Cormier, V. F., 1987, Slab diffracted S-waves (abstract), IASPEI Abstracts, Vol. 1, p 308, XIX General Assembly, IUGG.

Creager, K. C., and Jordan, T. H., 1984, Slab penetration into the lower mantle, J. Geophys. Res., 89:3031-3049.

Creager, K. C., and Jordan, T. H., 1986, Slab penetration into the lower mantle beneath the Mariana and other island arcs of the north west Pacific, J. Geophys. Res., 91:3573-3590.

Davies, O., and Julian, B., 1972, A study of short-period P-wave signals from Longshot, Geophys. J. R. Astron. Soc., 29:185-202.

Engdahl, E. R., and Gubbins, D., 1987, Simultaneous travel-time inversion for earthquake location and subduction zone structure in the central Aleutian Islands, J. Geophys. Res., 92:13,855-13,862.

Engdahl, E. R., and Kind, R., 1986, Interpretation of broad-band seismograms from central Aleutian earthquakes, Ann. Geophys., 4:233-240.

Fukao, Y., Kanjo, K., and Nakamura, I., 1978, Deep seismic zone as an upper mantle reflector of body waves, Nature, 272:606-608.

Matsuzawa, T., Umino, N., Hasegawa, A., and Takagi, A., 1986, Upper mantle velocity structure estimated from PS-converted wave beneath the north-eastern Japan arc, Geophys. J. R. Astron. Soc., 86:767-787.

Minear, J. W., and Toksoz, M. N., 1973, Thermal regime of a downgoing slab and new global tectonics, J. Geophys. Res., 78:6009-6020.

Owens, T. J., and Crosson, R. S., 1986, Teleseismic P-waveform modelling for deep structure in a subduction zone environment (abstract), EOS, Trans. Am. Geophys. Union, 67:1116.

Silver, P. G., and Chan, W. W., 1986, Observations of body wave multi-pathing from broad band seismograms: evidence for lower-mantle slab penetration beneath the Sea of Okhotsk, J. Geophys. Res., 91:13,787-13,802.

Sleep, N. H., 1973,, Teleseismic P-wave transmission through slabs, Bull. Seism. Soc. Am., 63:1349-1373.

Vidale, J. E., and Helmberger, D. F., 1986, Waveform effects of a high-velocity, subducted slab (abstract), EOS, Trans. Am. Geophys. Union, 44:1114.

Vidale, J. E., 1987, Waveform effects of a high-velocity, subducted slab, Geophys. Res. Lett., 14:542-545.

Vidale, J. E., and Garcia-Gonzalez, D., 1987, Observations of body wave distortion caused by subducting slabs (abstract), EOS, Trans. Am. Geophys. Union, 68:1379.

White, B. S., Burridge, R., Noriss, A., and Bayliss, A., 1987, Some remarks on the Gaussian beam summation method, Geophys. J. R. Astron. Soc., 89:579-646.

Witte, D., 1987, Numerical simulations of seismic waves distorted by subducting slabs (abstract), EOS, Trans. Am. Geophys. Union, 16:352.

THEORETICAL BACKGROUND FOR THE INVERSION OF SEISMIC

WAVEFORMS, INCLUDING ELASTICITY AND ATTENUATION (*)

Albert Tarantola

Institut de Physique du Globe
4, place Jussieu
F-75252 Paris Cedex 05

ABSTRACT

To account for elastic and attenuating effects in the elastic wave equation, the stress-strain relationship can be defined through a general, anisotropic, causal relaxation function $\psi^{ijkl}(\mathbf{x},\tau)$. Then, the wave equation operator is not necessarily symmetric ('self-adjoint'), but the reciprocity property is still satisfied. The representation theorem contains a term proportional to the history of strain. The dual problem consists in solving the wave equation with final time conditions and an anti-causal relaxation function. The problem of interpretation of seismic waveforms can be set as the nonlinear inverse problem of estimating the matter density $\rho(\mathbf{x})$ and all the functions $\psi^{ijkl}(\mathbf{x},\tau)$. This inverse problem can be solved using iterative gradient methods, each iteration consisting in the propagation of the actual source in the current medium, with causal attenuation, the propagation of the residuals — acting as if they were sources — backwards in time, with anti-causal attenuation, and the correlation of the two wavefields thus obtained.

1: INTRODUCTION

The problem of interpretation of seismic waveforms can be set as the problem of obtaining the Earth model which bests predicts the actually observed seismograms. This opens two questions: i) given an earth model, how to solve the forward problem of predicting seismograms?, and ii) how to solve the inverse problem of obtaining the optimum earth model?

The tools for predicting seismograms are the elastic wave equation and the numerical methods developed to obtain solutions, as for instance finite-difference approximations to derivatives. Finite-difference approximations to the wave equation have the advantages of having enough flexibility to be almost model-independent and of accounting, in principle, for all sorts of waves. The are expensive, but nicely adaptable to the newly emerging class of massively parallel computers.

The inverse problem is essentially an optimization problem in a functional space. Difficulties arise because the problem is large sized, and the functional to be minimized is non-quadratic. The modest capabilities of present day computers prevent the use of true nonqua-

---

(*) Reproduced with permission from **Pageoph**, 1988, special issue on **Scattering and Attenuation of Seismic Waves.**

dratic methods of optimization, like Monte Carlo methods. Gradient methods can be used which lead to elegant results.

In this paper I first review the mathematics of the forward problem that are useful for the inverse problem: wave equation, Green's function, and representation theorem. Second, I review the mathematics of functional least squares. Finally, I give the solution to the seismic inverse problem, with more generality than in my previous papers, because here I take into account attenuation.

In the problem of interpretation of seismic reflection data the model space has many degrees of freedom (millions to billions), and methods of inversion based on a naïve use of least-squares formulas do not work. In particular, matrix algebra must not be used, and partial (or Fréchet) derivatives of data with respect to model parameters should not be computed. Much work has to be done analytically, in order to interpret the final formulas of least squares as operations involving only wave propagations, and no linear algebra computations.

For developping a theory for inversion including attenuation we must first choose a model. If in the perfectly elastic approximation it is clear that density $\rho(\mathbf{x})$ and elastic stiffnesses $c^{ijkl}(\mathbf{x})$ (or related quatities) are the right earth parameters to choose, for a more realistic approximation including attenuation, the choice is not so clear, as many models for attenuation exist. I take here the most optimistic point of view: that data sets exist which contain enough information for not needing a particular model of attenuation, and that the more general parameterization can be chosen: an arbitrary relaxation function $\psi^{ijkl}(\mathbf{x},\tau)$. The goal of inversion is then to obtain the density $\rho(\mathbf{x})$ and the functions $\psi^{ijkl}(\mathbf{x},\tau)$. Of course, some constraints have to be imposed to the relaxation functions, as for instance causality and symmetry conditions. If necessary, some soft constraints can also be imposed, as for instance spatial or temporal smoothness.

Let $u^i(\mathbf{x},t)$ be the i-th component of displacement at point $\mathbf{x}$ and time t . If $\mathbf{x}_r$ (r=1,2,...) denote the receiver locations, a possible criterion of goodness of fit between observed and computed seismograms is the minimization of

$$S = \sum_r \int_{t_0}^{t_1} dt \left| u^i(\mathbf{x}_r,t)_{obs} - u^i(\mathbf{x}_r,t)_{cal} \right| , \qquad (1a)$$

where 'obs' and 'cal' respectively represent the observed and the calculated displacements from a given earth model. Although results given by this criterion are fairly good, they are dificult to obtain, and this criterion is replaced by the least-squares criterion of minimization of

$$S = \sum_r \int_{t_0}^{t_1} dt \left[u^i(\mathbf{x}_r,t)_{obs} - u^i(\mathbf{x}_r,t)_{cal} \right]^2 , \qquad (1b)$$

which gives less robust results but computations which are manageable with present-day computers.

In Section 2 the rate-of-relaxation function is defined, which will be at the center of our mathematical developments. Section 3 rapidly reviews the fundamental equation: the elastic wave equation with attenuation. In Section 4 I recall the definition of the transposed (and adjoint) of an operator, and in Section 5 the transposed of the wave equation operator is given. The Green function is introduced in Section 6, leading to the general representation theorems of Section 7, which are independent of the recipricity relations shown in Section 8. Section 9 reviews rapidly the Born approximation, useful for computing the gradient of the least-squares misfit function. A very formal version of the least-squares theory in functional spaces is developed in Section 10, and is applied to the problem of interpretation of seismic waveforms in Section 11. The formula allowing to inverse waveforms for obtaining the rate-of-relaxation function is, to my knowledge, original. The last Section comments the results obtained.

2: THE RATE-OF-RELAXATION FUNCTION

The most general linear relationship between stress , $\sigma^{ij}(\mathbf{x},t)$, and strain, $\epsilon^{ij}(\mathbf{x},t)$, can be described using a kernel $\Psi_1^{ijkl}(\mathbf{x},t;\mathbf{x}',t')$:

$$\sigma^{ij}(\mathbf{x},t) = \int_V dV(\mathbf{x}') \int_{-\infty}^{+\infty} dt' \; \Psi_1^{ijkl}(\mathbf{x},t;\mathbf{x}',t') \; \epsilon^{kl}(\mathbf{x}',t') \; , \tag{2a}$$

which has to be causal, must have some symmetries, and may be a distribution (containing the delta 'function' and/or its derivatives). This is too general for most seismic purposes. Assuming that the stress-strain relationship is local,

$$\Psi_1^{ijkl}(\mathbf{x},t;\mathbf{x}',t') = \Psi_0^{ijkl}(\mathbf{x};t,t') \; \delta(\mathbf{x}-\mathbf{x}') \; ,$$

gives

$$\sigma^{ij}(\mathbf{x},t) = \int_{-\infty}^{+\infty} dt' \; \Psi_0^{ijkl}(\mathbf{x};t,t') \; \epsilon^{kl}(\mathbf{x},t') \; . \tag{2b}$$

If the medium properties do not depend on time,

$$\Psi_0^{ijkl}(\mathbf{x};t,t') = \Psi^{ijkl}(\mathbf{x},t-t') \; ,$$

and

$$\sigma^{ij}(\mathbf{x},t) = \int_{-\infty}^{+\infty} dt' \; \Psi^{ijkl}(\mathbf{x},t-t') \; \epsilon^{kl}(\mathbf{x},t') \; . \tag{2c}$$

The function $\Psi^{ijkl}(\mathbf{x},\tau)$ has to satisfy causality,

$$\Psi^{ijkl}(\mathbf{x},\tau) = 0 \; \text{ for } \; \tau < 0 \; , \tag{3}$$

and is assumed to have the symmetries:

$$\Psi^{ijkl}(\mathbf{x},\tau) = \Psi^{jikl}(\mathbf{x},\tau) = \Psi^{klij}(\mathbf{x},\tau) \; , \tag{4}$$

(from where it follows $\Psi^{ijkl}(\mathbf{x},\tau) = \Psi^{ijlk}(\mathbf{x},\tau) = \Psi^{jikl}(\mathbf{x},\tau)$). Notice that the causality property allows to write (2c) as

$$\sigma^{ij}(\mathbf{x},t) = \int_{-\infty}^{t^+} dt' \; \Psi^{ijkl}(\mathbf{x},t-t') \; \epsilon^{kl}(\mathbf{x},t') \; .$$

Instead of the function $\Psi^{ijkl}(\mathbf{x},\tau)$ it is customary to use the creep function $\phi^{ijkl}(\mathbf{x},\tau)$ defined by

$$\epsilon^{ij}(\mathbf{x},t) = \int_{-\infty}^{t} dt' \; \phi^{ijkl}(\mathbf{x},t-t') \; \dot{\sigma}^{kl}(\mathbf{x},t') \; , \tag{5}$$

or its inverse, the relaxation function $\psi^{ijkl}(\mathbf{x},\tau)$ defined by

$$\sigma^{ij}(\mathbf{x},t) = \int_{-\infty}^{t} dt' \; \psi^{ijkl}(\mathbf{x},t-t') \; \dot{\epsilon}^{kl}(\mathbf{x},t') \; , \tag{6}$$

where a dot denotes time differentiation. As we have

$$\Psi^{ijkl}(\mathbf{x},\tau) = \dot{\psi}^{ijkl}(\mathbf{x},\tau) \; , \tag{7}$$

$\Psi^{ijkl}(\mathbf{x},\tau)$ can be named the rate-of-relaxation function.

Example 1: Perfect elasticity. Choosing

$$\Psi^{ijkl}(\mathbf{x},\tau) = c^{ijkl}(\mathbf{x}) \, \delta(\tau) , \tag{8}$$

where $\delta(.)$ is the delta 'function', gives Hooke's law

$$\sigma^{ij}(\mathbf{x},t) = c^{ijkl}(\mathbf{x}) \, \epsilon^{kl}(\mathbf{x},t) . \tag{9}$$

For isotropic media,

$$c^{ijkl}(\mathbf{x}) = \lambda_e(\mathbf{x}) \, \delta^{ij} \, \delta^{kl} + \mu_e(\mathbf{x}) \, (\delta^{ik} \, \delta^{jl} + \delta^{il} \, \delta^{jk}) , \tag{10}$$

where $\lambda_e(\mathbf{x})$ and $\mu_e(\mathbf{x})$ are the (elastic) parameters of Lamé, related with the elastic bulk modulus by

$$3\kappa_e(\mathbf{x}) = 3\lambda_e(\mathbf{x}) + 2\mu_e(\mathbf{x}) . \tag{11}$$

As a first approximation, for usual rocks, $\lambda_e \simeq \mu_e$.

Example 2: Elasticity with viscosity. Choosing

$$\Psi^{ijkl}(\mathbf{x},\tau) = c^{ijkl}(\mathbf{x}) \, \delta(\tau) + d^{ijkl}(\mathbf{x}) \, \dot{\delta}(\tau) , \tag{12}$$

gives the Kelvin-Voigt law

$$\sigma^{ij}(\mathbf{x},t) = c^{ijkl}(\mathbf{x}) \, \epsilon^{kl}(\mathbf{x},t) + d^{ijkl}(\mathbf{x}) \, \dot{\epsilon}^{kl}(\mathbf{x},t) , \tag{13}$$

which corresponds, in a 1D problem, to a perfectly elastic spring and a perfectly viscuous dashpot in parallel. For isotropic viscosity,

$$d^{ijkl}(\mathbf{x}) = \lambda_v(\mathbf{x}) \, \delta^{ij} \, \delta^{kl} + \mu_v(\mathbf{x}) \, (\delta^{ik} \, \delta^{jl} + \delta^{il} \, \delta^{jk}) . \tag{14}$$

The viscous bulk modulus is defined by

$$3\kappa_v(\mathbf{x}) = 3\lambda_v(\mathbf{x}) + 2\mu_v(\mathbf{x}) . \tag{15}$$

As a first approximation, for usual rocks, $3\kappa_v = 3\lambda_v + 2\mu_v \simeq 0$.

Example 3: Constant Q. In a one-dimensional example, Kjartansson (1979) shows that the rate-of-relaxation function

$$\Psi(\tau) \propto \frac{1}{\tau^{1+2\gamma}} \qquad \text{for } \tau > 0 \tag{16}$$

$$\Psi(\tau) = 0 \qquad \text{for } \tau \leq 0 \tag{17}$$

implies a quality factor Q strictly independent on the frequency, and, for a sufficiently small value of the positive parameter γ , fits most of seismic data.

3: THE FUNDAMENTAL EQUATION

Let us be interested in the description of elastic waves propagating inside a volume $\mathbf{V}$, surrounded by a surface $\mathbf{S}$. Points inside $\mathbf{V}$ will be denoted $\mathbf{x}$, $\mathbf{x}'$,... while points on $\mathbf{S}$ will be denoted ξ, ξ',... From the fundamental laws of physics we can show (e.g. Dautray and Lions, 1984) that if we impose to a medium with matter density $\rho(\mathbf{x})$ and rate-of-relaxation function $\Psi^{ijkl}(\mathbf{x},\tau)$, a volume density of force $\phi^i(\mathbf{x},t)$, a moment density $M^{ij}(\mathbf{x},t)$, and a surface traction $\tau^i(\xi,t)$, then the displacement field $u^i(\mathbf{x},t)$ satisfies at any point inside $\mathbf{V}$ the relationship

$$\rho(\mathbf{x})\ \frac{\partial^2 u^i}{\partial t^2}(\mathbf{x},t)\ -\ \frac{\partial \sigma^{ij}}{\partial x^j}(\mathbf{x},t)\ =\ \phi^i(\mathbf{x},t) \tag{18}$$

where

$$\sigma^{ij}(\mathbf{x},t)\ =\ M^{ij}(\mathbf{x},t)\ +\ \int_{-\infty}^{+\infty} dt'\ \Psi^{ijkl}(\mathbf{x},t-t')\ \frac{\partial u^k}{\partial x^l}(\mathbf{x},t')\ , \tag{19}$$

and, at the surface,

$$n^j(\xi)\ \sigma^{ij}(\xi,t)\ =\ \tau^i(\xi,t)\ . \tag{20}$$

Notice that $\sigma^{ij}(\mathbf{x},t)$ is the total stress (internal plus external origine) and that in the equations of the previous section the external stress $M^{ij}(\mathbf{x},t)$ was implicitly assumed to be zero.

4: MATHEMATICAL PRELIMINARY: SYMMETRIC AND SELF-ADJOINT OPERATORS

Let $\mathbf{U}$ be a linear space, whose elements, denoted $\mathbf{u}$, are named vectors. For instance each element of $\mathbf{U}$ may be a displacement field $u^i(\mathbf{x},t)$. A linear form over $\mathbf{U}$ is a linear application from $\mathbf{U}$ into a space of scalars (identical to the real numbers of $\mathbf{R}$ excepted in that the scalars may have a physical dimension). We will say that a linear space $\hat{\mathbf{U}}$ is a dual of $\mathbf{U}$ if each element of $\hat{\mathbf{U}}$ defines a linear form over $\mathbf{U}$ (I prefer that definition to the traditional definition of mathematicians where the dual is the space of all linear forms). Let $\hat{\mathbf{u}}_0 \in \hat{\mathbf{U}}$. The scalar associated to any $\mathbf{u} \in \mathbf{U}$ by $\hat{\mathbf{u}}_0$ is denoted by $\langle \hat{\mathbf{u}}_0\ ,\ \mathbf{u} \rangle_{\mathbf{U}}$. Alternatively, each element $\mathbf{u}_0 \in \mathbf{U}$ defines a linear form over $\hat{\mathbf{U}}$ through

$$\langle \mathbf{u}_0\ ,\ \hat{\mathbf{u}} \rangle_{\hat{\mathbf{U}}}\ =\ \langle\ \hat{\mathbf{u}}\ ,\ \mathbf{u}_0 \rangle_{\mathbf{U}}\ . \tag{21}$$

Let $\mathbf{U}$ and Φ be two linear spaces, and $\mathbf{L}$ a linear operator from $\mathbf{U}$ into Φ. For instance, $\mathbf{L}$ may be the wave equation operator defined in Section 5. An operator $\mathbf{L}^t$ mapping $\hat{\Phi}$ into $\hat{\mathbf{U}}$ will be named the transposed of $\mathbf{L}$ if for any $\hat{\phi} \in \hat{\Phi}$ and any $\mathbf{u} \in \mathbf{U}$

$$\langle\ \hat{\phi}\ ,\ \mathbf{L}\,\mathbf{u} \rangle_{\Phi}\ =\ \langle\ \mathbf{L}^t\,\hat{\phi}\ ,\ \mathbf{u} \rangle_{\mathbf{U}}\ . \tag{22}$$

This definition of transposed operators may be compared with the definition of adjoint operators. The adjoint of an operator may only be defined if the spaces into consideration have an scalar product. Let for instance the linear operator $\mathbf{L}$ map the linear space $\mathbf{U}$ into the linear space Φ, and let $(\mathbf{u}_1\ ,\ \mathbf{u}_2)_{\mathbf{U}}$ and $(\phi_1\ ,\ \phi_2)_{\Phi}$ denote respectively the scalar products over $\mathbf{U}$ and Φ. An operator $\mathbf{L}^*$ mapping Φ into $\mathbf{U}$ will be named the adjoint of $\mathbf{L}$ if for any ϕ and $\mathbf{u}$,

$$(\phi\ ,\ \mathbf{L}\,\mathbf{u})_{\Phi}\ =\ (\mathbf{L}^*\,\phi\ ,\ \mathbf{u})_{\mathbf{U}}\ . \tag{23}$$

Details on the mathematical definition of transposed and adjoint operators may be found for instance in Taylor and Lay (1980).

Transposed operators have an important and very simple property which follows immediately from the definition of the kernel of an operator: if two operators are mutually transposed, their integral kernels are identical, excepted in that their variables are 'transposed'.

It can easily be seen that if $\mathbf{W_U}$ and $\mathbf{W_\Phi}$ are the operators defining the scalar products over $\mathbf{U}$ and $\mathbf{\Phi}$ respectively:

$$(\mathbf{u}_1 , \mathbf{u}_0)_U = \langle \mathbf{W_U} \, \mathbf{u}_1 , \mathbf{u}_0 \rangle_U \tag{24a}$$

$$(\phi_1 , \phi_0)_\Phi = \langle \mathbf{W_\Phi} \, \phi_1 , \phi_0 \rangle_\Phi , \tag{24b}$$

then transposed and adjoint are related by

$$\mathbf{L}^* = \mathbf{W_U}^{-1} \, \mathbf{L}^t \, \mathbf{W_\Phi} . \tag{25}$$

By definition, if $\mathbf{L}$ maps $\mathbf{U}$ into $\mathbf{\Phi}$, then $\mathbf{L}^t$ maps $\hat{\mathbf{\Phi}}$ into $\hat{\mathbf{U}}$. In the particular case where $\mathbf{\Phi}$ can be identified with a dual of $\mathbf{U}$, both $\mathbf{L}$ and $\mathbf{L}^t$ map $\mathbf{U}$ into $\hat{\mathbf{U}}$. In that case, the definition of transpose (22) can be rewritten

$$\langle \overleftarrow{\mathbf{u}} , \mathbf{L} \overrightarrow{\mathbf{u}} \rangle_{\hat{U}} = \langle \mathbf{L}^t \overleftarrow{\mathbf{u}} , \overrightarrow{\mathbf{u}} \rangle_U , \tag{26}$$

with $\overrightarrow{\mathbf{u}}$ and $\overleftarrow{\mathbf{u}}$ elements of $\mathbf{U}$. If, in that case, $\mathbf{L}$ and $\mathbf{L}^t$ have the same domain of definition, and if for any $\overrightarrow{\mathbf{u}}$ and $\overleftarrow{\mathbf{u}}$ we can replace in (26) $\mathbf{L}^t$ by $\mathbf{L}$:

$$\langle \overleftarrow{\mathbf{u}} , \mathbf{L} \overrightarrow{\mathbf{u}} \rangle_{\hat{U}} = \langle \mathbf{L} \overleftarrow{\mathbf{u}} , \overrightarrow{\mathbf{u}} \rangle_U , \tag{27}$$

we say that $\mathbf{L}$ is symmetric , and we write

$$\mathbf{L} = \mathbf{L}^t . \tag{28}$$

By definition, if $\mathbf{L}$ maps $\mathbf{U}$ into itself, then $\mathbf{L}^*$ also maps $\mathbf{U}$ into itself. In that case, the definition of adjoint (23) can be rewritten

$$(\overleftarrow{\mathbf{u}} , \mathbf{L} \overrightarrow{\mathbf{u}})_U = (\mathbf{L}^* \overleftarrow{\mathbf{u}} , \overrightarrow{\mathbf{u}})_U . \tag{29}$$

If, in that case, $\mathbf{L}$ and $\mathbf{L}^*$ have the same domain of definition, and if for any $\overrightarrow{\mathbf{u}}$ and $\overleftarrow{\mathbf{u}}$ we can replace in (29) $\mathbf{L}^*$ by $\mathbf{L}$:

$$(\overleftarrow{\mathbf{u}} , \mathbf{L} \overrightarrow{\mathbf{u}})_U = (\mathbf{L} \overleftarrow{\mathbf{u}} , \overrightarrow{\mathbf{u}})_U , \tag{30}$$

we say that $\mathbf{L}$ is self-adjoint, and we write

$$\mathbf{L} = \mathbf{L}^* . \tag{31}$$

In the problem of wave propagation later studied, there is no natural scalar product, so the general concept of transposed operator will be preferred to the more particular concept of adjoint operator.

Practically, when we have an operator $\mathbf{L}$, we can use the two follwing rules to obtain the formal transposed: i) a derivative operator is anti-symmetric, i.e., its transposed equals its opposite. ii) if $\mathbf{L}$ is an integral operator, we can introduce its integral kernel; the transposed operator is also an integral operator and the kernel of the transposed operator is the transposed of the original kernel.

Once we have the formal transposed, we compute the difference $\langle \hat{\phi} , \mathbf{L}\, \mathbf{u} \rangle_{\Phi} - \langle \mathbf{L}^t\, \hat{\phi} , \mathbf{u} \rangle_U$, and if necessary, we impose restrictions on the spaces U and $\hat{\Phi}$ for this difference to vanish. These restrictions are named dual conditions, and we will see some examples in the next section. Typically, for a differential operator, the dual conditions are boundary conditions. For an integral operator, there are no dual conditions to impose if the integrals defining the linear forms have the same bounds as the integrals defining the operators. I not, we have to impose conditions on the functions outside the bounds.

5: THE WAVE EQUATION OPERATOR AND ITS TRANSPOSE

Let U be the space of all conceivable displacement fields $\mathbf{u} = \{u^i(\mathbf{x},t)\}$, and Φ the space of all conveivable source fields $\phi = \{\phi^i(\mathbf{x},t)\}$. The wave equation operator (with attenuation) L is the operator L mapping U into Φ defined by

$$(\mathbf{L}\, \mathbf{u})^i(\mathbf{x},t) = \rho(\mathbf{x}) \frac{\partial^2 u^i}{\partial t^2}(\mathbf{x},t) - \frac{\partial}{\partial x^j} \int_{-\infty}^{+\infty} dt' \; \overrightarrow{\Psi}_0^{\,jikl}(\mathbf{x};t,t') \; \frac{\partial u^k}{\partial x^l}(\mathbf{x},t') , \tag{32}$$

where $\overrightarrow{\Psi}_0^{\,jikl}(\mathbf{x};t,t')$ is causal (the right arrow is to distinguish this function from an anti-causal function to be introduced later). This allows to write the wave equation with attenuation defined by equations (18)-(19) as:

$$\mathbf{L}\, \mathbf{u} = \hat{\phi} , \tag{33}$$

where

$$\hat{\phi}^i(\mathbf{x},t) = \phi^i(\mathbf{x},t) + \frac{\partial M^{ij}}{\partial x^j}(\mathbf{x},t) . \tag{34}$$

In the definition (32) we choose the function $\overrightarrow{\Psi}_0^{\,jikl}(\mathbf{x};t,t')$ defined in (2b), rather than the rate-of-relaxation function $\overrightarrow{\Psi}^{\,jikl}(\mathbf{x},\tau)$ defined in (2c) because it is not more difficult to handle and will later help to clarify the properties of the Green function. To simplify notations, the index $(_0)$ in $\overrightarrow{\Psi}_0^{\,jikl}(\mathbf{x};t,t')$ will be dropped in what follows.

For a given $\phi \in \Phi$ we can define, for any $\mathbf{u} \in U$, the scalar

$$\langle \phi , \mathbf{u} \rangle_U = \int_V dV(\mathbf{x}) \int_{t_0}^{t_1} dt \; \phi^i(\mathbf{x},t) \; u^i(\mathbf{x},t) , \tag{35}$$

which has the dimension of an action (energy x time). As each element of Φ defines a linear form over U, we can say that Φ is a dual of U. Alternatively, for a given $\mathbf{u} \in U$ we can define, for any $\phi \in \Phi$, the action

$$\langle \mathbf{u} , \phi \rangle_{\Phi} = \langle \phi , \mathbf{u} \rangle_U = \int_V dV(\mathbf{x}) \int_{t_0}^{t_1} dt \; \phi^i(\mathbf{x},t) \; u_0^{\,i}(\mathbf{x},t) . \tag{36}$$

The spaces U and Φ are mutually duals.

Equations (32)-(33) define the linear operator L, mapping U into Φ. As U and Φ are mutually duals, the transposed operator L^t also maps U into Φ. Let us demonstrate that the transposed of L is the operator L^t defined by

$$(\mathbf{L}^t\, \mathbf{u})^i(\mathbf{x},t) = \rho(\mathbf{x}) \frac{\partial^2 u^i}{\partial t^2}(\mathbf{x},t) - \frac{\partial}{\partial x^j} \int_{-\infty}^{+\infty} dt' \; \overleftarrow{\Psi}_0^{\,ijkl}(\mathbf{x};t,t') \; \frac{\partial u^k}{\partial x^l}(\mathbf{x},t') , \tag{37}$$

where $\overleftarrow{\Psi}_0^{\,ijkl}(\mathbf{x};t,t')$ is the anti-causal rate-of-relaxation function defined by

$$\overleftarrow{\Psi}_0{}^{ijkl}(\mathbf{x};t,t') = \overrightarrow{\Psi}_0{}^{ijkl}(\mathbf{x};t',t) , \qquad (38)$$

which means that the operator $\mathbf{L}^t$ corresponds to a wave equation with negative attenuation.

For we have to verify that (26) is satisfied. We have

$$\langle \overleftarrow{\mathbf{u}} , \mathbf{L} \overrightarrow{\mathbf{u}} \rangle_\Phi - \langle \mathbf{L}^t \overleftarrow{\mathbf{u}} , \overrightarrow{\mathbf{u}} \rangle_U =$$

$$= \int_V dV(\mathbf{x}) \int_{t_0}^{t_1} dt\, \overleftarrow{u}{}^i(\mathbf{x},t)\, (\mathbf{L}\overrightarrow{\mathbf{u}})^i(\mathbf{x},t) - \int_V dV(\mathbf{x}) \int_{t_0}^{t_1} dt\, (\mathbf{L}^t\overleftarrow{\mathbf{u}})^i(\mathbf{x},t)\, \overrightarrow{u}{}^i(\mathbf{x},t) .$$

Inserting (32) and (37). integrating per parts, and using the divergence theorem gives (see Appendix A):

$$\langle \overleftarrow{\mathbf{u}} , \mathbf{L} \overrightarrow{\mathbf{u}} \rangle_\Phi - \langle \mathbf{L}^t \overleftarrow{\mathbf{u}} , \overrightarrow{\mathbf{u}} \rangle_U =$$

$$+ \int_V dV(\mathbf{x})\, (\overleftarrow{u}{}^i(\mathbf{x},t)\, \overrightarrow{p}{}^i(\mathbf{x},t) - \overleftarrow{p}{}^i(\mathbf{x},t)\, \overrightarrow{u}{}^i(\mathbf{x},t)) \Big|_{t=t_0}^{t=t_1}$$

$$+ \int_S dS(\xi) \int_{t_0}^{t_1} dt\, (\overleftarrow{\tau}{}^i(\xi,t)\, \overrightarrow{u}{}^i(\xi,t) - \overleftarrow{u}{}^i(\xi,t)\, \overrightarrow{\tau}{}^i(\mathbf{x},t))$$

$$+ \int_V dV(\mathbf{x}) \int_{t_0}^{t_1} dt\, (\overleftarrow{\epsilon}{}^{ij}(\mathbf{x},t)\, \overrightarrow{\Sigma}{}^{ij}(\mathbf{x},t) - \overleftarrow{\Sigma}{}^{ij}(\mathbf{x},t)\, \overrightarrow{\epsilon}{}^{ij}(\mathbf{x},t)) , \qquad (39)$$

where

$$\overrightarrow{\Sigma}{}^{ij}(\mathbf{x},t) = \int_{-\infty}^{t_0} dt'\, \overrightarrow{\Psi}{}^{ijkl}(\mathbf{x};t,t')\, \overrightarrow{\epsilon}{}^{kl}(\mathbf{x},t') \qquad (40a)$$

$$\overleftarrow{\Sigma}{}^{ij}(\mathbf{x},t) = \int_{t_1}^{+\infty} dt'\, \overleftarrow{\Psi}{}^{ijkl}(\mathbf{x};t,t')\, \overleftarrow{\epsilon}{}^{kl}(\mathbf{x},t') ,$$

and, where for each sense of the arrows $\rightarrow$ and $\leftarrow$,

$$p^i(\mathbf{x},t) = \rho(\mathbf{x})\, \frac{\partial u^i}{\partial t}(\mathbf{x},t) \qquad (40b)$$

$$\epsilon^{ij}(\mathbf{x},t) = \frac{1}{2}\left[\frac{\partial u^i}{\partial x^j}(\mathbf{x},t) + \frac{\partial u^j}{\partial x^i}(\mathbf{x},t) \right] \qquad (40c)$$

$$\sigma^{ij}(\mathbf{x},t) = \int_{-\infty}^{+\infty} dt'\, \Psi^{ijkl}(\mathbf{x};t,t')\, \epsilon^{kl}(\mathbf{x},t') \qquad (40d)$$

and

$$r^i(\xi,t) = n^j(\xi)\, \sigma^{ij}(\xi,t) \qquad \text{for}\, \xi \in S . \qquad (40e)$$

$\overrightarrow{\Sigma}{}^{ij}(\mathbf{x},t)$ corresponds to the stress at time t due to the history of the strain $\overrightarrow{\epsilon}{}^{ij}(\mathbf{x},t)$ before t_0 , while $\overleftarrow{\Sigma}{}^{ij}(\mathbf{x},t)$ corresponds to the future of the strain $\overleftarrow{\epsilon}{}^{ij}(\mathbf{x},t)$ after t_1 .

If we restrict the domains of definition of L and L^t respectively to subspaces $\vec{U}$ and $\overleftarrow{U}$ such that for any $\vec{u} \in \vec{U}$ and $\overleftarrow{u} \in \overleftarrow{U}$ the right-hand side terms in (39) vanish, then L^t is the transposed of L. The elements of $\vec{U}$ and $\overleftarrow{U}$ satisfy then dual conditions.

First example of dual conditions: If the field $\vec{u}$ has a quiescent past:

$$\vec{u}^i(x,t) = 0 \qquad \text{for } t \leq t_0 \;, \tag{41a}$$

$$\frac{\partial \vec{u}^i}{\partial t}(x,t_0) = 0 \;, \tag{41b}$$

and satisfies a condition of free surface (computed with positive attenuation):

$$n^j(\xi) \int_{-\infty}^{+\infty} dt' \; \vec{\Psi}_0^{ijkl}(\xi;t,t') \; \frac{\partial \vec{u}^k}{\partial x^l}(\xi,t') = 0 \qquad \text{for } \xi \in S \;, \tag{41c}$$

and the field $\overleftarrow{u}$ has a quiescent future:

$$\overleftarrow{u}^i(x,t_1) = 0 \qquad \text{for } t \geq t_1 \;, \tag{42a}$$

$$\frac{\partial \overleftarrow{u}^i}{\partial t}(x,t_1) = 0 \;, \tag{42b}$$

and satisfies a condition of free surface (computed with negative attenuation):

$$n^j(\xi) \int_{-\infty}^{+\infty} dt' \; \overleftarrow{\Psi}_0^{ijkl}(\xi;t,t') \; \frac{\partial \overleftarrow{u}^k}{\partial x^l}(\xi,t') = 0 \qquad \text{for } \xi \in S \;, \tag{42c}$$

then, the integrals in (39) vanish: conditions (41)-(42) are dual conditions.

Second example of dual conditions: If in the example above we assume conditions of rigid, instead of free surface:

$$\vec{u}^i(\xi,t) = 0 \qquad \text{for } \xi \in S \;, \tag{43}$$

$$\overleftarrow{u}^i(\xi,t) = 0 \qquad \text{for } \xi \in S \;, \tag{44}$$

the integrals in (40) also vanish: conditions (43)-(44), toguether with the conditions (41a,b)-(42a,b) are dual conditions.

6: THE GREEN FUNCTION

Consider the operator L_{free} defined by the restriction of the formal definition of L (32) to the subspace $\vec{U} \subset U$ of fields satisfying the homogeneous initial conditions and the conditions of free surface defined by equations (41). Then, for any ϕ belonging to the source space Φ, the equation

$$L_{free} \, \vec{u} = \phi \tag{45}$$

has one solution $\vec{u}$, and only one (if we limit our consideration to functions regular enough).

This allows to define L_{free}^{-1}, the inverse of L_{free}. It is named the Green operator, and is denoted $\vec{G}_{free}$:

$$\vec{G}_{free} = L_{free}^{-1} . \tag{46}$$

Equation (45) is then solved formally by

$$\vec{u} = \vec{G}_{free} \phi . \tag{47}$$

An integral representation of (47) is written

$$\vec{u}^i(x,t) = \int_V dV(x') \int_{t_0}^{t_1} dt' \, \vec{G}_{free}^{ij}(x,t;x',t') \, \phi^j(x',t') , \tag{48}$$

and $\vec{G}_{free}^{ij}(x,t;x',t')$, the kernel of the Green operator, is named the Green function.

Consider now the operator L_{free}^t defined by the restriction of the formal definition (32) to the subspace $\overleftarrow{U} \subset U$ of fields satisfying the homogeneous final conditions and the conditions of free surface (with negative attenuation) defined by equations (42). Again, the equation

$$L_{free}^t \overleftarrow{u} = \phi \tag{49}$$

has one solution, and only one. Let us denote $\overleftarrow{G}_{free}$ the inverse of L_{free}^t :

$$\overleftarrow{G}_{free} = (L_{free}^t)^{-1} . \tag{50}$$

Equation (49) is solved formally by

$$\overleftarrow{u} = \overleftarrow{G}_{free} \phi , \tag{51}$$

or, introducing the kernel of $\overleftarrow{G}_{free}$,

$$\overleftarrow{u}^i(x,t) = \int_V dV(x') \int_{t_0}^{t_1} dt' \, \overleftarrow{G}_{free}^{ij}(x,t;x',t') \, \phi^j(x',t') . \tag{52}$$

By definition, we have

$$L_{free} \vec{G}_{free} = I , \tag{53a}$$

$$L_{free}^t \overleftarrow{G}_{free} = I . \tag{53b}$$

Using the definitions of L_{free} and L_{free}^t , and the kernels of $\vec{G}_{free}$ and $\overleftarrow{G}_{free}$, equations (53) take the explicit form

$$\rho(x) \frac{\partial^2 \vec{G}_{free}^{ip}}{\partial t^2}(x,t;x',t') - \frac{\partial}{\partial x^j} \int_{-\infty}^{+\infty} dt'' \, \vec{\Psi}^{ijkl}(x;t,t'') \frac{\partial \vec{G}_{free}^{kp}}{\partial x^l}(x,t'';x',t') = \delta^{ip} \, \delta(x-x') \, \delta(t-t')$$

$$\tag{54a}$$

$$n^j(\xi) \int_{-\infty}^{+\infty} dt'' \, \vec{\Psi}^{ijkl}(\xi;t,t'') \frac{\partial \vec{G}_{free}^{kp}}{\partial x^l}(\xi,t'';x',t') = 0 \quad \xi \in S \tag{54b}$$

$$\vec{G}_{free}^{ip}(x,t;x',t') = 0 \quad \text{for} \quad t \leq t' \tag{54c}$$

$$\frac{\partial \vec{G}_{free}{}^{ip}}{\partial t}(\mathbf{x},t;\mathbf{x}',t') = 0 \quad \text{for} \quad t \leq t' \,, \tag{54d}$$

and

$$\rho(\mathbf{x})\frac{\partial^2 \overset{\leftarrow}{G}_{free}{}^{ip}}{\partial t^2}(\mathbf{x},t;\mathbf{x}',t') - \frac{\partial}{\partial x^j}\int_{-\infty}^{+\infty} dt'' \; \overset{\leftarrow}{\Psi}^{ijkl}(\mathbf{x};t,t'')\frac{\partial \overset{\leftarrow}{G}_{free}{}^{kp}}{\partial x^l}(\mathbf{x},t'';\mathbf{x}',t') = \delta^{ip}\,\delta(\mathbf{x}-\mathbf{x}')\,\delta(t-t') \tag{55a}$$

$$n^j(\xi)\int_{-\infty}^{+\infty} dt'' \; \overset{\leftarrow}{\Psi}^{ijkl}(\xi;t,t'')\frac{\partial \overset{\leftarrow}{G}_{free}{}^{kp}}{\partial x^l}(\xi,t'';\mathbf{x}',t') = 0 \quad \xi \in S \tag{55b}$$

$$\overset{\leftarrow}{G}_{free}{}^{ip}(\mathbf{x},t;\mathbf{x}',t') = 0 \quad \text{for} \quad t \geq t' \tag{55c}$$

$$\frac{\partial \overset{\leftarrow}{G}_{free}{}^{ip}}{\partial t}(\mathbf{x},t;\mathbf{x}',t') = 0 \quad \text{for} \quad t \geq t' \,. \tag{55d}$$

As the inverse of the transposed equals the transposed of the inverse,

$$\overset{\leftarrow}{G}_{free} = (\vec{G}_{free})^t \,, \tag{56a}$$

and, as the kernel of the transposed has transposed variables,

$$\vec{G}_{free}{}^{ip}(\mathbf{x},t;\mathbf{x}',t') = \overset{\leftarrow}{G}_{free}{}^{pi}(\mathbf{x}',t';\mathbf{x},t) \,. \tag{56b}$$

Notice that this is not a reciprocity relation: it relates $\vec{G}_{free}$ to $\overset{\leftarrow}{G}_{free}$, but it do not expresses an internal symmetry of $\vec{G}_{free}$. The reciprocity relationships are analyzed in Section 8.

If instead of L_{free} we define L_{rigid} as the restriction of L to the subspace of fields satisfying the homogeneous initial conditions and the conditions of rigid surface defined by equations (48), and $L_{rigid}{}^t$ as the restriction of L^t to the subspace of fields satisfying the dual conditions defined by equations (41a,b) and (43), we can introduce $\vec{G}_{rigid}$ and $\overset{\leftarrow}{G}_{rigid}$ as above. The equivalent of equations (54)-(55) is

$$\rho(\mathbf{x})\frac{\partial^2 \vec{G}_{rigid}{}^{ip}}{\partial t^2}(\mathbf{x},t;\mathbf{x}',t') - \frac{\partial}{\partial x^j}\int_{-\infty}^{+\infty} dt'' \; \overset{\rightarrow}{\Psi}^{ijkl}(\mathbf{x};t,t'')\frac{\partial \vec{G}_{rigid}{}^{kp}}{\partial x^l}(\mathbf{x},t'';\mathbf{x}',t') = \delta^{ip}\,\delta(\mathbf{x}-\mathbf{x}')\,\delta(t-t') \tag{57a}$$

$$\vec{G}_{rigid}{}^{ip}(\xi,t;\mathbf{x}',t') = 0 \quad \xi \in S \tag{57b}$$

$$\vec{G}_{rigid}{}^{ip}(\mathbf{x},t;\mathbf{x}',t') = 0 \quad \text{for} \quad t \leq t' \tag{57c}$$

$$\frac{\partial \vec{G}_{rigid}{}^{ip}}{\partial t}(\mathbf{x},t;\mathbf{x}',t') = 0 \quad \text{for} \quad t \leq t' \,, \tag{57d}$$

and

$$\rho(\mathbf{x})\frac{\partial^2 \overset{\leftarrow}{G}_{rigid}{}^{ip}}{\partial t^2}(\mathbf{x},t;\mathbf{x}',t') - \frac{\partial}{\partial x^j}\int_{-\infty}^{+\infty} dt'' \; \overset{\leftarrow}{\Psi}^{ijkl}(\mathbf{x};t,t'')\frac{\partial \overset{\leftarrow}{G}_{rigid}{}^{kp}}{\partial x^l}(\mathbf{x},t'';\mathbf{x}',t') = \delta^{ip}\,\delta(\mathbf{x}-\mathbf{x}')\,\delta(t-t') \tag{58a}$$

$$\overset{\leftarrow}{G}_{rigid}{}^{ip}(\mathbf{x},t;\mathbf{x}',t') = 0 \quad \mathbf{x} \in S \tag{58b}$$

$$\overset{\leftarrow}{G}_{rigid}{}^{ip}(\mathbf{x},t;\mathbf{x}',t') = 0 \quad \text{for} \quad t > t' \tag{58c}$$

$$\frac{\partial \overleftarrow{G}_{rigid}{}^{ip}}{\partial t}(\mathbf{x},t;\mathbf{x}',t') = 0 \quad \text{for} \quad t>t' , \tag{58d}$$

and we also have

$$\overrightarrow{G}_{rigid}{}^{ip}(\mathbf{x},t;\mathbf{x}',t') = \overleftarrow{G}_{rigid}{}^{pi}(\mathbf{x}',t';\mathbf{x},t) . \tag{59}$$

7: REPRESENTATION THEOREMS

Let $\overleftarrow{G}^{ij}(\mathbf{x},t;\mathbf{x}',t')$ by any Green's function satisfying the wave equation associated with the dual problem (i.e., with negative attenuation in our case):

$$\rho(\mathbf{x}) \, \frac{\partial^2 \overleftarrow{G}^{ip}}{\partial t^2}(\mathbf{x},t;\mathbf{x}',t') - \frac{\partial}{\partial x^j} \int_{-\infty}^{+\infty} dt'' \, \overleftarrow{\Psi}^{ijkl}(\mathbf{x};t,t'') \, \frac{\partial \overleftarrow{G}^{kp}}{\partial x^l}(\mathbf{x},t'';\mathbf{x}',t') = \delta^{ip} \, \delta(\mathbf{x}-\mathbf{x}') \, \delta(t-t') , \tag{60a}$$

and with final conditions of rest:

$$\overleftarrow{G}^{ip}(\mathbf{x},t;\mathbf{x}',t') = 0 \quad \text{for} \quad t \geq t' \tag{60b}$$

$$\frac{\partial \overleftarrow{G}^{ip}}{\partial t}(\mathbf{x},t;\mathbf{x}',t') = 0 \quad \text{for} \quad t \geq t' . \tag{60c}$$

As no surface conditions have yet been specified, there exists an infinity of such Green's functions.

Let $u^i(\mathbf{x},t)$ be an arbitrary field, not necessarily satisfying a wave equation. For any t', $t_0 < t' < t_1$, we have

$$u^p(\mathbf{x}',t') = \int_V dV(\mathbf{x}) \int_{t_0}^{t_1} dt \, \delta^{ip} \, \delta(\mathbf{x}-\mathbf{x}') \, \delta(t-t') \, u^i(\mathbf{x},t) . \tag{61}$$

Inserting (60a) into (61), using the final conditions (60b-c), integrating per parts, and using the divergence theorem gives (see Appendix B)

$$u^p(\mathbf{x}',t') = \int_V dV(\mathbf{x}) \int_{t_0}^{t_1} dt \, \overleftarrow{G}^{ip}(\mathbf{x},t;\mathbf{x}',t') \, \overset{\circ}{\phi}{}^i(\mathbf{x},t)$$

$$+ \int_S dS(\boldsymbol{\xi}) \int_{t_0}^{t_1} dt \, \overleftarrow{G}^{ip}(\boldsymbol{\xi},t;\mathbf{x}',t') \, \overset{\circ}{\tau}{}^i(\boldsymbol{\xi},t)$$

$$- \int_S dS(\boldsymbol{\xi}) \int_{t_0}^{t_1} dt \left[n^j(\boldsymbol{\xi}) \int_{-\infty}^{+\infty} dt'' \, \overleftarrow{\Psi}^{ijkl}(\boldsymbol{\xi};t,t'') \, \frac{\partial \overleftarrow{G}^{kp}}{\partial x^l}(\boldsymbol{\xi},t'';\mathbf{x}',t') \right] u^i(\boldsymbol{\xi},t)$$

$$+ \int_V dV(\mathbf{x}) \, \rho(\mathbf{x}) \left[\overleftarrow{G}^{ip}(\mathbf{x},t_0;\mathbf{x}',t') \, \frac{\partial u^i}{\partial t}(\mathbf{x},t_0) - \frac{\partial \overleftarrow{G}^{ip}}{\partial t}(\mathbf{x},t_0;\mathbf{x}',t') \, u^i(\mathbf{x},t_0) \right]$$

$$- \int_V dV(\mathbf{x}) \int_{t_0}^{t_1} dt \, \frac{\partial \overleftarrow{G}^{ip}}{\partial x^j}(\mathbf{x},t;\mathbf{x}',t') \, \Sigma^{ij}(\mathbf{x},t) \tag{62}$$

where

$$\phi^i(\mathbf{x},t) \;=\; \rho(\mathbf{x})\,\frac{\partial^2 u^i}{\partial t^2}(\mathbf{x},t) \;-\; \frac{\partial}{\partial x^j}\int_{-\infty}^{+\infty} dt''\;\overleftarrow{\Psi}^{ijkl}(\mathbf{x};t,t'')\,\frac{\partial u^k}{\partial x^l}(\mathbf{x},t'')\;, \tag{63}$$

$$\Sigma^{ij}(\mathbf{x},t) \;=\; \int_{-\infty}^{t_0} dt''\;\overleftarrow{\Psi}^{ijkl}(\xi;t,t'')\,\frac{\partial u^k}{\partial x^l}(\mathbf{x},t'')\;, \tag{64}$$

and

$$\hat\tau^i(\xi,t) \;=\; n^j(\xi)\int_{-\infty}^{+\infty} dt''\;\overleftarrow{\Psi}^{ijkl}(\xi;t,t'')\,\frac{\partial u^k}{\partial x^l}(\xi,t'')\;. \tag{65}$$

Equation (62) is a quite general representation theorem, but remember that the Green function is not defined uniquely. Notice that it is because we have imposed a negative attenuation in the definition (60) of $\overleftarrow{G}$ that we have in (63) and (65) expressions which correspond to the usual source field and surface tractions of a field propagating with positive attenuation.

Example: Let us be interested in a field $\vec{u}^i(\mathbf{x},t)$ defined, for $t\in[t_0,t_1]$, by

$$\rho(\mathbf{x})\,\frac{\partial^2 \vec{u}^{\,i}}{\partial t^2}(\mathbf{x},t) \;-\; \frac{\partial \vec\sigma^{\,ij}}{\partial x^j}(\mathbf{x},t) \;=\; \phi^i(\mathbf{x},t) \tag{66a}$$

$$\vec\sigma^{\,ij}(\mathbf{x},t) \;=\; M^{ij}(\mathbf{x},t) \;+\; \int_{-\infty}^{+\infty} dt'\;\vec\Psi^{ijkl}(\mathbf{x};t,t')\,\frac{\partial \vec u^{\,k}}{\partial x^l}(\mathbf{x},t') \tag{66b}$$

$$n^j(\xi)\,\vec\sigma^{\,ij}(\xi t) \;=\; \tau^i(\xi,t) \qquad \text{for } \xi\in S \tag{66c}$$

and assume that the history of the field is known for $t<t_0$. Then, choosing for $\overleftarrow{G}$ the operator $\overleftarrow{G}_{\text{free}}$, satisfying free boundary conditions (55), introducing the transposed operator $\vec{G}_{\text{free}}$, using (62) we obtain, using equation (56b) and the divergence theorem, and relabelling variables,

$$
u^i(\mathbf{x},t) \;=\; \int_V dV(\mathbf{x}') \int_{t_0}^{t_1} dt'\; \vec{G}_{\text{free}}^{\,ij}(\mathbf{x},t;\mathbf{x}',t')\,\phi^j(\mathbf{x}',t')
$$

$$
+\; \int_S dS(\xi') \int_{t_0}^{t_1} dt'\; \vec{G}_{\text{free}}^{\,ij}(\mathbf{x},t;\mathbf{x}',t')\,\tau^j(\xi',t')
$$

$$
-\; \int_V dV(\mathbf{x}') \int_{t_0}^{t_1} dt'\; \frac{\partial \vec{G}_{\text{free}}^{\,ij}}{\partial x'^k}(\mathbf{x},t;\mathbf{x}',t')\,(M^{jk}(\mathbf{x}',t') + \Sigma^{jk}(\mathbf{x}',t')) \tag{67}
$$

$$
+\; \int_V dV(\mathbf{x}')\,\rho(\mathbf{x}')\left[\vec{G}_{\text{free}}^{\,ij}(\mathbf{x},t;\mathbf{x}',t'{=}t_0)\,\frac{\partial \vec u^{\,j}}{\partial t'}(\mathbf{x}',t'{=}t_0) - \frac{\partial \vec{G}_{\text{free}}^{\,ij}}{\partial t'}(\mathbf{x},t;\mathbf{x}',t'{=}t_0)\,\vec u^{\,j}(\mathbf{x}',t'{=}t_0)\right],
$$

where $\Sigma^{ij}(\mathbf{x},t)$ is the stress due to the strain for $t<t_0$ not already relaxed:

$$\Sigma^{ij}(\mathbf{x},t) \;=\; \int_{-\infty}^{t_0} dt'\;\vec\Psi^{ijkl}(\mathbf{x};t,t')\,\frac{\partial \vec u^{\,k}}{\partial x^l}(\mathbf{x},t')\;. \tag{68}$$

Notice that the partial derivatives of the Green function are with respect to the source space-time coordinates.

In equation (67) the fields ϕ , τ , M , Σ , u_0 , and v_0 , can be named 'generalized sources' of the field u . Then this equation defines a linear operator from the space of generalized sources into the space of displacements. This operator can be named the 'generalized Green operator'.

8: RECIPROCITY THEOREMS

In the previous definition of Green functions, we have used a function $\Psi^{ijkl}(x;t,t')$: the medium parameters may depend on time. Then there is no reciprocity relation satisfied.

If we assume that the medium parameters do not depent on time,

$$\Psi^{ijkl}(x;t,t') = \Psi^{ijkl}(x;t-t',0) , \tag{69}$$

then, changing t by $-t$ switches from the dual problem defined by (55) into the primal problem (54), and the dual problem (58) into the primal problem (57). Then

$$\vec{G}^{ij}_{free}(x,t;x',t') = \overleftarrow{G}^{ji}_{free}(x,-t;x',-t') , \tag{70}$$

and

$$\vec{G}^{ij}_{rigid}(x,t;x',t') = \overleftarrow{G}^{ji}_{rigid}(x,-t;x',-t') . \tag{71}$$

As the density $\rho(x)$ is also independent on time, the whole wave equation is invariane by translation on time. Then

$$\vec{G}^{ij}_{free}(x,t;x',t') = \vec{G}^{ij}_{free}(x,t-t';x',0) , \tag{72}$$

and

$$\vec{G}^{ij}_{rigid}(x,t;x',t') = \vec{G}^{ij}_{rigid}(x,t-t';x',0) . \tag{73}$$

From (70) and (72) it follows the reciprocity relation for G_{free} :

$$\boxed{\vec{G}^{ij}_{free}(x,\tau;x',0) = \vec{G}^{ji}_{free}(x',\tau;x,0) ,} \tag{74}$$

while from (71) and (73) if follows the reciprocity relation for G_{rigid} :

$$\boxed{\vec{G}^{ij}_{rigid}(x,\tau;x',0) = \vec{G}^{ji}_{rigid}(x',\tau;x,0) ,} \tag{75}$$

The response at point x along the i-th axis for a source at point x' along the j-axis, equals the response at point x' along the j-axis for a source at point x along the i-th axis. For both experiments, the source starts at 0 and we record at τ .

Notice that to any Green's function $\vec{G}^{ij}(x,t;x',t')$ we can associate the Green function of the dual problem, $\overleftarrow{G}^{ij}(x,t;x',t')$, that they satisfy necessarily the property

$$\vec{G}^{ij}(x,t;x',t') = \overleftarrow{G}^{ji}(x',t';x,t) , \tag{76}$$

but that to satisfy a recipricity property we need more structure: an equivalence between primal and dual problems under some change of variables.

9: THE BORN APPROXIMATION

Let us consider the field $u^i(x,t)$ defined by the system of equations

$$\rho(x) \, \frac{\partial^2 u^i}{\partial t^2}(x,t) - \frac{\partial \sigma^{ij}}{\partial x^j}(x,t) = \phi^i(x,t) \, , \tag{77a}$$

$$\sigma^{ij}(x,t) = M^{ij}(x,t) + \int_{-\infty}^{+\infty} dt' \, \Psi^{ijkl}(x,t-t') \, \frac{\partial u^k}{\partial x^l}(x,t') \, , \tag{77b}$$

$$n^j(\xi) \, \sigma^{ij}(\xi,t) = \tau^i(\xi,t) \qquad \text{for } \xi \in S \, , \tag{77c}$$

$$u^i(x,t_0) = \alpha^i(x) \, , \tag{77d}$$

$$\frac{\partial u^i}{\partial t}(x,t_0) = \beta^i(x) \, , \tag{77e}$$

$$u^i(x,t) = \gamma^i(x,t) \qquad \text{for } t < t_0 \, . \tag{77f}$$

A perturbation of the model parameters

$$\rho(x) \; \rightarrow \; \rho(x) + \delta\rho(x) \tag{78a}$$

$$\Psi^{ijkl}(x,\tau) \; \rightarrow \; \Psi^{ijkl}(x,\tau) + \delta\Psi^{ijkl}(x,\tau) \tag{78b}$$

leads to a perturbation of the displacement field

$$u^i(x,t) \; \rightarrow \; u^i(x,t) + \delta u^i(x,t) \, . \tag{78c}$$

For the use of gradient methods, we need the first order approximation to $\delta u^i(x,t)$. Inserting (78) into (77), substracting (77) and dropping second order terms we arrive easily at

$$\rho(x) \, \frac{\partial^2 \delta u^i}{\partial t^2}(x,t) - \frac{\partial \delta\sigma^{ij}}{\partial x^j}(x,t) = \delta\phi^i(x,t) \, , \tag{79a}$$

$$\delta\sigma^{ij}(x,t) = \delta M^{ij}(x,t) + \int_{-\infty}^{+\infty} dt' \, \Psi^{jikl}(x,t-t') \, \frac{\partial \delta u^k}{\partial x^l}(x,t') \, , \tag{79b}$$

$$n^j(\xi) \, \delta\sigma^{ij}(\xi,t) = 0 \qquad \text{for } \xi \in S \, , \tag{79c}$$

$$\delta u^i(x,t_0) = 0 \, , \tag{79d}$$

$$\frac{\partial \delta u^i}{\partial t}(x,t_0) = 0 \, , \tag{79e}$$

$$\delta u^i(x,t) = 0 \qquad \text{for } t < t_0 \, , \tag{79f}$$

where $\delta\phi$ and δM are the 'secondary Born sources'

$$\delta\phi^i(x,t) = - \, \delta\rho(x) \, \frac{\partial^2 u^i}{\partial t^2}(x,t) \, , \tag{80a}$$

and

$$\delta M^{ij}(\mathbf{x},t) = \int_{-\infty}^{+\infty} dt' \, \delta\Psi^{ijkl}(\mathbf{x};t,t') \, \frac{\partial u^k}{\partial x^l}(\mathbf{x},t') \,, \tag{80b}$$

The field $\delta\mathbf{u}$ defined by this system of equations corresponds to the Born approximation to displacement perturbation. The intuitive interpretation of equations (79) is as follows. The field $\delta\mathbf{u}$ propagates in the unperturbed medium (because $\rho(\mathbf{x})$ and $\Psi^{ijkl}(\mathbf{x};t,t')$ appear in the left-hand side, but not $\delta\rho(\mathbf{x})$ and $\delta\Psi^{ijkl}(\mathbf{x};t,t')$. Sources for this field exist where the medium has been perturbed. They are proportional to the perturbations $\delta\rho(\mathbf{x})$ and $\delta\Psi^{ijkl}(\mathbf{x};t,t')$, and to the reference field $u^i(\mathbf{x},t)$. By comparison with equations (18) to (20), we see that the source corresponding to the density perturbation is a force density, while the one corresponding to the perturbation of the visco-elastic parameters is a moment density.

Using the representation theorem (67) we obtain

$$\delta u^p(\mathbf{x}',t') = \int_V dV(\mathbf{x}) \int_{t_0}^{t_1} dt \, \vec{G}_{free}^{pi}(\mathbf{x}',t';\mathbf{x},t) \, \delta\phi^i(\mathbf{x},t)$$

$$- \int_V dV(\mathbf{x}) \int_{t_0}^{t_1} dt \, \frac{\partial \vec{G}_{free}^{pi}}{\partial x^j}(\mathbf{x}',t';\mathbf{x},t) \, \delta M^{ij}(\mathbf{x},t) \,. \tag{81}$$

Equations (80)-(81) give the explicit expression to the Born approximation.

10: LEAST SQUARES IN FUNCTIONAL SPACES

Assume that using sources $\phi^i(\mathbf{x},t)$, $\tau^i(\xi,t)$, and $M^{ij}(\mathbf{x},t)$ (usually of only one type) we generate a displacement field $\vec{u}^i(\mathbf{x},t)$ in a medium described by the parameters $\rho(\mathbf{x})$ and $\Psi^{ijkl}(\mathbf{x},\tau)$, and that we measure the field $\vec{u}^i(\mathbf{x},t)$ at some receiver locations $\mathbf{x}_r$ (r=1,2,...). We wish to use the observations $\vec{u}^i(\mathbf{x}_r,t)_{obs}$ to infer the values of the parameters $\rho(\mathbf{x})$ and $\Psi^{ijkl}(\mathbf{x},\tau)$ describing the medium.

We assume here that the field $\vec{u}^i(\mathbf{x},t)$ satisfies homogeneous initial conditions, and propagates with a free surface. In this section, the notations $\mathbf{L}$ and $\mathbf{G}$ will stand respectively for $\mathbf{L}_{free}$ and $\vec{G}_{free}$ (or the corresponding generalized operators introduced in Section 7). The field $\vec{u}$ is then defined by the equation $\mathbf{L}\,\vec{u} = \psi$, where ψ denotes the generalized sources (representing $\phi^i(\mathbf{x},t)$, $\tau^i(\xi,t)$, and/or $M^{ij}(\mathbf{x},t)$). The operator $\mathbf{L}$ is a function of the medium parameters. To make this dependence explicit, we write $\mathbf{L}[\mathbf{m}]$, where $\mathbf{m}$ represents a model of the medium, i.e., a set of functions $\{\rho(\mathbf{x}), \Psi^{ijkl}(\mathbf{x},\tau)\}$. The models of the medium belong to the 'model space' $\mathbf{M}$.

Then, $\vec{u}$ is defined by

$$\mathbf{L}[\mathbf{m}]\,\vec{u} = \psi \,. \tag{82}$$

The observed values $\vec{u}^i(\mathbf{x}_r,t)$ will be denoted by $\mathbf{d}_{obs}$. The values $\vec{u}^i(\mathbf{x}_r,t)$ calculated from a model $\mathbf{m}$ will be denoted by $\mathbf{d}_{cal}$ or $\mathbf{d}[\mathbf{m}]$. The data vectors $\mathbf{d}$ belong to a 'data space' $\mathbf{D}$.

The aim of least-squares inversion (Tarantola & Valette, 1982a, 1982b, Tarantola, 1987) is to obtain the model $\mathbf{m}$ minimizing the misfit function

$$S(\mathbf{m}) = \frac{1}{2} \left(\|\mathbf{d}[\mathbf{m}] - \mathbf{d}_{obs}\|^2 + \|\mathbf{m} - \mathbf{m}_{prior}\|^2 \right) \tag{83}$$

$$= \frac{1}{2} \left[\langle\, C_D^{-1} \, (d[m]-d_{obs}) \, , \, (d[m]-d_{obs}) \,\rangle \;+\; \langle\, C_M^{-1} \, (m-m_{prior}) \, , \, (m-_{prior}) \,\rangle \right] ,$$

where C_D is the covariance operator describing data uncertainties, m_{prior} is some a priori model, and C_M is the covariance operator describing uncertainties in m_{prior}.

The gradient $\hat{\gamma}$ of the misfit function is defined by the first order development

$$S(m+\delta m) \;=\; S(m) + \langle\, \hat{\gamma} \, , \, \delta m \,\rangle \;+\; O(\|\delta m\|^2) . \tag{84}$$

It is an element of the dual of the model space (identified with the model space weighted by C_M^{-1}).

The direction of steepest ascent is then (Tarantola, 1987)

$$\gamma \;=\; C_M \, \hat{\gamma} , \tag{85}$$

and the algorithm of steepest descent for the minimization of $S(m)$ is

$$m_{n+1} \;=\; m_n - \alpha_n \, \gamma_n , \tag{86}$$

where α_n is a constant sufficiently small to ensure

$$S(m_{n+1}) \;<\; S(m_n) . \tag{87}$$

Let us now formally compute the gradient of the misfit function. As the term $\langle\, C_M^{-1} \, (m-m_{prior}) \, , \, (m-m_{prior}) \,\rangle$ is quadratic in m, it makes no problem, and is dropped (the reader will easily correct for it).

Formally, d is obtained by projecting the field $\vec{u}^i(x,t)$ into the observation points x_r:

$$d \;=\; P \, \vec{u} , \tag{88}$$

where P is the projector ($P^2 = P$) defined by

$$(P \, \vec{u})^i(x_r,t) \;=\; \vec{u}^i(x_r,t) . \tag{89}$$

The reader may easily verify that the transposed of P is the operator defined by

$$(P^t \, \hat{d})^i(x,t) \;=\; \sum_r \delta(x-x_r) \, \hat{d}^i(x_r,t) , \tag{90}$$

where $\hat{d}$ is an element of the dual of the data space (identified with the data space weighted by C_D^{-1}).

We have

$$S(m) \;=\; \frac{1}{2} \langle\, C_D^{-1} \, (P \, \vec{u} - d_{obs}) \, , \, (P \, \vec{u} - d_{obs}) \,\rangle , \tag{91}$$

where

$$L[m] \, \vec{u} \;=\; \psi . \tag{92}$$

A perturbation of the medium parameters

$$m \;\to\; m + \delta m \tag{93}$$

leads to

$$S(m+\delta m) = \frac{1}{2} \langle C_D^{-1} (P (\vec{u}+\vec{\Delta u}) - d_{obs}), (P (\vec{u}+\vec{\Delta u}) - d_{obs}) \rangle, \tag{94}$$

where $\vec{\Delta u}$ is defined by

$$L[m+\delta m] (\vec{u}+\vec{\Delta u}) = \psi, \tag{95}$$

and depends (nonlinearly) on δm. Let $\vec{\delta u}$ denote the first order approximation to $\vec{\Delta u}$:

$$\vec{\Delta u} = \vec{\delta u} + O(\|\delta m\|^2). \tag{96}$$

Then

$$S(m+\delta m) = \frac{1}{2} \langle C_D^{-1} (P (\vec{u}+\vec{\delta u}) - d_{obs}), (P (\vec{u}+\vec{\delta u}) - d_{obs}) \rangle + O(\|\delta m\|^2)$$

$$= S(m) + \frac{1}{2} \langle C_D^{-1} (P \vec{u} - d_{obs}), P \vec{\delta u} \rangle + \frac{1}{2} \langle C_D^{-1} P \vec{\delta u}, (P \vec{u} - d_{obs}) \rangle$$
$$+ O(\|\delta m\|^2);$$

as covariance operators are symmetric,

$$S(m+\delta m) = S(m) + \langle C_D^{-1} (P \vec{u} - d_{obs}), P \vec{\delta u} \rangle + O(\|\delta m\|^2);$$

and, introducing the transpose of the projector P,

$$S(m+\delta m) = S(m) + \langle P^t C_D^{-1} (P \vec{u} - d_{obs}), \vec{\delta u} \rangle + O(\|\delta m\|^2). \tag{98}$$

The field $\vec{\delta u}$, first order approximation to $\vec{\Delta u}$ corresponds to the Born approximation to $\vec{\Delta u}$ (see Section 9). It corresponds then to the field created by some Born secondary generalized sources $\delta \psi$ and propagated into the unperturbed medium m :

$$\vec{\delta u} = G[m] \delta \psi. \tag{98}$$

Equation (97) then becomes

$$S(m+\delta m) = S(m) - \langle P^t C_D^{-1} (P \vec{u} - d_{obs}), G \delta \psi \rangle + O(\|\delta m\|^2)$$

$$= S(m) - \langle \overleftarrow{u}, \delta \psi \rangle + O(\|\delta m\|^2), \tag{99}$$

where $\overleftarrow{u}$ is defined by

$$\overleftarrow{u} = G^t P^t C_D^{-1} (P \vec{u} - d_{obs}), \tag{100a}$$

i.e.,

$$L^t \overleftarrow{u} = P^t C_D^{-1} (P \vec{u} - d_{obs}). \tag{100b}$$

The field $\overleftarrow{u}$ is created by the sources $P^t C_D^{-1} (P \vec{u} - d_{obs})$, and satisfies conditions dual to those satisfied by $\vec{u}$.

The Born secondary sources $\delta \psi$ depend linearly on $\vec{u}$ and on δm. Introducing the notation

$$\delta\psi \;=\; (A \,\overrightarrow{u}) \,\delta m \tag{101}$$

leads to

$$S(m+\delta m) \;=\; S(m) - \langle\, \overleftarrow{u}\, ,\, (A\,\overrightarrow{u})\,\delta m\,\rangle \;+\; O(\|\delta m\|^2)\,,$$

and properly introducing the transpose of the operator $(A\,\overrightarrow{u})$,

$$S(m+\delta m) \;=\; S(m) - \langle\, (A\,\overrightarrow{u})^t\,\overleftarrow{u}\, ,\, \delta m\,\rangle \;+\; O(\|\delta m\|^2)\,. \tag{102}$$

By comparison with (87), this last equation gives the gradient of the least squares misfit functional S :

$$\boxed{\;\hat{\gamma} \;=\; (A\,\overrightarrow{u})^t\,\overleftarrow{u}\;.\;} \tag{103}$$

Equation (88) gives then

$$\gamma \;=\; C_M\,(A\,\overrightarrow{u})^t\,\overleftarrow{u}\,, \tag{104}$$

and (89) finally gives

$$m_{n+1} \;=\; m_n - \alpha_n\, C_M\,(A\,\overrightarrow{u}_n)^t\,\overleftarrow{u}_n\,. \tag{105}$$

All the partial steps needed for an iteration of the steepest descent algorithm are:

$L[m_n]\,\overrightarrow{u}_n \;=\; \psi$	(solve for u_n)
$\delta d_n \;=\; P\,\overrightarrow{u}_n - d_{obs}$	(compute data residuals)
$C_D\,\delta\hat{d}_n \;=\; \delta d_n$	(solve to obtain the weighted residuals)
$\delta\phi_n \;=\; P^t\,\delta\hat{d}_n$	(consider these as sources)
$L^t\,\overleftarrow{u}_n \;=\; \delta\phi_n$	(solve for $\overleftarrow{u}_n$, i.e., propagate the sources with dual conditions, solving the dual problem)
$\hat{\gamma}_n \;=\; (A\,\overrightarrow{u}_n)^t\,\overleftarrow{u}_n$	(compute the gradient $\hat{\gamma}_n$, where A has been defined in (106))
$\gamma_n \;=\; C_M\,\hat{\gamma}_n$	(unweight the gradient)
$m_{n+1} \;=\; m_n - \alpha_n\,\gamma_n$	(update the model)

$$\tag{106}$$

The above formulas correspond to a crude steepest descent method. Current implementations of gradient methods for the inversion of seismic waveforms (Gauthier et al., 1986; Kolb, 1986; Mora, 1987; Pica, 1987) are rather based in conjugate gradients (e.g., Fletcher, 1981; Scales, 1985), which converge more rapidly.

11: THE INVERSE PROBLEM OF INTERPRETATION OF SEISMIC WAVEFORMS

This section applies the results of the previous section to the problem of interpretation of seismic reflection data. Typically, a source is fired consecutively at different locations x_s (s=1,2,...) , and, for each source position, the displacement u^i is observed at some locations x_r (r=1,2,...) . In all rigor, the time variable t runs frim $-\infty$ to $+\infty$ and there is only one source, concentrated at different points at different times. More intuitively, we can consider that the time variable is reset to $t = t_0$ at each new shot, and we record the Earth's surface displacements until $t = t_1$. In that case, it can easily be seen that the gradient of the misfit function for an experiment with different sources is simply the sum of the gradients corresponding to each source. We see thus that, with any of the two points of view, we can limit our consideration to an experiment with a single source.

The observed seismograms are denoted by $u^i(x,t)_{obs}$, while the computed seismograms corresponding to the n-th Earth model are denoted $u^i(x,t)_n$. The recording time belongs to the interval $[t_0,t_1]$.

Let us take in order all the steps (130)-(137).

Equation (106a): $L[m_n] \vec{u}_n = \psi$. Let $\rho(x)_n$ and $\Psi^{ijkl}(x,\tau)_n$ denote the current Earth model. If the sources of seismic waves are described by the force density $\phi^i(x,t)$ the surface traction $\tau^i(x,t)$ and/or the moment density $M^{ij}(x,t)$ then the current displacement field $u^i(x,t)_n$ is defined by

$$\rho(x)_n \frac{\partial^2 \vec{u}^i}{\partial t^2}(x,t)_n - \frac{\partial \vec{\sigma}^{ij}}{\partial x^j}(x,t)_n = \phi^i(x,t) , \tag{107a}$$

$$\vec{\sigma}^{ij}(x,t)_n = M^{ij}(x,t) + \int_{-\infty}^{+\infty} dt' \; \vec{\Psi}^{ijkl}(x,t-t')_n \frac{\partial \vec{u}^k}{\partial x^l}(x,t')_n , \tag{107b}$$

$$n^j(\xi) \vec{\sigma}(\xi,t)_n = \tau^i(\xi,t) \qquad \text{for } \xi \in S , \tag{107c}$$

$$\vec{u}^i(x,t_0)_n = 0 , \tag{107d}$$

$$\frac{\partial \vec{u}^i}{\partial t}(x,t_0)_n = 0 , \tag{107e}$$

$$\vec{u}^i(x,t)_n = 0 \qquad \text{for } t<t_0 , \tag{107f}$$

where a free surface and homogeneous initial conditions are assumed. The computation of the field $\vec{u}^i(x,t)_n$ can be performed using any numerical method, as, for instance, finite-differences (Alterman and Karal, 1968; Virieux, 1986).

Equation (106b): $\delta d_n = P \vec{u}_n - d_{obs}$. This simply corresponds to the definition of the residuals at the receiver locations:

$$\delta d^i(x_r,t)_n = \vec{u}^i(x_r,t)_n - \vec{u}^i(x_r,t)_{obs} . \tag{108}$$

Equation (106c): $C_D \; \delta \hat{d}_n = \delta d_n$. As an example, assume independent and uniforn uncertainties. Then,

$$\delta \hat{d}^i(x_r,t)_n = \frac{1}{v^0} \delta d^i(x_r,t)_n , \tag{109}$$

Of course, other more realistic choices of the covariance operator describing experimental uncertainties can be made.

Equation (106d): $\delta\phi_n = \mathbf{P}^t\,\delta\hat{\mathbf{d}}_n$. From equation (90) we obtain

$$\delta\phi^i(\mathbf{x},t)_n = \sum_r \delta(\mathbf{x}-\mathbf{x}_r)\,\delta\hat{d}^i(\mathbf{x}_r,t)_n \ . \tag{110}$$

This corresponds to a composite source, one point source at each receiver location, radiating the weighted residuals in phase.

Equation (106e): $\mathbf{L}^t\,\overleftarrow{\mathbf{u}}_n = \delta\phi_n$. As demonstrated Section 5, to compute the field $\overleftarrow{u}^i(\mathbf{x},t)_n$, solution of $\mathbf{L}^t\,\overleftarrow{\mathbf{u}}_n = \delta\phi_n$, means to solve the differential system

$$\rho(\mathbf{x})_n\,\frac{\partial^2\overleftarrow{u}^i}{\partial t^2}(\mathbf{x},t)_n - \frac{\partial\overleftarrow{\sigma}^{ij}}{\partial x^j}(\mathbf{x},t)_n = \delta\phi^i(\mathbf{x},t)_n \ , \tag{111a}$$

$$\overleftarrow{\sigma}^{ij}(\mathbf{x},t)_n = \int_{-\infty}^{+\infty}dt'\,\overleftarrow{\Psi}^{ijkl}(\mathbf{x},t-t')_n\,\frac{\partial\overleftarrow{u}^k}{\partial x^l}(\mathbf{x},t')_n \ , \tag{111b}$$

$$n^j(\xi)\,\overleftarrow{\sigma}^{ij}(\xi,t) = 0 \qquad \text{for } \xi\in S \ , \tag{111c}$$

$$\overleftarrow{u}^i(\mathbf{x},t_1)_n = 0 \ , \tag{111d}$$

$$\frac{\partial\overleftarrow{u}^i}{\partial t}(\mathbf{x},t_1)_n = 0 \ , \tag{111e}$$

$$\overleftarrow{u}^i(\mathbf{x},t)_n = 0 \qquad \text{for } t>t_1 \ , \tag{111f}$$

where instead of a quiescent past, the field has a quiescent future, and where the attenuation is negative (anti-causal). To solve this problem, we can use the same computer code needed to solve the system (107), running the time backwards (see Gauthier et al., 1986) and changing $\overrightarrow{\Psi}^{ijkl}(\mathbf{x},\tau)$ by $\overleftarrow{\Psi}^{ijkl}(\mathbf{x},\tau) = \overrightarrow{\Psi}^{ijkl}(\mathbf{x},-\tau)$. If the medium is not highly attenuating, the introduction of negative attenuation should not cause numerical troubles.

Equation (106f): $\hat{\gamma}_n = (\mathbf{A}\,\overrightarrow{\mathbf{u}}_n)^t\,\overleftarrow{\mathbf{u}}_n$. Once the field $\overleftarrow{u}^i(\mathbf{x},t)_n$ has neem computed, we can turn to the computation of the gradient. The operator $\mathbf{A}$ is defined by equation (101):

$$\delta\psi_n = (\mathbf{A}\,\overrightarrow{\mathbf{u}}_n)\,\delta\mathbf{m} \ , \tag{101 again}$$

where $\delta\psi_n$ is the Born secondary generalized source corresponding to the perturbation $\delta\mathbf{m} = \{\delta\rho\,,\,\delta\Psi^{ijkl}\}$. Using the results of Section 9 gives

$$\langle\,\overleftarrow{\mathbf{u}}_n\,,(\mathbf{A}\,\overrightarrow{\mathbf{u}}_n)\,\delta\mathbf{m}\,\rangle = \langle\,\overleftarrow{\mathbf{u}}_n\,,\delta\psi_n\,\rangle =$$

$$+ \int_V dV(\mathbf{x})\left[\int_{t_0}^{t_1}dt\,\overleftarrow{v}^i(\mathbf{x},t)\,\overrightarrow{v}^i(\mathbf{x},t)\right]\delta\rho(\mathbf{x})$$

$$- \int_V dV(\mathbf{x})\int_{-\infty}^{+\infty}dt'\left[\int_{t_0}^{t_1}dt\,\overleftarrow{\epsilon}^{ij}(\mathbf{x},t)\,\overrightarrow{\epsilon}^{kl}(\mathbf{x},t-t')\right]\delta\Psi(\mathbf{x},t') \ . \tag{112}$$

As, by definition of transpose,

$$\langle\,\overleftarrow{\mathbf{u}}_n\,,(\mathbf{A}\,\overrightarrow{\mathbf{u}}_n)\,\delta\mathbf{m}\,\rangle = \langle\,(\mathbf{A}\,\overrightarrow{\mathbf{u}}_n)^t\,\overleftarrow{\mathbf{u}}_n\,,\delta\mathbf{m}\,\rangle \ , \tag{113}$$

the components of the gradient follow using (103):

$$\hat{\gamma}_\rho(\mathbf{x})_n = \overset{\leftarrow}{v}{}^i(\mathbf{x},t)_n \otimes \vec{v}{}^i(\mathbf{x},t)_n \Big|_{t=0} ,$$

(114)

and

$$\hat{\gamma}_\Psi{}^{ijkl}(\mathbf{x},t)_n = - \overset{\leftarrow}{\epsilon}{}^{ij}(\mathbf{x},t)_n \otimes \overset{\rightarrow}{\epsilon}{}^{kl}(\mathbf{x},t)_n ,$$

(115)

where the time correlation between two functions $f(t)$ and $g(t)$ has been defined by

$$f(t) \otimes g(t) = \int_{t_0}^{t_1} dt' \, f(t') \, g(t'-t) .$$

(116)

In Appendix C, I give a more direct demonstration of formulas (114)-(115).

If we are not interested in the attenuating properties of the medium, but only in the elastic properties, we make the hypothesis

$$\Psi^{ijkl}(\mathbf{x},\tau) = c^{ijkl}(\mathbf{x}) \, \delta(\tau) .$$

(8 again)

Instead of modifying the theory to compute the gradient with respect to the elastic stiffnesses, it is clear that only the correlations (155) at zero lag can contribute. Then,

$$\hat{\gamma}_c{}^{ijkl}(\mathbf{x})_n = \overset{\leftarrow}{\epsilon}{}^{ij}(\mathbf{x},t)_n \otimes \overset{\rightarrow}{\epsilon}{}^{kl}(\mathbf{x},t)_n \Big|_{t=0} ,$$

(117)

which corresponds to the result demonstrated by Tarantola (1987b).

Equation (106g): $\gamma_n = \mathbf{C}_M \, \hat{\gamma}$. The operator $\mathbf{C}_M$ describes the confidence we have on our a priori model. Assuming for instance that the uncertainties on density are independent on the uncertainties on the rate-of-relaxation function,

$$\mathbf{C}_M = \begin{bmatrix} \mathbf{C}_\rho & 0 \\ 0 & \mathbf{C}_\Psi \end{bmatrix} ,$$

(118)

gives, introducing the corresponding covariance functions,

$$\gamma_\rho(\mathbf{x})_n = \int_V dV(\mathbf{x}') \, C_\rho(\mathbf{x},\mathbf{x}') \, \hat{\gamma}_\rho(\mathbf{x}')_n ,$$

(119a)

and

$$\gamma_\Psi{}^{ijkl}(\mathbf{x},\tau)_n = \int_V dV(\mathbf{x}') \int_{-\infty}^{+\infty} dt' \, C_\Psi{}^{ijklpqrs}(\mathbf{x},\tau;\mathbf{x}',\tau') \, \hat{\gamma}_\Psi{}^{pqrs}(\mathbf{x}',\tau')_n .$$

(119b)

Equation (106h): $\mathbf{m}_{n+1} = \mathbf{m}_n - \alpha_n \, \gamma_n$. This gives, finally,

$$\rho(\mathbf{x})_{n+1} = \rho(\mathbf{x})_n - \alpha_n \, \gamma_\rho(\mathbf{x})_n ,$$

(120)

and

$$\Psi^{ijkl}(\mathbf{x},\tau)_{n+1} = \Psi^{ijkl}(\mathbf{x},\tau)_n - \alpha_n \gamma_{\Psi}^{ijkl}(\mathbf{x},\tau)_n \; , \tag{121}$$

12: DISCUSSION AND CONCLUSION

For the purposes of inversion os seismic waveforms it seems better to introduce a completely general rate-of-relaxation function $\Psi^{ijkl}(\mathbf{x},\tau)$ rather than to use a particular model (Standard linear solid, Constant Q, etc.). Some seismic data sets at least seem to contain enough information to give useful constraints on $\Psi^{ijkl}(\mathbf{x},\tau)$, as for instance seismic reflection data.

The representation theorems are as simple with this general linear model as for the particular perfectly elastic model. Only a supplementary stress term appears. which is due to the history of deformation before the initial time. Reciprocity is also satisfied.

The formulas given in Section 11 correspond to a steepest descent method. Practically, a lot of modification must be introduced. For instance, conjugate gradients may be used instead of steepest descent. But more fundamentally, I suggest not to use from the begining all the components of the gradient that may be computed. First invert for elastic parameters, and, after convergence, allow attenuation to be introduced. This means that, at the begining, only the values for zero lag of the correlations (115) should be taken into account.

In fact, as suggested by Tarantola (1986) the elastic parameters themselves are highly hierachisized. For instance, with seismic reflection surface data, the long wavelengths of the P-wave velocity have to be first computed, then the P-wave impedance (product of density by velocity). When a good model has been obtained, the S-wave velocity and S-wave impedance can be computed.

When all these things have satisfactorily converged, then formula (115) allows to go one step further, and to obtain the best model of attenuation.

ACKNOWLEDGEMENTS

I thank my colleague G. Jobert for many helpful suggestions. This research has been sponsored by the European Commission, the French CNRS, Amoco, Aramco, Elf-Aquitaine, C.G.G., Convex Computers, Merlin Geophysical, Shell, and Total-C.F.P.

REFERENCES AND REFERENCES FOR GENERAL READING

Aki, K., and Richards, P.G., 1980. Quantitative seismology (2 volumes), Freeman and Co.

Alterman, Z.S., and Karal, F.C., Jr., 1968. Propagation of elastic waves in layered media by finite difference methods, Bull. Seismological Soc. of America, 58, 367-398.

Backus, G., 1970. Inference from inadequate and inaccurate data: I, Proceedings of the National Academy of Sciences, 65, 1, 1-105.

Backus, G., 1970. Inference from inadequate and inaccurate data: II, Proceedings of the National Academy of Sciences, 65, 2, 281-287.

Backus, G., 1970. Inference from inadequate and inaccurate data: III, Proceedings of the National Academy of Sciences, 67, 1, 282-289.

Backus, G., and Gilbert, F., 1967. Numerical applications of a formalism for geophysical inverse problems, Geophys. J. R. astron. Soc., 13, 247-276.

Backus, G., and Gilbert, F., 1968. The resolving power of gross Earth data, Geophys. J. R. astron. Soc., 16, 169-205.

Backus, G., and Gilbert, F., 1970. Uniqueness in the inversion of inaccurate gross Earth data, Philos. Trans. R. Soc. London, 266, 123-192.

Backus, G., 1971. Inference from inadequate and inaccurate data, Mathematical problems in the Geophysical Sciences: Lecture in applied mathematics, 14, American Mathematical Society, Providence, Rhode Island.

Balakrishnan, A.V., 1976. Applied functional analysis, Springer-Verlag.

Bamberger, A., Chavent, G, Hemon, Ch., and Lailly, P., 1982. Inversion of normal incidence seismograms, Geophysics, 47, 757-770.

Ben-Menahem, A., and Singh, S.J., 1981. Seismic waves and sources, Springer-Verlag.

Bender, C.M., and Orszag, S.A., 1978. Advanced mathematical methods for scientists and engineers, McGraw-Hill.

Berkhout, A.J., 1980. Seismic migration, imaging of acoustic energy by wave field extrapolation, Elsevier.

Beydoun, W., 1985. Asymptotic wave methods in heterogeneous media, Ph.D Thesis, Massachusetts Institute of Technology.

Bland, D.R., 1960. The theory of linear viscoelasticity, Pergamon Press.

Bleistein, N., 1984. Mathematical methods for wave phenomena, Academic Press.

Bourbie, T., Coussy O., and Zinszner, B., 1986. Acoustique des milieux poreux, Technip, 1986.

Boschi, E., 1973. Body force equivalents for dislocations in visco-elastic media, J. Geophys. Res., Vol. 78, No. 32, 7727-8590.

Claerbout, J.F., 1971. Toward a unified theory of reflector mapping, Geophysics, **36**, 467-481.

Claerbout, J.F., 1976. Fundamentals of Geophysical data processing, McGraw-Hill.

Claerbout, J.F., 1985. Imaging the Earth's interior, Blackwell.

Claerbout, J.F., and Muir, F., 1973. Robust modelling with erratic data, Geophysics, **38**, 5, 826-844.

Courant, R., and Hilbert, D., 1953. Methods of mathematical physics, Interscience Publishers.

Dautray, R., and Lions, J.L., 1984 and 1985. Analyse mathématique et calcul numérique pour les sciences et les techniques (3 volumes), Masson, Paris.

Davidon, W.C., 1959. Variable metric method for minimization, AEC Res. and Dev., Report ANL-5990 (revised).

Devaney, A.J., 1984. Geophysical diffraction tomography, IEEE trans. Geos. remote sensing, Vol. GE-22, No. 1.

Fletcher, R., 1980. Practical methods of optimization, Volume 1: Unconstrained optimization, Wiley.

Fletcher, R., 1981. Practical methods of optimization, Volume 2: Constrained optimization, Wiley.

Franklin, J.N., 1970. Well posed stochastic extensions of ill posed linear problems, J. Math. Anal. Applic., **31**, 682-716.

Gauthier, O., Virieux, J., and Tarantola, A., 1986. Two-dimensional inversion of seismic waveforms: numerical results, Geophysics, **51**, 1387-1403.

Hamilton, E.L., 1972. Compressional-wave attenuation in marine sediments, Geophysics, **37**, 620-646.

Herman, G.T., 1980. Image reconstruction from projections, the fundamentals of computerized tomography, Academic Press.

Hsi-Ping Liu, Anderson, D.L., and Kanamori, H., 1976, Velocity dispersion due to anelasticity; implications for seismology and mantle composition, Geophys. J. R. astr. Soc., **47**, 41-58.

Ikelle, L.T., Diet, J.P., and Tarantola, A., 1986. Linearized inversion of multi offset seismic reflection data in the ω-k domain, Geophysics, **51**, 1266-1276.

Jacobson, R.S., 1987. An investigation into the fundamental relationships between attenuation, phase dispersion, and frequency using seismic refraction profiles over sedimentary structures, Geophysics, **52**, 72-87.

Johnson, L.R., 1974, Green's function for Lamb's problem, Geophys. J.R. astr. Soc., **37**, 99-131.

Kennett, B.L.N., 1974. On variational principles and matrix methods in elastodynamics, Geophys. J.R. astr. Soc., **37**, 391-405.

Kennett, B.L.N., 1978. Some aspects of non-linearity in inversion, Geophys. J. R. astr. Soc., **55**, 373-391.

Kjartansson, E., 1979. Constant Q-Wave propagation and attenuation, J. Geoph. Res., Vol. 84, No. B9, 4737-4748.

Licknerowicz, A., 1960. Eléments de Calcul Tensoriel, Armand Collin, Paris.

Lions, J.L., 1968. Contrôle optimal de systèmes gouvernés par des équations aux dérivées partielles, Dunod, Paris. English translation: Optimal control of systems governed by partial differential equations, Springer, 1971.

Liu, H.P., Anderson, D.L., and Kanamori, H., 1976. Velocity dispersion due to anelasticity; implications for seismology and mantle composition, Geophys. J. R. astr. Soc., **47**, 41-58.

Milne, R.D., 1980. Applied functional Analysis, Pitman Advanced Publishing Program, Boston.

Misner, Ch.W., Thorne, K.S., and Wheeler, J.A., 1973. Gravitation, Freeman.

Moritz, H., 1980. Advanced physical geodesy, Herbert Wichmann Verlag, Karlsruhe, Abacus Press, Tunbridge Wells, Kent.

Morse, P.M., and Feshbach, H., 1953. Methods of theoretical physics, McGraw-Hill.

Nercessian, Al., Hirn, Al., and Tarantola, Al., 1984. Three-dimensional seismic transmission prospecting of the Mont-Dore volcano, France, Geophys. J.R. astr. Soc., 76, 307-315.

Nolet, G., 1985. Solving or resolving inadequate and noisy tomographic systems, J. Comp. Phys., 61, 463-482.

O'Connell, R.J., and Budiansky, B., 1978. Measures of dissipation in viscoelastic media, Geophysical Research Letters, Vol. 5, No. 1, 5-8.

Park, J., and Gilbert, F., 1986. Coupled free oscillations of an aspherical, dissipative, rotating Earth: Galerkin theory, J. Geoph. Res., Vol. 91, No. B7, 7241-7260.

Parker, R.L., 1975. The theory of ideal bodies for gravity interpretation, Geophys. J. R. astron. Soc., 42, 315-334.

Parker, R.L., 1977. Understanding inverse theory, Ann. Rev. Earth Plan. Sci., 5, 35-64.

Polack, E. et Ribière, G., 1969. Note sur la convergence de méthodes de directions conjuguées, Revue Fr. Inf. Rech. Oper., 16-R1, 35-43.

Powell, M.J.D., 1981. Approximation theory and methods, Cambridge University Press.

Razavy, M., and Lenoach, B., 1986. Reciprocity principle and the approximate solution of the wave equation, Bull. Seism. Soc. of America, 76, 1776-1789.

Ricker, N.H., 1977. Transient waves in visco-elastic media, Elsevier.

Scales, L. E., 1985. Introduction to non-linear optimization, Macmillan.

Schoenberger M., and Levin, F.K., 1974. Apparent attenuation due to intrabed multiples, Geophysics, 39, 278-291.

Schwartz, L., 1965. Méthodes mathématiques pour les sciences physiques, Hermann, Paris.

Schwartz, L., 1966. Théorie des distributions, Hermann, Paris.

Schwartz, L., 1970. Analyse (topologie générale et analyse fontionelle), Hermann, Paris.

Tanimoto, A., 1985. The Backus-Gilbert approach to the three-dimensional structure in the upper mantle. - I. Lateral variation of surface wave phase velocity with its error and resolution., Geophys. J. R. astr. Soc., 82, 105-123.

Tarantola, A., 1984. Linearized inversion of seismic reflection data, Geophysical Prospecting, 32, 998-1015.

Tarantola, A., 1984. Inversion of seismic reflection data in the acoustic approximation, Geophysics, 49, 1259-1266.

Tarantola, A., 1984. The seismic reflection inverse problem, in: Inverse Problems of Acoustic and Elastic Waves, edited by: F. Santosa, Y.-H. Pao, W. Symes, and Ch. Holland, SIAM, Philadelphia.

Tarantola, A., 1986. A strategy for nonlinear elastic inversion of seismic reflection data, Geophysics, 51, 1893-1903.

Tarantola, A., 1987. Inverse problem theory; methods for data fitting and model parameter estimation, Elsevier.

Tarantola, A., 1987. Inversion of travel time and seismic waveforms, in: Seismic tomography, edited by G. Nolet, Reidel.

Tarantola, A., Jobert, G., Trézéguet, D., and Denelle, E., 1988. Inversion of seismic waveforms can either be performed by time or by depth extrapolation, Geophysical Prospecting (in press).

Tarantola, A. and Nercessian, A., 1984. Three-dimensional inversion without blocks, Geophys. J. R. astr. Soc., 76, 299-306.

Tarantola, A., and Valette, B., 1982a. Inverse Problems = Quest for Information, J. Geophys., 50, 159-170.

Tarantola, A., and Valette, B., 1982b. Generalized nonlinear inverse problems solved using the least-squares criterion, Rev. Geophys. Space Phys., 20, No. 2, 219-232.

Taylor, A.E., and Lay, D.C., 1980. Introduction to functional analysis, Wiley.

Virieux, J., 1986, P-SV wave propagation in heterogeneous media; velocity-stress finite-difference method, Geophysics, 51, 889-901.

Walsh, G.R., 1975. Methods of optimization, Wiley.

Watson, G.A., 1980. Approximation theory and numerical methods, Wiley.

Woodhouse, J.H., 1974. Surface waves in a laterally varying layered structure, Geophys. J.R. astr. Soc., 37, 461-490.

Woodhouse, J. H., and Dziewonski, A.M., 1984. Mapping the upper mantle: Three-dimensional modeling of Earth structure by inversion of seismic waveforms, Journal of Geophysical Research, Vol. 89, No. B7, 5953-5986.

Wu, R-S, and Ben-Menahem, A., 1985. The elastodynamic near field. Geophys. J.R. astr. Soc., **81**, 609-621.

Appendix A: TRANSPOSED OF THE WAVE EQUATION OPERATOR. DUAL CONDITIONS

The operators L and L^t have been defined by:

$$(L\,\mathbf{u})^i(\mathbf{x},t) = \rho(\mathbf{x})\,\frac{\partial^2 u^i}{\partial t^2}(\mathbf{x},t) - \frac{\partial}{\partial x^j}\int_{-\infty}^{+\infty}dt'\,\overrightarrow{\Psi}^{jikl}(\mathbf{x};t,t')\,\frac{\partial u^k}{\partial x^l}(\mathbf{x},t') \ , \tag{A-1}$$

and

$$(L^t\,\mathbf{u})^i(\mathbf{x},t) = \rho(\mathbf{x})\,\frac{\partial^2 u^i}{\partial t^2}(\mathbf{x},t) - \frac{\partial}{\partial x^j}\int_{-\infty}^{+\infty}dt'\,\overleftarrow{\Psi}^{ijkl}(\mathbf{x};t,t')\,\frac{\partial u^k}{\partial x^l}(\mathbf{x},t') \ , \tag{A-2}$$

where $\overrightarrow{\Psi}^{jikl}(\mathbf{x};t,t')$ is a causal function,

$$\overrightarrow{\Psi}^{ijkl}(\mathbf{x};t,t') = 0 \qquad \text{for } t<t' \ , \tag{A-3}$$

and

$$\overleftarrow{\Psi}^{ijkl}(\mathbf{x};t,t') = \overrightarrow{\Psi}^{ijkl}(\mathbf{x};t',t) \ . \tag{A-4}$$

We have

$$\langle \overleftarrow{\mathbf{u}} , L\,\overrightarrow{\mathbf{u}} \rangle_\Phi - \langle L^t\,\overleftarrow{\mathbf{u}} , \overrightarrow{\mathbf{u}} \rangle_U =$$

$$= \int_V dV(\mathbf{x})\int_{t_0}^{t_1}dt\,\overleftarrow{u}^i(\mathbf{x},t)\,(L\,\overrightarrow{\mathbf{u}})^i(\mathbf{x},t) - \int_V dV(\mathbf{x})\int_{t_0}^{t_1}dt\,(L^t\,\overleftarrow{\mathbf{u}})^i(\mathbf{x},t)\,\overrightarrow{u}^i(\mathbf{x},t) \ ,$$

and, inserting (A-1) and (A-2),

$$\langle \overleftarrow{\mathbf{u}} , L\,\overrightarrow{\mathbf{u}} \rangle_\Phi - \langle L^t\,\overleftarrow{\mathbf{u}} , \overrightarrow{\mathbf{u}} \rangle_U = A + B \ , \tag{A-5}$$

where

$$A = \int_V dV(\mathbf{x})\,\rho(\mathbf{x})\int_{t_0}^{t_1}dt\left[\overleftarrow{u}^i(\mathbf{x},t)\,\frac{\partial^2\overrightarrow{u}^i}{\partial t^2}(\mathbf{x},t) - \frac{\partial^2\overleftarrow{u}^i}{\partial t^2}(\mathbf{x},t)\,\overrightarrow{u}^i(\mathbf{x},t)\right] \ , \tag{A-6}$$

and

$$B = -\int_V dV(\mathbf{x})\int_{t_0}^{t_1}dt\,\overleftarrow{u}^i(\mathbf{x},t)\left[\frac{\partial}{\partial x^j}\int_{-\infty}^{+\infty}dt'\,\overrightarrow{\Psi}^{ijkl}(\mathbf{x};t,t')\,\frac{\partial\overrightarrow{u}^k}{\partial x^l}(\mathbf{x},t')\right]$$

$$+ \int_V dV(\mathbf{x})\int_{t_0}^{t_1}dt\left[\frac{\partial}{\partial x^j}\int_{-\infty}^{+\infty}dt'\,\overleftarrow{\Psi}^{ijkl}(\mathbf{x};t,t')\,\frac{\partial\overleftarrow{u}^k}{\partial x^l}(\mathbf{x},t')\right]\overleftarrow{u}^i(\mathbf{x},t) \ . \tag{A-7}$$

As

$$\frac{\partial}{\partial t}\left[\overleftarrow{u}{}^i\,\frac{\partial\overrightarrow{u}{}^i}{\partial t} - \frac{\partial\overleftarrow{u}{}^i}{\partial t}\,\overrightarrow{u}{}^i\right] = \overleftarrow{u}{}^i\,\frac{\partial^2\overrightarrow{u}{}^i}{\partial t^2} - \frac{\partial^2\overleftarrow{u}{}^i}{\partial t^2}\,\overrightarrow{u}{}^i\,,$$

we have

$$A = \int_V dV(\mathbf{x})\,\rho(\mathbf{x})\left[\overleftarrow{u}{}^i(\mathbf{x},t)\,\frac{\partial^2\overrightarrow{u}{}^i}{\partial t^2}(\mathbf{x},t) - \frac{\partial^2\overleftarrow{u}{}^i}{\partial t^2}(\mathbf{x},t)\,\overrightarrow{u}{}^i(\mathbf{x},t)\right]\Bigg|_{t=t_0}^{t=t_1}. \tag{A-8}$$

Equation (A-7) can be rewritten

$$B = C + D\,, \tag{A-9}$$

where

$$C = -\int_V dV(\mathbf{x})\int_{t_0}^{t_1} dt\,\frac{\partial}{\partial x^j}\left[\overleftarrow{u}{}^i(\mathbf{x},t)\left(\int_{-\infty}^{+\infty} dt'\,\overrightarrow{\Psi}{}^{ijkl}(\mathbf{x};t,t')\,\frac{\partial\overrightarrow{u}{}^k}{\partial x^l}(\mathbf{x},t')\right)\right]$$

$$+ \int_V dV(\mathbf{x})\int_{t_0}^{t_1} dt\,\frac{\partial}{\partial x^j}\left[\left(\int_{-\infty}^{+\infty} dt'\,\overleftarrow{\Psi}{}^{ijkl}(\mathbf{x};t,t')\,\frac{\partial\overleftarrow{u}{}^k}{\partial x^l}(\mathbf{x},t')\right)\overrightarrow{u}{}^i(\mathbf{x},t)\right], \tag{A-10}$$

and

$$D = +\int_V dV(\mathbf{x})\int_{t_0}^{t_1} dt\,\frac{\partial\overleftarrow{u}{}^i}{\partial x^j}(\mathbf{x},t)\left(\int_{-\infty}^{+\infty} dt'\,\overrightarrow{\Psi}{}^{ijkl}(\mathbf{x};t,t')\,\frac{\partial\overrightarrow{u}{}^k}{\partial x^l}(\mathbf{x},t')\right)$$

$$- \int_V dV(\mathbf{x})\int_{t_0}^{t_1} dt\left(\int_{-\infty}^{+\infty} dt'\,\overleftarrow{\Psi}{}^{ijkl}(\mathbf{x};t,t')\,\frac{\partial\overleftarrow{u}{}^k}{\partial x^l}(\mathbf{x},t')\right)\frac{\partial\overrightarrow{u}{}^i}{\partial x^j}(\mathbf{x},t)\,. \tag{A-11}$$

Using the divergence theorem gives

$$C = -\int_S dS(\boldsymbol{\xi})\int_{t_0}^{t_1} dt\,\overleftarrow{u}{}^i(\boldsymbol{\xi},t)\left[n^j(\boldsymbol{\xi})\int_{-\infty}^{+\infty} dt'\,\overrightarrow{\Psi}{}^{ijkl}(\boldsymbol{\xi};t,t')\,\frac{\partial\overrightarrow{u}{}^k}{\partial x^l}(\boldsymbol{\xi},t')\right]$$

$$+ \int_S dS(\boldsymbol{\xi})\int_{t_0}^{t_1} dt\left[n^j(\boldsymbol{\xi})\int_{-\infty}^{+\infty} dt'\,\overleftarrow{\Psi}{}^{ijkl}(\boldsymbol{\xi};t,t')\,\frac{\partial\overleftarrow{u}{}^k}{\partial x^l}(\boldsymbol{\xi},t')\right]\overrightarrow{u}{}^i(\boldsymbol{\xi},t)\,. \tag{A-12}$$

The term D can be rewritten

$$D = +\int_V dV(\mathbf{x})\int_{t_0}^{t_1} dt\,\frac{\partial\overleftarrow{u}{}^i}{\partial x^j}(\mathbf{x},t)\left[\left(\int_{-\infty}^{t_0} dt' + \int_{t_0}^{t_1} dt' + \int_{t_1}^{+\infty} dt'\right)\overrightarrow{\Psi}{}^{ijkl}(\mathbf{x};t,t')\,\frac{\partial\overrightarrow{u}{}^k}{\partial x^l}(\mathbf{x},t')\right]$$

$$- \int_V dV(\mathbf{x})\int_{t_0}^{t_1} dt\left[\left(\int_{-\infty}^{t_0} dt' + \int_{t_0}^{t_1} dt' + \int_{t_1}^{+\infty} dt'\right)\overleftarrow{\Psi}{}^{ijkl}(\mathbf{x};t,t')\,\frac{\partial\overleftarrow{u}{}^k}{\partial x^l}(\mathbf{x},t')\right]\frac{\partial\overrightarrow{u}{}^i}{\partial x^j}(\mathbf{x},t)\,,$$

or, using the causality and anti-causality properties of $\overrightarrow{\Psi}{}^{ijkl}(\mathbf{x};t,t')$ and $\overleftarrow{\Psi}{}^{ijkl}(\mathbf{x};t,t')$ respectively,

$$D = + \int_V dV(\mathbf{x}) \int_{t_0}^{t_1} dt\, \frac{\overleftarrow{\partial u}^i}{\partial x^j}(\mathbf{x},t) \left(\left[\int_{-\infty}^{t_0} dt' + \int_{t_0}^{t_1} dt' \right] \overrightarrow{\Psi}^{ijkl}(\mathbf{x};t,t') \frac{\overrightarrow{\partial u}^k}{\partial x^l}(\mathbf{x},t') \right)$$

$$- \int_V dV(\mathbf{x}) \int_{t_0}^{t_1} dt \left(\left[\int_{t_0}^{t_1} dt' + \int_{t_1}^{+\infty} dt' \right] \overleftarrow{\Psi}^{ijkl}(\mathbf{x};t,t') \frac{\overleftarrow{\partial u}^k}{\partial x^l}(\mathbf{x},t') \right) \frac{\overleftarrow{\partial u}^i}{\partial x^j}(\mathbf{x},t) \, ,$$

i.e.,

$$D = E + F \, , \tag{A-13}$$

where

$$E = + \int_V dV(\mathbf{x}) \int_{t_0}^{t_1} dt \int_{t_0}^{t_1} dt'\, \frac{\overleftarrow{\partial u}^i}{\partial x^j}(\mathbf{x},t)\, \overrightarrow{\Psi}^{ijkl}(\mathbf{x};t,t')\, \frac{\overrightarrow{\partial u}^k}{\partial x^l}(\mathbf{x},t'))$$

$$- \int_V dV(\mathbf{x}) \int_{t_0}^{t_1} dt \int_{t_0}^{t_1} dt'\, \overleftarrow{\Psi}^{ijkl}(\mathbf{x};t,t')\, \frac{\overleftarrow{\partial u}^k}{\partial x^l}(\mathbf{x},t'))\, \frac{\overleftarrow{\partial u}^i}{\partial x^j}(\mathbf{x},t) \, , \tag{A-14}$$

and

$$F = + \int_V dV(\mathbf{x}) \int_{t_0}^{t_1} dt\, \frac{\overleftarrow{\partial u}^i}{\partial x^j}(\mathbf{x},t) \left(\int_{-\infty}^{t_0} dt'\, \overrightarrow{\Psi}^{ijkl}(\mathbf{x};t,t')\, \frac{\overrightarrow{\partial u}^k}{\partial x^l}(\mathbf{x},t') \right)$$

$$- \int_V dV(\mathbf{x}) \int_{t_0}^{t_1} dt \left(\int_{t_1}^{+\infty} dt'\, \overleftarrow{\Psi}^{ijkl}(\mathbf{x};t,t')\, \frac{\overleftarrow{\partial u}^k}{\partial x^l}(\mathbf{x},t') \right) \frac{\overleftarrow{\partial u}^i}{\partial x^j}(\mathbf{x},t) \, . \tag{A-15}$$

Using (A-4) and the symmetries between indexes ij and kj gives

$$E = 0 \, . \tag{A-15}$$

This ends the proof for equation (39) of the text.

Appendix B: REPRESENTATION THEOREM

Let $\overleftarrow{G}^{ij}(\mathbf{x},t;\mathbf{x}',t')$ be defined by

$$\rho(\mathbf{x})\, \frac{\partial^2 \overleftarrow{G}^{ip}}{\partial t^2}(\mathbf{x},t;\mathbf{x}',t') - \frac{\partial}{\partial x^j} \int_{-\infty}^{+\infty} dt''\, \overleftarrow{\Psi}^{ijkl}(\mathbf{x};t,t'')\, \frac{\partial \overleftarrow{G}^{kp}}{\partial x^l}(\mathbf{x},t'';\mathbf{x}',t') = \delta^{ip}\, \delta(\mathbf{x}-\mathbf{x}')\, \delta(t-t') \, , \tag{B-1a}$$

$$\overleftarrow{G}^{ip}(\mathbf{x},t;\mathbf{x}',t') = 0 \quad \text{for } t \geq t' \tag{B-1b}$$

$$\frac{\partial \overleftarrow{G}^{ip}}{\partial t}(\mathbf{x},t;\mathbf{x}',t') = 0 \quad \text{for } t \geq t' \, , \tag{B-1c}$$

where $\overleftarrow{\Psi}^{ijkl}(\mathbf{x};t,t')$ is anti-causal:

$$\overleftarrow{\Psi}^{ijkl}(\mathbf{x};t,t') = 0 \qquad \text{for } t > t' \, . \tag{B-2}$$

For an arbitrary function $u^i(x,t)$ we have, for any t', $t_0 < t' < t_1$,

$$u^p(x',t') = \int_V dV(x) \int_{t_0}^{t_1} dt \, \delta^{ip} \, \delta(x-x') \, \delta(t-t') \, u^i(x,t) . \tag{B-3}$$

Inserting (B-1a), into (B-3) gives

$$u^p(x',t') = A + B , \tag{B-4}$$

where

$$A = \int_V dV(x) \, \rho(x) \int_{t_0}^{t_1} dt \, \frac{\partial^2 \overleftarrow{G}^{ip}}{\partial t^2}(x,t;x',t') \, u^i(x,t) , \tag{B-5}$$

and

$$B = - \int_V dV(x) \int_{t_0}^{t_1} dt \left[\frac{\partial}{\partial x^j} \int_{-\infty}^{+\infty} dt'' \, \overleftarrow{\Psi}^{ijkl}(x;t,t'') \, \frac{\partial \overleftarrow{G}^{kp}}{\partial x^l}(x,t'';x',t') \right] u^i(x,t) . \tag{B-6}$$

As

$$\frac{\partial^2 \overleftarrow{G}^{ip}}{\partial t^2} u^i = \frac{\partial}{\partial t}\left[\frac{\partial \overleftarrow{G}^{ip}}{\partial t} u^i - \overleftarrow{G}^{ip} \frac{\partial u^i}{\partial t} \right] + \overleftarrow{G}^{ip} \frac{\partial^2 u^i}{\partial t^2} ,$$

and, as $\overleftarrow{G}^{ip}(x,t;x',t')$ is anti-causal, we have

$$A = \int_V dV(x) \int_{t_0}^{t_1} dt \, \overleftarrow{G}^{ip}(x,t;x',t') \, \rho(x) \, \frac{\partial^2 u^i}{\partial t^2}(x,t)$$

$$- \int_V dV(x) \, \rho(x) \left[\frac{\partial \overleftarrow{G}^{ip}}{\partial t}(x,t_0;x',t') \, u^i(x,t_0) - \overleftarrow{G}^{ip}(x,t_0;x',t') \, \frac{\partial u^i}{\partial t}(x,t_0) \right] . \tag{B-7}$$

Equation (B-6) can be rewritten

$$B = C + D , \tag{B-8}$$

where

$$C = - \int_V dV(x) \int_{t_0}^{t_1} dt \, \frac{\partial}{\partial x^j} \left[\left(\int_{-\infty}^{+\infty} dt'' \, \overleftarrow{\Psi}^{ijkl}(x;t,t'') \, \frac{\partial \overleftarrow{G}^{kp}}{\partial x^l}(x,t'';x',t') \right) u^i(x,t) \right] , \tag{B-9}$$

and

$$D = + \int_V dV(x) \int_{t_0}^{t_1} dt \left[\left(\int_{-\infty}^{+\infty} dt'' \, \overleftarrow{\Psi}^{ijkl}(x;t,t'') \, \frac{\partial \overleftarrow{G}^{kp}}{\partial x^l}(x,t'';x',t') \right) \frac{\partial u^i}{\partial x^j}(x,t) \right] . \tag{B-10}$$

Using the divergence theorem gives

$$C = - \int_S dS(\xi) \int_{t_0}^{t_1} dt \left[n^j(\xi) \int_{-\infty}^{+\infty} dt'' \, \overleftarrow{\Psi}^{ijkl}(x;t,t'') \, \frac{\partial \overleftarrow{G}^{kp}}{\partial x^l}(x,t'';x',t') \right] u^i(x,t)) . \tag{B-11}$$

Using

$$\overleftarrow{\Psi}^{ijkl}(x;t,t'') = \overrightarrow{\Psi}^{klij}(x;t'',t) , \tag{B-12}$$

changing $t \longleftrightarrow t''$, $ij \longleftrightarrow kl$, and reordering gives

$$D = \int_V dV(x) \int_{-\infty}^{+\infty} dt \frac{\partial \overleftarrow{G}^{ip}}{\partial x^j}(x,t;x',t') \int_{t_0}^{t_1} dt'' \ \overrightarrow{\Psi}^{ijkl}(x;t,t'') \frac{\partial u^k}{\partial x^l}(x,t'') \ . \tag{B-13}$$

The anti-causality property of $\overleftarrow{G}^{ip}(x,t;x',t')$ and the causality property of $\overrightarrow{\Psi}^{ijkl}(x;t,t')$ allow to change the bounds of integration:

$$D = \int_V dV(x) \int_{t_0}^{t_1} dt \frac{\partial \overleftarrow{G}^{ip}}{\partial x^j}(x,t;x',t') \int_{t_0}^{+\infty} dt'' \ \overrightarrow{\Psi}^{ijkl}(x;t,t'') \frac{\partial u^k}{\partial x^l}(x,t'') \ , \tag{B-14}$$

i.e.,

$$D = \int_V dV(x) \int_{t_0}^{t_1} dt \frac{\partial \overleftarrow{G}^{ip}}{\partial x^j}(x,t;x',t') \left(\int_{-\infty}^{+\infty} dt'' - \int_{-\infty}^{t_0} dt'' \right) \overrightarrow{\Psi}^{ijkl}(x;t,t'') \frac{\partial u^k}{\partial x^l}(x,t'') \ . \tag{B-15}$$

This gives

$$D = E + F \ , \tag{B-16}$$

where

$$E = + \int_V dV(x) \int_{t_0}^{t_1} dt \frac{\partial \overleftarrow{G}^{ip}}{\partial x^j}(x,t;x',t') \int_{-\infty}^{+\infty} dt'' \ \overrightarrow{\Psi}^{ijkl}(x;t,t'') \frac{\partial u^k}{\partial x^l}(x,t'') \ , \tag{B-17}$$

and

$$F = - \int_V dV(x) \int_{t_0}^{t_1} dt \frac{\partial \overleftarrow{G}^{ip}}{\partial x^j}(x,t;x',t') \int_{-\infty}^{t_0} dt'' \ \overrightarrow{\Psi}^{ijkl}(x;t,t'') \frac{\partial u^k}{\partial x^l}(x,t'') \ . \tag{B-18}$$

Equation (B-17) can be written

$$E = G + H \ , \tag{B-19}$$

where

$$G = + \int_V dV(x) \int_{t_0}^{t_1} dt \frac{\partial}{\partial x^j} \left[\overleftarrow{G}^{ip}(x,t;x',t') \int_{-\infty}^{+\infty} dt'' \ \overrightarrow{\Psi}^{ijkl}(x;t,t'') \frac{\partial u^k}{\partial x^l}(x,t'') \right] \ , \tag{B-20}$$

and

$$H = - \int_V dV(x) \int_{t_0}^{t_1} dt \ \overleftarrow{G}^{ip}(x,t;x',t') \left[\frac{\partial}{\partial x^j} \int_{-\infty}^{+\infty} dt'' \ \overrightarrow{\Psi}^{ijkl}(x;t,t'') \frac{\partial u^k}{\partial x^l}(x,t'') \right] \ . \tag{B-21}$$

Using the divergence theorem gives

$$G = + \int_S dS(\xi) \int_{t_0}^{t_1} dt \ \overleftarrow{G}^{ip}(\xi,t;x',t') \left[n^j(\xi) \int_{-\infty}^{+\infty} dt'' \ \overrightarrow{\Psi}^{ijkl}(\xi;t,t'') \frac{\partial u^k}{\partial x^l}(\xi,t'') \right] \ . \tag{B-22}$$

This ends the proof for equation (62) of the text.

Appendix C: ALTERNATIVE COMPUTATION OF THE GRADIENT

The calculated displacement at the receiver locations, $\mathbf{d} = \{ \vec{u}^i(x_r,t) \}$, is a nonlinear function of the model parameters $\mathbf{m} = \{ \rho(x), \Psi^{ijkl}(x,\tau) \}$. A perturbation of the model parameters

$$\rho(x) \rightarrow \rho(x) + \delta\rho(x) \tag{C-1a}$$

$$\Psi^{ijkl}(x,\tau) \rightarrow \Psi^{ijkl}(x,\tau) + \delta\Psi^{ijkl}(x,\tau) \tag{C-1b}$$

leads to a perturbation of the displacement field

$$\vec{u}^i(x_r,t) \rightarrow \vec{u}^i(x_r,t) + \delta\vec{u}^i(x_r,t). \tag{C-1c}$$

Writting the first order approximation to $\delta\mathbf{d} = \{ \delta\vec{u}^i(x_r,t) \}$ as

$$\delta\mathbf{d} = \mathbf{A} \, \delta\rho + \mathbf{B} \, \delta\Psi, \tag{C-2a}$$

or, explicitly,

$$\delta\vec{u}^i(x_r,t) = \int_V dV(x) \, A^i(x_r,t;x) \, \delta\rho(x) + \int_V dV(x) \int_{-\infty}^{+\infty} d\tau \, B^{ijklm}(x_r,t;x,\tau) \, \delta\Psi^{jklm(x,\tau)}, \tag{C-2b}$$

defines the derivative operators of calculated displacements with respect to model parameters and their kernels.

As demonstrated in Section 9,

$$\delta\vec{u}^P(x',t') = - \int_V dV(x) \int_{t_0}^{t_1} dt \, \vec{G}_{free}^{\,pi}(x',t';x,t) \, \delta\rho(x) \, \frac{\partial^2\vec{u}^i}{\partial t^2}(x,t)$$

$$- \int_V dV(x) \int_{t_0}^{t_1} dt \, \frac{\partial \vec{G}_{free}^{\,pi}}{\partial x^j}(x',t';x,t) \int_{-\infty}^{+\infty} dt' \, \delta\Psi^{ijkl}(x;t,t') \, \frac{\partial\vec{u}^k}{\partial x^l}(x,t'), \tag{C-3}$$

which can also be written, at the receiver locations,

$$\delta\vec{u}^i(x_r,t) = \int_V dV(x) \left[- \int_{t_0}^{t_1} dt' \, \vec{G}_{free}^{\,ij}(x_r,t;x,t') \, \frac{\partial^2\vec{u}^j}{\partial t'^2}(x,t') \right] \delta\rho(x)$$

$$+ \int_V dV(x) \int_{-\infty}^{+\infty} d\tau \left[- \int_{t_0}^{t_1} dt' \, \frac{\partial\vec{G}_{free}^{\,ij}}{\partial x^k}(x_r,t;x,t') \, \frac{\partial\vec{u}^l}{\partial x^m}(x,t'-\tau) \right] \delta\Psi^{jklm}(x,\tau). \tag{C-4}$$

By comparison with (C-2) this gives

$$A^i(x_r,t;x) = - \int_{t_0}^{t_1} dt' \, \vec{G}_{free}^{\,ij}(x_r,t;x,t') \, (\partial^2\vec{u}^{\,j}\frac{}{\partial t'^2}(x,t'), \tag{C-5a}$$

and

$$B^{ijklm}(x_r,t;x,\tau) = - \int_{t_0}^{t_1} dt' \, \frac{\partial\vec{G}_{free}^{\,ij}}{\partial x^k}(x_r,t;x,t') \, (\partial\vec{u}^{\,l}\frac{1}{\partial x^m}(x,t'-\tau). \tag{C-5b}$$

The operators **A** and **B** defined by (C-5) map the density and rate-of-relaxation spaces into the data space. Their transposes map the dual of the data space into the duals of the density and rate-of-relaxation spaces, i.e., for fixed $\delta\hat{\mathbf{d}}$ they give

$$\delta\hat{\rho} = \mathbf{A}^t\,\delta\hat{\mathbf{d}}\,,\tag{C-6a}$$

and

$$\delta\hat{\Psi} = \mathbf{B}^t\,\delta\hat{\mathbf{d}}\,,\tag{C-6b}$$

or, introducing their kernels,

$$\delta\hat{\rho}(\mathbf{x}) = \sum_r \int_{t_0}^{t_1} dt\,(A^t)^i(\mathbf{x},\mathbf{x}_r,t)\,\delta\hat{u}^i(\mathbf{x}_r,t)\,,\tag{C-7a}$$

and

$$\delta\hat{\Psi}^{jklm}(\mathbf{x},\tau) = \sum_r \int_{t_0}^{t_1} dt\,(B^t)^{jklmi}(\mathbf{x},\tau,\mathbf{x}_r,t)\,\delta\hat{u}^i(\mathbf{x}_r,t)\,,\tag{C-7b}$$

but, as the kernels of the transposes equal the transposes of the kernels, we can write

$$\delta\hat{\rho}(\mathbf{x}) = \sum_r \int_{t_0}^{t_1} dt\,A^i(\mathbf{x}_r,t,\mathbf{x})\,\delta\hat{u}^i(\mathbf{x}_r,t)\,,\tag{C-8a}$$

and

$$\delta\hat{\Psi}^{jklm}(\mathbf{x},\tau) = \sum_r \int_{t_0}^{t_1} dt\,B^{ijklm}(\mathbf{x}_r,t,\mathbf{x},\tau)\,\delta\hat{u}^i(\mathbf{x}_r,t)\,,\tag{C-8b}$$

i.e.,

$$\delta\hat{\rho}(\mathbf{x}) = -\sum_r \int_{t_0}^{t_1} dt \int_{t_0}^{t_1} dt'\,\vec{G}_{\text{free}}{}^{ij}(\mathbf{x}_r,t;\mathbf{x},t')\,\frac{\partial^2\vec{u}^j}{\partial t'^2}(\mathbf{x},t')\,\delta\hat{u}^i(\mathbf{x}_r,t)\,,\tag{C-9a}$$

and

$$\delta\hat{\Psi}^{jklm}(\mathbf{x},\tau) = -\sum_r \int_{t_0}^{t_1} dt \int_{t_0}^{t_1} dt'\,\frac{\partial \vec{G}_{\text{free}}^{ij}}{\partial x^k}(\mathbf{x}_r,t;\mathbf{x},t')\,\frac{\partial \vec{u}^l}{\partial x^m}(\mathbf{x},t'-\tau)\,\delta\hat{u}^i(\mathbf{x}_r,t)\,.\tag{C-9b}$$

Defining

$$\overleftarrow{u}^j(\mathbf{x},t') = \sum_r \int_{t_0}^{t_1} dt\,\vec{G}^{ij}(\mathbf{x}_r,t;\mathbf{x},t')\,\delta\hat{u}^i(\mathbf{x}_r,t)\tag{C-10}$$

gives

$$\delta\hat{\rho}(\mathbf{x}) = -\int_{t_0}^{t_1} dt'\,\overleftarrow{u}^j(\mathbf{x},t')\,\frac{\partial^2\vec{u}^j}{\partial t'^2}(\mathbf{x},t')\,,\tag{C-11a}$$

and

$$\delta\hat{\Psi}^{jklm}(\mathbf{x},\tau) = -\int_{t_0}^{t_1} dt'\,\frac{\partial\overleftarrow{u}^j}{\partial x^k}(\mathbf{x},t')\,\frac{\partial\vec{u}^l}{\partial x^m}(\mathbf{x},t'-\tau)\,.\tag{C-11b}$$

Using (56b) of the text, the definition (C-10) can be rewritten

188

$$\overleftarrow{u}^j(x,t') = \sum_r \int_{t_0}^{t_1} dt \ \overleftarrow{G}^{ji}(x,t';x_r,t) \ \delta\hat{u}^i(x_r,t) \ , \tag{C-12}$$

i.e.,

$$\overleftarrow{u}^i(x,t) = \int_V dV(x') \int_{t_0}^{t_1} dt' \ \overleftarrow{G}^{ij}(x,t;x',t') \ \delta\phi^j(x',t') \ , \tag{C-13}$$

where

$$\delta\phi^i(x,t) = \sum_r \delta(x-x_r) \ \delta\hat{u}^i(x_r,t) \ . \tag{C-14}$$

The representation theorem allows then the interpretation of $\overleftarrow{u}^i(x,t)$ as a field satisfying final conditions of rest, propagating with anti-causal attenuation, and due to the sources (C-13).

Using an integration per parts, the causality of $\overrightarrow{u}^i(x,t)$ and the anti-causality of $\overleftarrow{u}^i(x,t)$, equation (C-11a) can also be written

$$\delta\hat{\rho}(x) = \int_{t_0}^{t_1} dt' \ \frac{\partial \overleftarrow{u}^j}{\partial t'}(x,t') \ \frac{\partial u^j}{\partial t'}(x,t') \ . \tag{C-11c}$$

Formulas (C-11b) and (C-11c) correspond to formulas (114)-(115) of the text.

SOME PROBLEMS OF INTEGRATED ACTIVE SEISMIC

METHODS FOR CRUSTAL EXPLORATION

R. Cassinis

Sezione Geofisica
Dipartimento di Scienze della Terra
Università di Milano, Italy

1. INTRODUCTION

During the two last decades crustal exploration, especially on the continents, experienced a very fast growth: both the techniques and the objectives were deeply transformed. Though other improved geophysical methods have been successfully applied, only the active seismic methods can face the problems brought up by the new objective, i.e., the exploration of the "fine" structure of the crust and, at a minor extent, of the lower lithosphere. In the sixties and early seventies, the refraction-wide angle reflection technique (generally referred to as DSS - Deep seismic soundings), was able to subdivide the crust into several "types" and to build distribution models of compressional velocity much more detailed than the ones derives from passive seismology.

The analysis was almost exclusively cinematic, the waveforms and the amplitude being considered qualitatively only to identify the type of the event and to assess the correlations along the various "branches" of the time-distance plot.

In DSS, the dynamic approach was attempted only when the petroleum geophysicists began to apply the near vertical reflection (NVR) to crustal problems, i.e., in the late seventies.

Therefore, the digital revolution reached very late DSS and only in the processing and interpretation phase. The data gathering is still mainly analog, the dynamic range being limited in most cases to about 45 db. The reason for this delay is due primarily to the fact the DSS has been considered as confined to the "research" field and therefore operated by the scientific, non profit, poorly funded, community.

But there are also physical justifications that lay primarily on the lower resolution and on less stringent need for wave analysis.

A slow and difficult process of integration between the two techniques is presently under way and, therefore, between their respective origin disciplines, i.e., passive seismology and oil exploration.

When vertical reflections were first received from the deep crust, several people considered that DSS had no reason to exist any longer. In

Table 1. A Comparison between NVR and DSS

	Near Vertical Reflection (~20°)	Deep Seismic Soundings (Refraction and Wide Angle Reflection)
Vertical resolution (for the intermediate crust)	Very high (Δh min ~ 50 m)	Low (Δh min ~ 200 m)
Penetration	Limited by the signal to noise ratio, type of formation and type of source	Less constraints (lower frequencies involved)
Ability to delineate the structures	Very good (in 2 dimensions, in some cases in 3 dimensions)	Limited due to the effect of lateral heterogeneities and to low horizontal resolution (large wavelength)
Propagation velocity measurement	The determination of the average velocity is impossible for large depths and in non-stratified media	The max velocity is measurable, with low accuracy, at all depths (except in inversion zones)
Sources of return energy in non-stratified media	Faults, lamellar structures, state and lithology changes, anisotropy, differential pressure	Gradients of velocity; discontinuities of 1st or higher order
Modelling tools	Synthetic seismograms (including multiple reflections and anelasticity effects). Modelling of dynamic behavior is relatively simple. Migration process is complex in non-stratified media	Synthetic seismograms (the dynamic approach to the complete seismograms is complex). Ray-tracing. Direct approach for the cinematic interpretation (trial and error)
Major interpretation problems	"Transparence" of the formation (lack of coherent signals). Lateral sources of reverberations	Low velocity zones. Types of transitions

fact, the COCORP program was based exclusively on NVR. Presently, nearly everybody is convinced that NVR and DSS are two complementary techniques and that in order to reach a better understanding of the fine structure of the lithosphere, they must be used jointly in every case.

It is perhaps useful for the benefit of those not deeply involved in seismic crustal techniques, to briefly discuss the peculiarities of NVR and DSS and, considering the present state-of-the-art, the requirements of an integrated exploration.

2.   NVR AND DSS: A COMPARISON

Up to the late seventies, except some tests, only DSS have been used for crustal exploration. There are several reasons for it, both economical and technical; some of the latter are summarized in Table 1. The main differences concern the horizontal and vertical resolution (and therefore the ability to define the structures), the measure and meaning of the velocity distribution with depth, the type of events used for the interpretation and their attenuation.

It is clear that NVR was borne for the exploration of the sedimentary basins, i.e., for sub-horizontal stratified sequences of laterally continuous layers. The exploration of non-sedimentary formations seems an objective contradictory to the physical principles of NVR. Nevertheless, not only back scattered but reflected energy from fairly continuous horizons is received in some instances. However, the coherence of the events is not comparable to that of regular reflections from stratified media (Figure 1). This is due to the possible nature of the sources (fault planes and mylonitic zones in the granitic crust, "laminated" structure of the transition in the lower crust, contacts and joints in granites, possible partial fusion or effects of pore pressure in lower crust, elastic anisotropy).

It must be noted that the influence of processing ought to be examined with care while interpreting NVR results. Especially the horizontal stacking (CDP) procedure destroys part of the diffraction pattern giving the impression of a lateral continuity more consistent than the actual one. When the S/N ratio is sufficiently high, it should be advisable to produce a single fold, constant offset, section to be compared with the final stacked version before attempting the first "line drawing" of the time, unmigrated, cross section. This is in order to try a better separation of reflected and diffracted energy, the latter being scattered by objects smaller than the Fresnel zone or by the edges of such objects.

As an average, one Fresnel zone is ten times the wave length (Figure 2). Therefore, in the granitic crust, considering a wavelength of 200 m, the first Fresnel zone is of about 2 km, while in the lower crust, assuming a wavelength of 400 m, reaches at least 4 km.

However, as it will be discussed later, broken or discontinuous horizons can also be produced (especially in the upper crust) by the influence of the overburden.

Another difficult problem is the migration process, especially at great depths. The effect of the 3D structure becomes very strong in the lower crust and migration can yield misleading results if the data are collected only along a single profile.

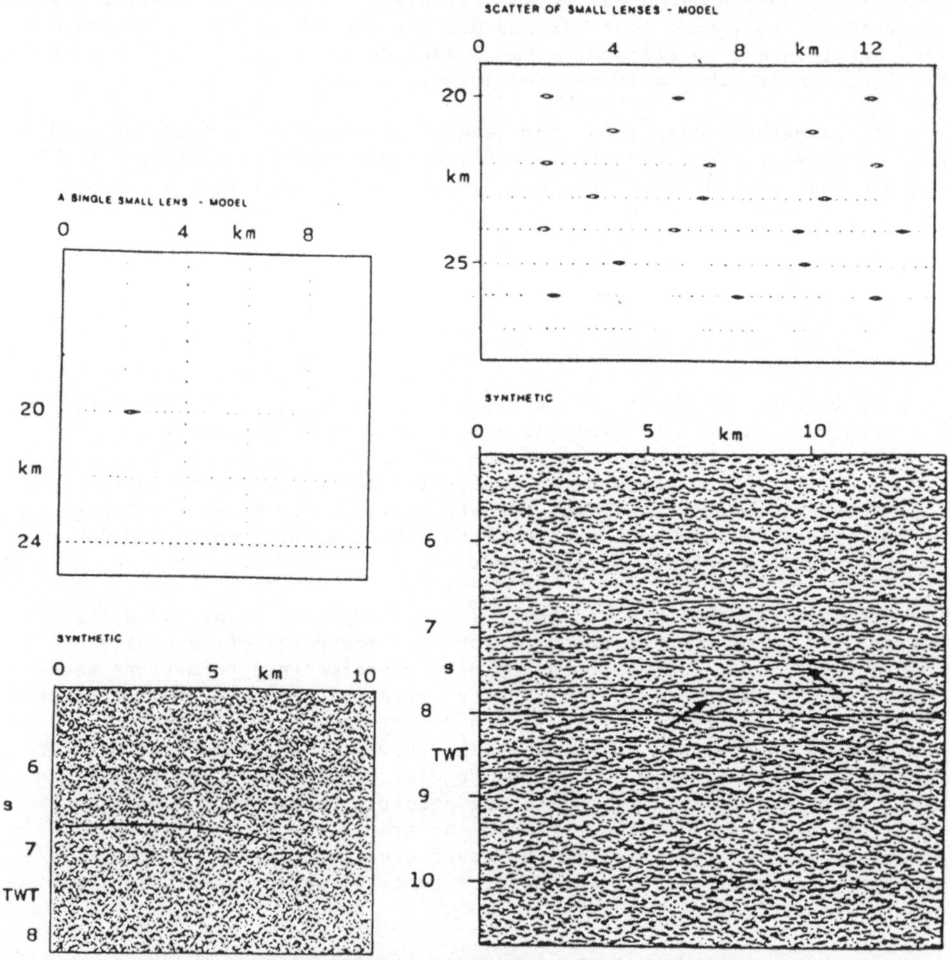

SCATTER OF SMALL LENSES - MODEL

A SINGLE SMALL LENS - MODEL

SYNTHETIC

SYNTHETIC

Fig. 1.   Response of NVR in non-sedimentary crust (lower crust):
diffraction hyperbolae from a lenticular structure can give the
impression of continuous reflecting horizons (Reston, 1987).

A very negative point for NVR is that it is impossible to get
information on the average velocity for TWT larger than 2 ÷ 2.5 s. This
is due not only to the small moveout, but also, for events coming from
non-stratified media, to the lack of regular reflection hyperbolae.

Penetration is a key factor for NVR; it depends on the decay with
time of S/N ratio and on the properties of the formations (Figure 3). It
is commonly used the term "transparent" to mean a formation where only
random energy is observed. However, one must be careful in using this
term, because the lack of signals can be originated by a fast energy
decay.

A clear example is the one reported by Meyer and Brown (1986). The
NVR profiles are crossing different geological and tectonic regions. In
the transition zone between the Basin and Range and the Colorado plateau,
the lower crust appears as "transparent", in contrast with the response
observed in the surrounding areas. However, considering the behavior of
the amplitude decay curve, it seems more likely that the lack of response
is due to insufficient penetration.

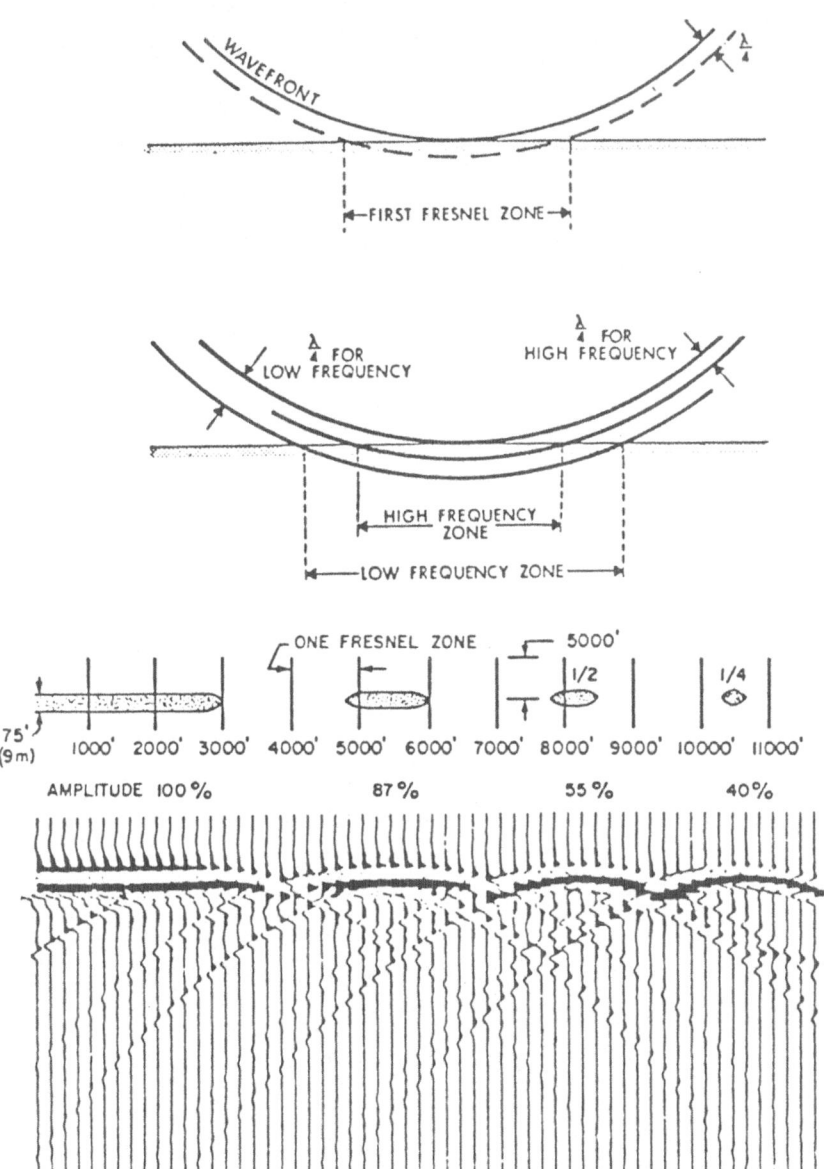

Fig. 2. Horizontal resolution: Fresnel zones (Sherif, 1980).

The advantage of DSS is, basically, the capability to yield a coarse description of the velocity distribution and to identify the main discontinuities under almost every condition even at great depth. The S/N ratio is less influenced by depth, or, it can be said that wide angle reflections can be better observed because they are less attenuated than NVR. In general, the main discontinuities seen by DSS and NVR seem to correspond. However, there are exceptions.

As there are more types of events than in NVR (transformed PS, P_MP and P_n) their identification is more complex than of vertical P reflections; however, the multiples are a less difficult problem. On the other hand, the low resolving power, the variable incidence angle, the

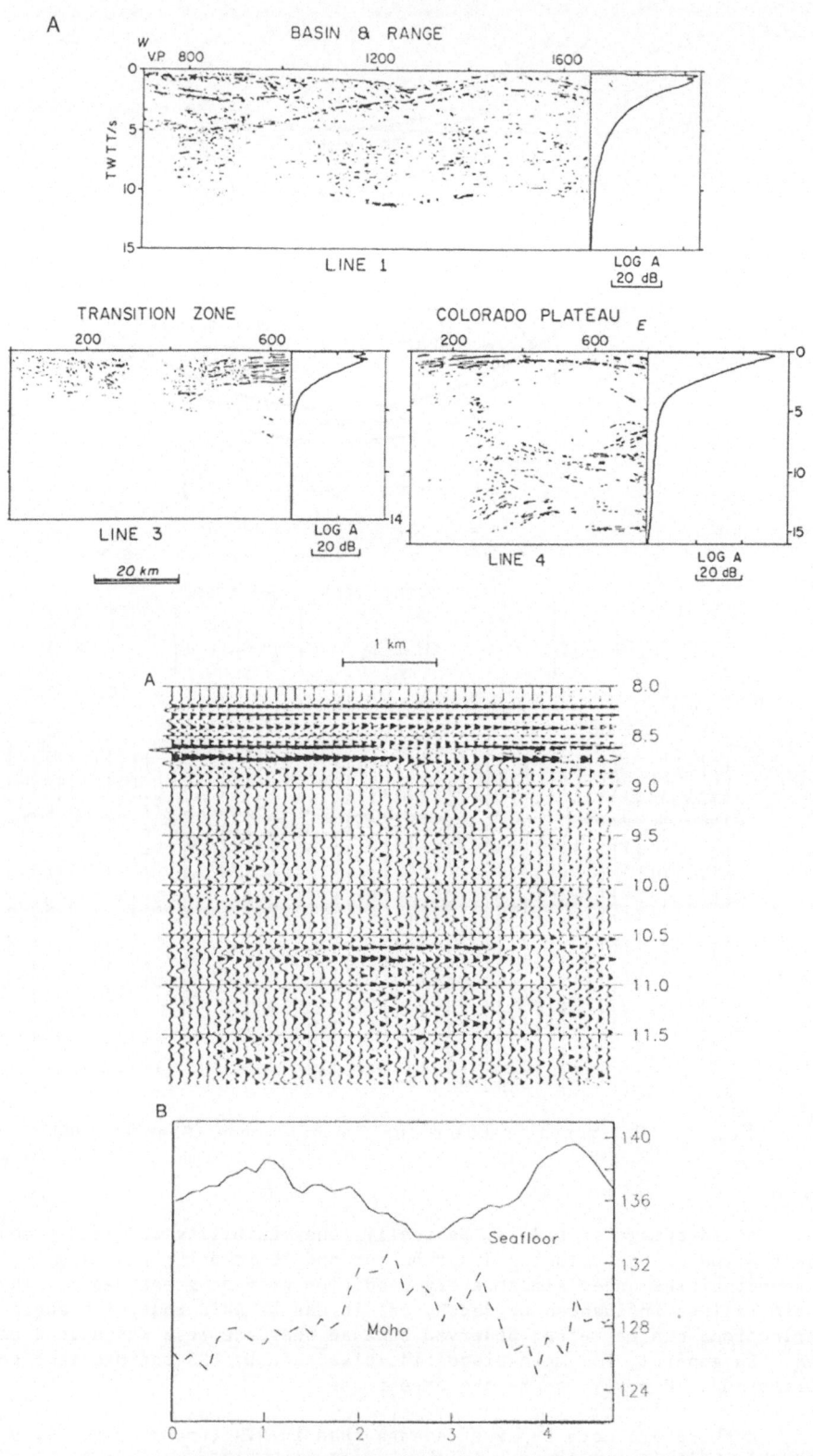

Fig. 3. See next page for legend.

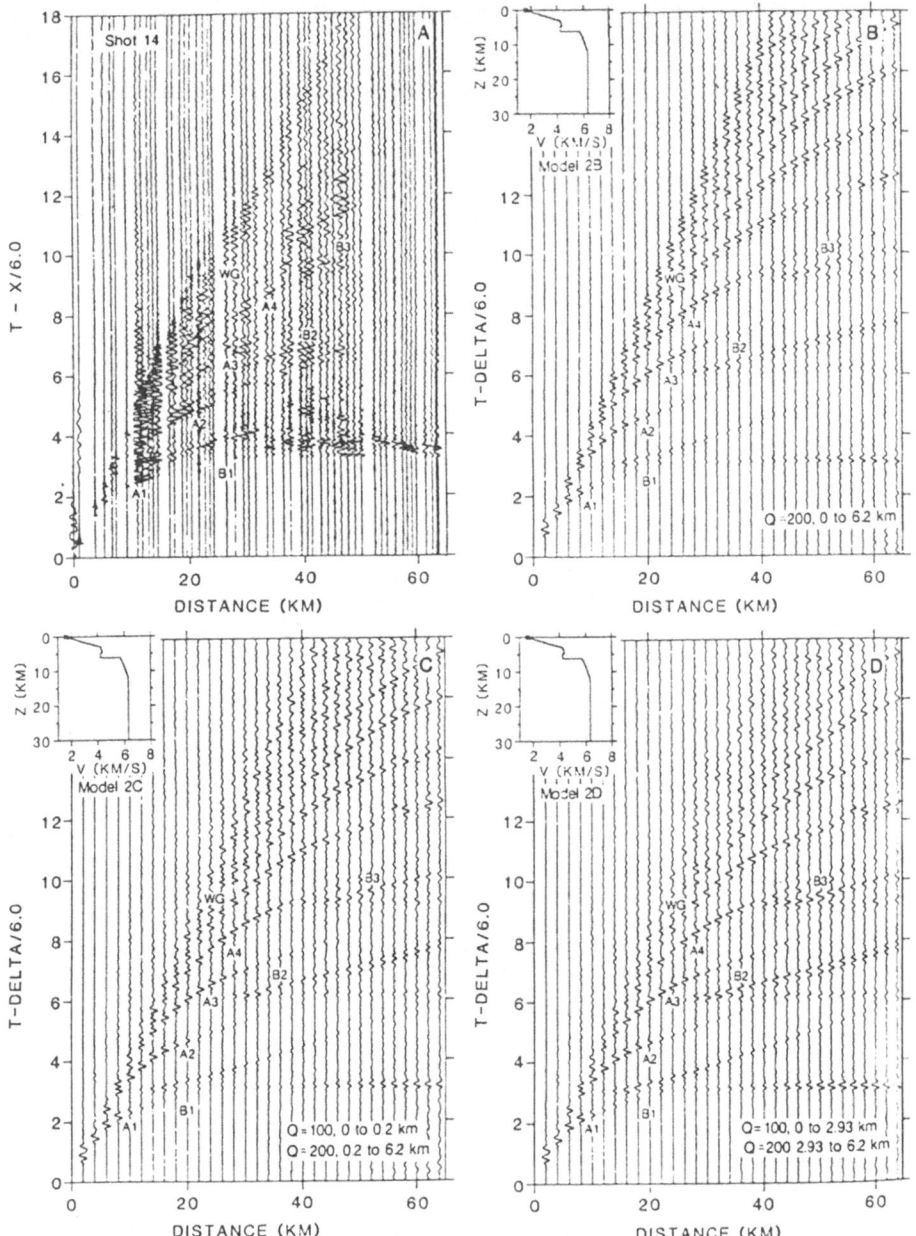

Fig. 4. Free surface multiples in a thick sedimentary basin (Hwang and
         Mooney, 1986) and synthetics computed on different assumptions of
         Q structure. Q evaluations are made looking for the model that
         fits better to the observed seismograms.

Fig. 3. Penetration in NVR profiles: a) in lines 1 and 4 the lower crust
         appears reflective while in line 3 it appears as "transparent".
         This is probably due to the very fast decay of the signal (Meyer
         and Brown, 1986); b) the reflection from the M boundary (10.7 s)
         is stronger when the reflection at 8.7 s (sea bottom) is weaker
         (Collins et al., 1986).

long and continuously changing wave paths are all well known limiting factors for DSS.

The accuracy of the recording and the sophistication of processing must be planned taking carefully into account these intrinsic limiting factors.

3.   PROBLEMS OF CRUSTAL EXPLORATION

In Table 2 some problems of crustal exploration are summarized.

a)   Upper Crust

The influence of sediments is an important factor to take into consideration while recording DSS profiles on several km thick sedimentary basins (Great Valley, California; Po Valley, Italy). The attenuation of deeper events can become very high and strong free-surface multiples are produced if the shot point lies in the basin (Figure 4). Where large and variable thicknesses of layered sediments are present, NVR is the optimum tool to evaluate their influence on DSS deep data.

A more difficult problem is represented by the blankets of allochtonous sediments of various lithology and of large, very variable thickness, sometimes of chaotic structure ("olistostromes").

This situation is well-known in Italy, because it is peculiar of the whole Apenninic range. The objectives (both sediments, like the mesozoic carbonatic sequence, or the crystalline basement) lay hidden under the overburden, and their exploration still remains one of the most difficult problems faced by geophysical methods.

It seems surprising that little improvement was obtained in NVR data using the most sophisticated techniques.

Figure 5 illustrates some of the reasons for these difficulties. Those due to the very rough topography and to the lateral etherogeneities can be overcome, to a certain extent.

However, they bring limitations in CDP and in the use of vibratory sources. But the influence of intermediate and deeper formations is much stronger. The allochtonous, often chaotic overburden, even of moderate thickness, produces a severe decay of S/N. The allochtonous itself can either generate broken events or behave as "transparent", depending on its structure, tectonic history, lithology.

A further decay of the quality of NVR is caused by the intermediate and deep structure itself (low angle faults, decollement tectonics, irregular basement top). The influence of these various factors can be observed in a typical cross section, published by Agip (Figure 6). The synthetic seismograms of Figure 7 illustrate the time distortions and the amplitude changes of NVR received from a flat reflector buried under a thick overburden with a typical decollement tectonics.

In conditions like the ones met under the Apennines, the only way to obtain some information on the structure of the upper crust is to integrate NVR and DSS. The crustal technique of DSS must be adapted to the shallower objectives, using shorter geophone spacing (about 1 km) and reducing the total length of the profiles (up to 30 - 35 km) and the radius of the fans (Figure 8).

Table 2. Problems of Crustal Exploration in Continental Crust

	Problems	NVR	DSS
Upper Crust	Surface conditions	Amplitude anomalies and attenuation	Hidden layers
	Type and thickness of sediments		
	Crystalline and sedimentary basement		Type of V functions
	Heterogeneities in quarzitic rocks	"Transparent" or reflective U.C. Reflections from LVZ, polarity	
	LVZ: p, T, pore pressure, anisotropy		Resolution and average V inside LVZ
Lower Crust and "M"	Transitional or sharp boundaries?	Reflective or transparent L.C. (problem of penetration and S/N)	Type of function. LVZ observed in L.C.
	Lamellar structure of "M"?		Strength of P_MP and of P_n
	Zone of "mélange"?	Nature of sub-Moho reflections	Repeated inverse branches of P_MP
	"Refracting and reflecting" Moho		Velocity at the M boundary from wide angle reflection (in case of weak P_n)

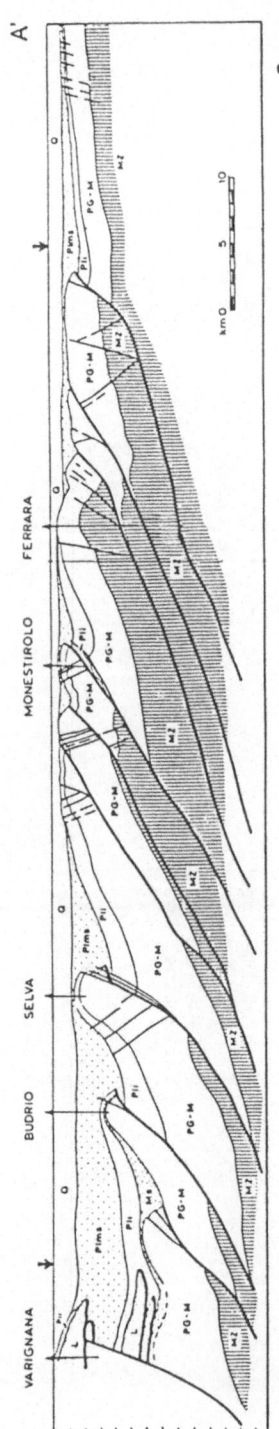

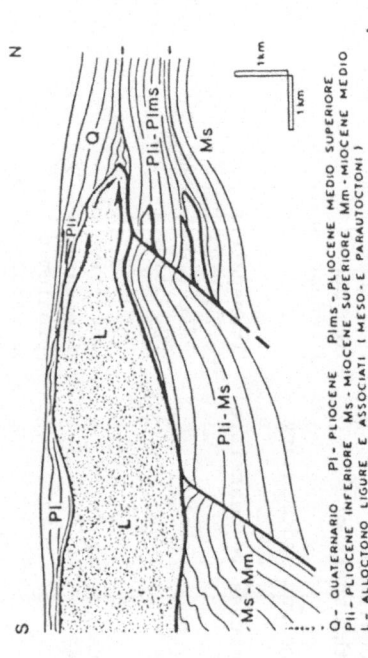

Fig. 5. Illustration of the structural situation along the Padanian front of the Apennines (northern Italy).
a) Schematic structure of the sedimentary crust across the Po Plain. The left end (A) corresponds to the Apenninic outcrops.
b) Structural sketch of the Apenninic front showing the emplacement of the Ligurides allochtonous nappes (L) over the tertiary and mesozoic formations as well as the decollement tectonics (Castellarin et al., 1985).

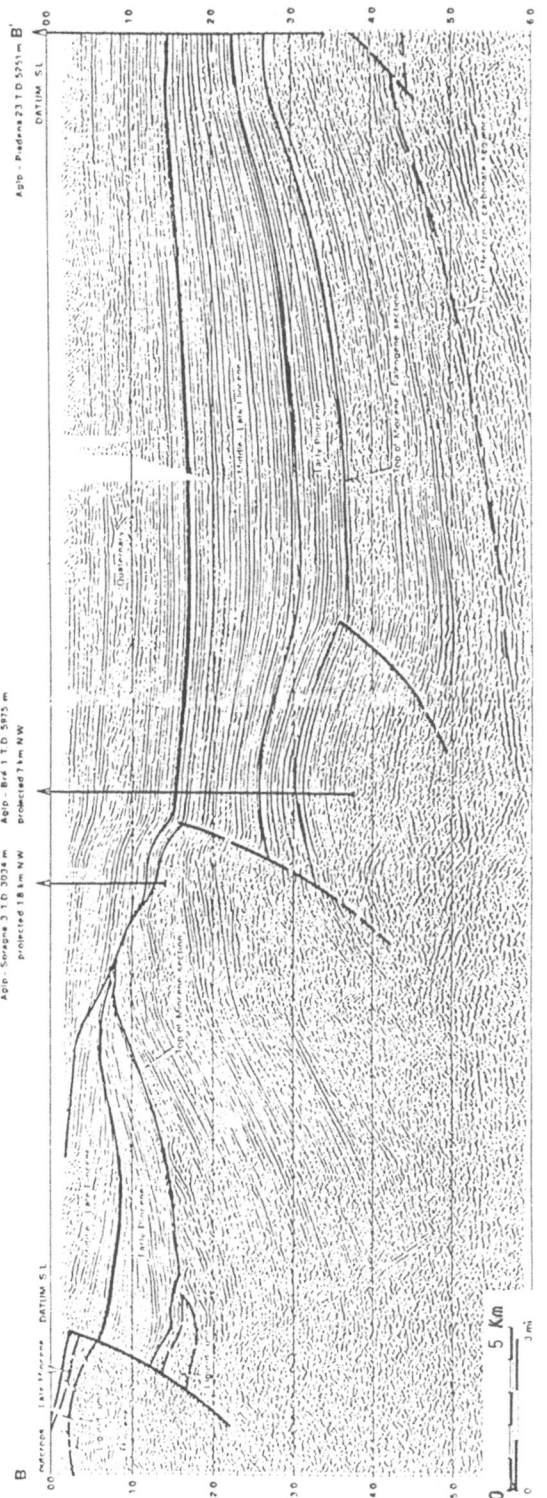

Fig. 6. Example of NVR line (unmigrated) at the Apenninic border. It is clear the influence of both the thrust tectonics and of the Ligurides nappes (left end) on the S/N ratio (Bally, 1983).

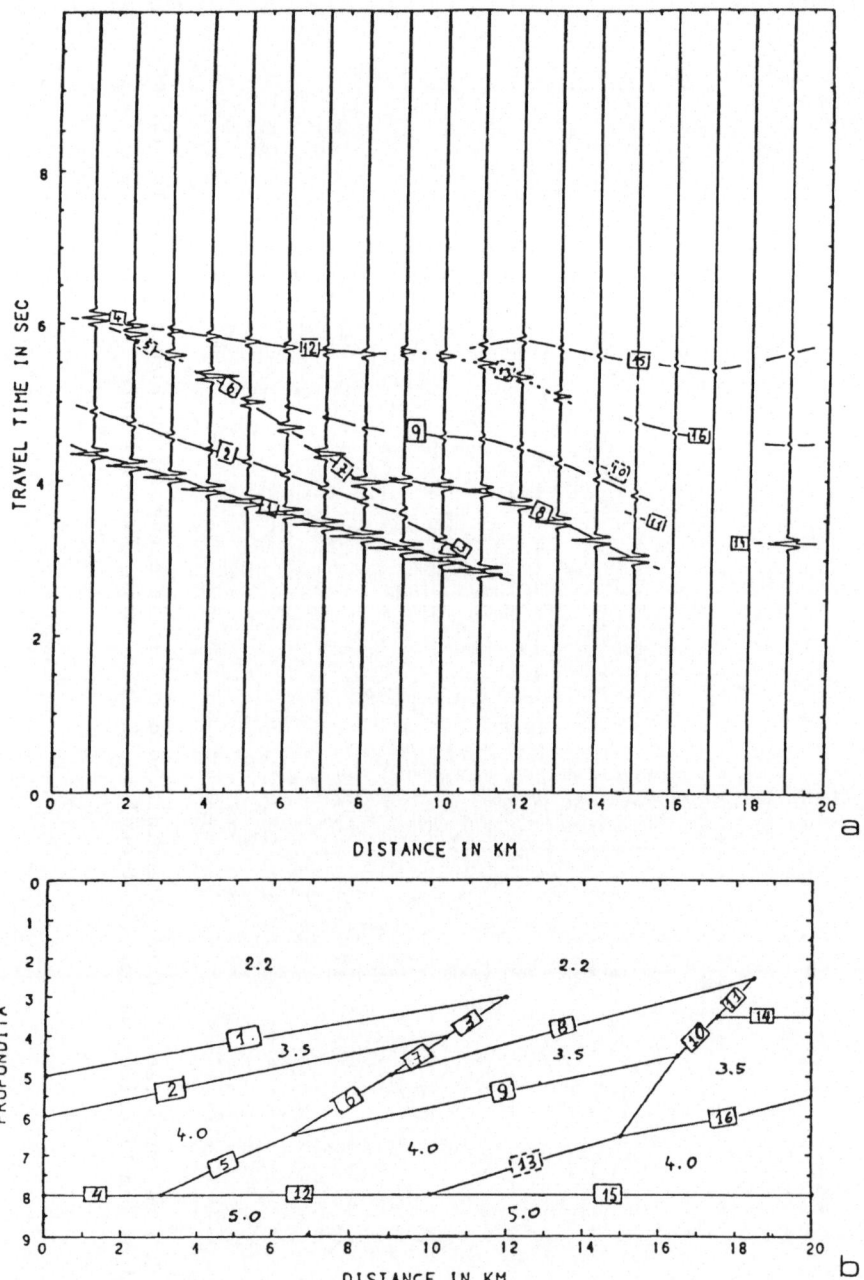

Fig. 7. Synthetic seismograms (ART) showing the effect on times and
amplitudes of vertical reflections of a decollement tectonics
overlaying a flat reflector (CNR-IGL, private communication).

The passband must be raised to 8 - 20 Hz and the presentation of the
time reduced cross sections prepared accordingly.

The examples of Figure 9 show the type of velocity functions obtained
in different cases, compared to the NVR. It is clear that, if a wide
angle reflection reveals a velocity discontinuity, the uncertain packages
of energy observed on NVR cross sections at comparable vertical times

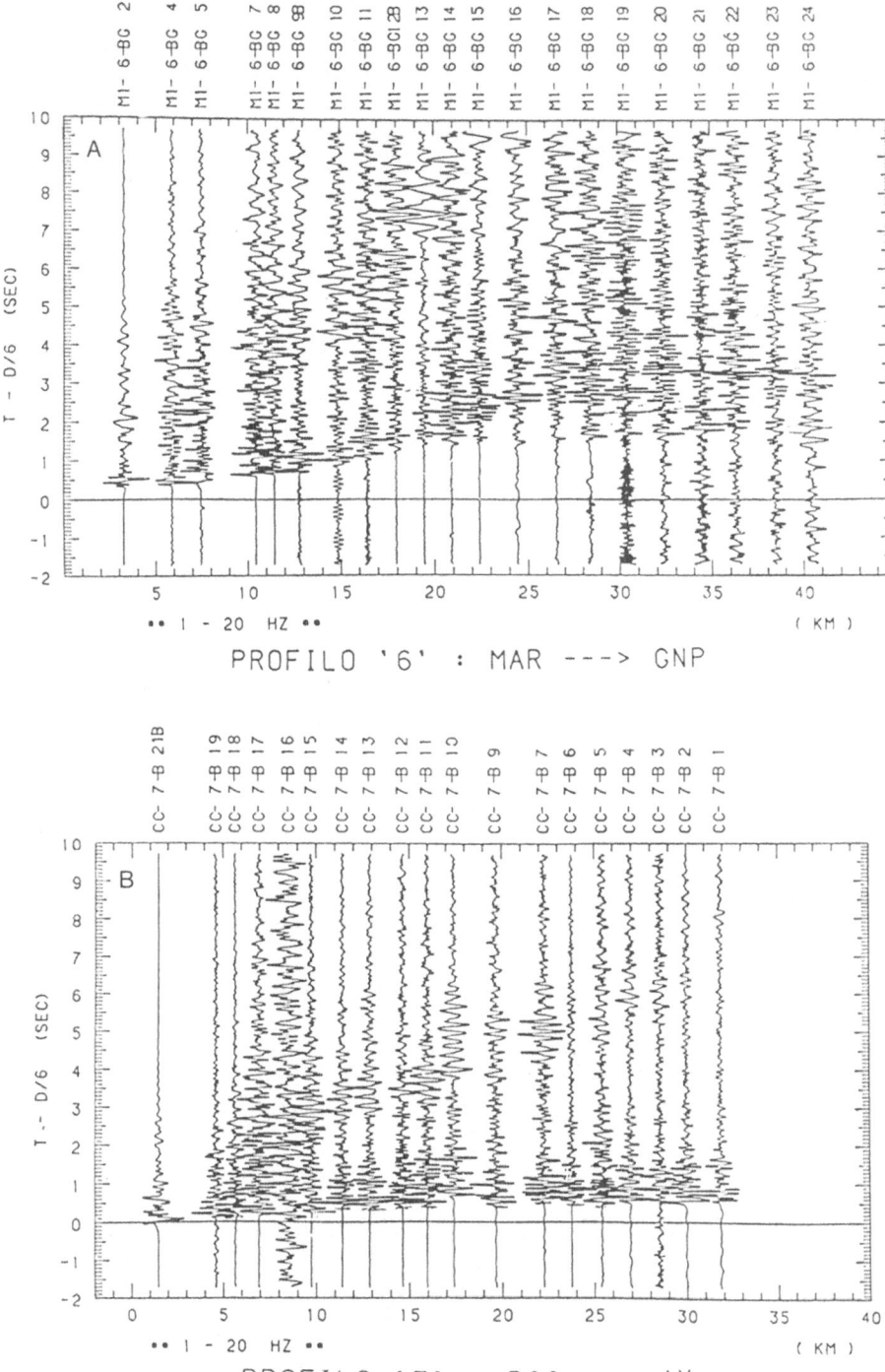

Fig. 8. Examples of DSS time sections for the exploration of the upper crust. a) On the northern slope of the Apenninic range; b) on the Ligurides formations (thick allochtonous overburden).

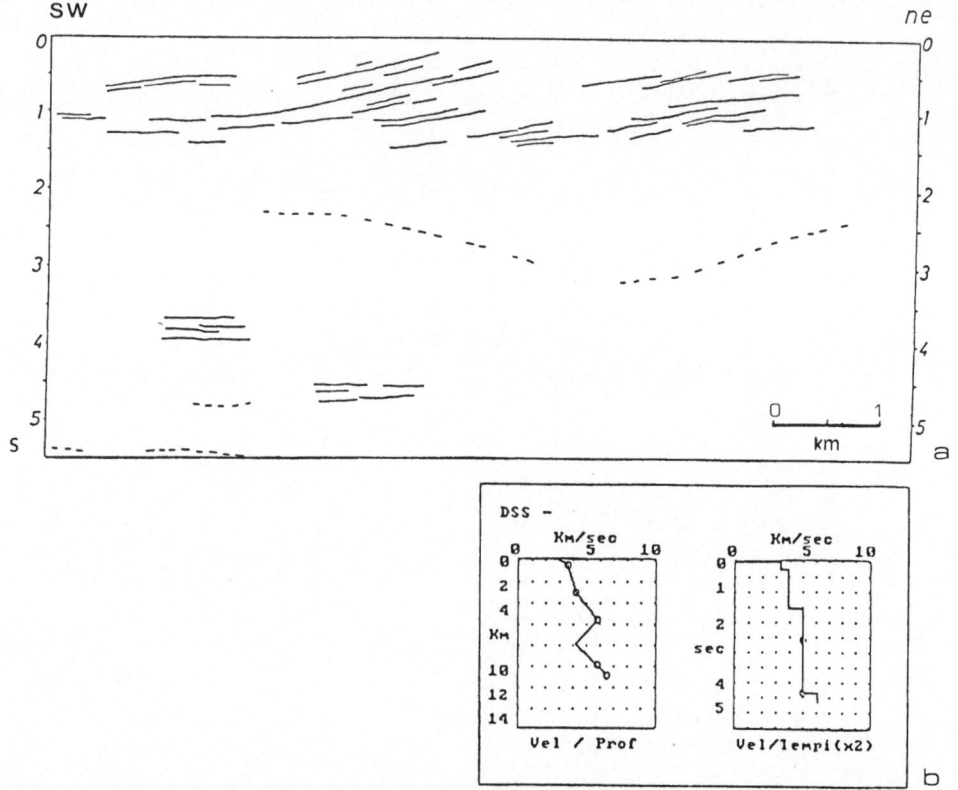

Fig. 9. Velocity functions derived from inversion of the DSS compared
with a "line drawing" of a NVR profile (a). Some correlation can
be found between the velocity and the return reflected energy.
(b) Vel. versus depth (left); Vel. versus TWT (right).

become more meaningful. However, when the allochtonous overburden is very
thick, there are problems even with DSS.

The prediction of the dynamic behavior of chaotic formations is
almost impossible by modelling, therefore the synthetic seismogram shown
in Figure 10 foresees only horizontal layering, taking account of an
allochtonous overburden exhibiting a continuous velocity function; at its
bottom, a LVZ of variable thickness is overlapping a basement of constant
velocity. The limitation in vertical resolution of wide angle reflection
is clearly shown.

The LVZ problem in upper crystalline crust is still one of the most
debated problems. It is generally accepted that at least one LVZ is
observed almost on every type of continental crust, its top being found at
depths ranging from about 5 to 10 Km. The main evidences for the
existence of such inversions are still of cinematic nature, i.e., a
sufficiently large time lag between the asymptotic branches of Pg and P*
or P_MP reflections (Figure 11). Very few evidences from NVR profiles are
available, the "granitic" crust giving very few coherent signals, and it
seems unlikely that phase inversions could be observed in these
conditions.

On the other hand, wide angle synthetics do not exhibit appreciable
amplitude anomalies due to the presence of LVZ.

204

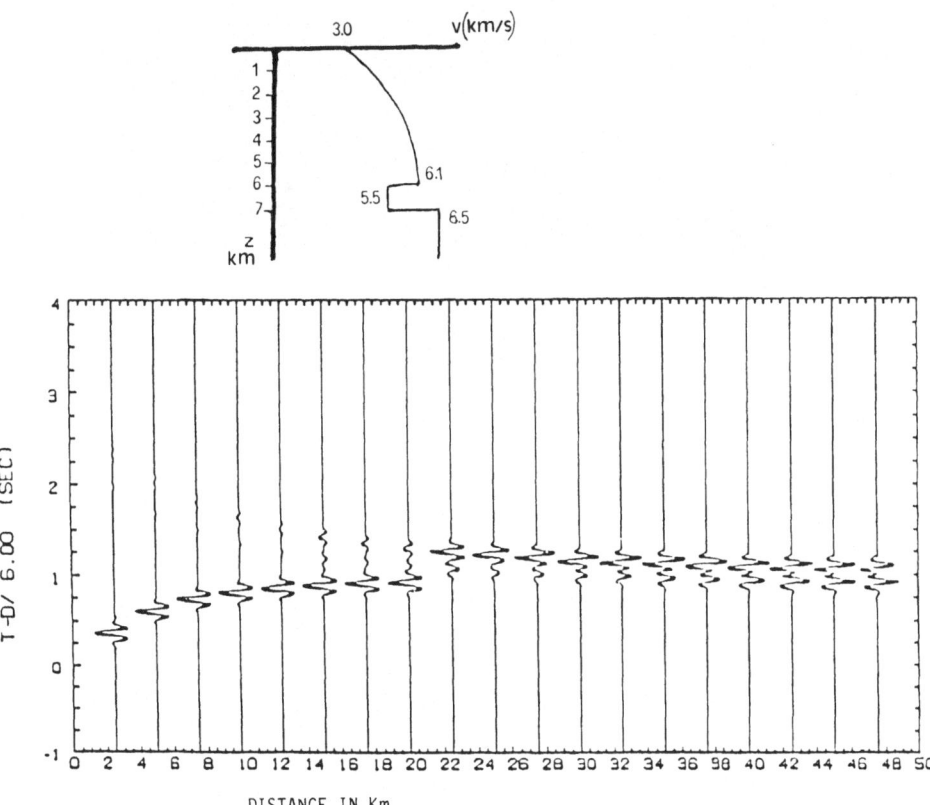

Fig. 10. Synthetic (ART, no attenuation) help in the correlation of
branches in DSS sections (compare with Figure 8b).

So, the problem of a unambiguous determination of velocity
distribution inside the LVZ and of its thickness so far remains unsolved.
According to Christensen (1979) the critical temperature gradient can be
reached only in areas of high heat flow.

Cassinis et al. (1985) compared, observed and calculated velocity
functions for different values of heat flow and assuming different
compositional models; a general remark that is made is that the average
velocity in the LVZ is always lower than in models. This suggests that
other factors are responsible for the inversions other than pressure and
temperature.

Some authors interpret as LVZ every small time lag, even if the
resulting thickness is less than the wavelength. To avoid
misinterpretation reverse profiles and lateral control are mandatory.

b)   Lower Crust and "M" Transition

Here the problems are even more controversial. Several authors (like
Meissner, 1986) while analyzing the seismic response in different types of
the continental crust, agree that its general character varies according
to the age:

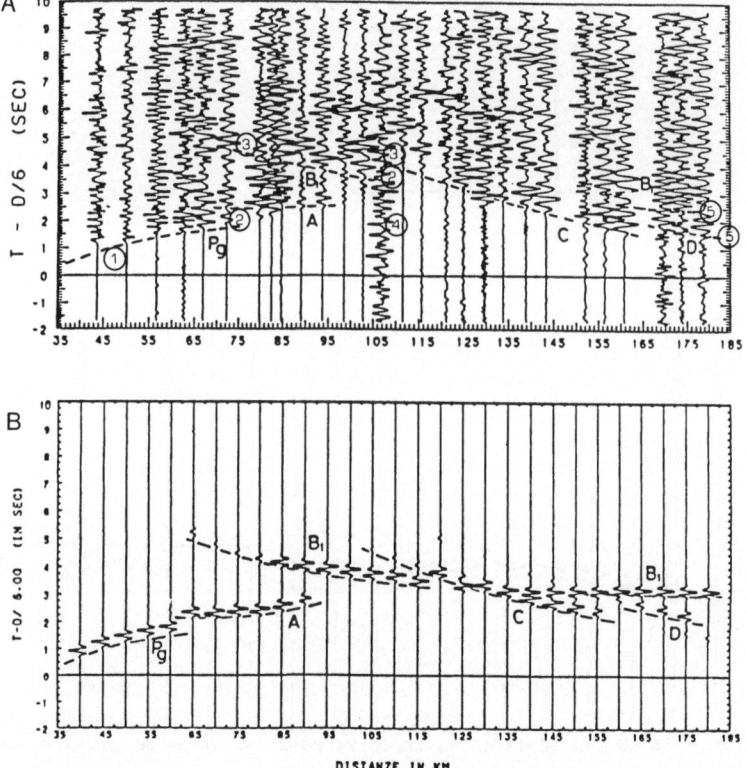

Fig. 11. A typical crustal profile (top: observed; bottom, synthetics) showing time lags interpreted as LVZ (when the reverse profile is available (from Scarascia et al., 1987). The major problems met while reading the record section are indicated:
1. Correlations of first arrivals. Velocity estimate (slope of branch).
2. Time lags and correlation of reflected events.
3. V_{max} read on reversed branch: the actual critical distance is uncertain. Generally V_{max} is over-estimated.
4. One disturbed trace makes critical the correlation of events for a spacing of 3 Km, like the one in the example.
5. Later events are masked by noise at large distances.

Young and Warm Crust	Old and Cold Crust (Shields)
Reflective lower crust; sharp transitions; thin crust (except when below orogens) P_n weak, P_MP strong Good NVR from Moho; LVZ only in Upper Crust	"Transparent" lower crust; smooth transitions; thick crust P_n strong, P_MP weak Poor NVR reflections from the gradual transition

However, there are many exceptions and a complex variability of cases is recorded.

LVZ seem to be observed by DSS even in the lower crust, especially in areas of high H.F. At the other hand, in the lower crust the wavelength of diving waves is of the order of one km, therefore only LVZ of large thickness can be observed.

Transitions seen by NVR in the lower crust are very differentiated. Sometimes (in areas of same age) the lower crust is "reflective", sometimes it is "transparent".

Though the similarity of synthetic and recorded traces cannot lead to a unique solution, the dynamic and cinematic models can considerably reduce the number of acceptable models for the "M" transition. In marginal or recent orogenic belts (like the Alpine-Himalayan) DSS results often led to the conclusion that a very complex crust-mantle transition can exist, whose thickness is sometimes of the same order as the one of the normal crust.

This complex transition was interpreted in several ways (thrusting, melange, flake tectonics, lamellar structure). The evidence of this situation as seen in DSS records are repeated inverse branches ($P_M P$), separated by a LVZ having an average velocity compatible with crustal material. At the "first M transition" the V_p is generally consistently lower than 8 km sec^{-1}.

An example of such a situation is shown along a S-N profile on the EGT, crossing the Po Valley, Italy. However, so far the interpretation is in the preliminary phase and this important problem is still open; NVR profiles have been recently recorded across the Alpine orogen and it is hoped that the results will allow a clearer view of the relationship between the lower crust and the upper crust.

A problem of general interest concerns the attenuation of seismic waves. A direct determination of Q from the amplitude studies, is generally not yet feasible. Only in some instances (attenuation of multiples in the sedimentary overburden) some results can be obtained. In general, some information on the most probable Q is gathered by the comparison of synthetic and recorded amplitudes (see Figure 4).

4. REQUIREMENTS OF INTEGRATED ACTIVE SEISMIC METHODS

a) Data Acquisition

The need of integration has been discussed. However, it seems that DSS is a regional technique and, therefore, still has the role of first exploration tool to better plan the more expensive, NVR, profiles. During the detailed exploration by NVR, DSS profiles and fans are useful in order to gather velocity information on the "transparent" upper crust, to look for lateral variations and to determine the velocity distribution at depth.

The critical points of data acquisition for DSS are: the geophone spacing (and, therefore, the number of recording points); the availability of reverse profiles: the standardization and calibration of the individual mobile stations. The spacing should be maintained within 1.5 km for crustal lines.

The problem of updating the equipment is not yet solved, the digital mobile stations being not yet developed to a sufficient degree to justify a very expensive substitution of the existing analog recorders.

Nevertheless, if the acquisition is carefully accomplished, in spite of the low dynamic range of analog equipment, the digitized traces can be conveniently processed.

NVR techniques and equipment are fully developed. The acquisition is quite expensive also because the need of integration of vibratory and explosive sources. The need of high order folding in CDP is a critical point, especially in areas with poor trafficability, where it is difficult to maintain a constant number of folds.

b)   Processing and Interpretation

In DSS, the digitalization of traces has become compulsory. It is a rather slow and cumbersome process, that must be applied also to old data, provided they were recorded with sufficient quality requirements. Techniques and softwares able to speed up this early part of the processing phase should be carefully studied, because the huge mass of data acquired in the past years still contains a large unexploited potential of information.

The presentation of cross sections is an almost standardized process: however, the choice of optimum band width and of gain is a problem to be solved individually for every case.

In Figure 11 the major problems encountered while reading a DSS cross section are illustrated.

The process of interpretation includes: correlation of the different branches (perhaps the most critical point of the interpretation); a first inversion of the travel time curves in order to build the preliminary velocity model assuming horizontal layering; ray tracing to take account of lateral variations, synthetics assuming different options. The whole process is iterative and, therefore, non standard and slow. The correlation can be improved by spectral analysis of the wave trains (on windows) and, sometimes, by trace coherence (or semblance) analysis.

All this (and especially the modelling) implies the need of dedicated computing and plotting devices. It suggests that the processing of DSS data should be concentrated in few well equipped Centers. However, as the process is not a standard one, it must be attended by the researchers responsible for each project.

As far as NVR is concerned, processing is, at a certain extent, more routinely carried out in order to produce the basic cross sections. However, special enhancements and migration, require very sophisticated programs and must be applied with care.

It is well-known that in NVR profiles not only the length of the spread but that of the whole line is an essential factor. Especially in transitional areas, lines shorter than about 100 km cannot give useful information. Therefore a program of integrated surveys must take into account a minimum annual budget.

ACKNOWLEDGEMENTS

The synthetic seismograms have been realized at the CNR, Institute for the Geophysics of the Lithosphere.

REFERENCES

Bally, A. W., 1983, "Seismic expression of structural styles", Vol. 3, A. W. Bally, ed., published by The American Association of Petroleum Geologists, Tulsa, USA.

Cassano E., Anelli, L., Fichera, R., and Cappelli, V., 1986, Pianura Padana. Interpretazione integrata di dati geofisici e geologici. Agip 73 Congresso Soc. Geol. It., Roma 29 Settembre - 4 Ottobre 1986.

Cassinis, R., Mazzoni, P., and Ranzoni, A., 1985, Active seismic layers and crustal structure in some Italian regions, J. Geophys., 56:153-159.

Castellarin, A., Eva, C., Giglia, G., and Vai, G. B., 1985, Analisi strutturale del fronte Appeninnico Padano, Giornale di Geologia, 3, vol. 47:1-2.

Cattaneo, U., De Franco, R., and Scarascia, S., 1985, Sismogrammi sintetici per strutture con variazioni laterali: una applicazione su un profilo sismico crostale (private communication).

Christensen, N. I., 1979, Compressional wave velocities in rocks at high temperatures and pressures, critical thermal gradients and crustal low-velocity zones, J. Geophys. Res., 84:6849-6857.

Collins, J. A., Brocher, T. M., and Karson, J. A., 1986, Two-dimensional seismic reflection modelling of the inferred fossil oceanic crust/mantle transition in the Bay of Islands ophiolite, J. Geophys. Res., vol. 91, n. B12:12520-12538.

Hwang, L. J., and Mooney, W. D., 1986, Velocity and Q structure of the Great Valley, California, based on synthetic seismogram modelling of seismic refraction data, Bull. Seismol. Soc. Am., 76, n. 4:1053-1067.

Meissner, R., 1986, The continental crust. A geophysical approach, W. B. Donn, ed., International Geophysics series, Vol. 34, Academic Press, Harcourt, Brace Jovanovich Publishers, London.

Meyer, J. R., and Brown, L. D., 1986, Signal penetration in the COCORP Basin and Range-Colorado plateau survey, Geophysics, Vol. 51, n. 5:1050-1055.

Reston, T. J., 1987, Spatial interference, reflection character and structure of the lower crust under extension, Results from 2-D seismic modelling, Annales Geophysicae, 5B(4):339-348.

Sheriff, R. E., 1980, Seismic stratigraphy, published by International Human Resources Development Corporation, Boston, pp 134-135.

ASPECTS OF ACQUISITION AND INTERPRETATION

OF DEEP SEISMIC SOUNDING DATA

J. Ansorge

Institute of Geophysics
ETH-Hönggerberg
CH-8093 Zurich, Switzerland

INTRODUCTION

In recent years an increasing number of large scale seismic
refraction and wide angle reflection profiling has been performed in
national and international projects over large parts of Europe to
determine the structure of the crust and upper mantle both onshore and
offshore. Although the observational techniques and mostly analogue
recording instrumentation is largely the same as 15 or 20 years ago some
major improvements were achieved over the years. These are a significant
reduction of intervals between shotpoints and receivers to minimum values
of about 50 km and 1 km, respectively, with the aim to increase the
lateral resolution of the shallow and deep structure. This is
accompanied by a considerable increase of the collected amount of data for
each experiment, which can only be handled and interpreted properly by
analog to digital conversion and further digital processing. In fact, at
present much more time has to be spent to digitize the analogue field
data, which are still collected in Europe mostly with the same type of
frequency-multiplex recording system MARS 66 or minor technical
modifications thereof (Berckhemer, 1970), than for the preparation and
realization of the field experiment. Therefore, the change from analogue
to digital field equipment for refraction and wide angle profiling is one
of the most important goals which have to be achieved in the near future.
Lightweight, low-power digital recording instruments have been designed
recently or are being developed at several institutes or by groups of
institutions (e.g., IRIS, 1984). Hopefully, this will lead to
instruments, which can be bought for a reasonable price and with which the
international cooperation can be continued.

Complementary information about the deep structure of the lithosphere
comes from large-scale deep near-vertical reflection profiling (e.g.,
Brewer and Oliver, 1980; Barazangi and Brown, 1986). The gap between both
methods with respect to coverage at depth and distance range of
observations may be closed eventually. However, there are significant
differences in the deep structural response to the two techniques of deep
seismic profiling, which have been worked out very carefully recently by
Mooney and Brocher (1987) by comparison of coincident reflection and
refraction studies on a global scale. It is the purpose of this short
review to present some aspects and properties of refraction and wide angle

reflection data which were obtained recently from different sources, field techniques and methods of interpretation.

SOURCE ASPECTS

Until now explosives are still the most widely used sources on land and at sea for deep refraction and wide angle reflection studies. These range in size from several tens of kilograms to several tons. However, also large airgun arrays have been used recently for combined onshore-offshore wide angle crustal surveys which reach distances well beyond 100 km. The amount of explosives, its geometrical distribution at the shotpoint and depth of deployment together with the immediate near surface geological structure determine the frequency contents of the seismic signals and consequently the structural resolution.

On land borehole shots have replaced almost completely the earlier widely used commercial quarry blasts. The size and location of borehole shots can be designed specifically according to the regional geological and tectonic structure of the area which is to be investigated at depth.

As an example, Figure 1 shows the trace-normalized, vertical component record section from a north-south refraction profile along the eastern boundary of the Black Forest in south-western Germany (Deichmann, 1984; Deichmann and Ansorge, 1983). This section is composed of the recordings from eight different quarry blasts with charge sizes of 700 kg (shot number 1) and about 2000 kg (shots 2 to 10) which were detonated instantaneously. The analysis of only the P_g wavelet indicated on Figure 2a by arrows together with several tenths of a second of the local background noise prior to the arrival of P_g gives dominant frequencies between 7 and 10 Hz. The analysis of the entire window shown here gives a somewhat broader spectrum with more scatter (Figure 2a upper part, right). The difference between the spectra shown in Figure 2a derived from the same shot is most probably caused by the local near surface sedimentary structure at the two recording sites 3 and 14. In Figure 2b the spectra of phases refracted and reflected from the lower crust and Moho are presented for shot S8 recorded at sites 31 and 33.

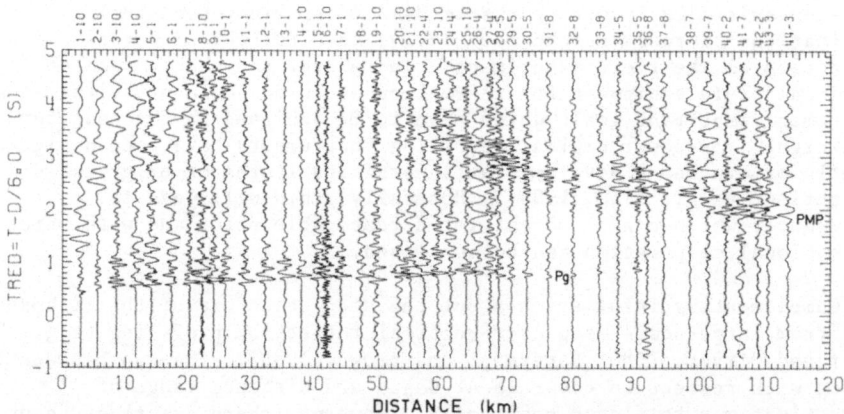

Fig. 1.   Trace-normalized, vertical component record section of 8 quarry blasts in Muschelkalk near Sulz at the eastern boundary of the Black Forest (FRG). Numbers above each trace indicate station and shot (reduction velocity 6 km/s, filtered with 32 Hz low pass) (after Deichmann and Ansorge, 1983).

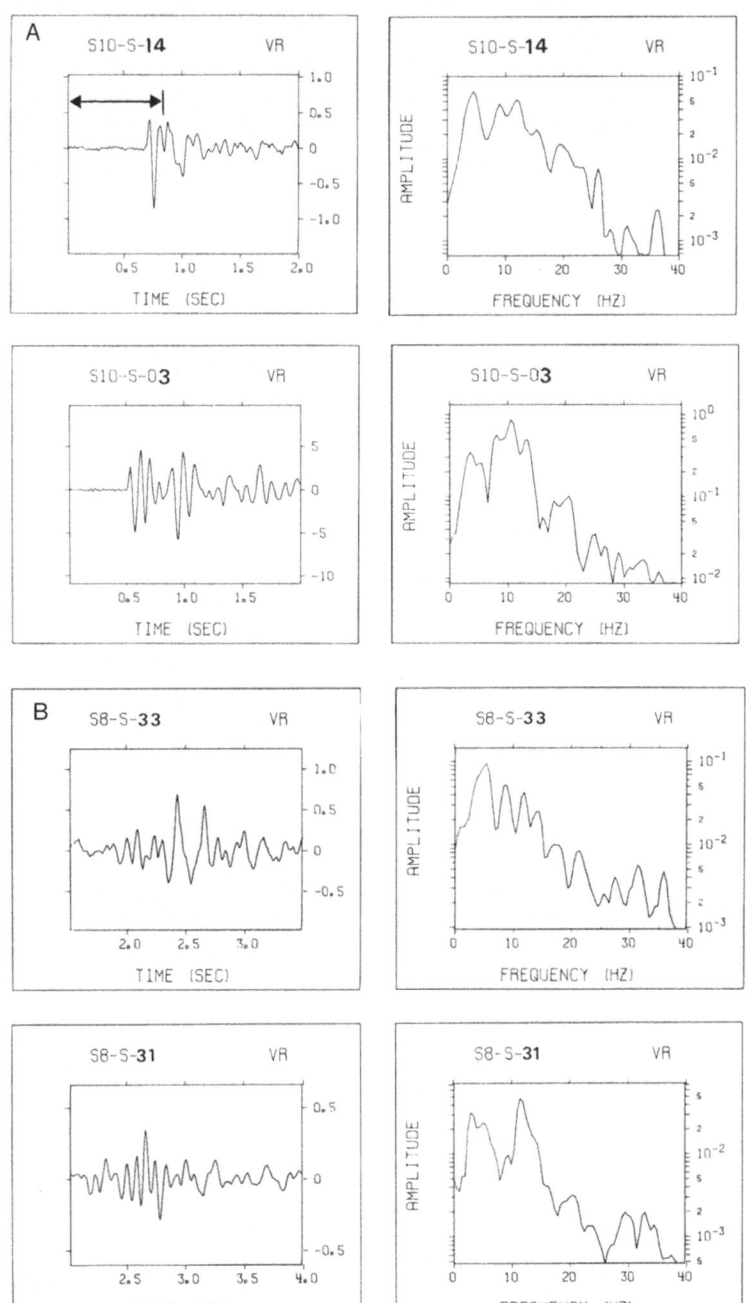

Fig. 2a. Seismograms and amplitude spectra of shot number 10 at stations 3 and 14 (see Figure 1).

2b. Seismograms and amplitude spectra of shot number 8 at stations 31 and 33; only phases from the lower crust are analyzed (see Figure 1).

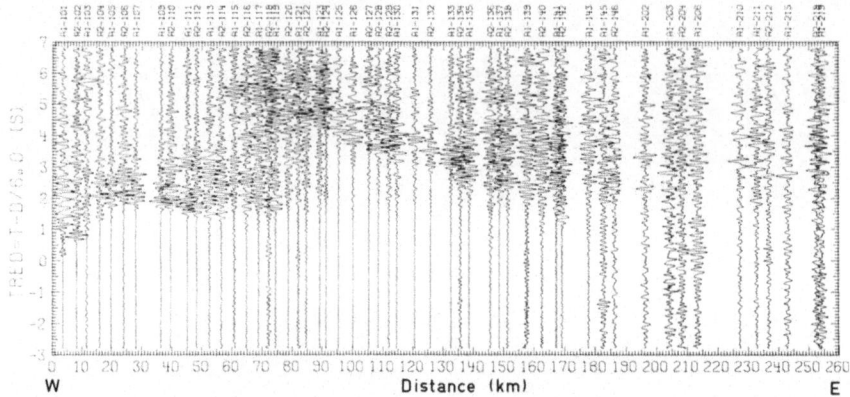

Fig. 3. Trace-normalized, vertical component record section, filtered
2-16 Hz, from borehole shots near the southern tip of Lago
Maggiore to the east. Numbers above each trace indicate shot and
station (after Deichmann et al., 1986).

Compared to quarry blasts, explosions at well selected borehole
locations are much more efficient. Figure 3 shows again the trace-
normalized record section of a west-east running profile in the southern
Alps (Deichmann, 1984; Deichmann, Ansorge and Mueller, 1986). This
profile was obtained from two explosions with charges of 800 kg (A 1) and
400 kg (A 2) distributed in six boreholes at a depth of about 50 m, near
the southern tip of Lago Maggiore in the bottom of a small valley which

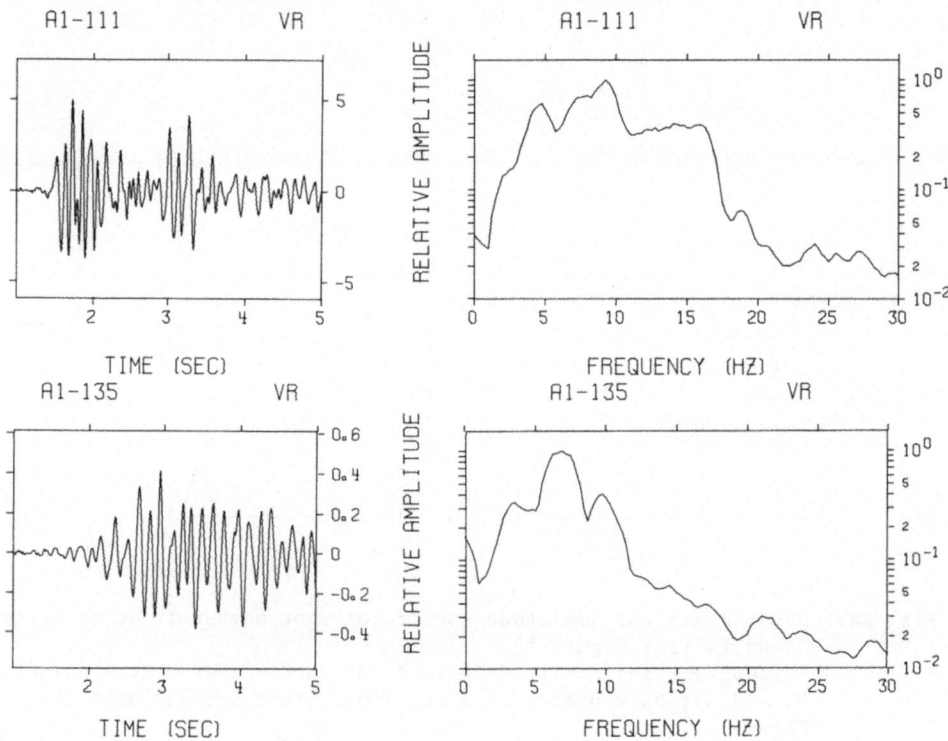

Fig. 4. Seismograms and amplitude spectra of shot A1 at stations 111 and
135 (see Figure 3).

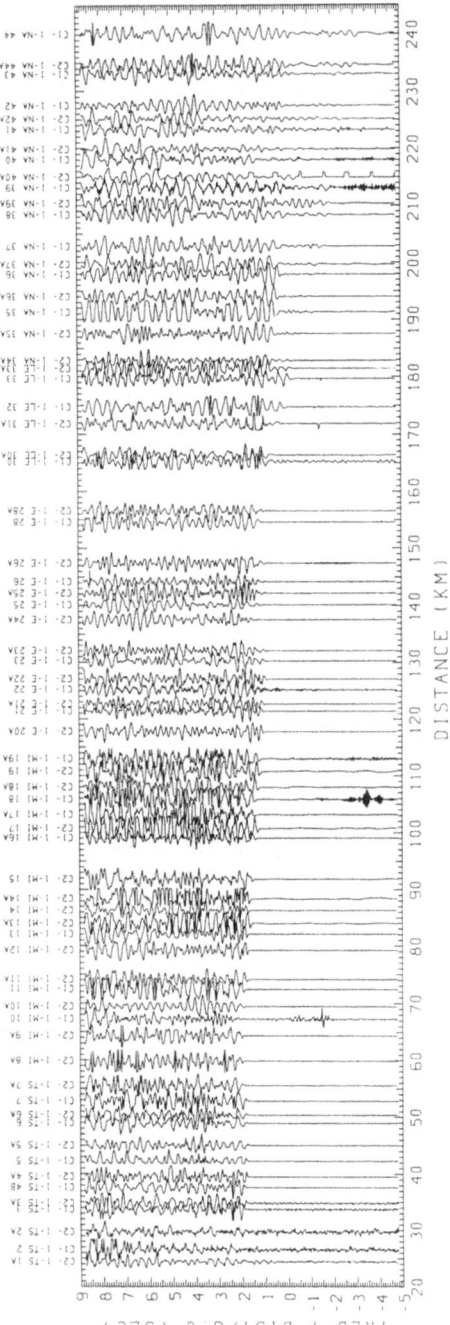

Fig. 5. Unfiltered trace-normalized, vertical component record section in Sardinia from two 1 to offshore shots in the Strait of Bonifacio to the south (after Egger, 1985).

consisted of water saturated Quaternary sediments. The frequency spectra are again characterized by a strong scatter between the recordings of the same explosion A1 at two different sites. The dominant frequencies range from 5 to 10 Hz (Figure 4) for two portions of the selected seismograms. Record number A1-111 includes the P_g phase and signals from the upper crust whereas seismogram number A1-135 is dominated by phases from the lower crust.

For some time it was fairly easy and cheap to use explosions in lakes as a very efficient seismic source. Because of the environmental implications this technique can no longer be accepted.

The controls of signal frequency and range of observation for a given charge size becomes much easier and clearer with offshore explosions recorded along profiles on land or by a set of ocean-bottom seismographs at sea for marine seismic profiles. The dominant period of the signal for a given charge is determined by the depth at which the shot is fired. For a given charge size an optimum water depth exists where the bubble pulse and the surface reflection are in phase. This is reached when the blubber period of the gas bubble equals four times the travel time of the explosive signal between the explosion at depth and the water surface (Wielandt, 1972).

As an example, Figure 5 shows the trace-normalized vertical component record section of a north-south profile on Sardinia with the shotpoint C, 24 km off the north coast of Sardinia in the Strait of Bonifacio (Egger, 1985; Egger et al., 1988). At this shotpoint two explosions of 1 ton each were set off at a depth of 80 m and a water depth of 240 m. The corresponding summed noise and signal spectra for 46 and 71 traces are shown in Figure 6. For the noise spectra a time window of 4 s preceding the first arrivals was taken. The signal plus noise spectra were derived from a time window of 11 s (-4 s to + 7 s reduced travel time). The peak frequency lies at 3 Hz within a band from 1.5 Hz to about 7 Hz. This peak value is slightly higher than the one for optimum depth, because the depth of the charge was only 80 m instead of 130 m.

Similarly, a series of offshore shots fired along a profile can be recorded by single fixed stations on land or by OBS. Figure 7 presents the unfiltered record section of 41 explosions set off in the Sardinia Channel and recorded at one site in southern Sardinia (Ye, 1986). In this specific case the charge sizes were 80 kg detonated at a depth between 60 and 70 m and with depths to the sea bottom ranging from 600 m to 2400 m. Because of the smaller charges and the shallower shooting depth the dominant frequencies of the signal have now increased to values around 5 Hz (Figure 8) which is still significantly lower than the frequency range generated in borehole explosions as was shown previously.

Higher signal frequencies allow a better resolution of structural details. The smooth velocity structure of the crust derived quite often from the recordings of offshore explosions may partly be caused by the smaller resolution power of the low-frequency signals. Large charges at optimum depth provide higher signal amplitudes and consequently long recording distances at the expense of structural resolution. In addition, the optimum depths for large charges require quite often long distances from the shoreline, i.e., to the nearest recording site which reduces the resolution of the upper crust still further. Jacob (1975) has shown that a set of small dispersed charges fired at shallow optimum depth provide amplitudes which are directly proportional to the number of individual charges without changing the waveform generated by a single unit. Besides, small dispersed charges can be handled more easily than a single large charge.

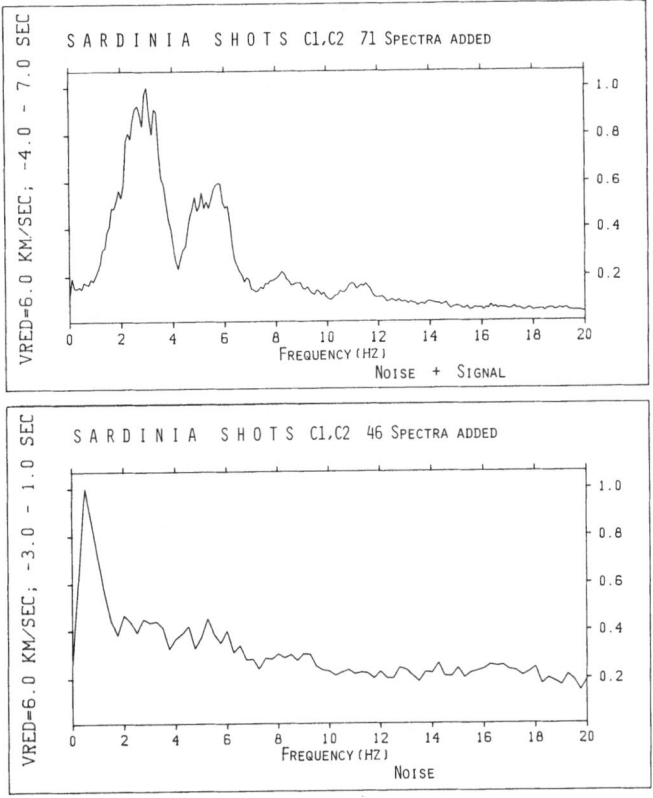

Fig. 6. Summed amplitude spectra of noise (lower part) and noise + signal (upper part) of 1 ton marine shots C1, C2 recorded in Sardinia. Time windows are indicated on the left for a reduction velocity of 6 km/s (after Egger, 1985).

EXPERIMENTAL DESIGN

Most of the refraction, wide angle reflection measurements are carried out along linear arrays or profiles with distances ranging from a few kilometers up to several thousand kilometers as for example on the Fennoscandian Long-Range experiment (FENNOLORA) across the Baltic Shield where almost 2000 km distance were reached (Guggisberg, 1986). This experimental set-up is still most appropriate for the determination of the main structural features. In reality, these profiles are seldom completely reversed which implies some uncertainty and assumptions for the interpretation.

The depth variation of distinct horizons perpendicular to the main profile can be determined by fan observations around the shotpoints at appropriate distances. Such surveys have been carried out successfully for example in the central Alps to outline the depth to the Moho on a north-south cross section (Behnke, 1969; Egloff, 1979). Figure 9 shows this fan section recorded at 140 km distance east of the shotpoint Lago Bianco in the Gotthard Massif with clear lateral variations of the wide

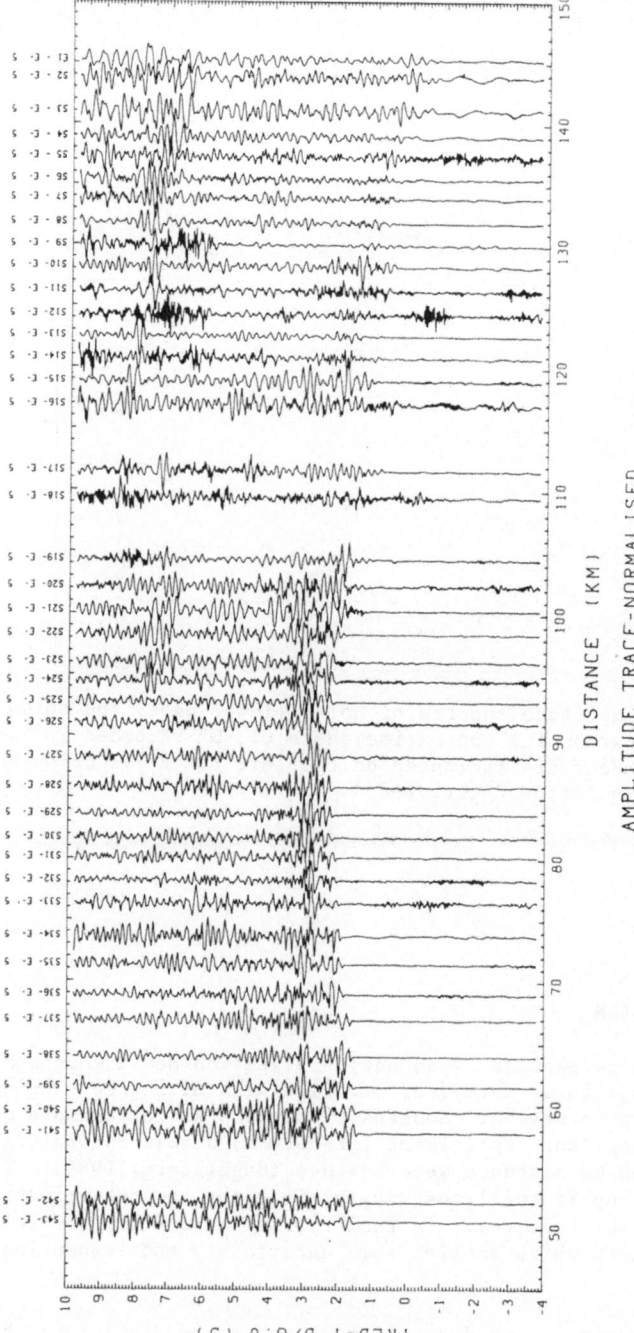

Fig. 7. Unfiltered trace-normalized, vertical component seismogram section from 43 marine shots in the Sardinian Channel recorded at a single station in southern Sardinia (after Ye, 1986).

218

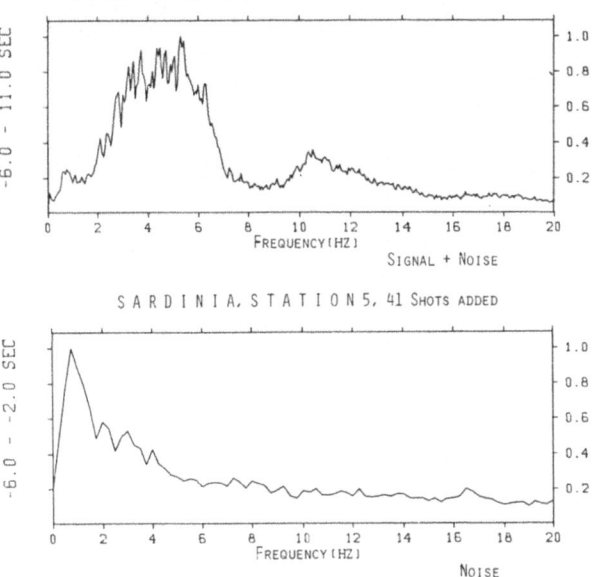

Fig. 8.   Summed amplitude spectra of noise (lower part) and noise + signal
(upper part) of 41 marine shots, 80 kg each, recorded in
Sardinia.   Time windows are indicated on the left for a reduction
velocity of 6 km/s (after Ye, 1986).

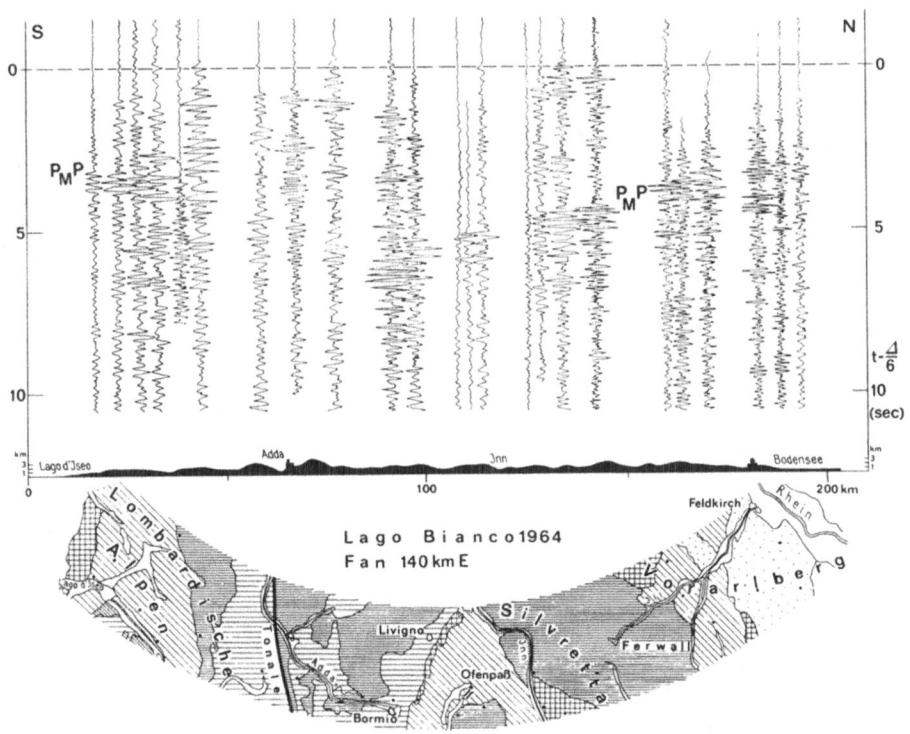

Fig. 9.   Fan section at 140 km distance east of shotpoint Lago Bianco
(1964) in the Gotthard Massif, Central Alps (reduction velocity 6
km/s).   Note delay of P_MP with increasing crustal thickness
(after Behnke, 1969).

angle reflection from the Moho (P_MP). Hirn et al. (1980) have determined a vertical throw of the Moho of more than 10 km from north to south across the Pyrenees over a distance of only a few kilometers. Care has to be taken to include proper corrections for the influence of near-surface low-velocity sediments on the travel times.

Station spacing is another important parameter which influences the resolution of structural details both with depth and along the profiles. Not too long ago the mean station spacing was only 5 km (for example see Giese, Prodehl and Stein, 1976). Now an average spacing between recording sites of 1 to 3 km is used mostly. A few hundred meters are desirable, but financial means and often also the topography set the limits. Jentsch (1979) has reprocessed a profile from the Ukrainian Shield with an original station spacing of 0.2 km. Figure 10 shows how certain seismic phases from within the crust are lost and details of lateral variation of travel times disappear when the spacing is increased to 1 km and 3 km. Behrens et al. (1986) published recently an excellent record section from a single seismic station which has recorded continuously a series of airgun shots over 120 km distance with a spacing of 300 m. Of course, whether or not data of this quality can be obtained depends very much on the local sedimentary structure under the recording site and cannot be expected everywhere.

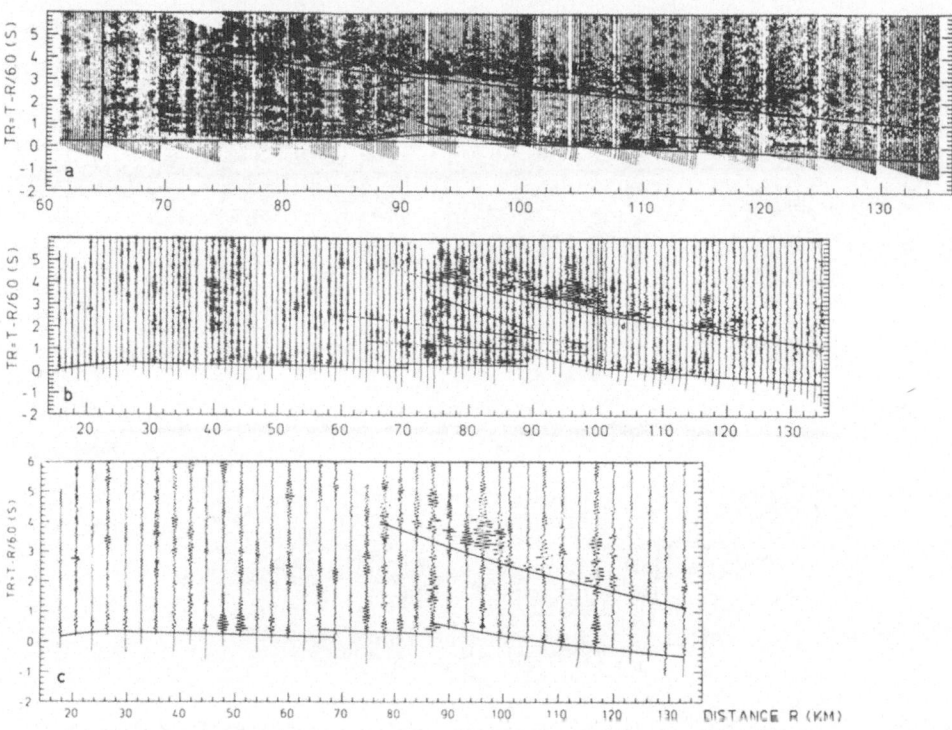

Fig. 10. Record section of a deep seismic sounding profile on the Ukrainian Shield with station spacing of (a) 200 m; (b) 1 km; and (c) 3 km, after Jentsch (1979), (reduction velocity 6 km/s).

220

ACCURACY OF TRAVEL TIME AND AMPLITUDE MEASUREMENTS

For any meaningful interpretation the knowledge of the accuracy of seismic data is an important prerequisite. Deichmann (1984) has carried out a careful study of the achievable accuracy of travel times and amplitudes based on a test experiment with 24 MARS stations equipped with MARK-L4-3D geophones. The instruments recorded a shot of 95 kg explosives at a distance of 1.4 km at the same location within a radius of about 3 m and the recordings were processed and analyzed routinely. The outcome is a standard deviation of the relatively sharp first arrivals of ± 22 ms. This includes the inaccuracy of picking the onsets, variations of the recording instruments, and the accuracy of the digitizing process. Errors in station location depend on the care of the field observers and quality of maps, and, therefore, are hard to estimate. Shot time and shot location are not considered because these are systematic errors. Also the contributions of elevation corrections are disregarded here, but of course, can be rather significant. Having this in mind, an arrival time accuracy of 30 ms can be considered as a realistic number with the presently used equipment and processing.

Deichmann (1984) has also used the same data to determine the uncertainty of the amplitude values. Considering the errors of the measurement of amplitudes on the plot, differences in the seismometer sensitivity and in the recording amplifier gains, and effects of processing the upper limit for the amplitude scatter is 20%. This does not include the differences due to ground coupling and to site geology between individual stations. In reality, these surface influences can change the amplitudes by a factor of two. But since the amplitudes of a refracted or reflected phase can vary systematically by factors between 4 and 30 over the entire profile depending on the velocity structure in the crust, theoretical model calculations are still the most reliable tool to determine this structure in detail.

INTERPRETATION

Instead of reviewing all current methods for the interpretation of deep seismic sounding data a few examples together with the interpretations by different workers are presented here. The Commission on Controlled Source Seismology (CCSS) in the International Association of Seismology and Physics of the Earth's Interior (IASPEI) organizes regularly one-week workshops. The objective is to have a group of people with different backgrounds and ideas work independently on common data sets to obtain an indication of constraints that various analysis methods can place on the resolution of such a structure. These workshops started with synthetic data from rather simple laterally homogeneous velocity-depth structures, included several sets of real data with the assumption of lateral homogeneity, proceeded to synthetic sections for laterally inhomogeneous structures and included recently also near-vertical reflection data from areas with pronounced lateral inhomogeneity.

Figure 11 presents a synthetic record section from the workshop in Karlsruhe, FRG, 1977, for a simple velocity-depth structure together with the corresponding travel time branches of refracted (a,c,f,h) and reflected phases (b,d,e,g) (Ansorge et al., 1982). The interpreters were free to use whatever inversion technique they wanted. Twelve models derived for this synthetic section are shown in Figure 12 together with the original structure. All the models show the main boundaries at almost identical depth. Differences stem from small amplitude phases which are caused by numerical noise or interferences and by differences in the identification of phases which decay rapidly in amplitude with increasing

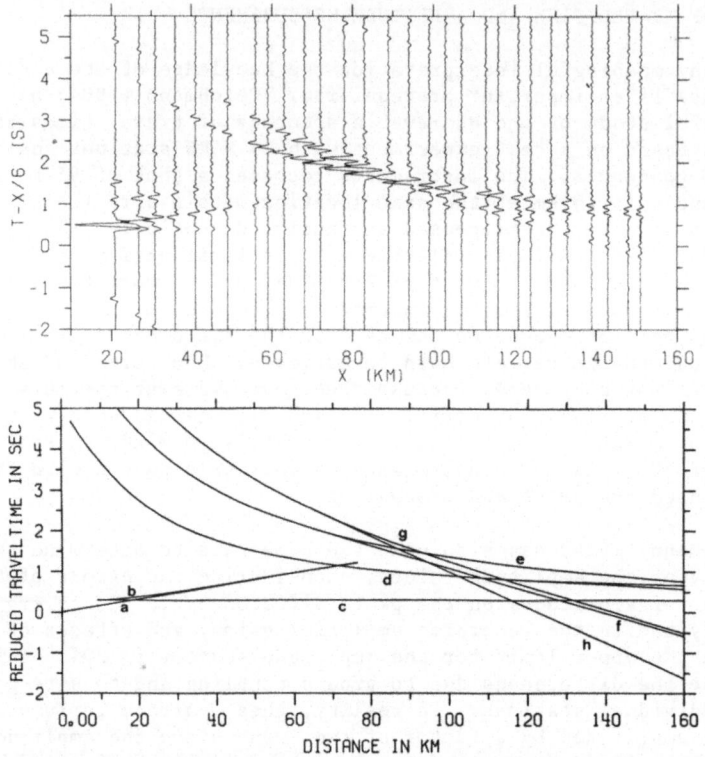

Fig. 11. Synthetic crustal section with true amplitudes and corresponding
          travel time curves: (a,c,f,h refracted phases; b,d,e,g reflected
          phases; reduction velocity 6 km/s) (after Ansorge et al., 1982).

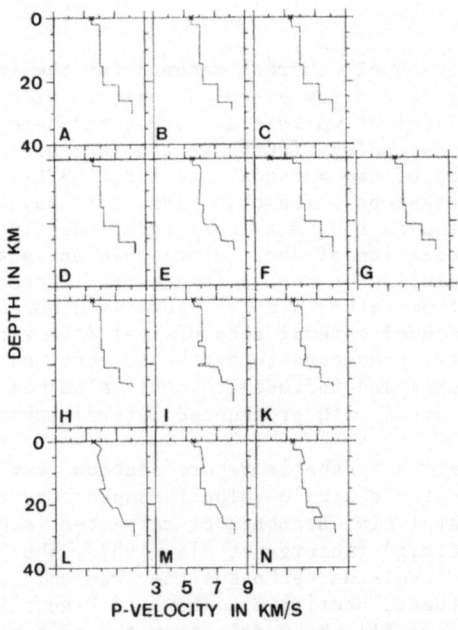

Fig. 12. Models derived from the synthetic record section in Figure 11 by
          various authors. (A) original model (after Ansorge et al.,
          1982).

distance. The variety of velocity-depth functions computed for these data reflects primarily the differences in the correlation of phases by the various workers. The different methods of inversion applied subsequently lead to structures with the same main features when the initial correlations are similar. Preconceived ideas about the velocity structure by the individual interpreter may bias the number and identification of correlations. Apart from clearly correlatable refracted arrivals (P_g, P_n), the most reliable waves for estimating the main parameters of the velocity model are the reflections, e.g., P_MP (Figures 1,3,7,10).

For comparison a record section of real data from the LISPB experiments in northern Britain 1974 (Bamford et al., 1976) was chosen for the same workshop (Figure 13). A representative selection of velocity-depth functions derived for this crustal record section by various participants is shown in Figure 14.

The variety of these results reflects primarily again the differences in phase correlations. All models show a clear crust-mantle boundary. Velocity inversions are of minor importance. The transition to a higher-velocity lower crust has been located by most authors between 20 and 25 km depth. While differences in details are quite obvious, the main characteristic properties, i.e., more or less constant velocity in the upper and a gradient zone in the lower crust, are well determined.

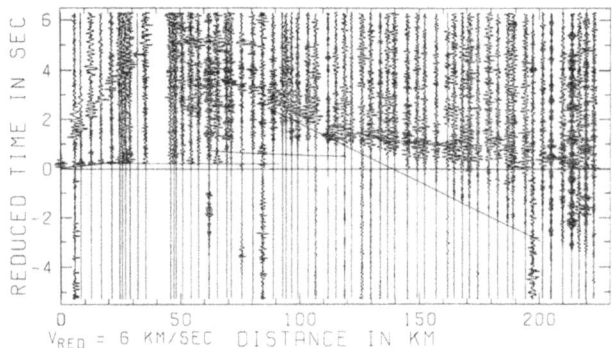

Fig. 13. Trace-normalized vertical component record section from LISPB profile ALPHA-1 (Bamford et al., 1976) (after Ansorge et al., 1982).

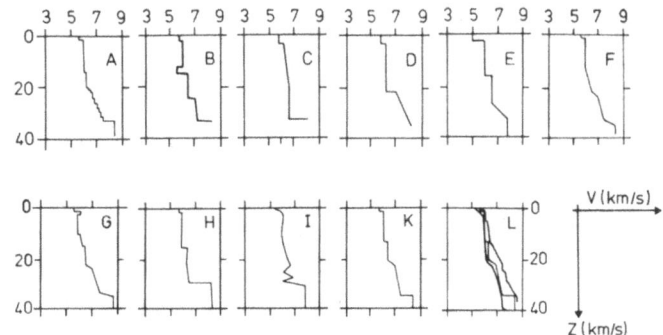

Fig. 14. Models derived from the record section in Figure 13 by various authors. (K) final model, (L) Tau-bounds together with final model (after Ansorge et al., 1982).

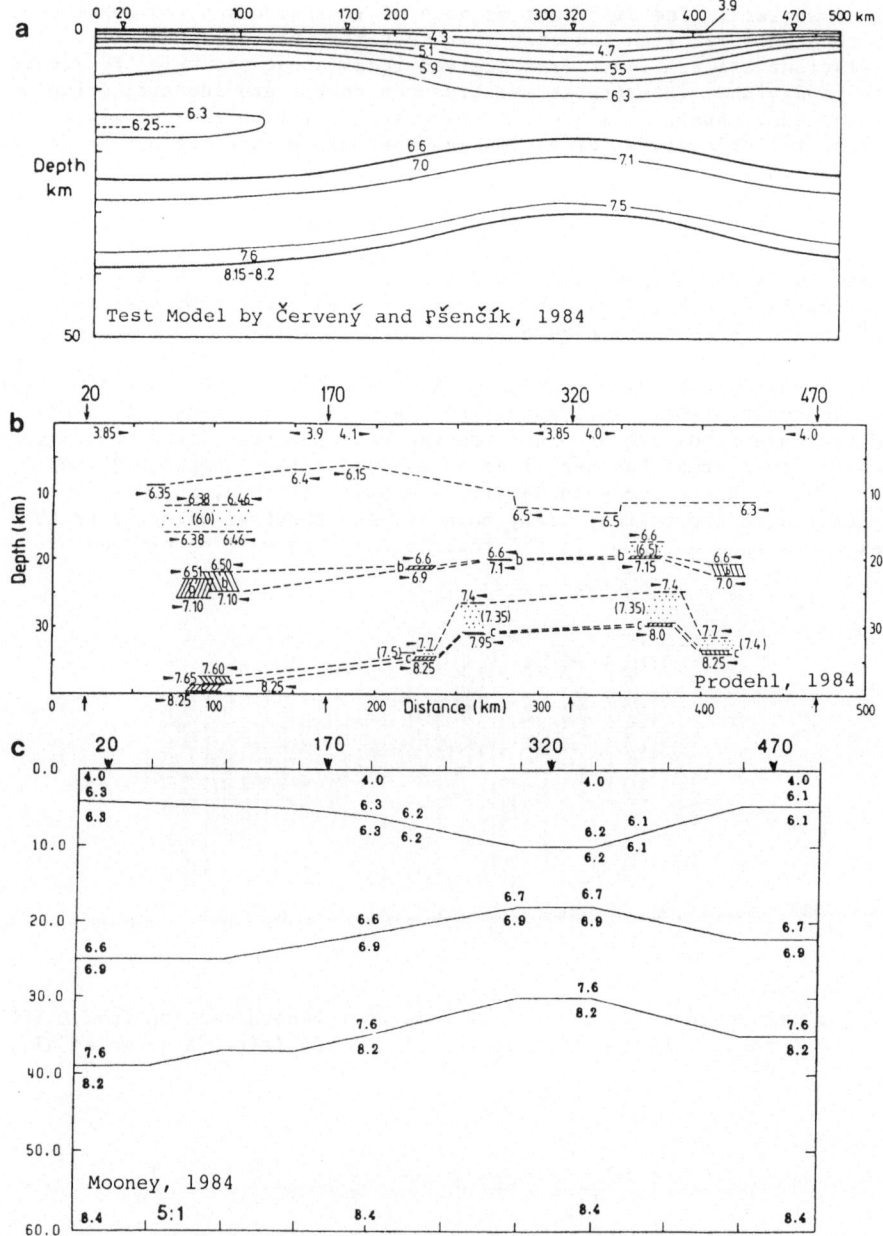

Fig. 15. (a) Velocity-depth model from which reversed synthetic record
sections were calculated at shotpoints 20 km, 170 km, 320
km, 470 km by Červený and Pšenčík (1984).
(b) Crustal structure derived by Prodehl (1984) from synthetic
data for (a). Arrows: direction of corresponding profile.
Hatched: transition zones. Dotted: low-velocity zones with
mean velocities.
(c) Crustal structure derived by Mooney (1984) from synthetic
data for (a). No low-velocity zone was found in this
interpretation.

A more recent workshop in Einsiedeln, Switzerland (1983) was dedicated to the interpretation of laterally inhomogeneous structures. A set of synthetic record sections was prepared by Červený and Psenčík (1984) for a 520 km long hypothetical crustal model with four shotpoints which provided reversed coverage. The model consisted of a crust varying in thickness between 32 and 40 km which included a sedimentary basin and a mid-crust low-velocity zone. Figure 15 shows the original model together with two interpretations. The derived model by Prodehl (1984) is based solely on a one-dimensional analysis. By concentrating on the cusps of reflections, critical points, and offsets between phases he was able to derive a model which comes close to the original structure. It even includes the low-velocity zone. Mooney (1984) used a two-dimensional ray-tracing technique which allowed him to derive a very close fit to the original data without inclusion of a low-velocity layer. However, one has to keep in mind that field data are rarely so clear as the synthetic set used for this exercise.

The interpretation is of real seismic data obtained in laterally inhomogeneous media were the main topics of the last two CCSS workshops in Susono, Shizuoka (Japan), 1985 (Walter and Mooney, 1987) and Whistler (Canada), 1987. Susono was dedicated to the interpretation of refraction and reflection data from a profile across the California Coast Ranges and the adjacent Great Valley. The participants agreed on the main features but important details for the understanding of the tectonic evolution could not be resolved. Most methods enable the determination of depths of boundaries and the corresponding average velocities if good data are available. The correlation of short and weak phases is problematic and can cause major differences in the derived structures. At the presence of strongly varying structures, differences in depth can be compensated by lateral velocity variations if good enough data are lacking. Therefore, more effort has to be put into the acquisition of dense and high quality data, i.e., further reduction of shotpoint and station intervals, which then can be subjected to careful processing with filtering and true amplitude recoverage as a basis for any inversion method.

ACKNOWLEDGEMENTS

The author would like to thank N. Deichmann, A. Egger and S. Ye for providing the final record sections and spectral analyses. Contribution No. 562, Institute of Geophysics, CH8093 Zurich, Switzerland.

REFERENCES

Ansorge, J., Prodehl, C., and Bamford, D., 1982, Comparative interpretation of explosion seismic data, J. Geophys., 51:69-84.
Bamford, D., Faber, S., Jacob, B., Kaminski, W., Nunn, K., Prodehl, C., Fuchs, K., King, R., and Willmore, P., 1976, A lithospheric seismic profile in Britain - I. Preliminary results, Geophys. J. R. A. S., 44:145-160.
Barazangi, M., and Brown, L., eds., 1986, Reflection seismology: a global perspective, Geodyn. Ser., 13 and 14, AGU, Washington, DC.
Behrens, K., Hansen, J., Flüh, E. R., Goldflam, S., and Hirschleber, H., 1986, Seismic investigations in the Skagerrak and Kattegat, Tectonophys., 128:209-228.
Behnke, C., 1969, Messdaten seismischer Untersuchungen in den Alpen 1959-1969. Bericht des Niedersaechsischen Landesamtes fuer Bodenforschung Hannover.

Brewer, J. A., and Oliver, J. E., 1980, Seismic reflection studies of deep crustal structure, Annu. Rev. Earth Planet Sci., 8:205-230.

Červený, V., andPsencik, I., 1984, Data Set I: Model Zurich. Computation of synthetic record section, in: "Workshop Proceedings: Interpretation of Seismic Wave Propagation in Laterally Heterogeneous Structures", D. M. Finlayson and J. Ansorge, eds., Bureau of Mineral Resources, Canberra (Australia), Report 258:15-39.

Deichmann, N., 1984, Combined travel time and amplitude interpretation of two seismic refraction studies in Europe, Ph.D. Thesis ETH Zürich, pp 144.

Deichmann, N., and Ansorge, J., 1983, Evidence for lamination in the lower continental crust beneath the Black Forest, J. Geophys., 52:109-118.

Deichmann, N., Ansorge, J., and Mueller, St., 1986, Crustal structure of the Southern Alps beneath the intersection with the European Geotraverse, Tectonophys., 126:57-83.

Egger, A., 1985, Struktur der Lithosphäre unter Korsika und Sardinien abgeleitet aus refraktionsseismischen Messungen, Diplom Thesis ETH Zürich, p 109.

Egger, A., Demartin, M., Ansorge, J., Banda, E., and Maistrello, M., 1988, The gross structure of the crust under Corsica and Sardinia, Tectonophys., 150:363-389.

Egloff, R., 1979, Sprengseismische Untersuchungen der Erdkruste in der Schweiz, Ph.D. Thesis ETH Zürich, p 167.

Giese, P., Prodehl, C., and Stein, A., eds., 1976, Explosion Seismology in Central Europe, Springer Co., Berlin, p 430.

Guggisberg, B. Ch., 1986, Eine zweidimensionale refraktionsseismische Interpretation der Geschwindigkeits-Tiefen-Struktur des oberen Erdmantels unter dem Fennoskandischen Schild (Projekt FENNOLORA), Ph.D. Thesis ETH Zurich, p 199.

Guggisberg, B., Sierro, N., Ansorge, J., Demartin, M., and Banda, E., 1984, Two-dimensional modelling of Data Set I, laterally inhomogeneous structure, in: "Workshop Proceedings: Interpretation of Seismic Wave Propagation in Laterally Heterogeneous Structures", D. M. Finlayson and J. Ansorge, eds., Bureau of Mineral Resources, Canberra, Australia, Report 258:152-158.

Hirn, A., Daignières, M., Gallart, J., and Vadell, M., 1980, Explosion seismic sounding of throws and dips in the continental Moho, Geophys. Res. Let., 7:263-266.

Incorporated Research Institutions for Seismology (IRIS), 1984, The Program for Array Seismic Studies of the Continental Lithosphere (PASSCAL), 1616 North Fort Myer Dr., Arlington, Virginia 22209, USA, p 169.

Jacob, A. W. B., 1975, Dispersed shots at optimum depth - an efficient seismic source for lithospheric studies, J. Geophys., 41:63-70.

Jentsch, M., 1979, Reinterpretation of a deep seismic sounding profile on the Ukrainian Shield, Z. Geophys., 45:355-372.

Mooney, W. D., 1984, An interpretation of synthetic seismograms for a two-dimensional structure, in: "Workshop Proceedings: Interpretation of Seismic Wave Propagation in Laterally Heterogeneous Structures", D. M. Finlayson and J. Ansorge, eds., Bureau of Mineral Resources, Canberra (Australia), Report 258:159-165.

Mooney, W. D., and Brocher, Th. M., 1987, Coincident seismic reflection-refraction studies of the continental lithosphere: a global review, Reviews of Geophysics, 20:723-742.

Prodehl, C., 1984, Interpretation of synthetic record sections for a 2-D laterally inhomogeneous structure, in: "Workshop Proceedings: Interpretation of Seismic Wave Propagation in Laterally Heterogeneous Structures", D. M. Finlayson and J. Ansorge, eds.,

Bureau of Mineral Resources, Canberra (Australia), Report 258:66-72.

Walter, A. W., and Mooney, W. D., eds., 1987, Interpretation of the SJ-6 seismic reflection-refraction profile, south central California, USA, Commission on Controlled Source Seismology, Proc. of 1985 Workshop in Susono, Shizuoka, Japan, p 132.

Wielandt, E., 1972, Generation of seismic waves by underwater explosions, Geophys. J. R. A. S., 40:421-439.

Ye, S., 1986, Die Struktur der Erdkruste unter Sudsardinien und dem sudlich angrenzenden Sardinischen Kanal aus refraktionsseismischen Messungen, Diplom Thesis ETH Zürich, p 117.

PROCESSING BIRPS DEEP SEISMIC REFLECTION DATA:

A TUTORIAL REVIEW

Simon L. Klemperer

British Institutions' Reflection Profiling
Syndicate (BIRPS), Bullard Laboratories
Madingley Rise, Madingley Road
Cambridge CB3 0EZ, UK

INTRODUCTION

Processing of deep reflection data involves many sequential
operations, with testing and then choice of processing parameters required
at several of these stages. This paper reviews the processing sequence
used for a recent BIRPS profile, and illustrates the complete range of
tests performed during that processing. Although every seismic line has
its own individual processing problems, the standard processes appropriate
to almost all marine deep reflection data are shown here, so that this
paper may serve both as a tutorial to workers entering this field and as a
reivew of a standarized processing sequence presently in use at BIRPS.

Deep seismic reflection profiling of the continents has been
developed, essentially since 1975 (Oliver et al., 1976), to become the
highest resolution technique available for study of the whole crust and
potentially of the whole lithosphere (McGeary and Warner, 1985). Perhaps
50,000 km of deep (whole crustal) continental profiles are now available,
principally recorded by academic researchers in Western Europe, North
America and Australia (Barazangi and Brown, 1986a and 1986b; Matthews and
Smith, 1987). The unique ability of reflection profiling to trace
geologic features mapped at the surface deep into the crust, and to image
structures less than one hundred meters thick and only a few kilometers
across as deep as or deeper than the Moho, ensures that deep profiling
will remain in the forefront of geophysical exploration for many years.

Study of the lithosphere by reflection profiling involves three major
component activities: data acquisition, data processing, and data
interpretation. Careful data acquisition is the sine qua non of
experimental work, but because of the high cost of both land and marine
recording some parameter choices may be dictated by equipment availability
and financial exigency rather than geophysical desirability. Acquisition
for hydrocarbon exploration is well reviewed by Sheriff and Geldart (1982)
while aspects pertinent to deep reflection profiling are touched on by
Brown (1986) for land work and by Warner (1986) for marine studies.
Because the major crustal structures of Britain all run offshore, BIRPS
has been able to work at sea, where profiling is cheaper and generally
gives better data quality than on land (Warner, 1986), without excluding
important geological targets.

Processing marine data is the subject of this paper. However, this paper does not aim to explain the theory of seismic analysis, for which see texts by, for example, Sheriff and Geldart (1982; 1983), Sheriff (1984) and Hatton et al. (1986). Rather, this paper is intended as a practical guide to processing deep reflection data, and concentrates on illustrating the parameter choices the geophysicist must make during processing. Processing land data utilizes the same general techniques as used in processing marine data but there are additional complexities, including corrections for variable source-receiver geometries and near-surface velocities, for a discussion of which see, for example, Sheriff and Geldart (1982) or, with special reference to deep profiling, Zhu and Brown (1986).

Data interpretation involves a range of activities. For the deep crust geological interpretations tend to rely on a synthesis of reflection data with other geophysical data (e.g., Clowes et al., 1984; Brewer and Smythe, 1986; and many others) or on compilation and generalized analysis of large numbers of data sets (e.g., Matthews, 1986; Smithson, 1986). Detailed interpretations of reflector geometries (e.g., Peddy et al., 1986) and geophysical properties such as reflection coefficients (e.g., Warner and McGeary, 1987), amplitude-with-offset variations (e.g., Louie and Clayton, 1987), and Q (e.g., Clowes and Kanasewich, 1970; Jannsen et al., 1985) are increasingly important.

PROCESSING PHILOSOPHY

The ultimate aim of seismic processing is to produce a true cross-section of the reflective properties of the earth with unlimited resolution and perfect signal-to-noise (S/N) ratio. The first step towards achieving this is the unmigrated time section or common mid-point (CMP) stack which is a reduction of the common source-point recording experiment to a coincident source-receiver geometry. Subsequent migration (e.g., Sheriff and Geldart, 1983) of the CMP stack to produce a true depth section requires detailed velocity and 3-D structural control that is not typically available for deep seismic profiles.

Appropriate processing is further constrained by necessary compromises between improving resolution and improving S/N ratios. The conflicts between resolution and noise reduction are exemplified by the process of deconvolution, which broadens the spectrum thus improving temporal resolution but reduces S/N by amplifying parts of the spectrum inherently low in signal. Another example is migration, which improves spatial resolution but which, for deep crustal data in particular, tends to give an increase in apparent noise levels in the form of migration "smiles" which arise due to artificial (non-geologic) discontinuities in the reflected wavefield (Warner, 1987). Conversely, array simulation or trace summing increases S/N but with a reduction in spatial resolution, while bandpass filtering is used to improve S/N but if applied too severely can lead to a loss of temporal resolution. For deep seismic data the choice is normally weighted towards improved S/N to the extent this can be achieved without reducing resolution below the ultimate physical limitations set by wavefront spreading (Fresnel zone criterion for lateral resolution) and earth absorption of high frequencies (limit on vertical resolution).

Practical processing is also constrained by the sheer volume of data: the 235 km long profile described in this paper contains over 5×10^8 data samples. Processing must therefore be highly automated with interactive decision making only possible for a restricted volume of the data set and at a few restricted points in the processing sequence. If processing

parameters selected for a few test sections are to be applied to the whole
data set, which may run from exposed basement to deep sedimentary basins,
and from noisy inshore waters to quieter deep marine environments, then
the processing methods must be robust enough to cope with these lateral
changes in data character and quality.

BIRPS NORTHEAST COAST LINE (NEC)

The profile NEC is a 235 km long seismic line shot off the northeast
coast of Britain to study a Paleozoic continental collision zone known as
the Iapetus Suture (Klemperer and Matthews, 1987; Freeman et al.,1988). A
thin veneer (< 1 km) of Mesozoic sedimentary rocks overlies deformed Lower
Paleozoic basement at the north end of the profile and moderately thick
Upper Paleozoic sedimentary rocks (> 4 km) at the southern end. The upper
crust contains few reflectors, but the lower crust is strongly laminated
from about 15 km down to the Moho at about 35 km. Although the type of
lower crust is variable along the seismic line, any velocity variation in
the deep crust is probably irrelevant to the appropriate choice of
stacking velocities for these data acquired with a short (3 km) streamer
(see below), though very relevant to the correct migration or depth
conversion of the data.

NEC was acquired with an airgun source that was large (nominal
7276 cu in or 119 l volume at 2000 psi or 13.8 MPa pressure), tuned (36
guns of size from 30 cu in (0.5 l) to 500 cu in (8.2 l), and areally
extensive (74 m wide, 37 m long). The source was towed at 7.5 m depth and
fired every 50 m. The 3 km recording cable contained 60 hydrophone groups
of length 50 m and was towed at a depth of 15 m. The near group offset
from the source was 273 m. Data length was 15 s recorded at 4 ms sampling
interval with filter settings of 5.3 Hz (roll-off 18 dB/octave) and 64 Hz
(72 dB/octave). Data was directly recorded in a standard demultiplex
format, SEG-D (SEG, 1980).

PROCESSING STRATEGY

Processing was carried out in two stages: preliminary quality-control
displays to identify any problems in the data and to guide further
processing, and production processing carried out along side and following
extensive parameter testing. Table 1 is a sequential list of the steps
used in processing, divided into the same subsections used in the text in
order to serve as an index to this paper. The quality-control displays
were used to appraise the data set, estimate processing needs and
potential problems, and to select test sections on which all processing
parameters would be tested and from which parameter selection would be
made. Three test sections (each 7.5 km in length, or in total about 10%
of the data) were selected, of which one is followed in detail in this
paper. Test sections were chosen to be of typical data quality and
representative surface geology, so that processing decisions could safely
be extrapolated from the test sections to the whole seismic line. For
each test several displays of the test section were prepared, each with a
different value of the processing parameter being tested. All the test
sections were displayed as stack sections. Processing steps that had
previously been tested and for which final stacking parameters had been
chosen were applied using these parameters selected for the final stack
(Table 2, column 4), while parameters for subsequent processing stages
were fixed to a standard (Table 2, column 3) based on the brute stack
(Table 2, column 2). For example, when testing the front-end mute
(process xi in Table 1) test stacks were prepared (Figure 9) with variable
mute functions but constant parameters for all the other processors.

Table 1. NEC Processing Sequence

Preliminary Processing & Quality Control (QC) Displays

(i)   Reformat from SEG-D to SEG-X SSL format for processing,
          and to SEG-Y format for archiving and data distribution
(ii)   Instrument phase compensation filter (IPCF)
(iii)   Anti-alias filter and resample to 8 ms
          [QC display: every 20th source-point gather]   (Figure 1)
(iv)   Delete bad records and noise records, and pad
          missing source-points with dummy records (Edit)
          [QC display: near-trace (constant-offset) section]   (Figure 2)
          [QC display: brute stack with guessed parameters]   (Figure 3)
(v)   Deletion of high-amplitude noise traces (Edit)

Pre-Stack Processing

*(vi)   Source- and receiver-array simulation   (Figure 4)
(vii)   Spherical divergence correction
(viii)   Sort into common-midpoint gathers
*(ix)   Predictive deconvolution before stack (DBS)   (Figure 5)
          [Display velocity-function gathers and mini-stacks] (Figs 6,7,8)
(x)   Normal moveout correction (NMO)
*(xi)   Front-end mute   (Figure 9)

Stack

(xii)   CMP stack   (Figure 10)
(xiii)   Apply gun and cable static correction

Post-Stack Processing

*(xiv)   F-k filter (dip filter)   (Figure 11)
*(xv)   Predictive deconvolution after stack (DAS)   (Figure 12)
*(xvi)   Time-variant bandpass filter (TVF)   (Figure 13)
*(xvii)   Time-variant equalization (AGC)   (Figure 14)
(xviii)   Display final stack   (Figure 15)

---

* = testing carried out at this stage to select parameters

Earlier processes in the sequence in Table 1 that had already been tested
(e.g., array simulation and deconvolution-before-stack) were applied with
the parameters chosen for the final stack while later processes in Table 1
(i.e., f-k filter, post-stack deconvolution, bandpass filtering and
amplitude equalization) were applied with the parameters used for all the
test stacks (Table 2, column 3). Thus at each stage a range of parameters
for a particular process is tested by their effect on the final stack
section in isolation from any variation in the parameters associated with
other processes. Note that most of the figures show data only from 0 to
6 s and from 8 to 14 s, even though the data was recorded, and all tests
displayed, continuously from 0 to 15 s. The display format in this paper
has been chosen purely to allow reproduction of the data at a reasonable
scale in this volume. Processing of the NEC line was carried out by
Seismograph Service Ltd under the direction of the BIRPS Core Group.

PRELIMINARY PROCESSING AND QUALITY CONTROL DISPLAYS

     Several initial processing steps are done automatically, without
testing. The field data tapes were reformatted from SEG-D to the
Seismograph Service Ltd house format within the SEG-X standard (SEG, 1980)

Table 2. NEC Processing Parameters

	1. Single-Trace Display	2. Bruce Stack	3. Test Stacks*	4. Final Stack
(i) Reformat	√	√	√	√
(ii) IPCF	√	√	√	√
(iii) Resample	√	√	√	√
(iv)(v) Edit	x	√	√	√
(vi) Array simulation	x	Sum adjacent traces and adjacent files	x	0.0-0.5 s no mixing; 0.5-2.5 s ramp; 2.5-15.0 s 1:3:1 mix
(vii) Spherical divergence	√	√	√	√
(ix) DBS: gap	x	32 ms	32 ms	32 ms/48 ms
operator		392 ms	392 ms	248 ms/352 ms
design window		near:0.5-2.5 s / far: 3.6-5.6 s	6.0-11.0 s	near: 0.5-3 s / 6-11 s / far: 3.6-6.1 s/6-11 s
apply window		0.0-15.0 s	0.0-15.0 s	near: 0-4 s/ramp/6-15 s / far: 0-5 s/ramp/7-15 s
(x) Velocities	x	Every 30 km; Constant velocity	Every 30 km; Constant velocity	Every 3 km; Variable velocity
(xi) Mute	x	x	x	± 27 ms/trace
(xiv) F-k filter	x	x	x	x
(xv) DAS: gap	x	32 ms	32 ms	32 ms/48 ms
operator		392 ms	392 ms	248 ms/352 ms
design window		0.5-2.5 s	6.0-11.0 s	0.5-3 s/6-11 s
apply window		0.0-15.0 s	0.0-15.0 s	0-4 s/ramp/6-11 s
(xvi) TVF	5-60 Hz	8-40 Hz	8-40 Hz	Variable, 0 s:15-45 Hz/15 s:5-30 Hz
(xvii) AGC	1 s	1 s	1 s	Variable, 0.25 s shallow/4 s deep

√ = process applied; x = process not applied.

* For each test stack all processing is held constant except for the parameter under test; earlier processes in sequence have final stack parameters applied (column 4); subsequent processes have test stack parameters applied (column 3).

for processing, and also to SEG-Y (SEG, 1980) for archiving and
distribution (step i in Table 1). In step ii, the data were digitally
filtered to compensate for distortions introduced by the analogue
recording filters. A low-pass, anti-alias digital filter was then applied
with a 57.5 Hz high-cut (5 Hz below the Nyquist frequency corresponding to
8 ms sample interval) low-pass digital filter and a sharp roll-off to
-18 dB at 62.5 Hz before resampling to 8 ms interval. The resampling is
done in order to reduce processing costs by 50%. There is little
reflected energy recorded above about 40 Hz except at very shallow levels
(see Figure 13b) so this procedure does not harm the data quality except
at the very top of the section. The data is recorded at 4 ms sample rate
because analogue low-pass filters have a slower roll-off than digital
filters and are normally set at half the Nyquist frequency (62.5 Hz for
4 ms sampling) to avoid aliassing. Use of an analogue anti-alias filter
at 31.25 Hz before 8 ms data sampling in the field would cause noticeable
bandwidth reduction in the reflected spectrum. Step iv in Table 1 is the
deletion of bad records (no shot fired or airguns fired but at incorrect
time ("misfires"), data corruption during writing to tape, etc.) and
padding of the missing shot points with blank traces. The padding of
missing files after deletion of bad records is done to ensure a uniform

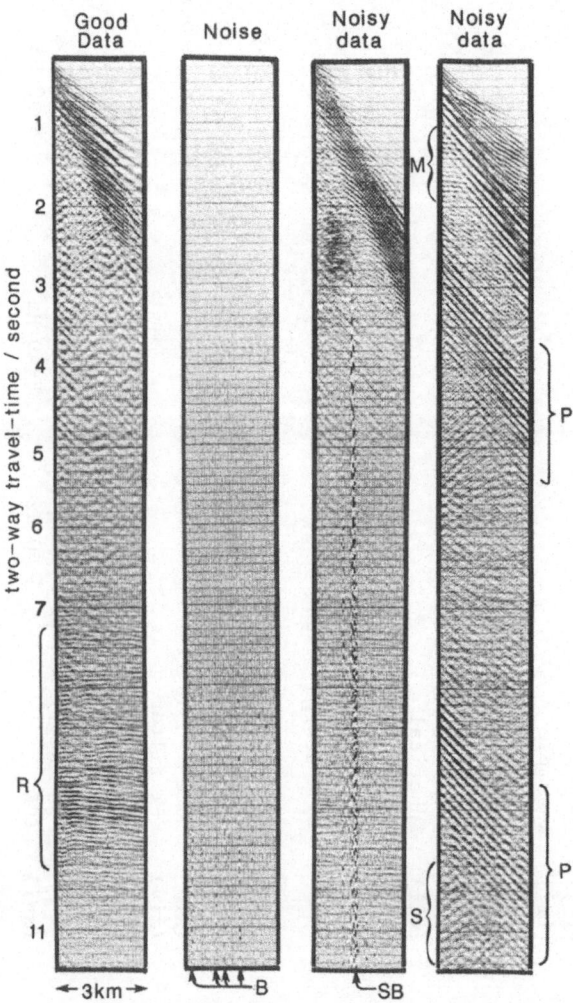

Fig. 1. Four source-point gathers displayed from 0.2 to 11.5 s. For
explanation, see text.

processing geometry, with equal numbers and ranges of traces in all common source-point and common mid-point gathers.

The first quality control (QC) display made (Figure 1) is a plot of every 20th source-point gather (one per kilometer of profile length for 50 m shot spacing), each consisting of 60 traces with a 3 km range of source-receiver offsets. Plots have a simple spherical divergence correction applied so that direct arrivals and deep reflections are visible on the same plot, but otherwise have no corrections or processing applied. This display is intended to allow measurement of the refraction velocities and hence to estimate the near-surface interval velocities; and also to allow a rapid appraisal of noise problems along the profile. From left to right in Figure 1, a good data file is shown; a noise record (airguns deliberately switched off); and two noisy source-point gathers. Deep primary reflections (R) are visible from 7 to 10 s on the good data file and also faintly on the left-hand noisy record. The amplitude of these reflections far exceeds that of the ambient noise level as seen on the noise record which is displayed with the same gain corrections as the other files shown in Figure 1. Low frequency noise on single traces (B) of this record is caused by the depth-controllers ("birds") that maintain constant streamer depth and are attached to the streamer at these points. Various kinds of seismic noise are also shown. "Swell breakout" (SB) occurs when the recording streamer flexes excessively due to deeply penetrating, long-period oceanic swell which sets up pressure waves in the oil-filled streamer. The amplitude is random, and only appears to increase with time in Figure 1 because of the spherical divergence applied to the plot. Water-borne propeller noise (S) from another ship broadside on to the streamer appears here as high-frequency hyperbolae. Reflections from seafloor topography, in this case two pipelines (P) on the sea bottom directly ahead of the recording ship, have the apparent velocity characteristic of water-borne waves (1.5 km.s^{-1}) and a reverbatory multiple character. In addition to ambient noise and side reflections there is normally also other systematic noise such as multiple reflections (M). For more discussion and examples of marine noise sources see Peardon and Cameron (1986) and Fulton (1985). Visual inspection of the noise patterns showed that traces 35 and 45 (trace 1 is the near trace) were consistently noisy due to birds on those streamer sections (trace 35 is the right-hand trace arrowed on the noise record in Figure 1). These traces were deleted (3% of the data). Random bad traces due principally to swell breakout were automatically zeroed if their average amplitude in the window from 13 to 15 s exceeded a reference value chosen by reference to the average amplitude of good traces. About 0.25% of the data were removed by this method. Because the frequency of occurrence and the amplitudes of noise sources are highly variable from profile to profile, the precise method and the degree of such trace editing must be chosen afresh for each line.

The second QC display is a single-trace display or constant-offset section (Figure 2: note that here, as elsewhere, only 0 to 6 s and 8 to 14 s travel time are shown), in which the same trace from successive source-point gathers is displayed side-by-side in a time section. A near trace is desirable so that shallow reflections are visible, but the closest traces to the ship are often the noisiest on the record, due to the ship's propeller noise, turbulence from the airguns and tugging of the ship on the streamer, and so are sometimes unsuitable for use in the single-trace display. For NEC, the near-trace offset (center of the gun array to center of the first receiver group) was 273 m, and the first trace was not significantly noisier than other traces (Figure 1), so trace 1 was used for the display in Figure 2. A simplified processing scheme was applied: predictive deconvolution with a derivation window 0.5 to 8.5 s, operator length 392 ms and gap length 32 ms, applied to the whole

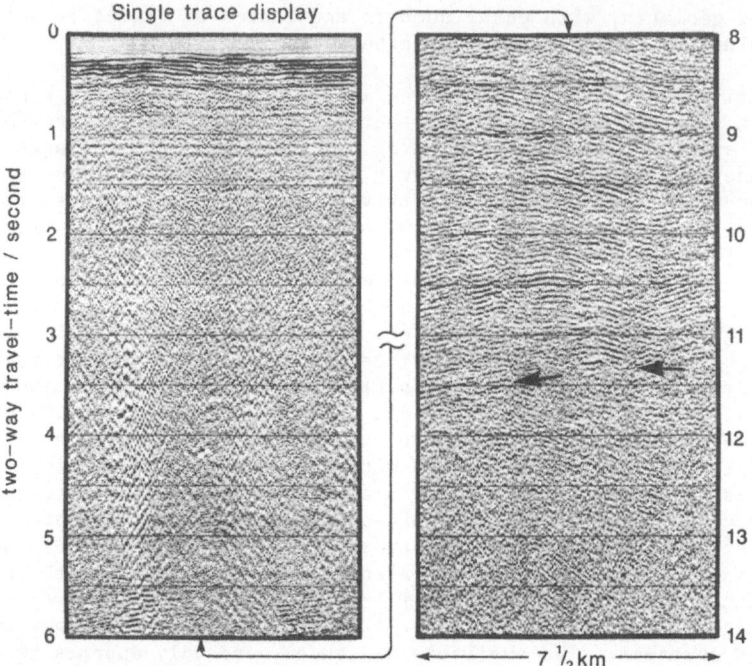

Single trace display

two—way travel—time / second

7 ½ km

Fig. 2. Constant-offset display of test section, 0 to 6 s and 8 to 14 s.

trace (the detailed significance of these parameters is explained later)
followed by a 5 to 60 Hz bandpass filter and a 1 s automatic gain control.
These parameters are a generalization of the final processing parameters
used for previous BIRPS surveys over similar geologic areas, but greatly
simplified to save money. Table 2 compares these parameters with those
used for the final stack of NEC. This simple processing allows some
geology to be interpreted: in Figure 2 many sub-horizontal reflections
("the reflective lower crust") are visible from 5.5 to 11.5 s. The base
of these reflections (marked by arrows in Figure 2) is defined as the
reflection Moho (Klemperer et al., 1986), and may well be equivalent to
the refraction Moho (Mooney and Brocher, 1987).

     The third stage of processing shown is the brute stack or raw stack
(Figure 3). The brute stack also uses a simplified processing sequence
(Table 2), based on prior experience modified slightly by examination of
the single-trace display (Figure 2). Adjacent traces in the source-point
gathers and adjacent source-point gathers were summed before stacking to
reduce the data volume (and processing cost of this step) by a factor of
four. Stacking velocity functions were derived from multi-velocity stacks
at 30 km intervals and published results of crustal seismic refraction
experiments. Deconvolution was applied both before and after stack, with
the same gap lengths and operator lengths as for the single-trace display,
but with derivation windows changed to 0.5 to 2.5 s on the near trace, 3.6
to 5.6 s on the far trace before stack, and 0.5 to 2.5 s after stack. An
8 to 40 Hz bandpass filter (more band-limited than that used for the
single-trace display because little very high or very low frequency energy
could be seen on Figure 2) and 1 s automatic gain control were used. The
brute stack is available soon after acquisition and allows preliminary
geologic analysis that is valuable both towards fulfilling the aims of the
profile and also for identifying the range of seismic events to be
preserved and enhanced during processing. The brute stack was used to
select the test sections on the basis of which all further parameters were

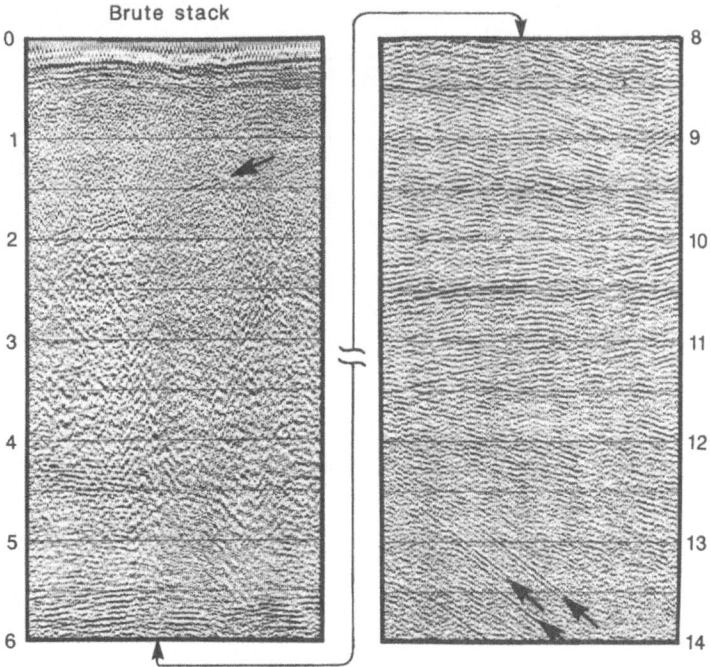

Fig. 3. Brute stack of same test section.

to be chosen. Figure 3 is one of those test sections, chosen because of the range of reflections visible at all travel times and because it seemed to represent average noise levels. On this section of the brute stack, in addition to the reflective lower crust seen on the constant-offset section (Figure 2) there are visible a steep reflection, probably a fault, dipping to the left from 1 to 2 s, and many diffraction tails, dipping right, at the base of the section (marked by arrows on Figure 3). These features may be recognized, and the improvement in their resolution and S/N noted through all the succeeding processing stages. Because the reflectivity in the lower crust is much more prominent than in the upper crust, the design window for the pre- and post-stack deconvolution was changed to a deep zone, 6 to 11 s, for all subsequent testing (except of course for the deconvolution tests themselves).

PRE-STACK PROCESSING

Not all seismic processors are commutative. The processing sequence may be broken into sections separated by changes in trace ordering: the gather from source-point to common mid-point format, and the stack or summation of the CMP-gathered traces. Techniques that depend on relative trace positioning, such as velocity analysis or rejection of coherent noise, must be inserted in the sequence appropriately, whereas trace-by-trace processes such as deconvolution or bandpass filtering can theoretically be done at almost any stage. The order of application of these later processes may be important for some real data-sets. For example: data overprinted by very strong noise outside the frequency bandwidth of interest may benefit from frequency filtering before deconvolution (so that the deconvolution operators are designed on signal not on noise) as well as after deconvolution, while good-quality data may not require prior filtering. However, it is not normally possible to test all possible orderings of the different processors and an expected processing sequence is normally selected after inspecting the brute stack.

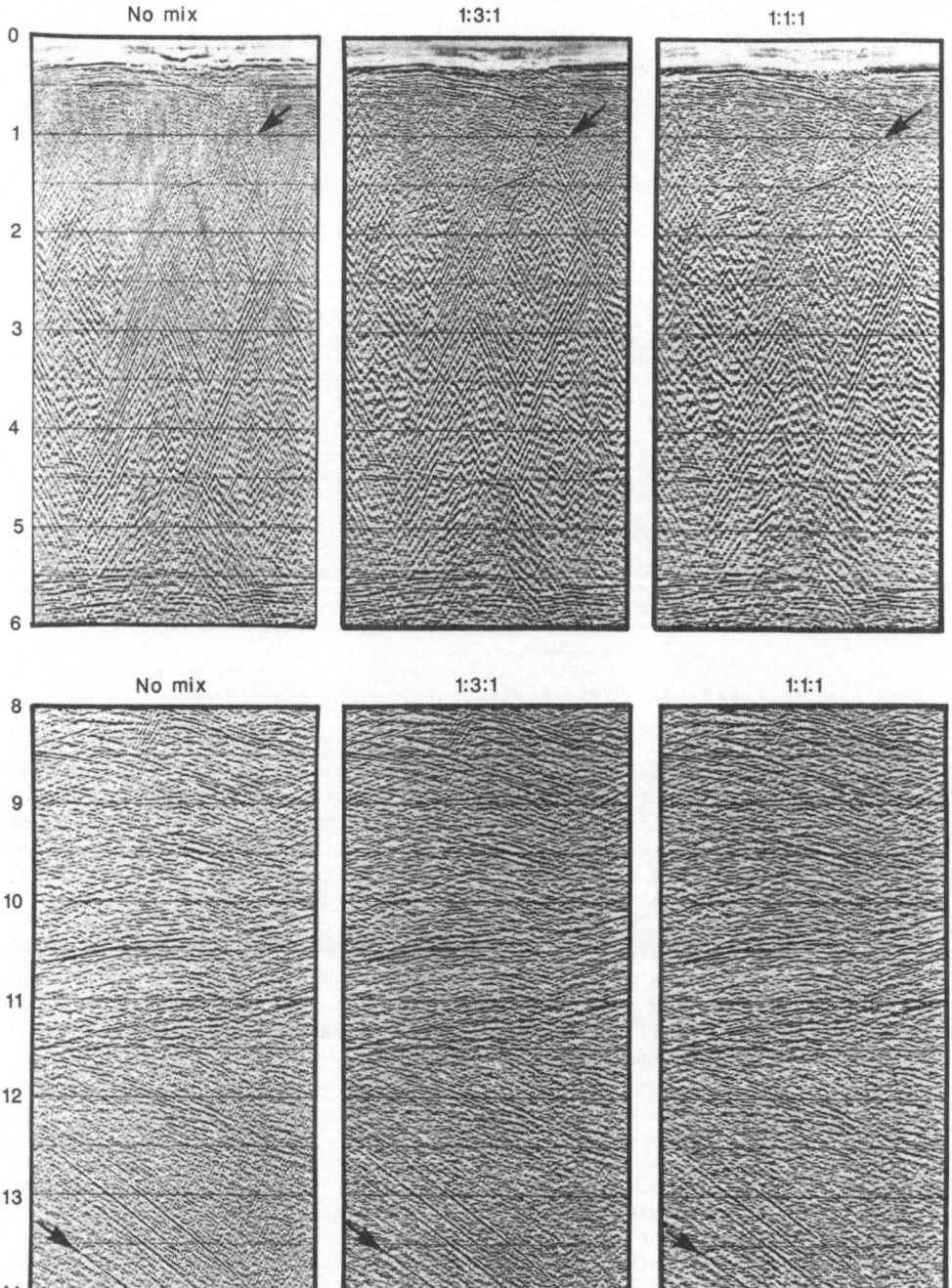

Fig. 4. Top: source- and receiver-array simulation tests.
Bottom: source- and receiver-array simulation tests.

The processing sequence used for NEC begins with source and receiver array simulation (Figure 4, top and bottom). Array simulation, as noted above, increases S/N at the expense of lateral resolution. The test shown compares no array simulation (50 m source and receiver arrays) with center-weighted (1:3:1) 150 m source and receiver arrays and with uniformly weighted (1:1:1) 150 m arrays. The source array is simulated by mixing three adjacent traces in the common-receiver domain and the receiver array by mixing in the common-source domain. For each test panel, array simulation is followed by an identical sequence of processing parameters, through CMP stack to display (Table 2, column 3). In addition to the panels shown, tests were run with 250 m simulated arrays, both with five adjacent traces mixed in the ratio 1:3:9:3:1 and with the approximately equivalent post-stack mixing of five adjacent traces in the ratio 1:2:3:2:1. Though post-stack mixing gives similar results to pre-stack array simulation and is far cheaper, pre-stack array simulation is generally preferred because the resulting enhanced S/N allows better pre-stack deconvolution and velocity analysis. The effect of increasing mixing (moving from left to right in Figure 4) is to increasingly attenuate steeply dipping water-borne side reflections and any random noise. Reflection S/N is generally enhanced because the arrivals are nearly flat on source-point and receiver-point gathers unless they are from steeply dipping reflectors or from distant diffractors, or unless they are long-offset reflections from shallow levels in the section. Over-mixing in the right-hand panel of Figure 4 results in smearing and slight loss of continuity of the shallow dipping reflection above 1 s and the diffraction tails at 12 to 14 s (marked by arrows in Figure 4), and to "worminess" of lower crustal reflections at 8 to 11 s. At the Moho the Fresnel-zone diameter is about 5km and array simulation on the scale described gives no loss in resolution. Loss of resolution is only significant at very shallow levels where the simulated array of 150 m length is large compared with the Fresnel zone, less than about 0.3 s travel time for the NEC velocity structure. This problem, and the danger of attenuating shallow, long-offset reflections by array simulation before normal moveout correction, are avoided by time-varying the application of array simulation. The choice of parameters is clearly somewhat subjective. Based on Figure 4, in the final stack no array formation was used from 0 to 0.5 s, then trace mixing was ramped in from 0.5 to 2.5 s, with full 1:3:1 array simulation applied from 2.5 to 15 s.

After array simulation parameters had been selected, these and a simple spherical divergence correction were applied (Table 1, vi, vii). Spherical divergence is used as a correction for amplitude decay before deconvolution, since an assumption made in deconvolution is that the signal is time-stationary. A data-independent scaling which can easily be removed for subsequent true-amplitude studies is preferable to data-dependent scalers. Correction for spherical divergence satisfies this condition and leads eventually to a "relative amplitude" stack. A fair approximation to the true spherical divergence correction is a gain curve proportional to travel-time multiplied by the square of the stacking velocity (Newman, 1973). To avoid differential trace-to-trace scaling along the final stack section, a constant gain function was applied to the whole profile. The chosen function, generalized from the expected velocity structure, was 1.5 km.s^{-1} to 0.2 s, then linearly interpolated to 3 km.s^{-1} at 1.5 s, 5 km.s^{-1} at 3 s, 6 km.s^{-1} at 6 s and 7 km.s^{-1} at 15 s.

CMP sorting (Table 1, viii) was carried out in this processing sequence after spherical divergence correction. Sorting is routine for marine data collected with regular geometry, and for the NEC profile yielded 30-trace CMP gathers at 25 m intervals, two per source-point. On land, crooked line geometries may require calculation of the mid-point of each source-receiver pair and binning of the data before selection of those traces to be used in each CMP gather.

Predictive deconvolution is the application of an inverse filter designed trace-by-trace to remove short-period multiples and to compress the source wavelet. Three parameters must be specified for predictive deconvolution: the limits of the auto-correlation derivation window that is used for the design of the inverse filter, the active operator length, and the gap. The gap, or minimum lag, determines the amount of wavelet compression and hence the output frequency spectrum. The operator length, or maximum minus minimum lag, determines the time extent of predictable amplitude removed. Because the frequency content of the signal changes substantially with depth for deep crustal data, two operators were designed for each trace, one for the sedimentary section and upper crust, the other for the lower crust and upper mantle. The active operator length was chosen, without testing but following previous experience with BIRPS data, to be 0.248 s and 0.352 s for the upper and lower operators respectively. These operator lengths are considerably greater than the first water-bottom multiple delay, which is never more than 0.13 s since the water depth on NEC is never more than 100 m. The design-window limits were selected by inspection of the source-point gathers (Figure 1) and the brute stack (Figure 3). For the first operator a sliding design window of 0.5 to 3.0 s on the near trace to 3.6 to 6.1 s on the far trace was chosen to exclude refracted noise trains, and for the second operator a fixed window of 6 to 11 s was chosen to include all the prominent lower crustal reflections. A long design window increases the statistical validity of the derived operator, particularly in zones of sparse reflections such as the upper crustal basement, and the design window is typically chosen to be at least ten times the operator length.

The deconvolution-before-stack (DBS) tests in Figure 5 compare the test section without deconvolution with the same section deconvolved with varying gap length. Note that gaps of 32 ms and 24 ms are shown for the top part of this figure, but gaps of 48 ms and 32 ms are shown for the bottom part. This is because gaps are usually lengthened for the deep section where the frequency content of the signal is lower than in the upper crust. At the base of Figure 5 are shown the auto-correlations of the whole of each of the test section after processing (note the symmetry about time zero). Clearly, both deconvolutions shown strongly suppress the water-bottom multiple (marked by arrows in the auto-correlation displays). Decreasing the gap length corresponds to greater whitening of the output spectrum by increasing the high-frequency content of the output wavelet. Comparing the right-hand two panels of Figure 5 (bottom), the data deconvolved with a 32 ms gap show subtly higher frequencies than the data deconvolved with a 48 ms gap, but this is due to amplification of high-frequency components of the spectrum that may have low S/N, and so there is some reduction in overall S/N. The lower S/N is seen for example in the poorer continuity of deep diffraction tails (marked by arrows on Figure 5) when DBS is applied with a 32 ms gap than when a 48 ms gap is used. An additional test panel (not shown) with 48 ms gap and 64 ms gap for the upper and lower operators respectively was examined, but the preferred gap lengths for the final stack were 32 ms and 48 ms. The shallow operator was applied from 0 to 4 s on the near trace and from 0 to 5 s on the far trace, while the deep operator was applied from 6 to 15 s and from 7 to 15 s respectively. The effects of the two operators were smoothly merged over the 2 s gap between their windows of full application. This very gradual variation in processing parameters with depth is extremely important in order to avoid abrupt artificial changes in spectrum, S/N or general character of the data that may subsequently confuse interpretations.

Array simulation enhances S/N, and both array simulation and DBS improve reflection clarity, thus making the process of picking stacking velocities simpler and more reliable. Various devices are available for picking stacking velocities, including moved-out CMPs, constant velocity or variable velocity mini-stacks, and velocity semblance spectra. Though velocity semblance spectra are convenient and reliable for analyzing sedimentary basin profiles with high-amplitude, flat-lying reflections they are less appropriate - certainly so if used in isolation - for

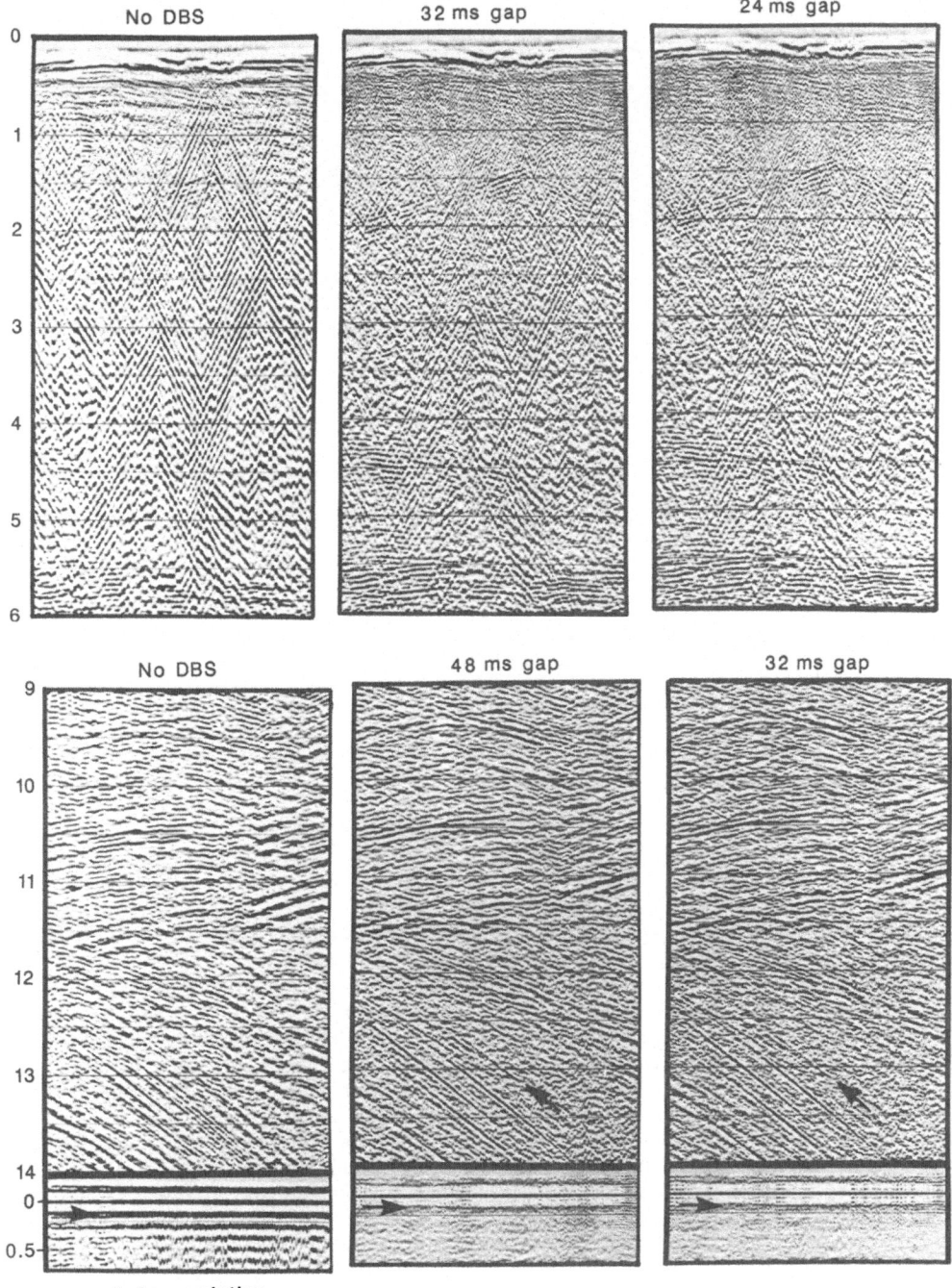

Fig. 5. Top: deconvolution-before-stack tests.
Bottom: deconvolution-before-stack tests and trace auto-
correlations.

picking velocities from sparse, weak and steeply dipping reflective
sequences because semblance peaks are as likely to represent multiples or
non-reflected energy as to be primary reflections.

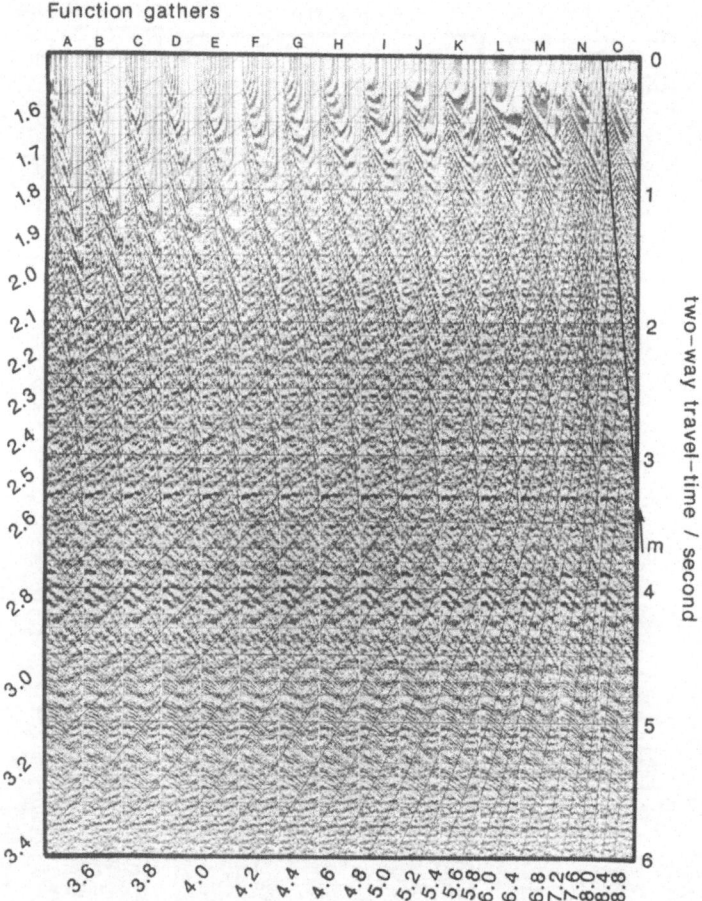

Fig. 6. CMP gathers with NMO applied for variable velocity functions.

Figure 6 is a display of function gathers, that is, a single CMP gather displayed with 15 separate velocity functions (A to O) defined by iso-velocity contours from 1.6 to 8.8 km.s^{-1}. The NMO correction that has been applied to a particular gather at any travel time is given by the iso-velocity contour crossing the center of that function gather at that travel time. Velocity function A was chosen as the minimum conceivable stacking velocity at any depth and function O as the maximum conceivable, i.e., in this area of the earth the velocity will not be less than 1.5 km.s^{-1} at the surface to 3.5 km.s^{-1} at 6 s (function A) nor greater than 6.8 km.s^{-1} at the surface to 8.8 km.s^{-1} at 6 s travel time (function O). Figure 7 shows fifteen stack sections, of which the center trace is the stack of the corresponding gather in Figure 6 after muting with the mute m shown in Figure 6. Note that the velocities vary slowly in the top left of Figures 6 and 7, allowing precise picks where such precision is possible due to the low velocities commonly present near the surface, but that the velocities change rapidly in the bottom right corner where reflections need little NMO correction and intrinsic velocity resolution is greatly reduced. Reflections at 5.5 to 6 s visually stack equally well at 5 km.s^{-1} (stack J) and at over 8 km.s^{-1} (stack O) because the short (3 km) recording array does not permit greater resolution. Although this lack of resolution means that reliable interval velocities cannot be calculated, it also implies that the precise stacking velocities chosen do not greatly affect the resultant section. When there is uncertainty it is

Function stacks

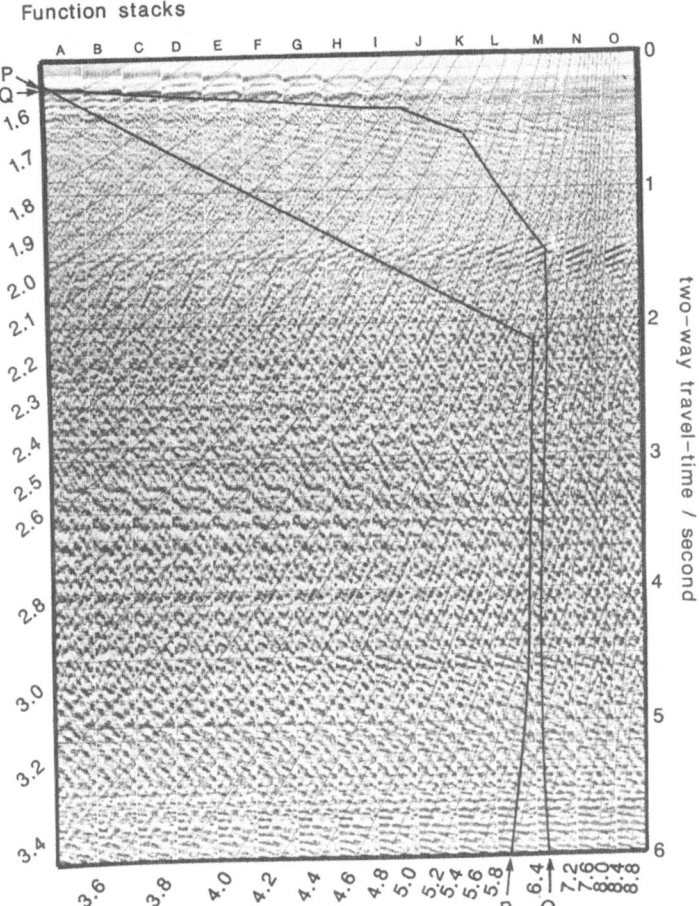

Fig. 7. Sections stacked with NMO corrections as in Figure 6.

better to pick too-high velocities since these will militate most strongly against multiples or side-swipes. This philosophy is evident in the comparison between the computer-selected stacking velocity function obtained from semblances calculated for the function stacks in Figure 7 - this machine pick is shown as the line PP - and the final stacking velocity function chosen, QQ. Note that unusually high velocities are required to properly stack the dipping reflection at 1.5 s - comparison with the brute stack (Figure 3) confirms that this is a real reflector - and that the machine pick did not select this but rather a steeply dipping noise pattern just below 2 s.

Displays of function gathers and stacks were repeated every 3 km (every spread length) along the profile, and augmented by constant-velocity mini-stacks displayed to 15 s every 15 km (detail shown in Figure 8). Though there is no meaningful velocity resolution for real reflections at 15 s travel time, these displays helped identify low-velocity multiples or side-swipes, for example the flat reflection at 8.2 s (Figure 8) which stacks best at 3.2 km.s^{-1}. Velocity functions were chosen using all these plots and also incorporating information from the refraction velocities obtained from the source-point gathers (Figure 1), then smoothed to avoid sharp lateral stacking-velocity changes, before being applied as NMO corrections to the profile (Table 1, x).

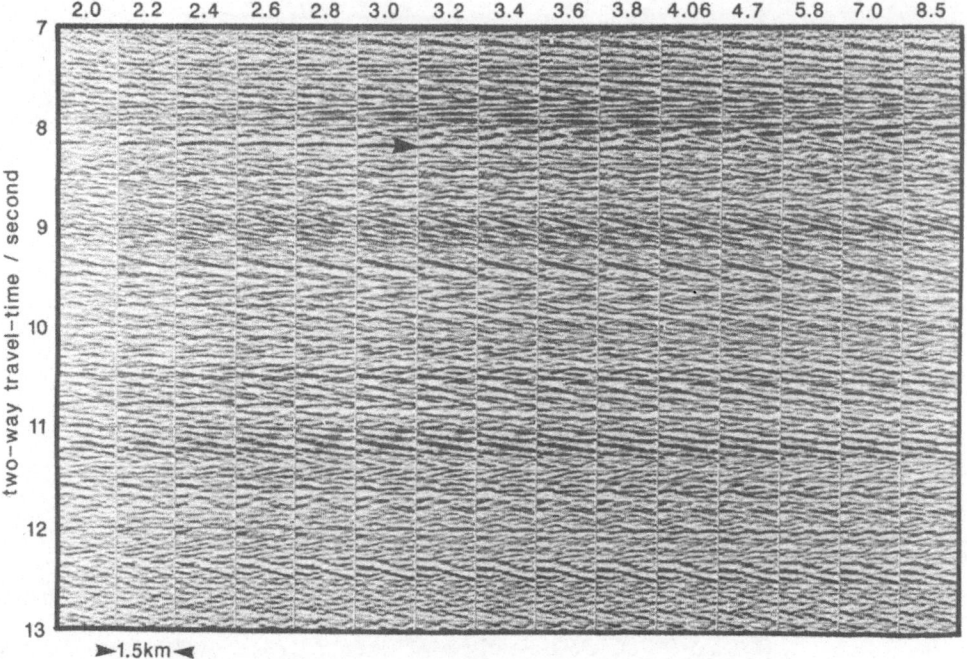

Constant – velocity mini – stacks

2.0 2.2 2.4 2.6 2.8 3.0 3.2 3.4 3.6 3.8 4.06 4.7 5.8 7.0 8.5

two-way travel-time / second

►1.5km◄

Fig. 8. Sections stacked at constant velocity.

Front-end mutes to zero that are part of the trace dominated by non-reflected energy, or distorted by NMO stretch, were tested by displaying panels of successively greater stacking fold, then selecting at each travel time the panel showing the best data quality. Panels with 1, 2, 4, 6, 10, 14, 18, 22, 26 and 30 fold data were examined; a subset of these and the section with the selected time-varying mute (on which is marked the chosen fold at each travel time) are shown in Figure 9. The greater the stacking fold at any travel time, the greater the reduction in random noise and the cancellation of multiples. However, the greater the fold the more the refracted or mode-converted energy allowed into the stack, for times less than 3.5 s. Thus a strong multiple is present at 1.2 s on the 1-trace display, because the fold is too low to effect cancellation, but is not present on the 4-trace display. Strong, stacked coherent noise is visible at 1.5 s in the center of the 18-trace test section and at 1.5 to 2.5 s on the 30-trace section; clearly the fold is too high at these times on these displays. The nature of the energy that is stacking-in as horizontal events is most easily determined by reference back to the moved-out function gathers of Figure 6. After selection of a mute a display was prepared (left panel of Figure 9a) to check its effect before application to the whole data set. In practice the required mute may vary considerably along a profile that crosses both deep basins and exposed basement, and mute tests at several locations may well be needed.

STACK

After NMO correction and front-end muting, the NEC data were stacked, and then scaled by the number of live traces (not muted or edited) present at each travel time in order to preserve relative amplitudes. For deep land profiles, in contrast, some trace-to-trace amplitude balancing before stack may be essential in order to overcome near-surface effects and

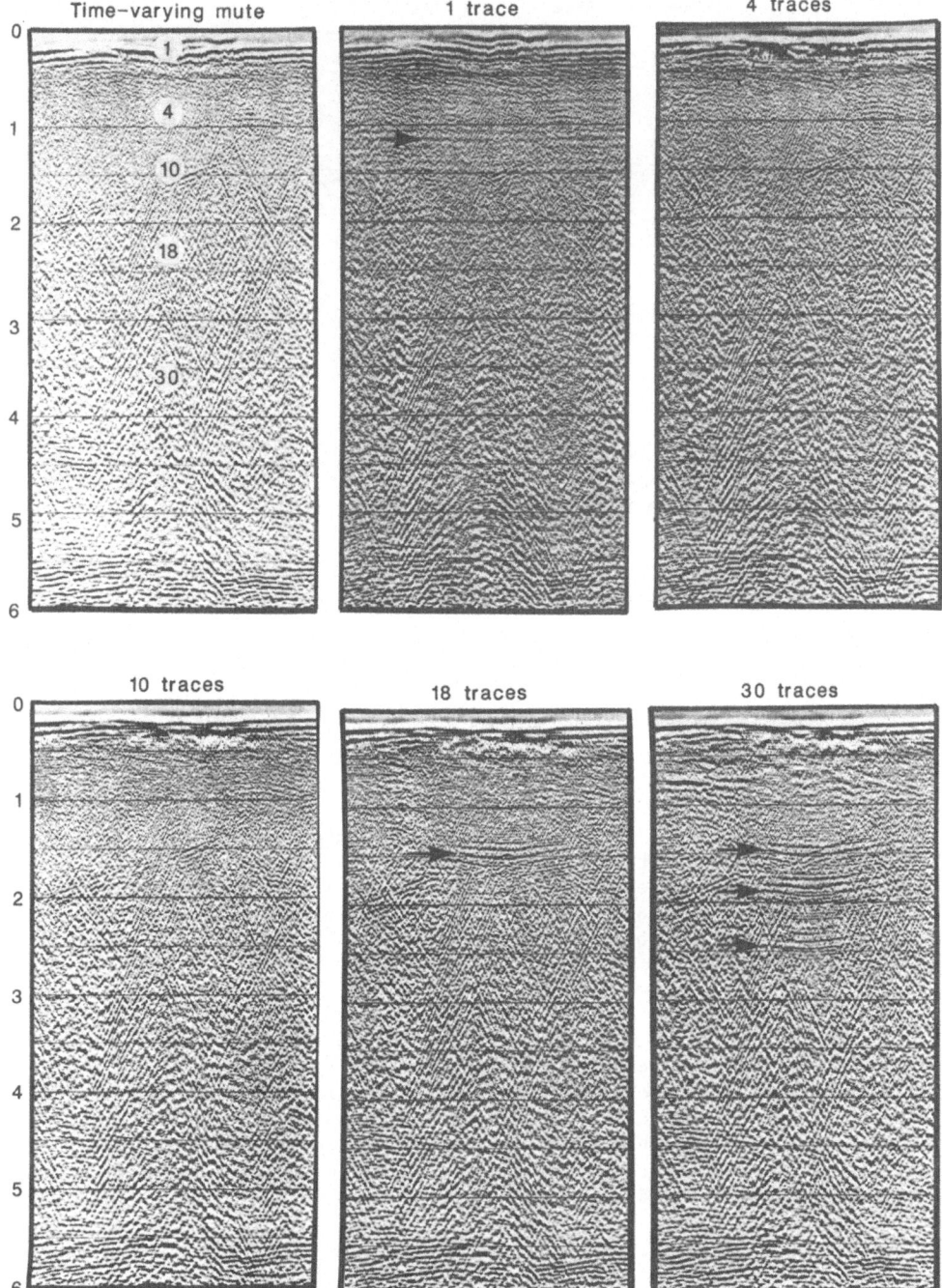

Fig. 9. (a) Muted panel and unmuted, variable-fold test sections.
(b) Unmuted, variable-fold test sections.

ambient noise variations, but such scaling precludes subsequent true-amplitude study. Figure 10, the intermediate stack, shows considerable enhancement over Figure 3, the brute stack, including notably greater coherence of shallow reflections (marked by arrows at 0.75 and 1.25 s). Some obvious problems still remain, such as the steeply dipping coherent noise from 1.5 to 4.5 s (marked by arrows at 3.75 and 5 s). The

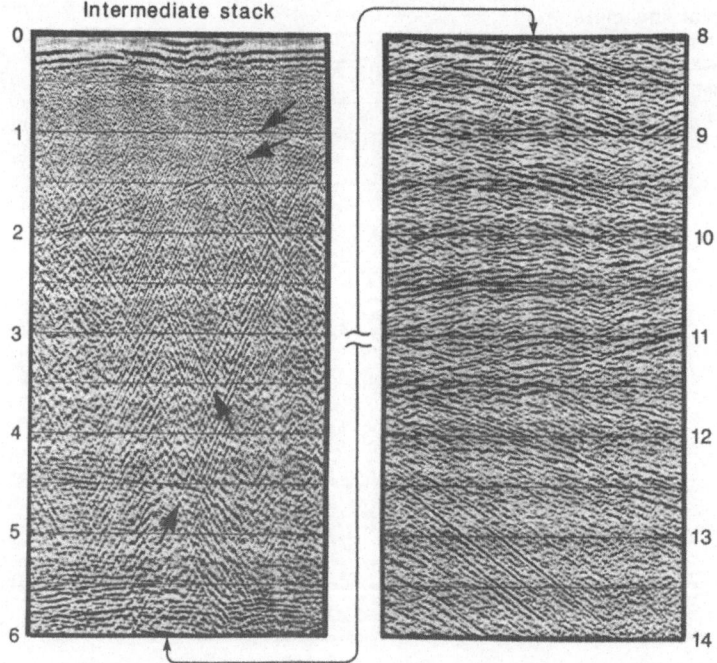

Intermediate stack

Fig. 10.   Intermediate stack.

intermediate stack was displayed for the whole profile, the first such
display since the QC displays, in order to confirm that the pre-stack
processing had achieved the intended results for all the data, not just
for the test sections. After stacking had reduced the volume of data to
be processed, a constant static correction of +16 ms was applied to every
trace to change the experimental datum from the actual gun and receiver
depth to sea-level. Though small, this correction may become important
when tying reflections between different seismic surveys.

POST-STACK PROCESSING

The post-stack processing sequence for NEC consisted of coherent-
noise filtering, deconvolution, frequency filtering and amplitude scaling.
Though not essential, this process ordering is common: deconvolution
modifies the spectrum which can then be re-adjusted by bandpass filtering;
and filtering modifies the trace amplitudes which may then be re-balanced
before the final stack display.

Frequency-wavenumber (f-k) filtering was used in the NEC processing
sequence as a dip-filter after stack to remove spurious events with true
dips greater than 45° (before migration). These events are typically
side-reflections through the water or shallow sedimentary layers, and are
clearly visible dipping both left and right from 0 to 6 s on the
unfiltered section in Figure 11. Test sections were prepared using dip
pass-bands in the ranges ±30, ±24, ±18 and ±12 ms/trace. These filters
all have bell-shaped (cosine-squared) tapers to prevent generation of
residual f-k noise. At ±12 ms/trace all dipping noise has been removed
but steep real reflections (the event dipping left from 0.5 to 1.5 s on
the right of the test panel) and diffractions (from 12 to 14 s) are
becoming noticeably attenuated, and the upper part of the section is
becoming "wormy". With a ±24 ms/trace pass-band most of the steepest

246

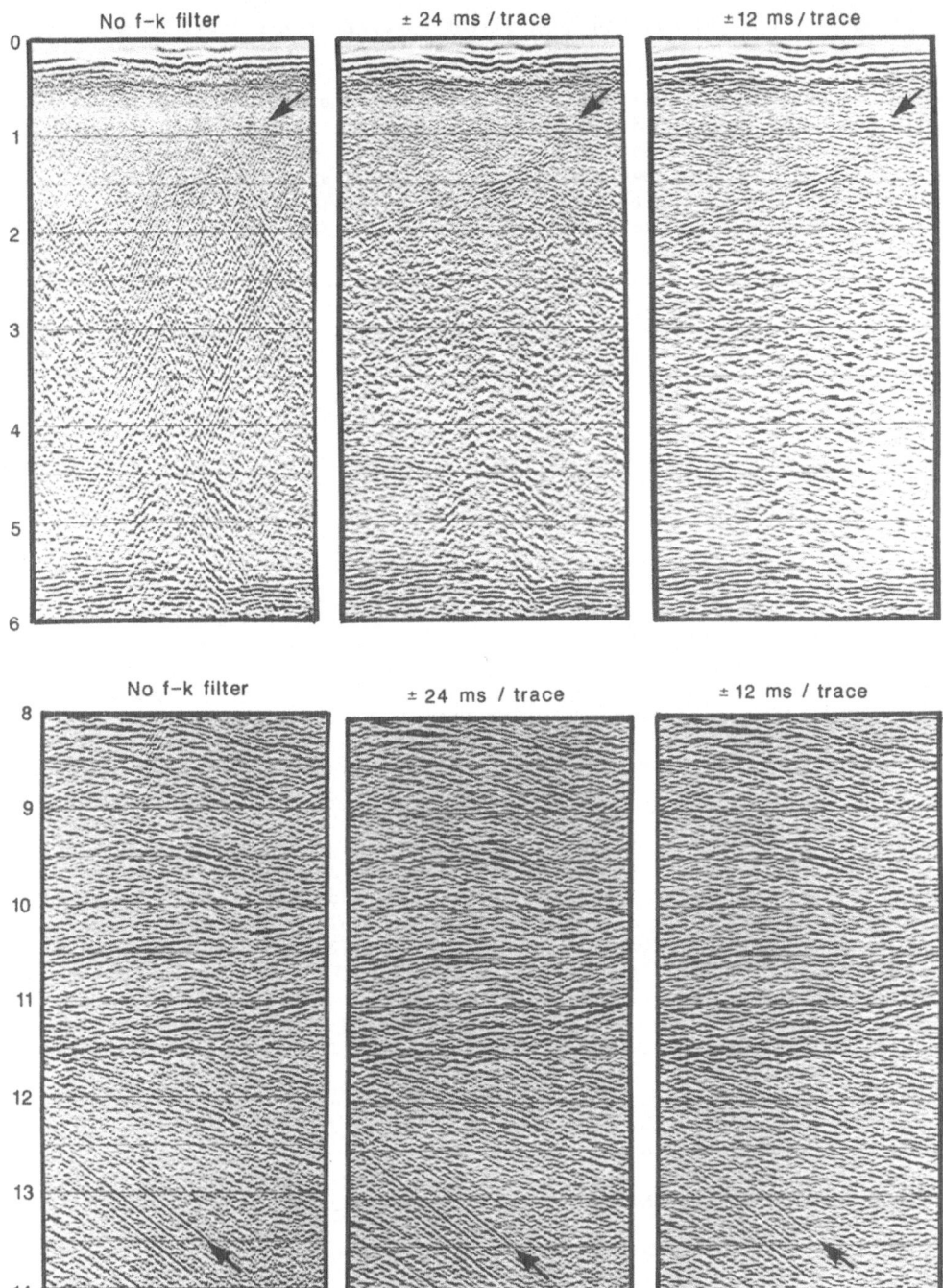

Fig. 11.   Top: test of frequency-wavenumber filtering.
           Bottom: test of frequency-wavenumber filtering.

noise is removed, but real reflections are hardly affected. To assist
interpretation of the final section, it is important to try to preserve
the "character" of the seismic data. In this instance it was judged
appropriate to allow some obvious noise to remain in the section (which
should be recognizable to the interpreter) rather than to over-filter the
data and give it an artificial appearance (which is less easily discounted

during interpretation), and so a slightly weaker f-k filter, ±27 ms/trace, was specified.

Deconvolution is typically applied both before stack (DBS; Table 1, ix) and after stack (DAS; Table 1, xv). Because 30-fold stacking improves S/N by a factor of $5\frac{1}{2}$ for random noise (the expected improvement is the square root of the number of traces summed) and also substantially attenuates multiples, predictive deconvolution gives substantial improvement when applied a second time after stack. The DAS test (Figure 12) used identical parameters to the DBS test (Figure 5 and discussion in text) with the exception of course of range-dependent design and application windows which are not applicable after stack. Comparison of the auto-correlations at the bottom of Figures 5 and 12 shows the extent to which multiple periodicity has been removed from the section by deconvolution. Though there is not much to choose between the effect of 48 ms and 32 ms gap length DAS, the middle section of Figure 12 (bottom) (48 ms gap) shows slightly greater continuity of reflections at 11.5 s and of diffractions at 12.5 to 13 s than does the right-hand panel. The DAS parameters thus chosen from this test, 32 ms and 48 ms gaps for the upper and lower operators respectively, were the same as those selected for DBS.

Time-variant frequency filtering is essential because earth absorption acts to reduce the high-frequency component of the signal with increasing depth. Therefore there will be a signal at higher frequencies in the shallow section than in the deep section, and a time-variant filter chosen to take advantage of this will have a high-frequency bandpass for the upper part of the stack and progressively lower-frequency pass-bands for later arriving signals. The unfiltered upper section in Figure 13 (top) shows low-frequency (< 15 Hz) noise, for example that dipping left to right from 2.5 to 6 s, which may be filtered out of the upper section, but which lies in the bandwidth of highest S/N for the lower section (Figure 13, bottom). There is no significant reflected energy at less than 14 Hz at 0 to 4 s but there is lots of signal at less than 7 Hz at 10 to 12 s. The test used consisted of successive one-octave bandpass filters, from which the processor selects the appropriate, or highest S/N, bandwidth at each travel time. Panels were displayed lowpass to 7 Hz, 5 to 10 Hz, 7 to 14 Hz, 10 to 20 Hz, 15 to 30 Hz, 20 to 40 Hz, 30 to 60 Hz, and 45 to 62.5 Hz, for the whole 15 s data length; only a selection of these are shown in Figure 13. Note that in Figure 13 test panels are shown in the range from 7 to 60 Hz for data from 0 to 6 s, but from 0 to 40 Hz for data from 8 to 14 s. The useful signal bandwidth of any survey will vary depending on recording characteristics (source signature, notches due to the source and receiver ghosts (here 50 Hz from the 15 m streamer depth and 100 Hz from the 7.5 m airgun depth), recording filters (here 5.3 Hz low-cut and 57.5 Hz anti-alias high-cut)), on the noise spectrum, and on the reflection spectrum. Based on the displays shown, the chosen filters for NEC were: pass 15 to 45 Hz (80% amplitude ($\simeq$ -2 dB) and 90% amplitude ($\simeq$ -1 dB) respectively), rolling off to 20% (-14 dB) at 10 Hz and 10% (-20 dB) at 55 Hz) at time zero, 10 to 40 Hz (20% at 7 Hz and 10% at 48 Hz) from 1 to 4 s, 5 to 40 Hz (20% at 3 Hz and 10% at 48 Hz) from 8 to 11 s, and 5 to 30 Hz (20% at 3 Hz and 10% at 36 Hz) at 15 s, with continuous variation in filter parameters from 0 to 1 s, 4 to 8 s, and 11 to 15 s.

Amplitude equalization (Table 1, xvii) is important to generate a stack section on which all the reflections are easily visible; but nonetheless it is desirable to try to retain some sense of relative reflection strength in order to assist interpretation. Automatic gain control with windows of 0.25, 0.5, 1, 2, 3, and 5 s was applied to the test section (Figure 14). Clearly, short equalization windows homogenize

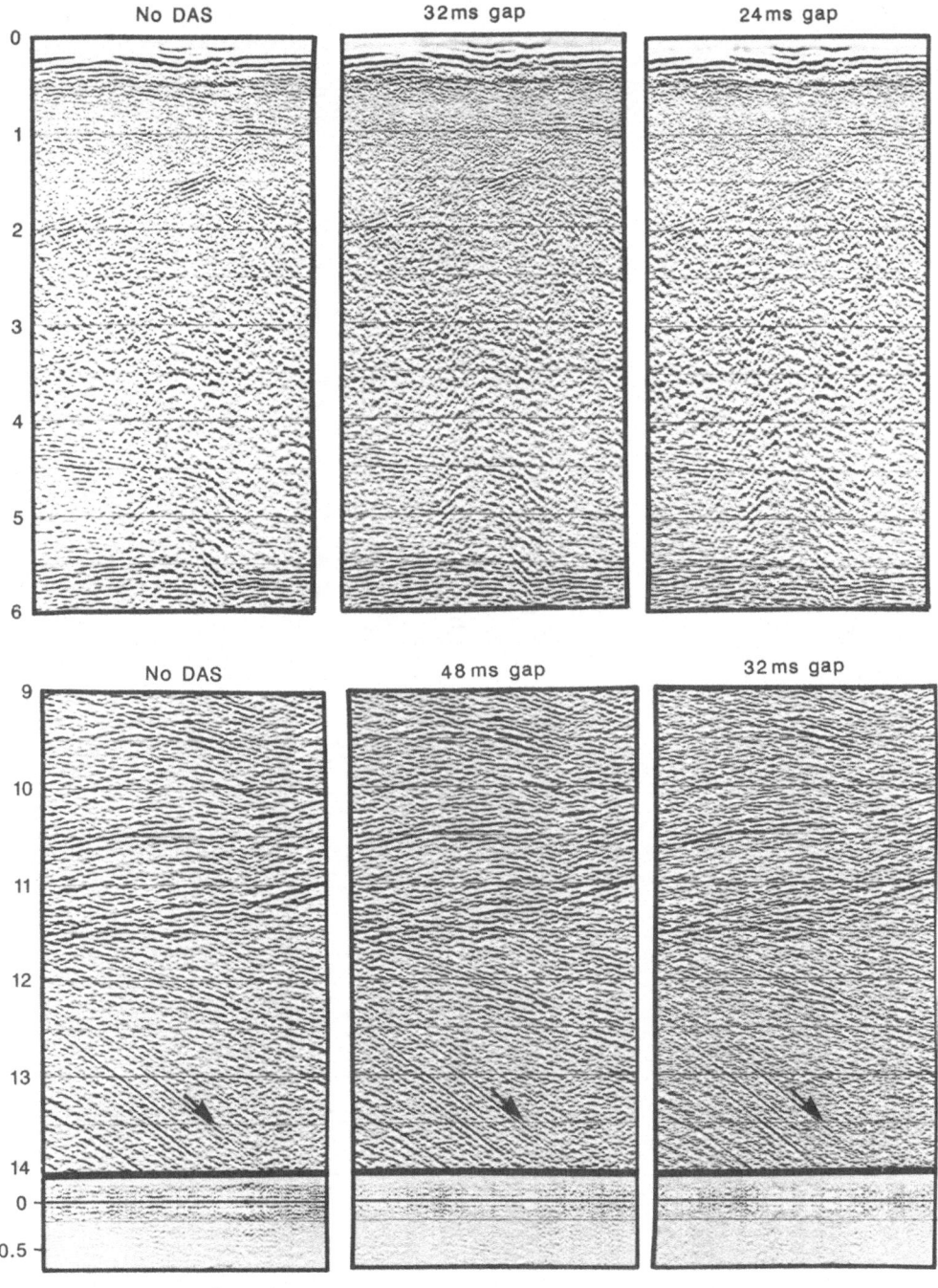

Autocorrelation

Fig. 12. Top: deconvolution-after-stack tests.
Bottom: deconvolution-after-stack tests and trace auto-
correlations.

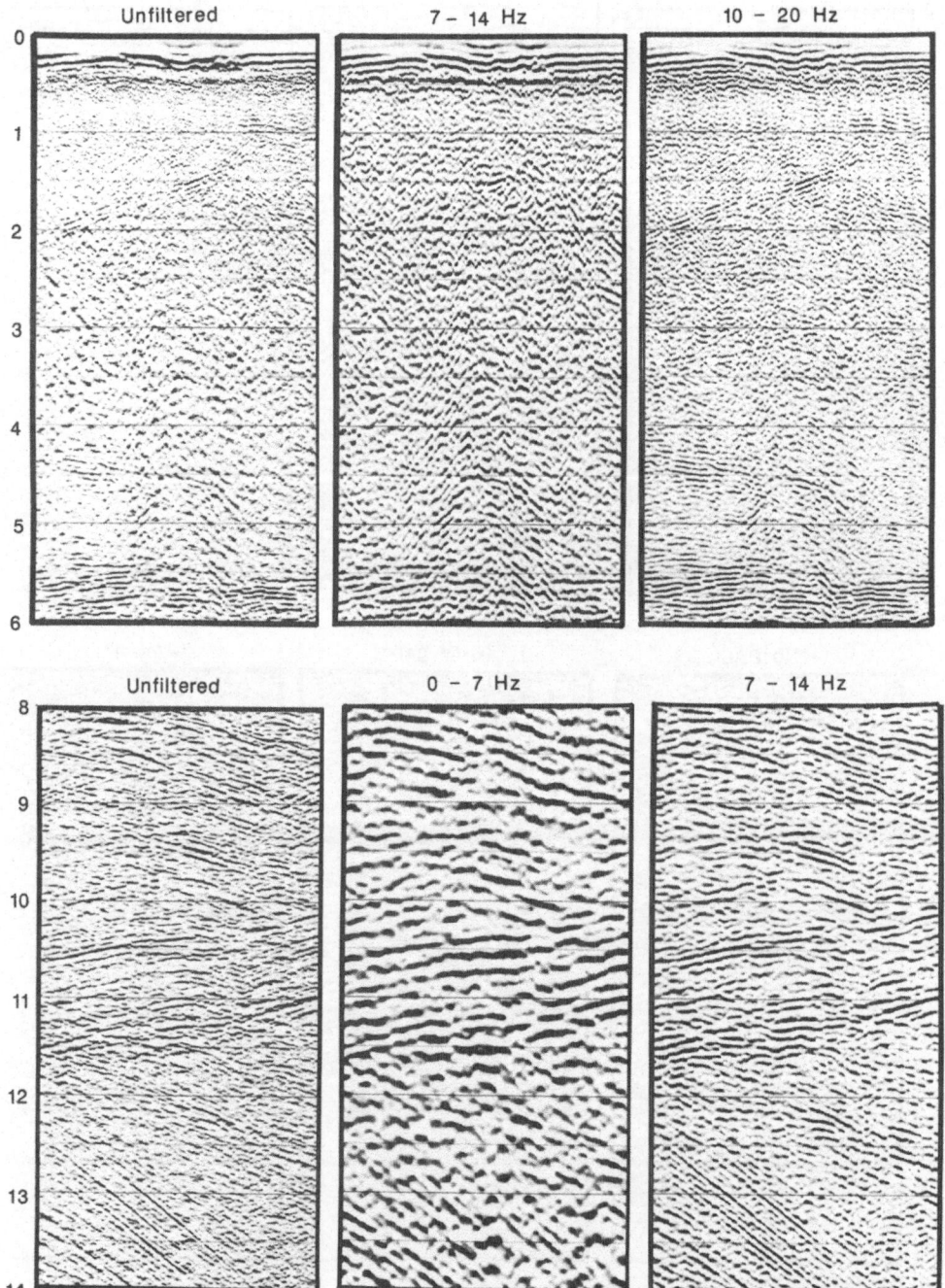

Fig. 13a. Top: bandpass filter test - unfiltered and low frequencies.
Bottom: bandpass filter test - unfiltered and low frequencies.

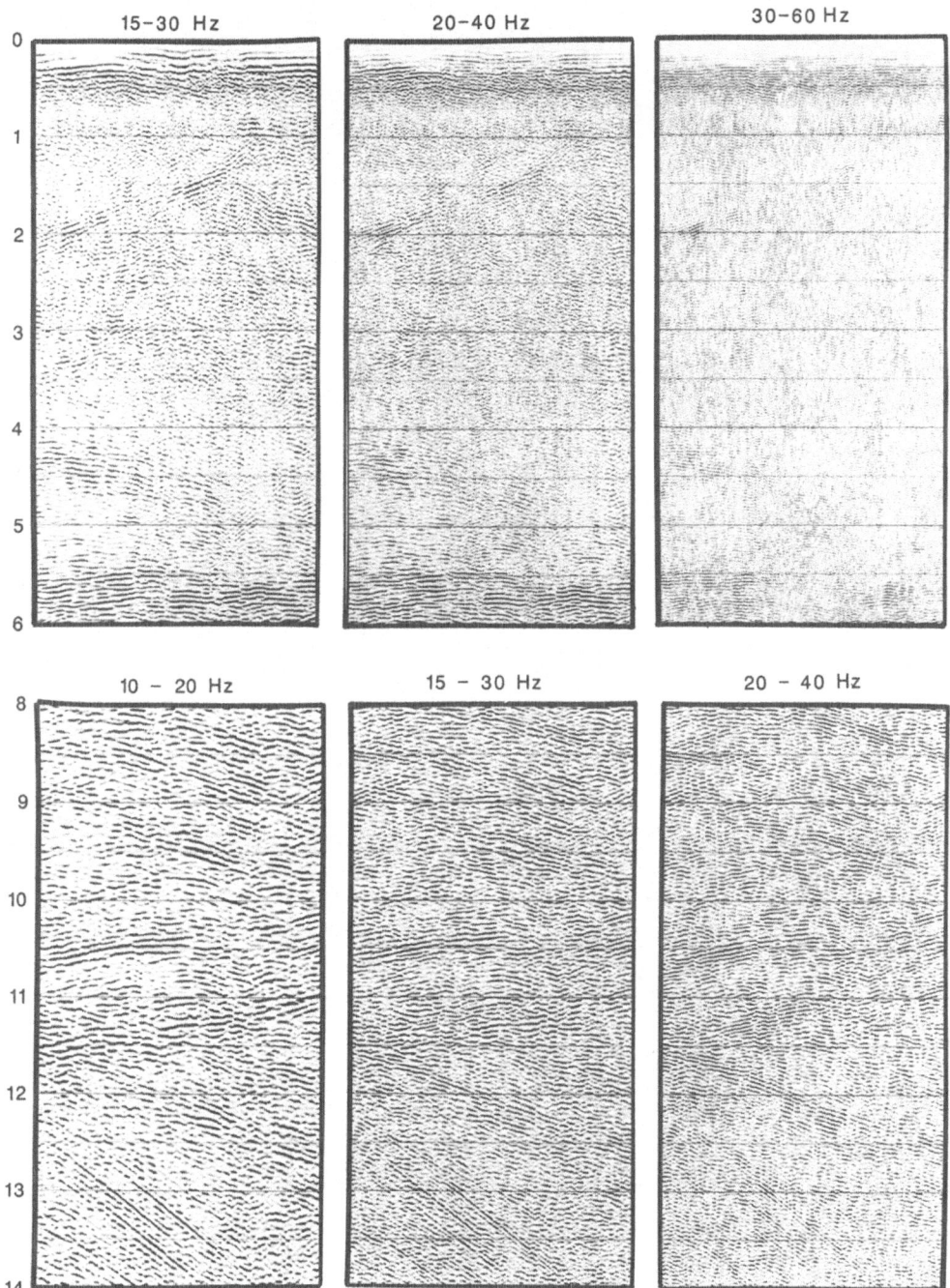

Fig. 13b. Top: bandpass filter test - high frequencies.
           Bottom: bandpass filter test - high frequencies.

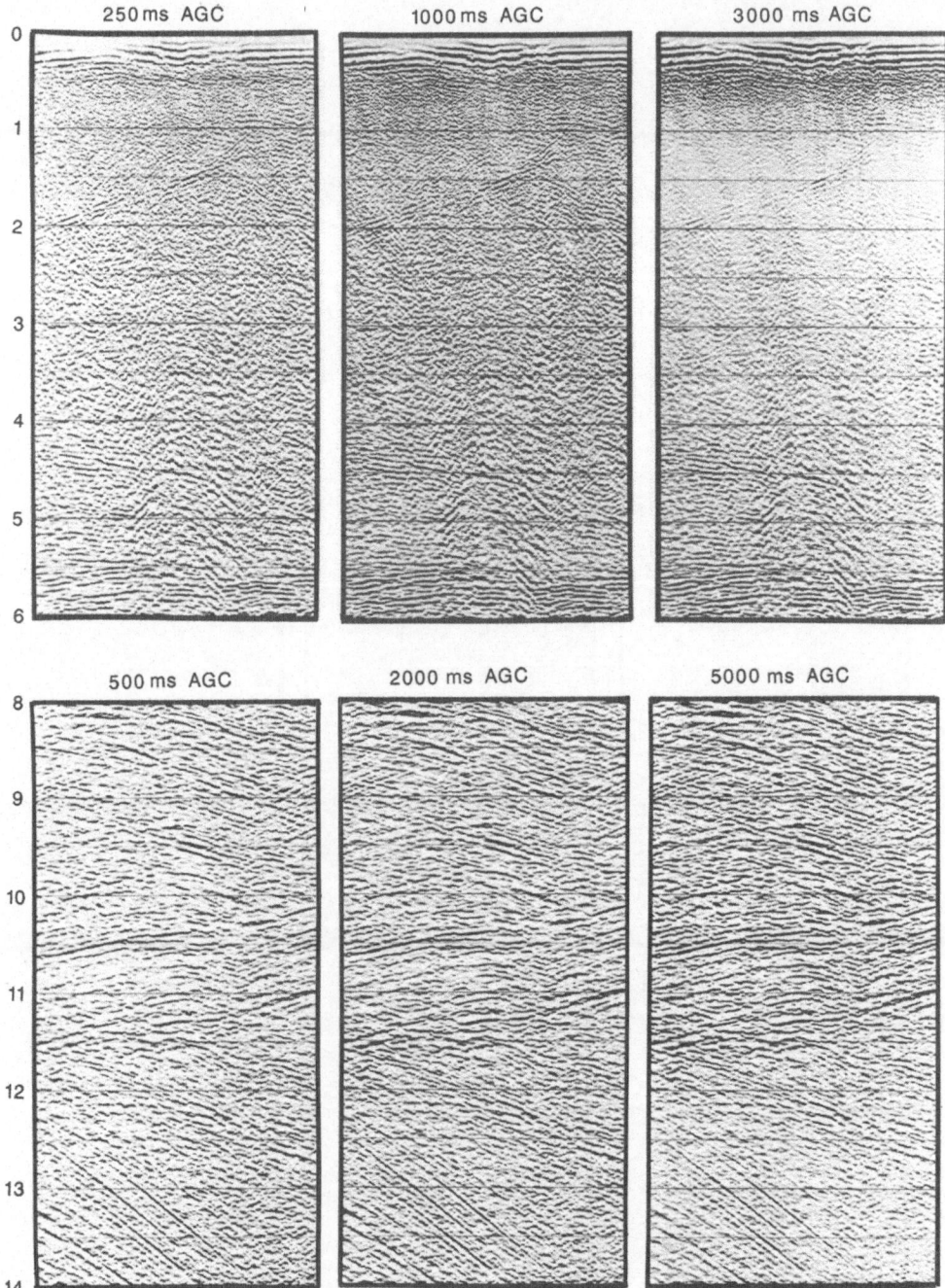

Fig. 14. Top: time-variant equalization test.
Bottom: time-variant equalization test.

the data while long windows can produce blank zones above and below strong reflections. Shorter time windows are generally more appropriate when signal levels are more variable, i.e., at shallow depths, than when signal amplitudes are generally more uniform as is typical at longer travel times. From these test panels, data equalization windows were selected with end times 0.25, 0.5, 1.0, 1.5, 2.5, 4, 6, 9, 12, and 15 s. The gain

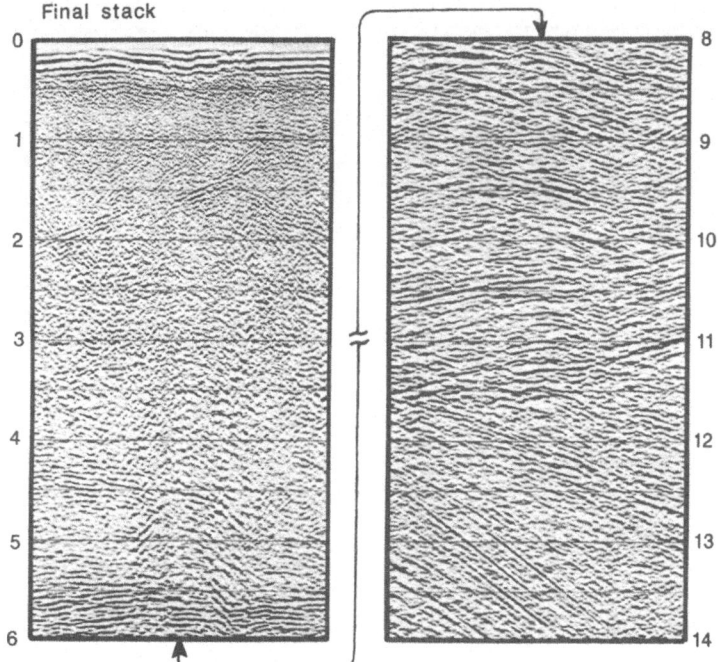

Final stack

Fig. 15. Final stack section.

factor at the center of each window of each trace is the reciprocal of the
mean amplitude within the window, while gain factors at intermediate times
are calculated by linear interpolation. The gain curves thus derived have
equalization windows that increase from 0.25 s at the top to 3 s at the
bottom of the section. When these gain curves had been applied to the
profile, the final stack, Figure 15, was complete.

The final appearance of the stack is controlled by the plotting
parameters, including gain level (determining the signal level at which
trace overlap occurs), bias (a constant amplitude added to each trace to
change the proportion of each wavelet that is shaded), and clip level (the
maximum amplitude plotted). A judicious choice of parameters will ensure
that background noise does not generally overlap from trace to trace, thus
avoiding a false coherency. Plot parameters will vary somewhat with size
of reproduction. All the figures in this paper are reductions of sections
originally displayed at 1:50,000 horizontally, 5 cm/s vertically (true
scale where interval velocity is 5 km.s^{-1}), and in consequence contain
almost too much information on too fine a scale to be resolved. Very
small scale sections, as for publication in journals, should in general be
low-pass filtered and if necessary have adjacent traces summed, so that
the spatial frequencies present in the final plot do not exceed the
resolution limit of the printing medium and can be easily resolved by the
reader.

SUMMARY

Figure 16 compares the three stacks prepared in this processing
sequence: brute, intermediate and final. The benefits of a well-tested
processing route are shown by the increasing geologic information
available, from the quality-control brute stack display prepared without
testing, to the intermediate stack formed after tested pre-stack signal

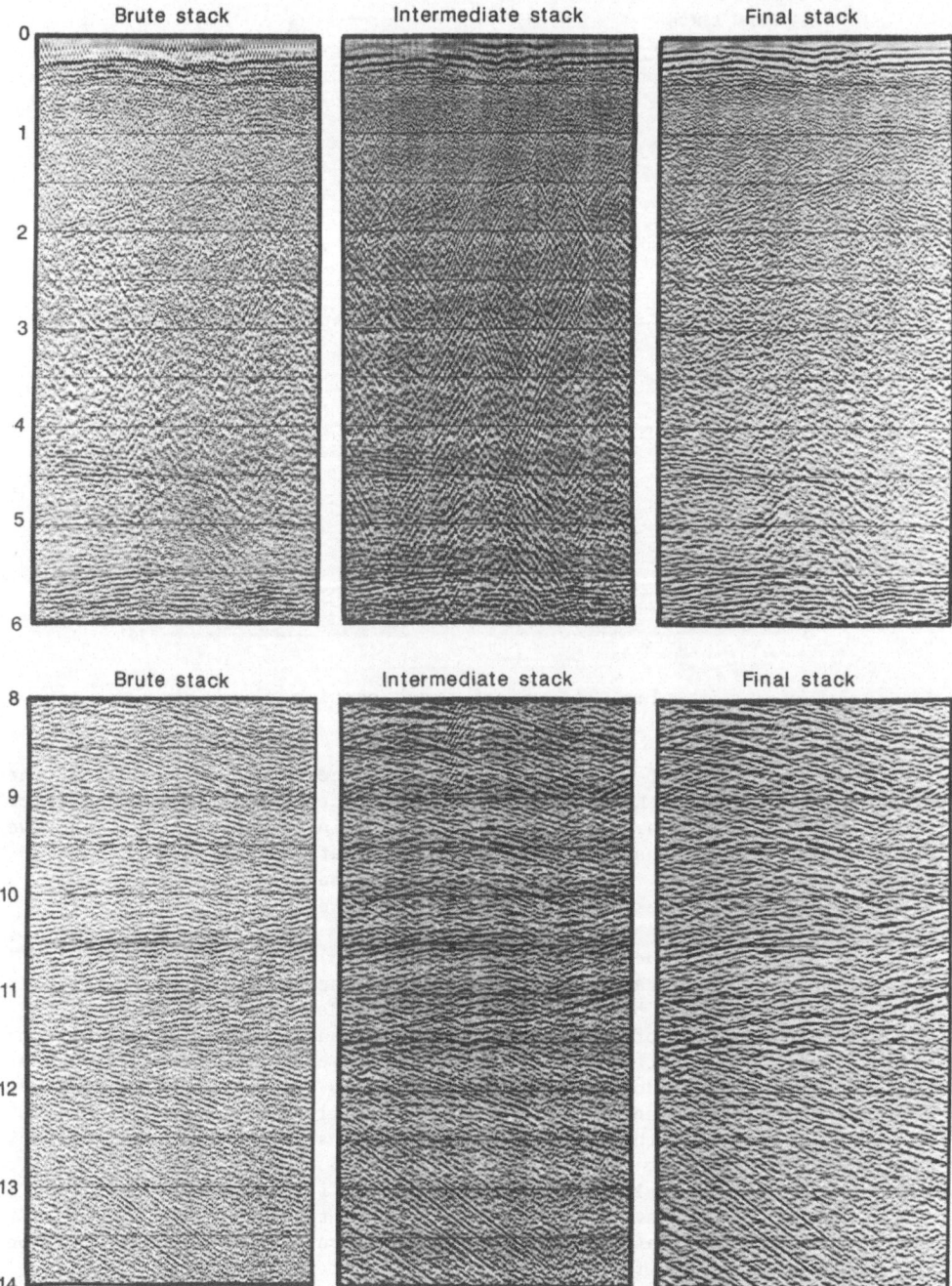

Fig. 16.   Top: brute stack, intermediate stack and final stack compared.
Bottom: brute stack, intermediate stack and final stack
compared.

enhancement but with untested post-stack parameters, to the final stack
incorporating tested post-stack filters. It is important that the testing
be done step-by-step for two reasons. First, if more than one processing
parameter is varied between two test panels it is often difficult to tell
which parameter change caused what effect. Second, if all five array

254

simulations were tested in combination with all four DBS trials, and these in all possible combinations with the five f-k filter trials and the four DAS tests, an overwhelming four hundred display panels would result (or 2400 display panels if process ordering is also varied). In contrast, fewer than twenty test panels were required for these tests since each parameter in turn was selected before testing the next parameter in sequence. Although the testing shown here suffices for a complete processing sequence many additions are possible, and perhaps necessary, for other data sets. In particular, the data discussed here cross an area with rather uniform velocity structure. In many areas with complex and/or laterally variable velocity functions, velocity and mute testing will be more important. Preliminary velocity analysis may be necessary for each test section if the pre-NMO testing is to have validity, and mutes may need testing at many places along the line. Perhaps the most important omission from this paper is the problem of migrating deep seismic data (Warner, 1986, 1987), neglected at least in part because choice amongst the available options - wavefront, f-k, or finite difference migration, pre- or post-stack migration (e.g., Sheriff and Geldart, 1983) - is often made as much on financial grounds as on geophysical desirability.

In processing deep crustal seismic data, the aim is to produce the most interpretable section. This does not always require the maximum possible S/N or highest possible resolution to be attained at every point on the section. It may be more important to preserve the intrinsic character of the recorded data, and to retain high-amplitude but recognizable noise than to risk introducing unrecognizable processing artefacts. The mapping of the deep crust is still in its infancy, and the environment and physical significance of the reflections is still largely unknown. The slow variation of processing parameters in time and space is therefore very necessary if the interpreter is to distinguish real lateral and vertical changes in the deep crust from changes in data-character due only to processing effects.

ACKNOWLEDGEMENTS

This paper describes practices adopted since 1981 by members of the BIRPS group, led by Drum Matthews. Particular thanks are due to Richard Hobbs who carefully reviewed the manuscript, to Mike Warner whose 1986 paper provides the framework for much of the material presented here, and to Jim Downes who was responsible for the processing of NEC at Seismograph Service Ltd. SLK is supported by a Royal Society University Research Fellowship. BIRPS is funded by the Deep Geology Committee of the Natural Environment Research Council. The NEC data were acquired for BIRPS by GECO, processed by Seismograph Service Ltd, and are now available for the cost of reproduction from the Marine Geophysics Programme Manager, British Geological Survey, Murchison House, West Mains Road, Edinburgh. This paper is Cambridge Earth Sciences contribution 1025.

REFERENCES

Barazangi, M., and Brown, L. D., eds., 1986a, "Reflection Seismology: A Global Perspective", American Geophysical Union, Washington, Geophysical Monograph 13.

Barazangi, M., and Brown, L. D., eds., 1986a, "Reflection Seismology: The Continental Crust", American Geophysical Union, Washington, Geophysical Monograph 14.

Brewer, J. A., and Smythe, D. K., 1986, Deep structure of the foreland to the Caledonian orogen, NW Scotland: results of the BIRPS WINCH profile, Tectonics, 5:171.

Brown, L. D., 1986, Aspects of COCORP deep seismic reflection profiling, in: "Reflection Seismology: A Global Perspective", M. Barazangi and L. D. Brown, eds., American Geophysical Union, Washington, Geophysical Monograph 13:209.

Clowes, R. M., and Kanasewich, E. R., 1970, Seismic attenuation and the nature of reflecting horizons within the crust, J. Geophys. Res., 75:6693.

Clowes, R. M., Green, A. G., Yorath, C. J., Kanasewich, E. R., West, G. F., and Garland, G. D., 1984, Lithoprobe - a national program for studying the third dimension of geology, J. Can. Soc. Explor. Geophys., 20:23.

Freeman, B., Klemperer, S. L., and Hobbs, R. W., 1988, The deep structure of northern England and the Iapetus Suture zone from BIRPS deep seismic reflection profiles, J. Geol. Soc. Lond., 145:727.

Fulton, T. K., 1985, Some interesting seismic noise, Geophysics: The Leading Edge of Exploration, 4(9):70.

Hatton, L., Worthington, M. H., and Jakin, J., 1986, "Seismic data processing: theory and practice", Blackwell Scientific Publications, Oxford.

Jannsen, D., Voss, J., and Theilen, F., 1985, Comparison of methods to determine Q in shallow marine sediments from vertical reflection seismograms, Geophys. Prospect, 33:479.

Klemperer, S. L., and Matthews, D. H., 1987, Iapetus suture located beneath the North Sea by BIRPS deep seismic reflection profiling, Geology, 15:195.

Klemperer, S. L., Hauge, T. A., Hauser, E. C., Oliver, J. E., and Potter, C. J., 1986, The Moho in the northern Basin and Range Province, Nevada, along the COCORP 40 N seismic reflection transect, Geol. Soc. Am. Bull., 97:603.

Louie, J. N., and Clayton, R. W., 1987, The nature of deep crustal structures in the Mojave Desert, California, Geophys. J. Roy. Astron. Soc., 89:125.

Matthews, D. H., 1986, Seismic reflections from the lower crust around Britain, in: "The nature of the lower continental crust", J. B. Dawson, D. A. Carswell, J. Hall, and K. H. Wedepohl, eds., Geological Society of London Special Publication 24:11.

Matthews, D. H., and Smith, C. A., eds., 1987, "Deep seismic reflection profiling of the continental lithosphere", Geophys. J. Roy. Astron. Soc., 89.

McGeary, S., and Warner, M. R., 1985, Seismic profiling the continental lithosphere, Nature, 317:795.

Mooney, W. D., and Brocher, T. M., 1987, Coincident seismic reflection/refraction studies of the continental lithosphere: a global review, Rev. Geophys., 25:723.

Newman, P., 1973, Divergence effects in a layered earth, Geophysics, 38:481.

Oliver, J., Dobrin, M., Kaufman, S., Meyer, R., and Phinney, R., 1976, Continuous seismic reflection profiling of the deep basement, Hardeman County, Texas, Geol. Soc. Am. Bull., 87:1537.

Peardon, L., and Cameron, N., 1986, Acoustic and mechanical design considerations for digital streamers, Society of Exploration Geophysicists Expanded Abstracts with Biographies, 56th Annual Meeting 1986:291.

Peddy, C. P., Brown, L. D., and Klemperer, S. L., 1986, Interpreting the deep structure of rifts with synthetic sesimic sections, in: "Reflection Seismology: A Global Perspective", M. Barazangi and L. D. Brown, eds., American Geophysical Union, Washington, Geophysical Monograph 13:301.

SEG, 1980, "Digital Tape Standards", Society of Exploration Geophysicists, Tulsa, USA.

Sheriff, R. E., 1984, "Encyclopedic Dictionary of Exploration Geophysics", 2nd edition, Society of Exploration Geophysicists, Tulsa, USA.

Sheriff, R. E., and Geldart, L. P., 1982, "Exploration Seismology, Volume 1: History, Theory and Data Acquisition", Cambridge University Press, UK.

Sheriff, R. E., and Geldart, L. P., 1983, "Exploration Seismology, Volume 2: Data Processing and Interpretation", Cambridge University Press, UK.

Smithson, S. B., 1986, A physical model of the lower crust from North America based on seismic reflection data, in: "The nature of the lower continental crust", J. B. Dawson, D. A. Carswell, J. Hall and K. H. Wedepohl, eds., Geological Society of London Special Publication 24:23.

Warner, M. R., 1986, Deep seismic reflection profiling the continental crust at sea, in: "Reflection Seismology: A Global Perspective", M. Barazangi and L. D. Brown, eds., American Geophysical Union, Washington, Geophysical Monograph 13:281.

Warner, M. R., 1987, Migration: why doesn't it work for deep continental data?, Geophys. J. Roy. Astron. Soc., 89:21.

Warner, M. R., and McGeary, S., 1987, Seismic reflection coefficients from mantle fault zones, Geophys. J. Roy. Astron. Soc., 89:223.

Zhu, T.-F., and Brown, L. D., 1986, Consortium for Continental Reflection Profiling Michigan Surveys: Reprocessing and results, J. Geophys. Res., 91:11477.

3-D REFLECTION SEISMIC DESIGN, ACQUISITION AND PROCESSING

Luigi Salvador

AGIP SpA, Direzione Generale dei Servizi
dell'Esplorazione, DES P.O. Box 12069
20120 Milano, Italy

INTRODUCTION

In conventional 2-D seismic the energy emitted from a source S and backscattered from a layer R towards the surface where it is picked up by a receiver G is supposed to travel only on a vertical plane passing through S and G (plane of the section).

In fact the energy emitted from S propagates in a spherical manner, in such a way that at time T the receiver detects the energy scattered both from reflectors that lie on the section plane and from reflectors outside it that belong to the ellipsoid for which the following relationship is valid:

$$T1 + T2 = T$$

where T1 is the descending time from source down to the reflector and T2 the ascending time back from the reflector to the receiver. Collected energy coming from points outside the section plane (lateral event) can seriously damage the 2-D seismic picture, and it is very difficult to cancel them in conventional processing. The 3-D technique has been designed to take advantage of this fact, in the sense that after a 3-D data processing a reconstruction of the three-dimensional shape of the layers can be carried out, thus getting vertical seismic sections free from spurious elements.

In a 3-D survey we collect the data so densely that all the subsurface to be investigated is covered, rather than some strips only as in 2-D. The subsurface is divided into equal elementary areas, usually called bins, in which the recorded seismic traces are collected. Gathering the bins into ordered sequences, for example parallel to the orthogonal axes of the survey, we can obtain a large number of crossing profiles. In practice a 3-D survey is a 2-D grid where the distance between profiles equals the bin size. Moreover, it is possible to cut horizontally (at constant time) the 3-D volume. Through the use of such horizontal slices the interpretation becomes easier, faster and more reliable than the one based upon vertical sections only.

Clearly only complex problems justify the big effort required to carry on a 3-D survey. In fact, the problems that geophysical research will meet in the future will become even more severe, and so we can expect a wider application of this method in the exploration stage. More recently, to reduce the cost, a new approach to the 3-D design was introduced.

Since the percentage of the total cost of a 3-D survey to ascribe to the acquisition ranges from 70 to 90%, significative saving can be achieved by being able to take advantage, through the use of more sophisticated processing techniques, of less information physically collected in the field. This means less spatial and temporal resolution, but sometimes, like in the early exploration stage of a complex area, it is enough to be able to run a 3-D migration interesting the lower spatial frequencies (structural solution) only. The quality of the final pictures will be reliable enough to prevent misplacements of the wells even though not the best possible.

The key processing allowing this new way of surveying is the interpolation of the collected profiles to produce a less coarse grid so that the 3-D migration can be run far away from the aliasing limit. The reduction of the spatial sampling in the cross-line direction to one third allows to save about 50% of the total cost.

3-D ACQUISITION TECHNIQUES

Basically, there are two methodologies for data acquisition:

- parallel profiling,
- cross-array profiling.

All other techniques are directly derived by applying, in various forms, such basic ideas.

1.1 Parallel Profiling

In the parallel profiling technique the survey is carried out simply by recording a series of conventional 2-D profiles, very close together, until all the area to be surveyed is covered.

This was the first method applied, since it represents nothing more than a conventional survey recorded by using a less coarse line interval. It is still used in marine environment, where the rigid acquisition geometry imposed by the streamers does not allow more interesting solutions.

During the last years a new generation of research vessels, characterized by larger dimensions, more powerful propellers and wider sterns, was launched. Such vessels offer the capability to tow more than one streamer and more than one source array. Consequently, more than one subsurface profile can be recorded per boat course, thus highly decreasing the costs.

1.2 Cross-Array Profiling

While in parallel profiling the shot points lie on the same trajectory followed by the receiver ensemble (the streamer in the marine case), in the cross-array technique the shots are aligned along a direction perpendicular to that of the receiver ensemble. Figure 1 shows some examples of implementation.

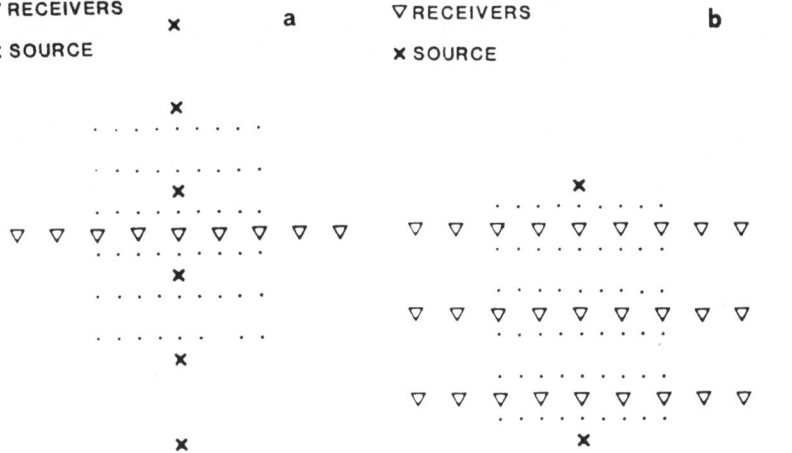

Cross-array technique, example (a) Cross-array technique, example (b)

Fig. 1. Examples of Cross-Array Layout. The same coverage can be obtained through different field layouts.

It is extremely flexible, and for this reason is used in land surveys. Looking at Figure 1, it is easy to recognize how the coverage of the same subsurface can be achieved by shooting many shot-points while recording through a few receivers or vice-versa, by means of many shots.

The direction of alignment of the receivers is usually called the "in-line" direction, while "cross-line" is used for the direction along which the shot-points are aligned.

The ensemble of the receivers belonging to one or more receiver-lines, and of the shot-points belonging to one or more shot-lines that insists on such receiver-lines is called "block". The block represents the modulus of the survey: by appropriately moving it along both the in-line and the cross-line direction it is possible to cover the area to be surveyed with a predetermined folding level.

The subsurface strip covered by moving the block only along the in-line direction is called "swath". A survey recorded by using the cross-array technique is very often referred to as "swath survey".

Swath surveying provides a high level of production and operative flexibility. It can be carried out both in land and shallow water surveys, provided that the streamers or the bay-cables can be fixed to the sea bed.

It is important to note that this technique allows the recording of the energy pertaining to subsurface points lying outside the vertical of both the receiver-lines and the shot-lines. This means the possibility to cover areas where the access are problematic.

1.3 Reconnaissance Techniques

In the reflection 2-D environment the basic element is the point in the subsurface lying midway between the source and the receiver (mid-point, or Common Mid Point if multiple folding is used).

Each CMP is associated to a seismic trace, and the ensemble of the CMPs is the seismic profile. CMP is a bi-dimensional concept, being the seismic profile associated to a vertical plane.

In 3-D environment, where we are dealing with a volume of data rather than with a single plane, the basic element is the elementary subsurface area lying again midway between the source and the receiver. Such area is usually called "bin".

By collecting the bins along one direction, we will produce a conventional profile, despite the fact the couples source-receiver which mid point lie on the considered bins are dispersed in positions apart from the track of the profile in the surface.

The cost of a 3-D survey is relatively high, and in fact it was first applied by oil companies to enhance the knowledge of the subsurface setting where hydrocarbon was always found.

Where the energy propagation conditions allows it, and mainly for exploratory purposes, a cheaper technique can be applied.

In fact, to be able to process in a three-dimensional fashion the data (in particular to migrate and display in a 3-D form), we very often do not need to record in the field all the bins used to process the data.

In the following sections all the computational aspects will be deeply investigated, as well as the design and checking procedures providing enough reliability to the final results. At this stage we can consider the economic advantage resulting from the use of a wider bin size during the field operations.

By halving only the cross-line dimension of the bin we reduce by 50% the cost of the survey. Since the acquisition cost represents 85% of the total cost of the project in marine environment and 70-75% in land, we can save from 40 to 45%.

By halving both the dimension of the bin we can save from 50 to 60%. There are different ways to reduce the number of recorded bins minimizing the loss of information.

One method, often used in marine operations, consists of the collection of couples of subsurface lines separated by a gap multiple of the distance between the paired lines. If there are two streamers or two sources, the operation is straightforward, and results in a substantial reduction of the sail-line number.

The availability of couples of close lines strongly helps the interpolation programs run to fill in the empty bins during the processing stage.

Another method, applied generally in land operations where problems of shortage of field material (receivers, cables or available channels) exists, considers an extension of the twin-streamer technique.

Since the recording is generally made through swath surveying the bin decimation is achieved by properly separating the shots along the shot-lines for the cross-line direction, and, again, by properly decimating the receivers along the receiver lines.

If the bin decimation has been correctly planned taking into account the geophysical characteristics of the area, there are minor differences between a "full" 3-D and a "reconnaissance" 3-D result.

3-D SURVEY DESIGN

The design procedure can be divided in four steps:

a) recognition of the geophysical problem to be solved in order to obtain the best result at target level;
b) determination of the acquisition parameters that satisfy at best the constraints;
c) survey layout modification to fit the environmental constraints;
d) overall check through simulation (shooting on paper).

If the final check is satisfactory the design stage is completed and the field operations can be started, otherwise it is necessary to iterate through step b) to d) until all the constraints are satisfied.

2.1 Recognition of the Geophysical Problem

When we are going to design our survey, we must remember that the acquisition problem to be solved in order to get the best image of the target can be different from the best acquisition of the target itself.

To clarify the sentence, let us consider the following problem. The target is a simple structure buried at 3 sec (twt), fairly continuous and gently dipping, lying below a thick, inhomogeneous and noisy level. The propagation velocity is relatively low up to the target, where it rapidly increases.

The diffraction hyperbolae generated in the low velocity noisy layer exhibit long and steeply dipping tails that mask the reflections coming from the target. If we collect at the best such hyperbolae we will be able to remove them completely by the 3-D seismic migration at the processing stage, evidencing at the end the target. On the contrary, if we concentrate on the target, neglecting the diffractions in which we are not interested, we will record spatially aliased data. The migration process will work very badly, producing a final section contaminated by strong and uneliminable noise.

Once the problem has been recognized, it can be translated in terms of couples dip-frequency (function of the reflection time) which we like to preserve on the final result.

In other words, the result of the first design step is the determination of the characteristics of the seismic signals appearing on the profiles that should be preserved in order to solve the geophysical problem that we are dealing with.

By imposing, at a known time, the values of dip and frequency we can compute, once the typical velocity function is available, the sampling intervals in time and in space, as well as the surface dimensions of the survey and of the block.

2.2 Acquisition Parameters

The elements we need to determine are the following:

- energy source and temporal sampling,
- in-line spatial sampling,
- cross-line spatial sampling,
- orientation of the survey,
- recording time,
- survey dimensions,
- in-line coverage,
- cross-line coverage,
- move-up.

Moreover, the survey design is affected by special requirements caused by the application of particular processing sequences. Typical are the problems related to the trace spacing for seismic migration and for interpolation through Dip Move Out programs.

2.2.1 <u>Energy source and temporal sampling</u>. The choice of the sample interval is linked to the following factors:

- the characteristics of the source,
- the typical seismic response of the area,
- the characteristics of the target.

The three factors are linked together, in the sense that once the target is defined, the choice of the source is generally directly related to the acoustic characteristics of the area to be investigated.

Sometimes many different energy sources can provide the wanted bandwidth of the signals, and in this case the choice of the temporal sampling is made simply by taking into account the maximum expected temporal frequency. More complicated is the choice of the most appropriate source when the acoustic response of the area does not fit with the requirements in terms of vertical resolution. As a general rule, in this case it is better to work at 2 msec of sampling interval, since some corrective actions, like de-phasing and theoretical minimum phase conversion, applied in the early stage of data processing produce improved results if run at lower sampling rate.

2.2.2 <u>Spatial sampling</u>. The choice of the spatial sampling is often the key for a successful survey, but also the most critical parameter both for the economic and technical implications.

The choice is made after considering the following factors:

- the Fresnel zone extension,
- the aliasing problem during acquisition,
- the aliasing problem during processing.

2.2.2.1 <u>Fresnel radius</u>. The concept of Fresnel zone is well-known in optic. In the following we will briefly examine the problem of the interference related to the problem of the spatial resolution in seismic.

Following the Huygens principle, the propagation of the energy is explained through the fact that each subsurface point becomes energy emitter once interested by the energy propagating from the surface. The energy emitted from these "secondary" sources reaches other points, thus creating new secondary sources and so on.

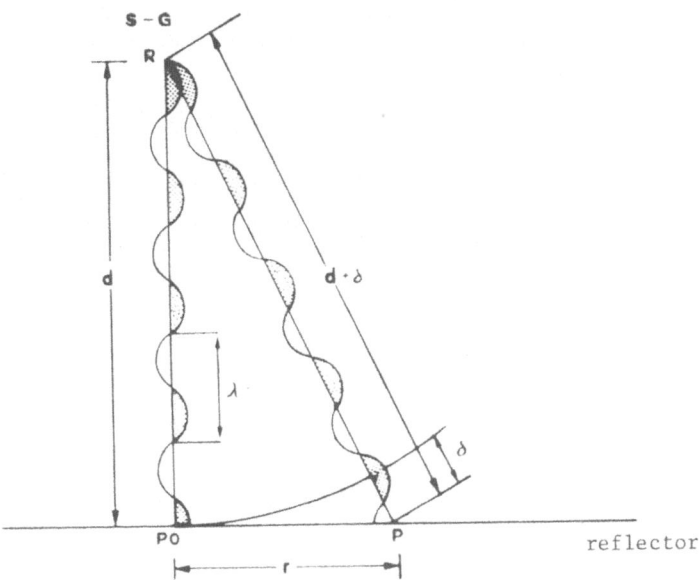

Fig. 2. Phase Difference due to Different Ray Paths. The phase of the
wave depends on the incidence angle.

Let us consider a simple model with only one reflecting surface
between two different homogeneous layers. The velocity is considered
constant inside each layer. Moreover the energy is monochromatic, that
means that the frequency is constant as well as the wavelength λ:

$$\lambda = \frac{V}{f} \tag{1}$$

With reference to Figure 2, we see that the phase of the wave emitted
from the source S when it reaches the reflecting surface changes as the
ray path followed changes.

In fact, the surfaces where the phase is constant are the spheres
with centre in S.

In general the energy emitted from the secondary sources activated
along the reflecting interface will present a phase different from that of
the primary source.

Let us consider what happens at a receiver location when it
intercepts the energy emitted from two different secondary sources
belonging to the reflecting interface. If the difference between the
length of the ray paths is an odd multiple of $\lambda/2$ the phase of the two
corresponding waves differs by odd multiples of π, that means totally
destructive interference.

Figure 3 shows how the relative amplitude of the signal recorded by
the geophone changes as function of the difference in ray path lengths
expressed in terms of wavelength λ.

It is clear that for differences less than $\lambda/2$ there is coherency in
the phases at the receiver location, or conditions of constructive
interference.

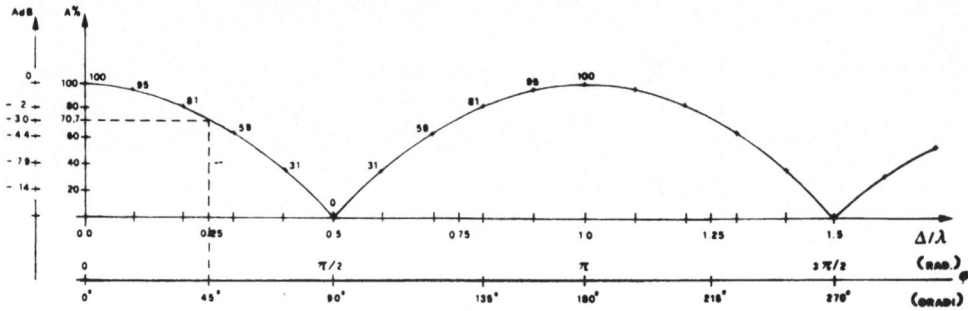

Fig. 3. Effects of Interference. The amplitude changes with the phase
displacement. The Fresnel radius corresponds to the distance of
the first zero crossing.

The horizontal distance corresponding to half the wavelength is
called Fresnel Radius.

In order to get the Fresnel radius in seismic environment we should
bear in mind that the energy captured by the receiver covers twice the
distance (down and up) between the source and the reflecting interface.
This means that the distance corresponding to the first destructive
interference is associated with a phase difference of $\lambda/4$.

If d denotes the vertical distance between the emitter and the
reflecting interface we see that the Fresnel radius F is:

$$F^2 = (d + \lambda/4)^2 - d^2$$

or

$$F = \frac{V}{2}\sqrt{\frac{t}{f}} \tag{2}$$

This solution is valid for constant velocity media only.

In fact, the propagation takes place generally in variable velocity
media. A more realistic approximation of the Fresnel radius, valid for
depth p many times higher than the wavelength $\lambda = V/f$, is the following
(Thatam et al., 1981):

$$F = \sqrt{\frac{VaVzt}{4f}} \tag{3}$$

where Va denotes the average velocity from the surface down to the
reflector and Vz the formational velocity just below the reflecting
surface.

Figure 4 shows a set of curves produced by applying (3) to a real
example for different values of the frequency f.

In optic the diffraction fringes appear as a series of circles
alternately bright and dark. The bright zones are associated with
constructive interference, whilst the dark are associated with
destructive interference. With white light (multi-component) only the
first zone (inside the Fresnel radius) is clearly recognizable (and
white), while outside it there is a superposition of colored fringes,
being the interference frequency dependent on λ.

In seismic the phenomenon is exactly the same: there is a central
zone of good hearing, since a constructive interference of all the

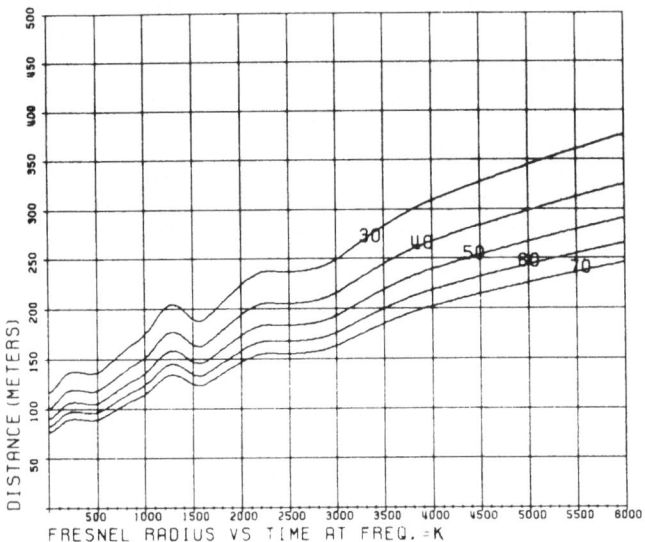

Fig., 4. Fresnel Radius. Examples of distributions of the Fresnel radius
with reflecting time when velocity changes with depth.

frequencies takes place, followed by circular zones of very low and
disturbed response.

It is well-known that the reconstruction of a signal from its samples
is possible if there are almost 2 samples available per period. This
condition is at the limit of aliasing, thus in seismic it is preferable to
operate with twice this precision.

If we accept for F the definition (3), implicitly we accept a phase
dispersion between -90 and +90 degrees for the various frequency
components of our band-limited acoustic signal. Since a phase change is
associated with an amplitude change, it is better to limit the phase
dispersion between -45 and +45 degrees only. The corresponding
variation in amplitude is limited to 3 dB. This condition allows for a
difference in the ray path of $\lambda/8$, and is equivalent to 30% reduction of
the Fresnel radius.

Concluding this discussion, we can say that the spatial sampling,
that is the distance between adjacent seismic traces in subsurface, should
satisfy:

$$\text{spatial sampling} < a\sqrt{\frac{VaVzt}{4f}} \tag{4}$$

where the corrective factor a is between 0.5 and 0.7, in order to avoid
destructive interference.

2.2.2.2 <u>Aliasing during acquisition</u>. The spatial aliasing problem
occurs whenever the acquisition system is unable to properly sample the
seismic events, and is generally associated to dipping or diffracted
events.

Let us introduce the concept of "apparent velocity". Along a seismic
profile two receivers are X meters apart. They record a signal coming
from the same buried interface with a difference in time (usually called
step-out) t. Regardless of the true propagation velocity, the apparent
velocity of the signal is:

$$Vapp = X/t.$$

The wave number associated to the propagation along the profile in the surface is:

$$K = f/Vapp$$

where f represents the temporal frequency of the signal.

The apparent velocity is the link between the temporal frequency and the spatial wave number of a seismic event.

In the monochromatic case the spatial aliasing occurs any time the step-out t is greater than half the period of the monochromatic temporal wave. In other words, any time the phase shift recognizable between two adjacent traces exceeds 180 degrees, it is impossible to reconstruct the seismic event starting from the recorded samples.

In Figure 5 we can see the traces of two wavefronts traveling upwards and emerging to the surface with the angle θ. The two wavefronts are shown when they cross the receivers. The distance between two adjacent receivers is X. If V is the propagation velocity and t the step-out (two way time), we see that:

$$\frac{V}{2} t = X \sin\theta \tag{6}$$

or:

$$t = \frac{2X \sin\theta}{V} . \tag{7}$$

The maximum step-out tmax allowed before aliasing is related to the maximum temporal frequency fmax. Introducing them we get:

$$tmax = \frac{1}{2fmax} = \frac{2X \sin\theta}{V} \tag{8}$$

or:

$$fmax = \frac{V}{4X \sin\theta} . \tag{9}$$

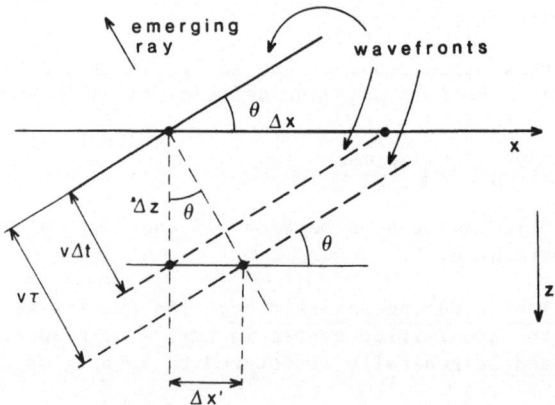

Fig. 5. Wavefront Propagation. Energy moves towards the surface. R is the distance between the receivers, V the propagation velocity and θ the angle of emergence.

268

Equation (9) gives the maximum unaliased frequency recorded by a field system having a trace distance of X m in presence of dip θ.

2.2.2.3 <u>Aliasing during migration</u>. In the previous section all the considerations refer to the acquisition, or to the early stage of the processing (before stack). Now we can repeat the considerations, but concentrating on what happens after stack, and particularly during the seismic migration process.

Seismic migration is a process that moves the seismic events from the apparent position where they were recorded to the true position where they lie.

During this process the apparent recorded dips are changed into true dips. The data input to migration are defined in the bi-dimensional space (x,t) (or, in 3-D environment, in the three-dimensional space (x,y,t)), while the output is defined in the space $(x,depth)$ (or, in 3-D, $(x,y,depth)$).

In fact, only in the last years, since the supercomputer appeared on the market, the outputs started to be of this kind. Formerly the output space was the theoretical $(x,migrated\ time)$ one, obtained from $(x,depth)$ dividing it by the velocity field used during migration. Anyway, the following considerations are correct, since the computation space was and is $(x,depth)$.

The migration process transforms the input dips, defined in (x,t) and so limited to a maximum of 45 degrees, into the output true dips defined in $(x,depth)$ space.

During this process it may happen that some dipping events, correctly recorded in the field and correctly processed up to stack level, fall into aliasing because of such dip change.

Figures 6 to 11 show what happens to some dipping layers when they are migrated with different trace spacing.

The relationships between the maximum temporal frequency that can be saved through the migration as a function of the trace spacing and of the true dip of the seismic horizons is easy to obtain. Referring to Figure 12, we denote again with X the subsurface trace spacing input to the migration, with V the interval velocity across the seismic horizon and with θ its true dip, measured starting parallel to the horizontal axis.

The maximum distance between two peaks lying on adjacent traces, in depth and at the limit of aliasing, is:

$$D = \frac{V}{2} * \frac{T}{2} . \tag{10}$$

The distance between the traces is X. The corresponding dip, that is the critical dip, is:

$$\theta = arc\ tan\left(\frac{VT}{4X} \right). \tag{11}$$

T is the period of the monochromatic wave we are considering. Thus, the corresponding frequency is $f = 1/T$ and (11) becomes:

$$f = \frac{V}{4X\ tan\theta} . \tag{12}$$

INPUT STACK MODEL – S/N RATIO 1:1

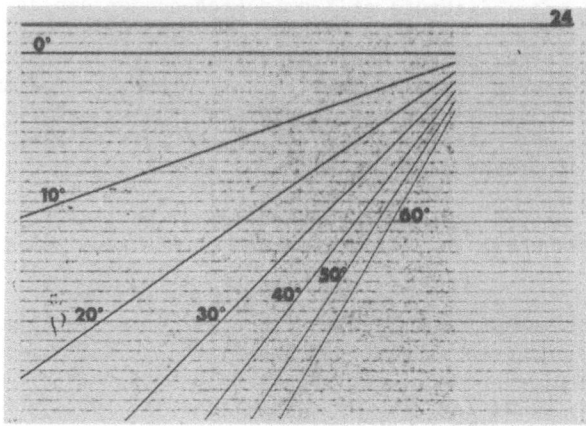

Fig. 6

FK MIGRATION TEST – GROUP INTERVAL 100 METERS

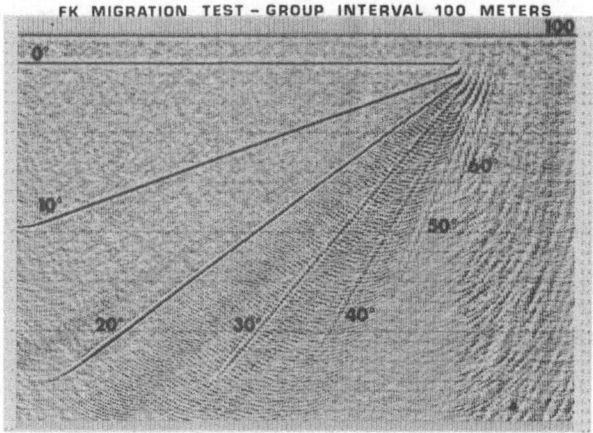

Fig. 7

FK MIGRATION TEST – GROUP INTERVAL 72 METERS

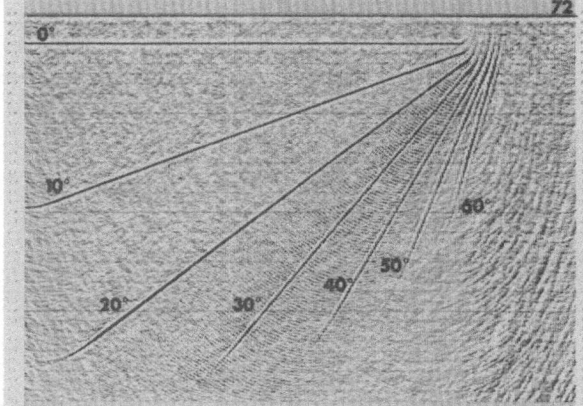

Fig. 8

Figs. 6-11. Examples of Migration. The synthetic stack is migrated using
            different trace intervals. The number on the upper right
            corner is the equivalent receiver distance in surface.

FK MIGRATION TEST – GROUP INTERVAL 48 METERS

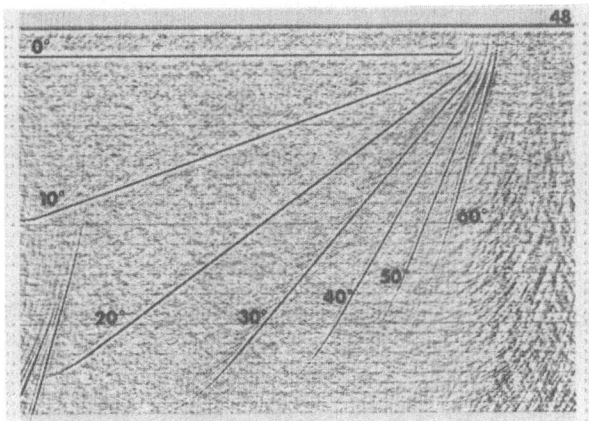

Fig. 9

FK MIGRATION TEST – GROUP INTERVAL 24 METERS

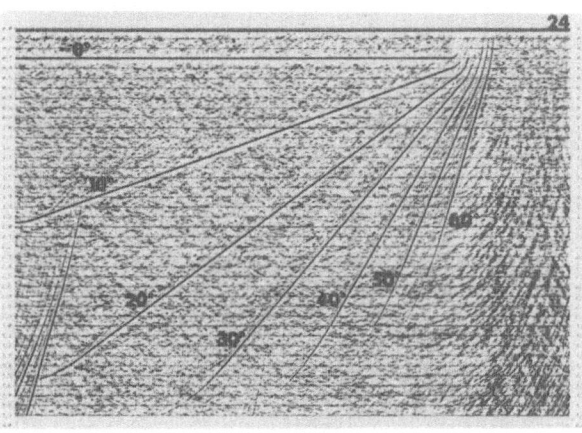

Fig. 10

FK MIGRATION TEST – GROUP INTERVAL 12 METERS

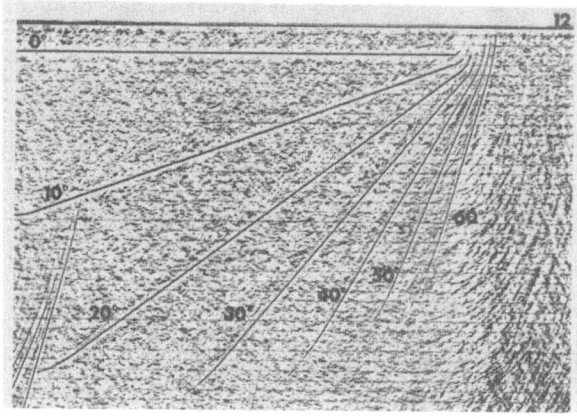

Fig. 11

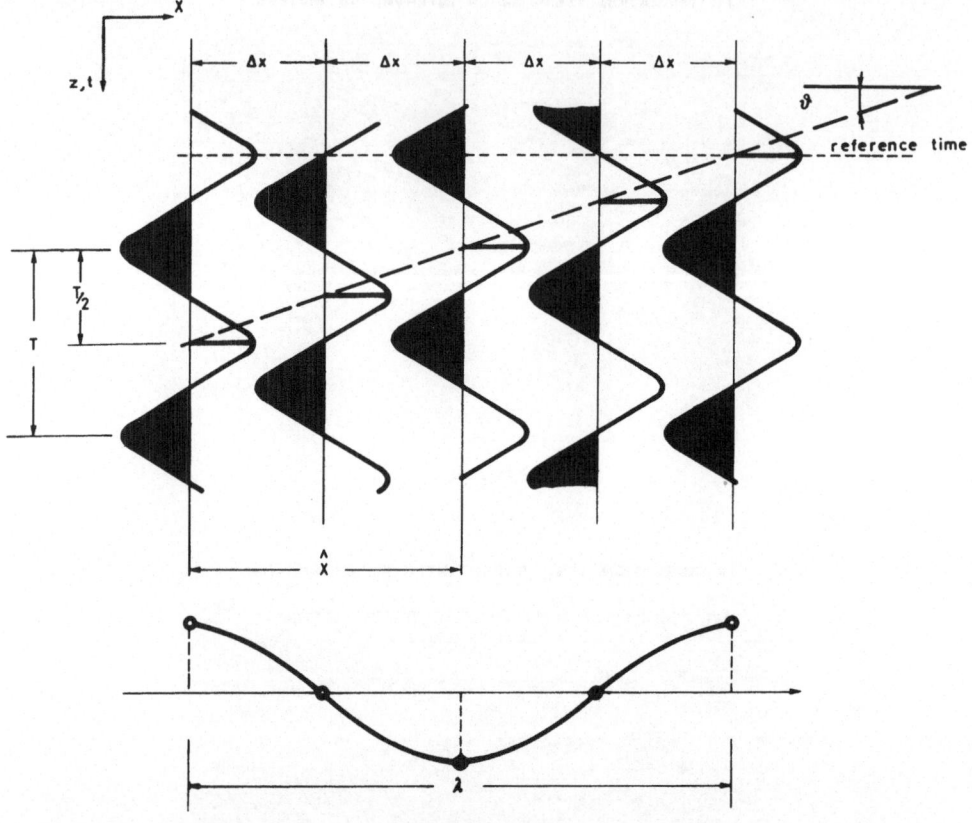

Fig. 12.   Spatial Aliasing during Migration.   The model shows how the
computations are carried out.

Apart from the fact that the sin changes into tan, Equation (12) is
equivalent to (9).   The tangent (before migration) is equivalenced to sin
(after migration) by migration process.

The relationship (11) is valid for a monochromatic wave propagating
through a constant velocity medium.

In order to utilize (12) for the estimation of the spatial sampling
during the design stage it is better to analyze what happens to the three
parameters involved (temporal frequency f, distance between receivers X
and dip of the target events) as a function of the recording time.

During the recognition of the geophysical problem the couples
significative dips-interesting frequencies as function of the two way
recording time were identified.

Introducing them in (12) and rearranging we obtain:

$$f(t) = \frac{V(t)}{4X \tan\theta(t)} \tag{13}$$

$$X(t) = \frac{V(t)}{4f(t) \tan\theta(t)} \tag{14}$$

$$\theta(t) = \text{arc tan}\left(\frac{V(t)}{4Xf(t)} \right). \tag{15}$$

The functions (13) and (14) are portrayed in Figures 13 and 14.

272

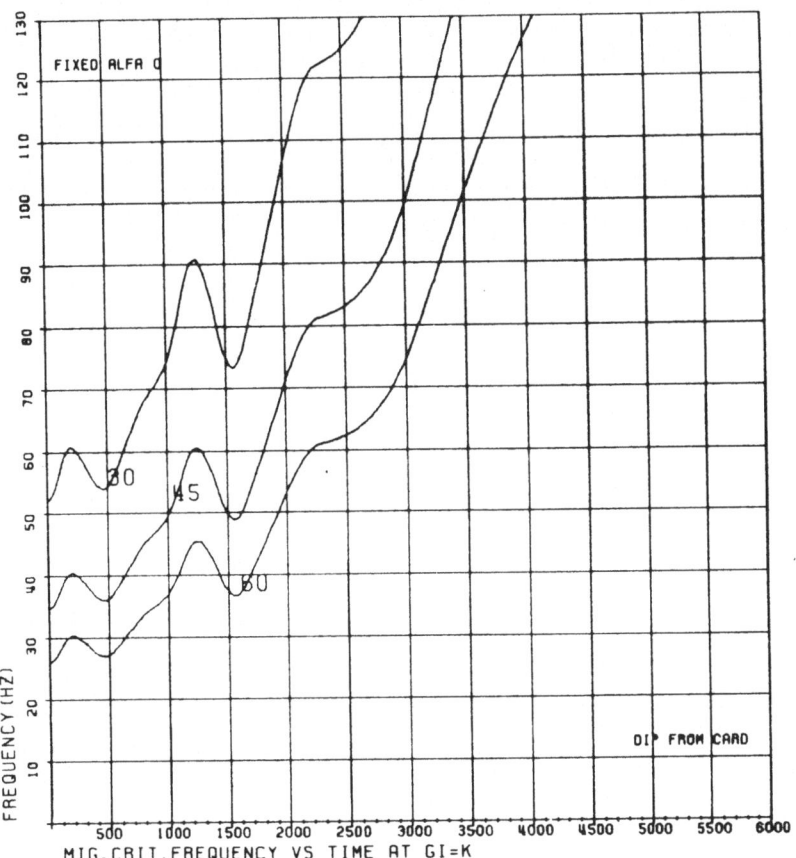

Fig. 13. Aliasing Control in Migration. The maximum unaliased frequency versus recording time is shown. Dips are externally defined.

By analyzing what happens as the recorded time changes we can evaluate the effects of a parameter choice on the capability of the recording system. In particular (14) allows the determination of the proper distance between the traces, while (13) and (15) provide the means to study the impact on the maximum frequency and on the maximum recordable dip respectively, of the choice of a particular value of X.

Concluding this section we can say that the choice of the sampling interval should be consistent with:

$$X \leqslant \frac{V}{4F \tan\theta} \tag{16}$$

where F is the upper bandwidth limit we need at target level, θ is the maximum dip of the target horizon and V the best estimate of the formational velocity across the target.

The three relationships (13), (14) and (15) provide the elements for the selection of the most appropriate subsurface spatial sampling. In 3-D environment very often the computations should be done twice, since the structural shape of the target horizons along two orthogonal directions is quite different.

273

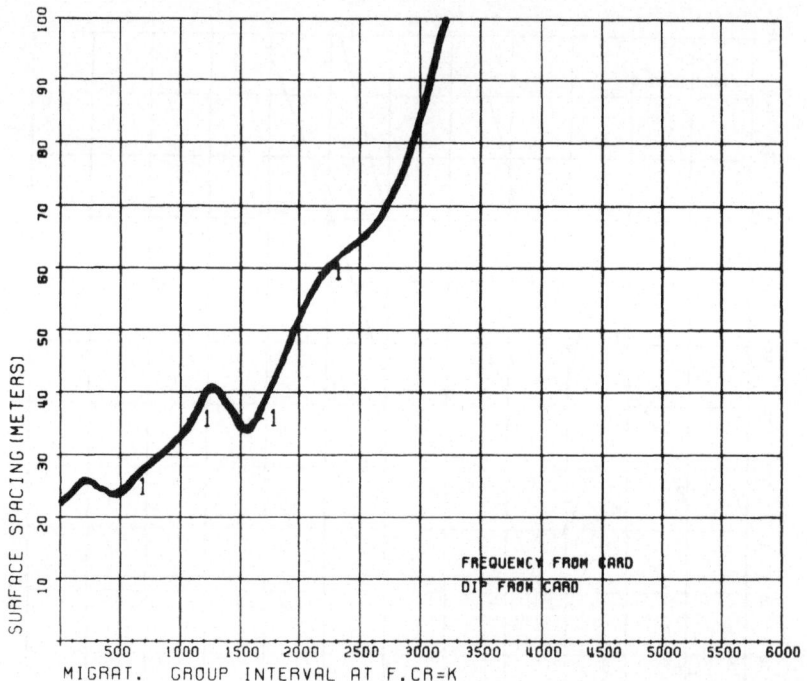

Fig. 14. Aliasing Control in Migration. The maximum trace distance
allowing the handling of the frequency f(t) in presence of the
dips O(t) is plotted as function of the recording time t.

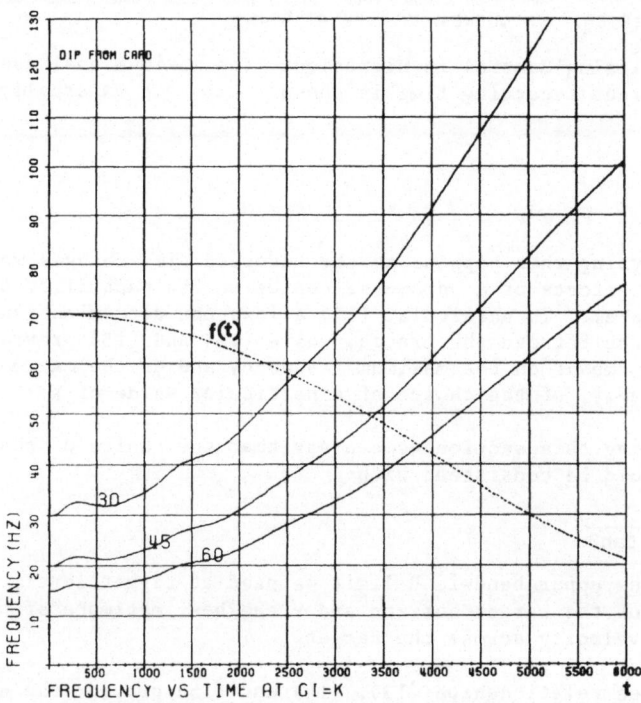

Fig. 15. Aliasing Control in Acquisition. The maximum unaliased
frequency versus recording time is shown. Dips are externally
defined.

274

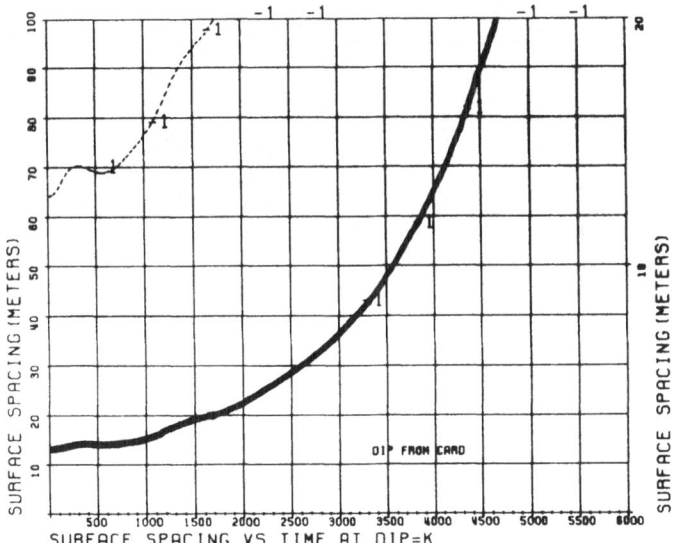

Fig. 16. Aliasing Control in Acquisition. The maximum trace distance
allowing the handling of the frequency f(t) in presence of the
dips θ(t) is plotted as function of the recording time t.

For comparison, Figures 15 and 16 show the aliasing curves computed
in the same conditions of Figures 13 and 14, but by applying the
relationship (9).

The spatial aliasing control is a fundamental task in 3-D processing.
If we get some aliasing, we record at the wave number Kr the energy that
was actually in the signal at the wave number K, where Kr = K - Ks and Ks
is the Nyquist wave number. Suppose the signal is sinusoidal with time
and with pulsation w = 2πf. Since the angle θ of emergence of the plane
wave is connected to the wave number, if we equivocate the wave number we
will incorrectly interpret as θr the emergence angle

$$\sin\theta = \frac{KV}{2w} \; ; \; \sin\theta r = \frac{Ks - K}{2w} V.$$

In conclusion, we will migrate in the wrong place the aliased plane
wavefront: in particular θr = 0 when Ks = K and w = Ks V/2.

We get to the same conclusion if we analyze a single reflecting
horizon, slanting with given slope, and we improperly sample the wavefield
with an insufficient distance between neighboring couples of source and
receiver.

If the sampling frequency was sufficiently high, the semi-circles,
outcome of the migration of the single spikes that compose the section,
would interfere destructively everywhere but on their envelope, that
coincides with the correctly migrated sections (Figure 17).

If, on the other hand, the spatial sampling frequency is too low,
this interference will not take place. We see that for instance, at some
frequencies, the aliased angle is zero, and thus part of the reflections
will remain at their places (Figures 6-11). Figure 18 shows a single
horizon and two migrations, the first at a proper spatial sampling, the
other at coarser trace spacing. Because of aliasing, a part of the energy

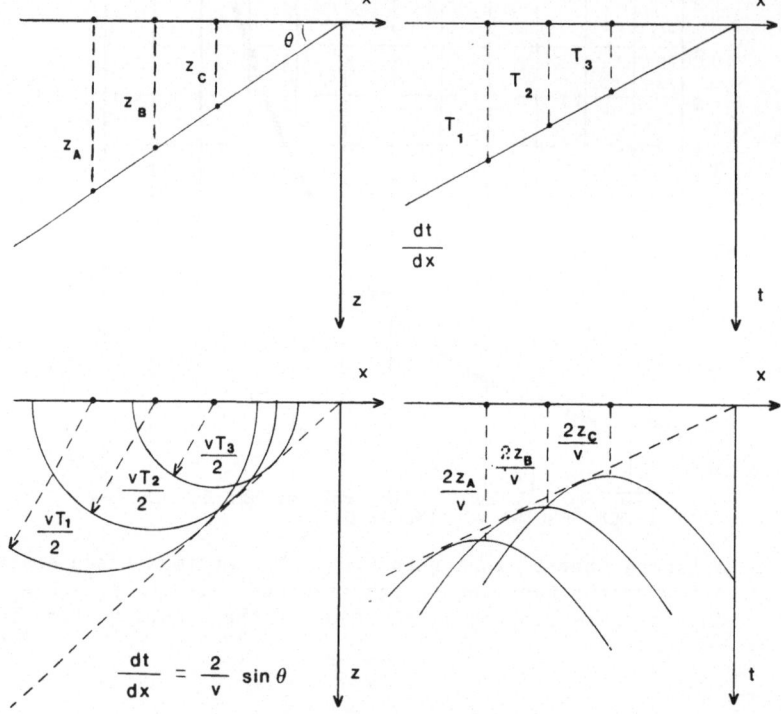

Fig. 17. Seismic Migration.

is scattered along circles, centered across the unmigrated horizon position. The sketch on the left side explains how and where the energy is scattered.

In 3-D the situation becomes worse, since such a result will be migrated again in the orthogonal direction, thus originating semisphere of noise strongly interfering with the signals.

2.2.3 <u>Dip limitation due to the size of the survey.</u> There are dip limitation effects due to the recording time and the size of the survey.

In order to evaluate such effects we can observe what happens when the stacked data are migrated. With reference to Figure 19, we can observe that the point S, belonging to a dipping horizon, appears on the vertical of point M at a distance z:

$$z = V \ t/2.$$

Due to the migration process, the dipping horizon will move up-dip, and the point S, moving along the circle centered in M with radius z, will move to the true position P. The vertical true depth (VP) is:

$$z' = V \ tz/2.$$

From Figure 19 we see that:

$$\theta = arc \ cos(tz/t).$$

The maximum recordable dip follows from the substitution of the generic time t with the recording time Tr:

276

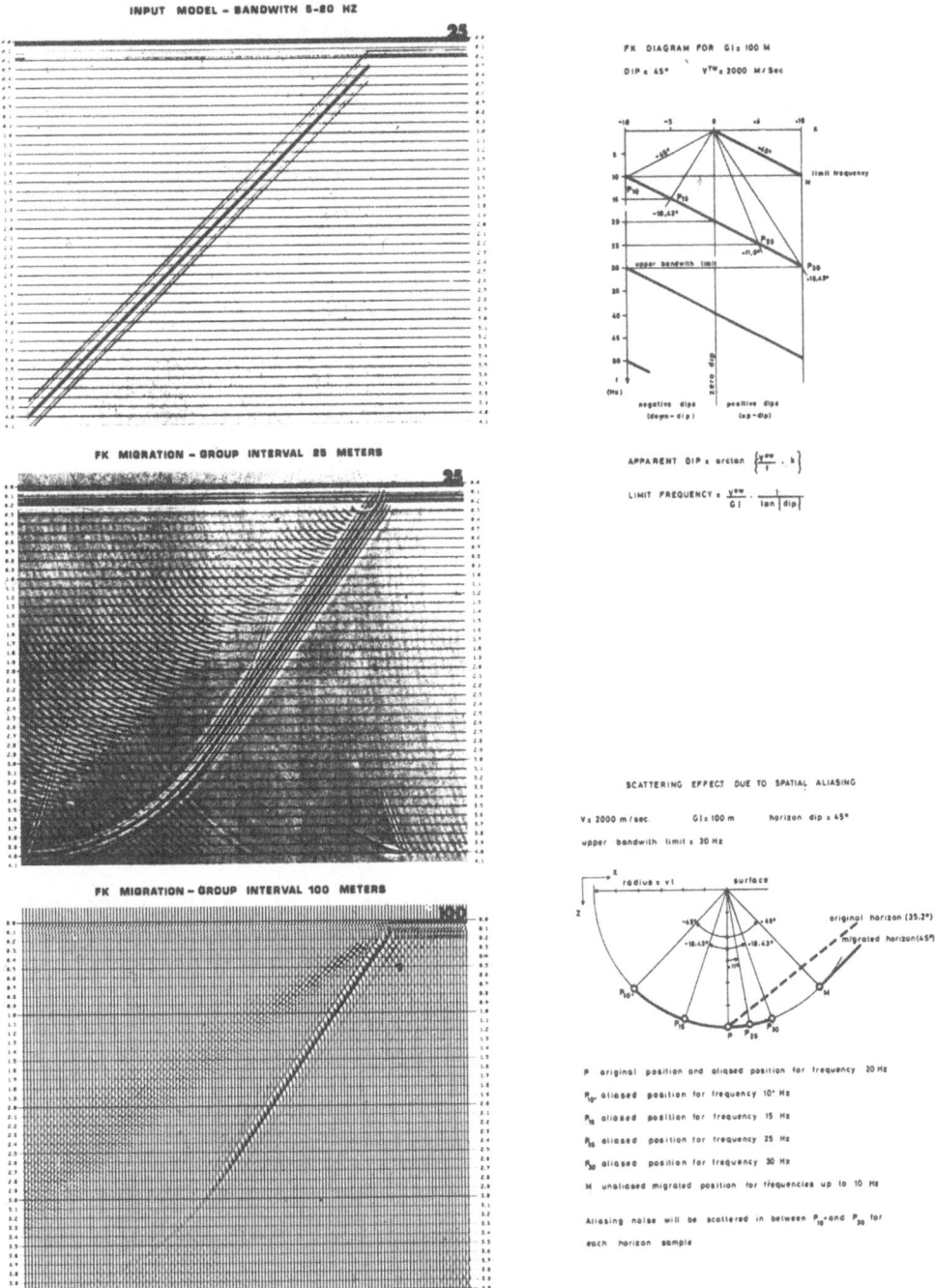

Fig. 18. Aliasing Noise in Migration. The effect of the spatial aliasing
is shown. Migration with coarse trace spacing disperses the
horizon.

$$\theta_{max} = arc \ cos(tz/Tr). \tag{17}$$

Referring again to Figure 19, we note that the lateral displacement
of the point S when migrated to the true position P is:

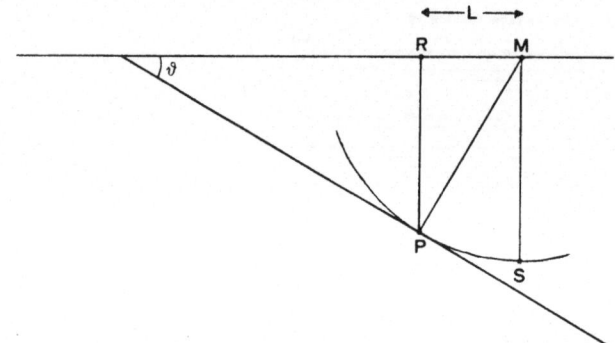

Fig. 19. Model for the Computation of Dip Limitation due to Survey.

displacement = true depth * tan θ.

Rearranging we get:

$$\theta_{max} = \text{arc } \tan\left(\frac{L}{tzV/2}\right) \tag{18}$$

where L represents the length of the line and max the maximum dip
angle that can be recorded through such a line.

The dip limitation for recording time is due to the fact that the
recording time is shorter than the two way time from the surface down to
the reflector and up to the surface again when the horizons are dipping.
Clearly if the layers are horizontal there is no dip limitation due to
this fact. As a rule of thumb the recording time is taken twice the two
way time of the target. This means the possibility to record up to 60
degrees.

The dip limitation due to finite length of the profile is due to the
fact that the migration can in fact move to the true position only on the
recorded data. In other words, if the lateral displacement, that
increases with the dip angle, is comparable with the profile length, many
S point lie outside the recorded section and cannot be moved by migration.

Figure 20 shows a dip field, obtained by applying (17) and (18). It
is easy to recognize the "edge effect".

The relationships found out in this section are very useful in
planning the data acquisition, since they provide an analytic means to
establish how many extra data we need to correctly image our target.

2.2.4 <u>Areal extension of the survey block</u>. Once the size of the
survey has been decided, we need to establish the size of the elementary
block of the survey.

In fact, it is generally impossible to complete all the survey at
once, and it has to be divided in more blocks.

To determination of the block size means to establish the size along
the in-line and the cross-line direction, as well as the more opportune
offset range and the azimuth distribution.

2.2.4.1 <u>Determination of the offset range</u>. The relationship giving
the Normal Move Out term is the well-known:

278

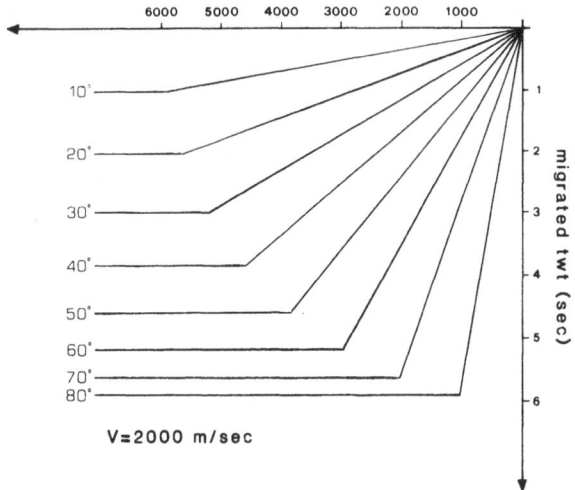

Fig. 20. Maximum Dip Field. The maximum dip handled by the acquisition system is shown along a profile. The edge effects are clearly recognizable.

$$t = \frac{S^2}{2 \text{ to } V^2} \tag{19}$$

where S represents the offset distance, to the vertical two way time and V the stacking velocity.

In the presence of multiple reflections superimposed to the primary events, when the data are corrected of NMO by using the velocity of the primaries, the multiples will be under-corrected of the quantity:

$$tmax = \frac{S^2}{2 \text{ to } Vm^2} - \frac{S^2}{2 \text{ to } Vp^2} . \tag{20}$$

If tmax is the longest period of the multiple, the relationship can be used to provide the maximum offset allowing a good multiple suppression through simple stacking of data. Such offset is:

$$Smax = \sqrt{tmax \ 2 \text{ to } \frac{Vp^2 \ Vm^2}{Vp^2 - Vm^2}} . \tag{21}$$

In 3-D environment Smax equals 2/3 of the target depth in the majority of the cases.

In fact, considering a 20 Hz multiple, that is tmax = 50 msec, a primary velocity of 3000 m/sec and a multiple velocity of 2500 m/sec at a vertical time to = 2 sec we obtain for the maximum offset Smax the value 2020 m. The target depth is 3000 m, thus the maximum offset equals 2/3 of the target depth.

Note that in 2-D environment, it is generally requested for the maximum offset a figure equal to the depth of the target.

Repeating the procedure, but imposing that the difference in the NMO correction between primary and multiple be less than half the sampling interval, we obtain the minimum useful offset (in terms of multiple suppression) Smin:

$$tmin = 1/(2 \text{ fsam}) \tag{22}$$

$$Smin = \sqrt{tmin \ 2 \ to \ \frac{Vp^2 \ Vm^2}{Vp^2 - Vm^2}} \ . \tag{23}$$

In the same conditions of the above example, once we fix the sampling interval at 4 msec, we obtain for Smin the value 400 m. This means that for offset less than 400 m there is no multiple suppression due to the stacking operation.

Another criterium to select the maximum offset is based upon considerations on the efficiency of the recording operations.

In fact, the recorded data do not consist of reflected energy only. The most evident "different" energy is the refracted one, that is eliminated from the record since there are difficulties to process it in a "reflection" environment.

The mute process eliminates all the energy, the recording time of which is less than:

$$t' = v' \ S/2 \tag{24}$$

where v' is the "mute velocity", very close to the refraction velocity, and S the offset from source to receiver.

Figure 21 shows the Normal Move Out curves at different offsets plotted starting with the first useful (unmuted) time. We can see that at 3000 m from source the first useful time is 2500 msec.

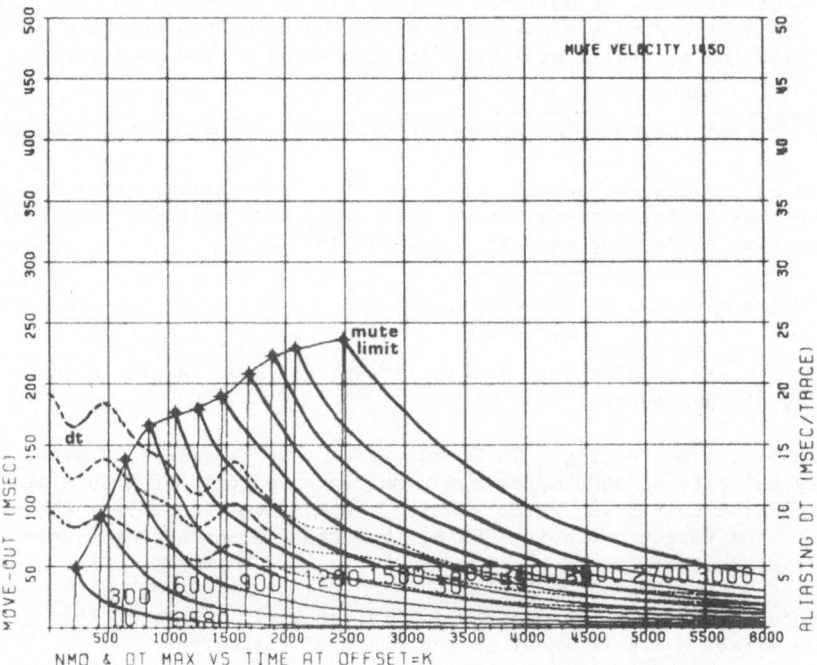

Fig. 21. Mute Plot. The NMO curves are plotted as function of time, starting from the first unmuted time as function of offsets.

280

It makes sense to limit the offset to a value allowing to record
energy, the reflection time of which, Tr, satisfies:

$$\text{Smax} < \frac{\text{Tr}}{v'} \, . \tag{25}$$

2.2.4.2 <u>Orientation of the survey</u>. The practice of 2-D surveying
states the need to shoot along the "dip" direction, that is to orientate
the profiles along the direction of maximum structural dip.

In 3-D environment the rule is to elongate the acquisition block
in the strike direction, that is exactly the opposite.

Suppose the structure we are dealing with be "cylindrical", that
means without dip component along one direction. In such a structure
the velocity field changes along the profiles recorded in the "dip"
direction, while it is constant along the orthogonal direction (usually
called "strike"). When NMO and stack are performed, some troubles will
be encountered if the dips are high or the structure severely
tectonized. If the recording and processing is carried out along the
strike direction, all these problems are avoided. The same kind of
considerations can be made in presence of surface obstacles or when
shooting close to the shorelines too.

The same conclusion comes out from the analysis that follows. Let us
introduce a local coordinate system where the x axis is parallel to the
strike direction, while the y axis is parallel to the dip direction.
Moreover, suppose the structure be cylindrical. The general definition of
the NMO correction is:

$$\text{NMO} = \frac{S^2}{2 \text{ to } Va^2} \tag{26}$$

where S is the offset, to the vertical time and Va the velocity to be used
for the best correction.

The offset S, in our coordinate system, is:

$$S^2 = ax^2 + ay^2 . \tag{27}$$

We are dealing with a structure presenting some dips along the y
direction only. For the correction along y we should apply a velocity Va
that is linked to the stacking velocity from the relationship:

$$Va = Vst/\cos \theta \tag{28}$$

while for the correction along x, since dip is zero, the Vst will be
applied.

By computing the difference between NMO computed in presence and
absence of dip we get:

$$\text{NMO} = \frac{ay^2}{2 \text{ to } Vst^2} (1 - \cos^2\theta) . \tag{29}$$

By imposing that the difference in NMO be less than half the period
of the upper bandwidth limit, we ensure that the error in the NMO
correction can be neglected. In other words, keeping ay small enough we do
not need to explicitly compensate the velocity for dip.

This fact simplifies the data processing, that, if such condition is not satisfied, will require the knowledge of the two components (along dip and along strike directions) of the velocity for NMO correction.

The condition can be formalized as follows:

$$\frac{ay^2 \sin^2\theta}{2 \text{ to } Vst^2} < \frac{1}{2 \text{ fmax}} \tag{30}$$

that means:

$$aymax < \frac{Vst}{\sin\theta} \sqrt{\frac{to}{fmax}} . \tag{31}$$

It is clear that ay decreases as dip increases, or, in other words, the elongation in the dip direction should decrease as the structural dip increases.

The in-line elongation further decreases in presence of high dips in the strike direction.

The NMO correction without dips is:

$$NMO1 = \frac{ax^2 + ay^2}{2 \text{ to } Vst^2} . \tag{32}$$

In presence of dip α along y and dip β along x, it becomes:

$$NMO2 = \frac{ax^2\cos^2\beta + ay^2\cos^2\alpha}{2 \text{ to } Vst^2} . \tag{33}$$

The condition (29) thus becomes:

$$NMO1-NMO2 = \frac{ax^2\sin^2\beta + ay^2\sin^2\alpha}{2 \text{ to } Vst^2} < \frac{1}{2 \text{ fmax}} \tag{34}$$

but:

$$ax^2 = S^2 - ay^2$$

so:

$$\frac{S^2\sin^2\beta - ay^2\sin^2\beta + ay^2\sin^2\alpha}{\text{to } Vst^2} < \frac{1}{fmax} \tag{35}$$

that finally gives:

$$aymax < \sqrt{\frac{\text{to } Vst^2}{fmax(\sin^2\alpha - \sin^2\beta)} \quad S^2 \frac{\sin^2\beta}{\sin^2\alpha - \sin^2\beta}} . \tag{36}$$

that limits the maximum offset of the block.

Figure 22 shows how the elongation in the dip direction is affected by the dip. Since (37) has the same structure as (31), the picture provides also an estimation of the maximum offset in the presence of cross-line dips.

Figure 23 shows the effect of the cross-dips in the value of the in-line extension.

$$S < \sqrt{\frac{\text{to } Vst^2}{fmax \sin^2\beta}} = \frac{Vst}{\sin\beta} \sqrt{\frac{to}{fmax}} \tag{37}$$

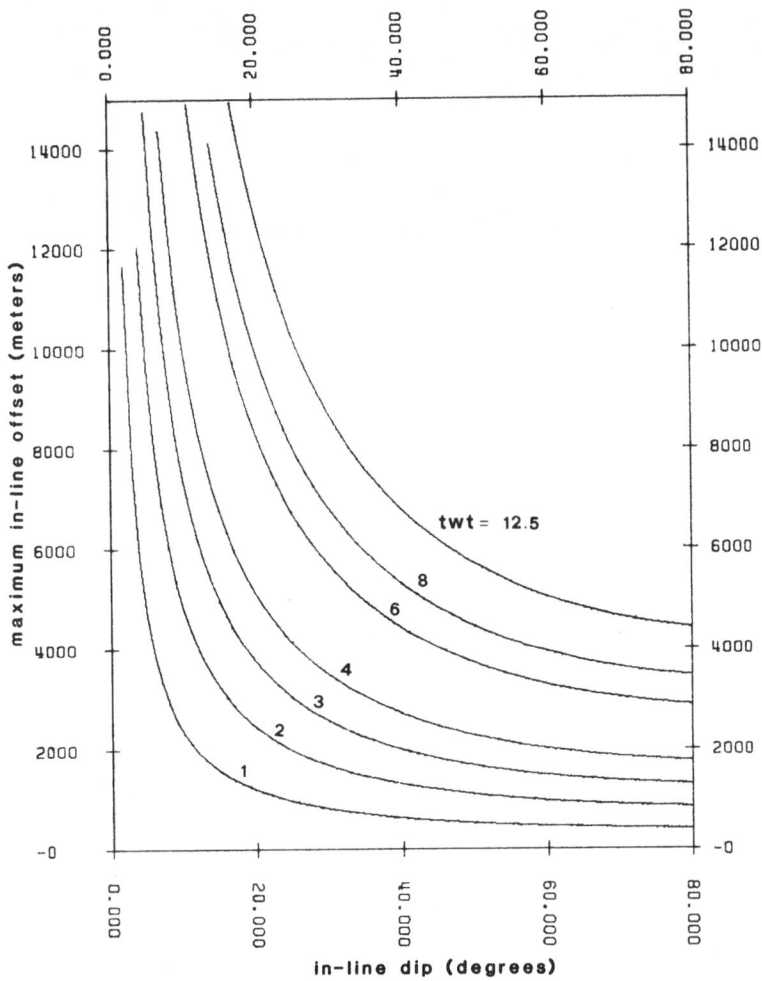

Fig. 22. In-Line Elongation. The maximum elongation (meters) along the in-line direction of the basic block is plotted versus the in-line dip. The displayed curves refer to the same survey, at target times 1, 2, 3, 4, 6, 8 and 12.5 s. The elongation increases with depth of the target and decreases with the dip of the horizons.

2.3 Survey Layout

Once the dimensions of the block has been established, next step is the set up of the field layout. This means to select:

- the total level of coverage,
- the in-line and cross-line folding,
- the number of receiver lines to connect,
- the maximum admissible displacements of the actual positions with respect to the theoretical ones,
- the tolerance on the positioning of shot and receivers.

This kind of choice depends on many factors:

- type of source,
- number of recording channels available,

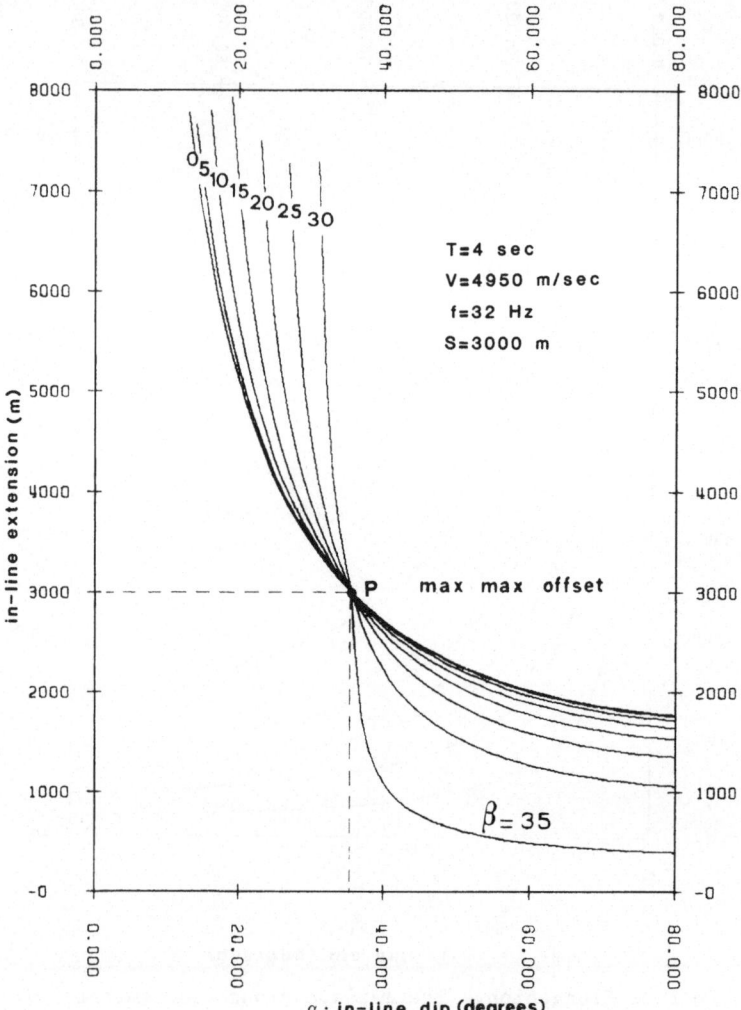

Fig. 23. In-Line Elongation in Presence of Cross-Line Dip. The maximum
elongation versus the in-line dip is shown for dips in the
cross-line direction from 0 to 35 deg. All the curves cross in
the point P, which abscissa is the maximum dip that can be seen
by the maximum spread when the cross-dip is zero and the
ordinate equals the maximum spread. On the left of P the
elongation increases with the cross-dip, whilst the opposite
happens on the right. The position of P moves rightwards as the
spread length decreases.

- accessibility of the positions,
- background noise level,
- static problem,
- economical constraints.

In the following, and through the analysis of some examples, we will
describe the basic guidelines.

2.3.1 Coverage. The total coverage of a 3-D survey is generally
less than 2-D. In fact, as pointed out by Marshall (1984), the signal to

noise ratio of a 3-D is from 1.5 to 2.0 times higher than in a 2-D survey recorded by using the same acquisition parameters of 3-D.

Since the 3-D is carried out in areas where formerly some seismic activity has already been done, by analyzing such results it is possible to identify an adequate level of coverage.

As a rule of thumb in land operation 12 to 16 fold seem to be adequate to most cases, while for marine cases the corresponding figures are 30 to 60.

2.3.1.1 <u>In-line and cross-line coverage</u>. The total level of coverage is the product of the in-line by the cross-line folding.

In parallel profiling there are no special comments to do, since the acquisition is 2-D like.

In cross-array profiling things are more complicated. In order to enhance the reliability of the data, that are recorded in separated blocks, it is imperative to overlap the subsurface mid points with records belonging to adjacent blocks.

This statement can be satisfied by overlapping some receiver lines between adjacent swaths, or by shearing the same field positions between shot points belonging to adjacent swaths, or, better, by taking both these actions.

Moreover, the interlacing of the coverage of the adjacent block becomes increasingly important as the static problem becomes more severe. Such interlacing calls for a minimum of 3 fold coverage in both directions.

The general expression of the coverage in the in-line direction (after Bading, 1982) is:

$$My = \frac{NS}{2n} \tag{38}$$

where:

- N is the number of detector stations in the spread,
- S is the number of shot positions per each shot,
- n is the number of stations by which the spread is advanced.

A similar expression for the cross-line coverage, to be used with care, is:

$$My = \frac{LS}{nm} \tag{39}$$

where:

- L is the number of receiver lines simultaneously recorded,
- S is the number of shot per shot-line,
- n is the number of shot positions between adjacent receiver lines,
- m is the number of receiver lines to be advanced for the adjacent swath.

The above relationships can be useful for drafting areal design.

Another aspect should be taken into account when deciding how to subdivide the total coverage in the two directions: the azimuthal distribution into each bin.

Through the multiple coverage technique we obtain, at each subsurface location, a number of traces which offsets cover the full offset range. This in 2-D, while in 3-D, since data are recorded by combining shot and receiver locations spread out areally, care should be taken in order to ensure both a proper distribution of offsets and azimuths into each bin.

The best way to properly distribute the azimuths is to shoot in-line and cross-line through the roll-in/roll-out technique rather than through a simple roll along one (2-D standard).

Figure 24 shows an example of such a technique. There are four receiver lines, each consisting of 48 receivers, and six shot-lines that insists on them. Along each shot-line there are six shots. The distribution of the offsets and azimuths is shown for two bins.

By moving this block both in-line and cross-line it is possible to provide the coverage of the area to be surveyed with a predetermined folding level, being at the same time sure to provide an optimum offset and azimuth distribution.

The azimuthal distribution provides an increased S/N ratio on the stacked sections, and better velocity analysis as well.

Figure 25 shows some velocity analysis, in the form of Constant Velocity Stack, carried out on data collected this way. The velocity increment between adjacent strips is 50 m/sec, and it is easy to recognize the high resolution of the analysis also at very long times.

As a rule of thumb for full 3-D survey it is better to ensure about the same folding level to the two main directions.

As a matter of fact it is preferable, from the operative and economical point of view, to increase the cross-line folding against the in-line one. The statement will be clarified in an example. The reason is to ascribe to the better utilization of the field material made possible by such a choice.

Figure 26 shows some possible cross-line configurations realized by connecting four receiver lines. On the picture the four lines are

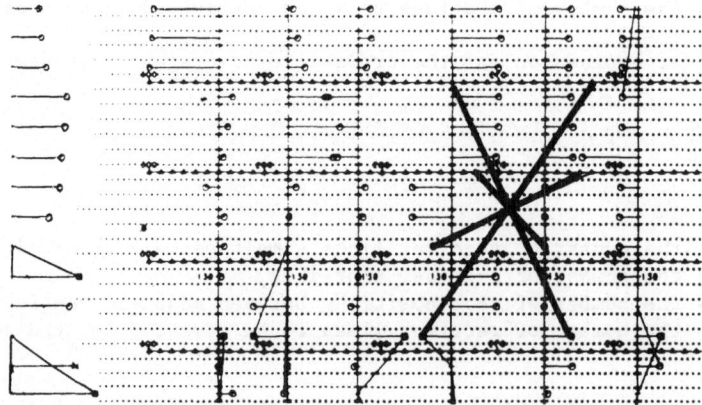

Fig. 24.   3-D Block. The surface layout, showing the receiver and the shot positions, and the corresponding center of the covered bins is displayed. For a couple of bins, the distribution of the offsets and azimuth has been evidenced.

286

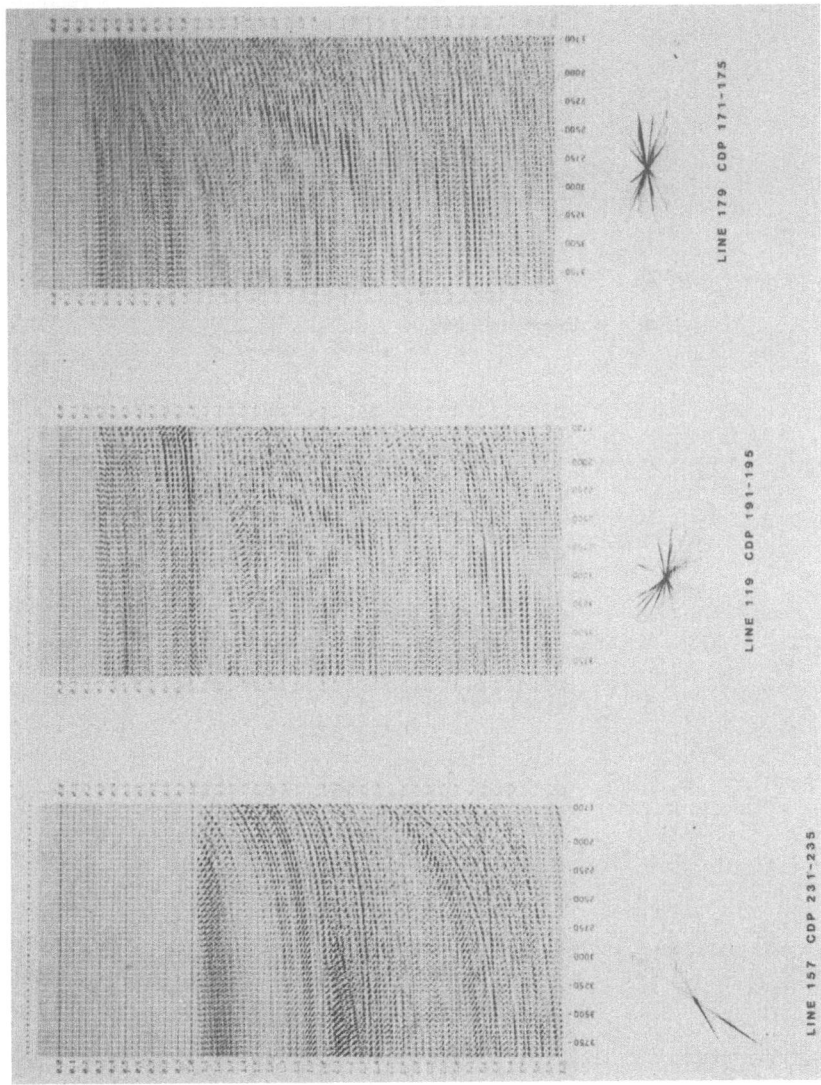

Fig. 25. 3-D Velocity Analysis. The display shows a Constant Velocity Stack with 50 m/sec of increment between the velocities. Note the high resolution of the analysis, due to the good azimuthal distribution of the traces involved (shown at the bottom).

indicated by the circles, and they are represented as seen from one side. As can be seen, by changing the number and the disposition of the shot points it is possible to change the cross-line folding and the lateral extension of the swath. All the distances are expressed in multiples of the subsurface distance among the covered points. In fact, to each point corresponds an entire subsurface line covered. To different dispositions pertain also a different interlacing of the adjacent swaths.

By studying the effect of the choice of a number of different configurations in-line and cross-line it is possible to optimize the survey. Generally, once the technical requirements are satisfied, the criterium followed for the optimization is the minimization of the total number of shots requested to complete the survey. Generally by moving the

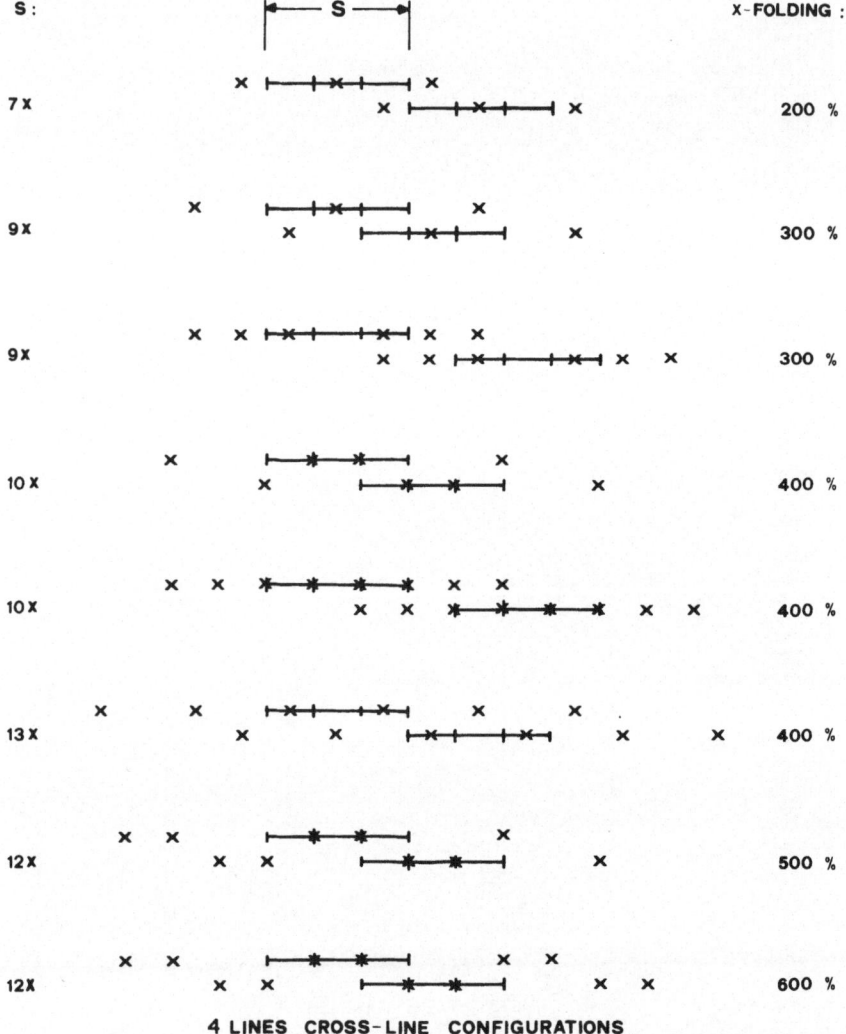

4 LINES CROSS-LINE CONFIGURATIONS

Fig. 26. Examples of Cross-line Layout. Some possible layouts built up
with 4 receiver lines are shown, as well as the maximum offset
(in units of subsurface spacing) and the cross-line folding
associated.

higher folding in the cross-line direction it is possible to reduce the
number of shot-lines to record for the fulfilling of the total folding
requirement, and this translates in a smaller total number of shots.

Each source presents particular problems, so that not all the
configurations can be profitably used.

Dynamite shooting does not allow the re-use of the shot positions.
In fact, for an efficient run of residual statics programs, it is
requested that only one static value be associated to a field position, a
condition not satisfied when using dynamite.

On the contrary, the re-use of the field positions is recommended
when the vibrators are used, since the shot static is the same for all the
shots fired on the same position.

Configurations providing a larger number of "far" positions are better for noisy sources like detonating chord, while these providing "close" firing are preferred for low energy sources when recording shallow target surveys.

The choice of the configuration is clearly affected by so many problems that it is evident the impossibility to fix a general rule.

2.3.2 <u>Displacement from the theoretical positions</u>. Quite often the layout theoretically drafted should be adapted to the environmental constraints. The most relevant effect is the displacement of the receiver (sometimes) and the shot (quite often) positions. The criteria to follow when taking such action can be derived from the former discussions.

In discussing about the size and orientation of the survey some rules were found out. The same apply to the displacements too, in the sense that they should not carry to situations for which the constraints are violated.

While planning the field operations, to each shot-line can be associated a maximum value of displacement both for in-line and cross-line movements.

In a general sense we can see that the displacements allowed decrease with the target time. Deep targets are less problematic to handle.

2.3.3 <u>Errors in positioning</u>. The problem of the accuracy of the positioning of source and receiver locations is a basic one. Certainly it is easier to control them in land operations than in marine.

In the following we will analyze the general criteria.

2.3.3.1 <u>Tolerance in positioning</u>. The analysis moves again from the consideration of the NMO relationship:

$$dt = \frac{S^2}{2 \text{ to } V^2} \tag{40}$$

from which, by partial differentiation, we get:

$$ds = \frac{\text{to } V^2}{S} d \ dt. \tag{41}$$

The maximum allowed error in time, due to positioning error, is the sample rate SR. This leads to:

$$RMS \ error = \frac{SR}{3} = \frac{1}{3} d \ dt. \tag{42}$$

By replacing S with Z in (41), since:

$$Z = \frac{\text{to } V}{2} \tag{43}$$

we get:

$$\frac{\text{to } V^2}{S} = 2V \tag{44}$$

and:

$$dS = V \frac{2}{3} SR. \tag{45}$$

We define the "tolerance radius" dr, valid for dips $< 10°$:

$$dr = \frac{dS}{2} = \frac{SR}{2} V. \tag{46}$$

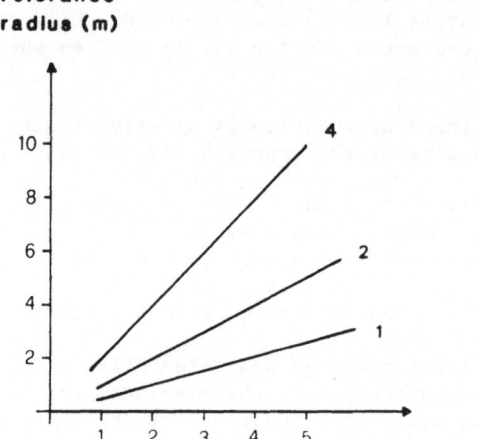

Fig. 27. Tolerance Radius. The tolerance radius is plotted as function of velocity and sample interval.

In the presence of higher dips, that influence the positioning of the mid point, the general relationship (Bading, 1980) is:

$$dr = \frac{SR}{2} V \frac{1}{2 - \cos^2 \theta} \, . \tag{47}$$

Figure 27 shows the tolerance radius plotted as function of the velocity and of the sampling rate. Figure 28 the effect of the dip on the tolerance radius.

2.3.3.2 <u>Tolerance in elevations</u>. Errors in elevations means static errors. As previously considered, the RMS error consequent to a maximum allowed error equal to the sample rate SR is:

$$dto = + \frac{SR}{3} \tag{48}$$

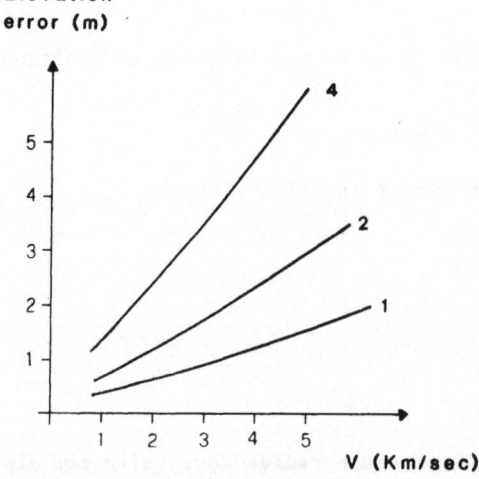

Fig. 28. Tolerance Radius. The tolerance radius is plotted as function of velocity and dip, for Sr 4 msec.

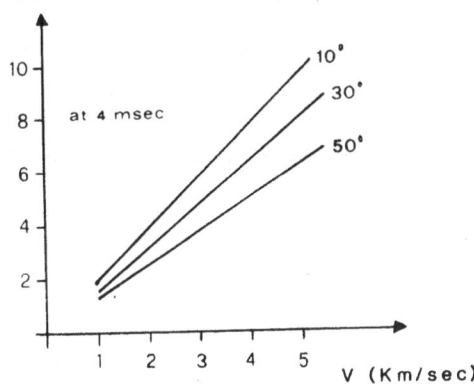

Fig. 29. Elevation Error. The elevation tolerance is plotted as function of the velocity and sample rate.

and the corresponding error in the elevation becomes:

$$Me = + \frac{SR}{3} \, Vsw \qquad\qquad (49)$$

where Vsw is the sub-weathering velocity used for static computations.

Figure 29 shows the elevation error plotted as function of the sub-weathering velocity and sampling rate.

2.3.3.3 <u>Positioning errors in marine surveys</u>. The positioning errors are of two kinds:

- errors due to the navigation system,
- errors due to the estimation of the streamer shape.

The navigation systems used for 3-D consist generally of 3 independent elements, helping each other in fixing the "antenna" position. The precision depends on the single element precision and, when used, on the number, as well as relative position and distance, of ground stations for triangulation.

The error figure is between 5 and 15 meters.

The errors in estimation of the streamer shape, and so its position, depend on the following factors:

- number of gyro-compasses along the streamer,
- number of depth controller along the streamer,
- type of tail buoy monitoring,
- type of streamer head monitoring,
- accuracy of the processing of field readings.

The number of gyro-compasses ranges from 8 to 12 on streamer 3000 meters long. Through standard calibration procedures the accuracy of the readings can be ensured within 1 degree.

Each contractor has its own tail buoy and monitoring system. The most sophisticated available on the market are based upon "mini-trisponders", that provide the possibility of measuring the distance from

fixed points of the vessels within some tenths of centimeters. In the
near future also some points of the streamer will be monitored this way.

With these precisions on the readings, the position of the single
traces of the streamer can be estimated within some meters if the
processing of the field data utilizes a Kalman filter.

These figures, when compared with those obtained from (47), look
still too high for very accurate seismic processing (processing at 1 or 2
msec of sampling rate with upper bandwidth limit greater than 120 Hz), but
precise enough for standard resolution surveys.

Apart from the errors introduced by the positioning errors, the
streamer feathering, because of its random figure, introduces some
additional problems. For this reason, generally the acquisition is
accepted only if the feathering is within 6-7 degrees.

Figure 30 shows the cable shape as determined while recording a line
belonging to a North Sea 3-D. The variations of the cable shape are
evident.

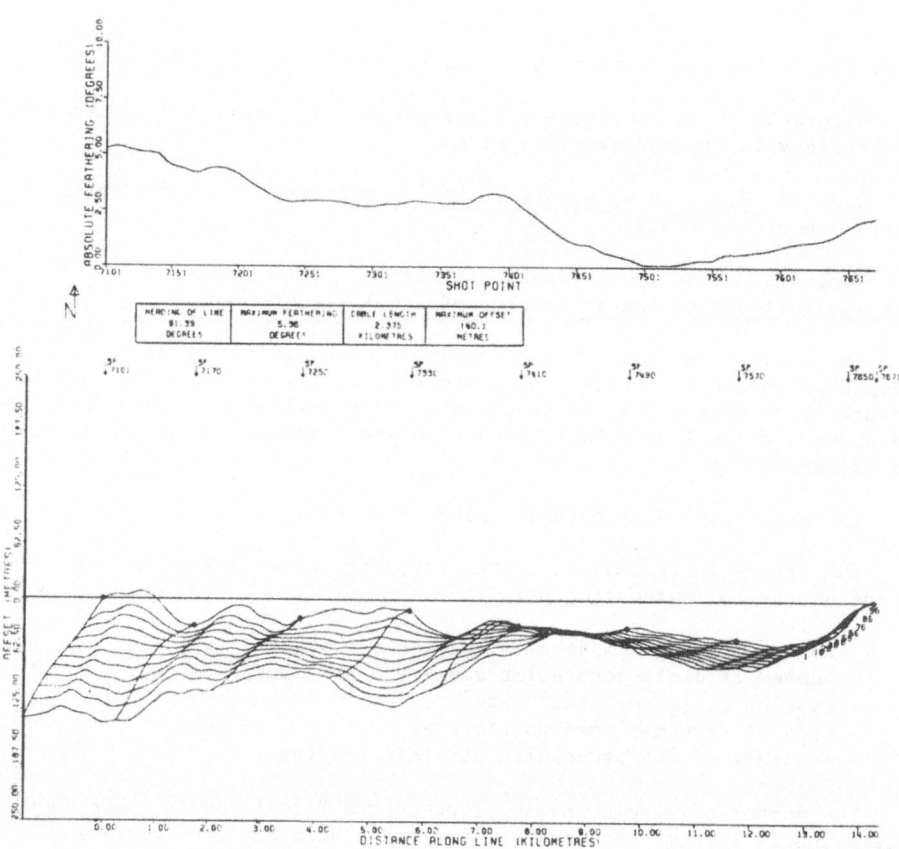

Fig. 30. Feathering Record. The plot shows the feathering of the
streamer. Offset is the perpendicular displacement from a
straight line drawn between first and last gun positions.

2.3.4 <u>Directivity of source and receiver patterns</u>. The directivity of the source and receiver patterns is used in 2-D for a reduction of the noise and of the "out of plane" energy. In 3-D only the noise has to be attenauted, since the method was conceived to take advantage of the side-swipes.

For this reason, the receivers are deployed in arrangements that present response curves very similar for different azimuths.

A very popular geophone setting is the one shown in Figure 31: it is called "3 arm windmill", and is almost omni-directional.

For the source pattern, when used, the considerations are the same, but while these patterns are easy to make with dynamite or dynamite-like sources, with vibrators moving along the roads it results are very difficult.

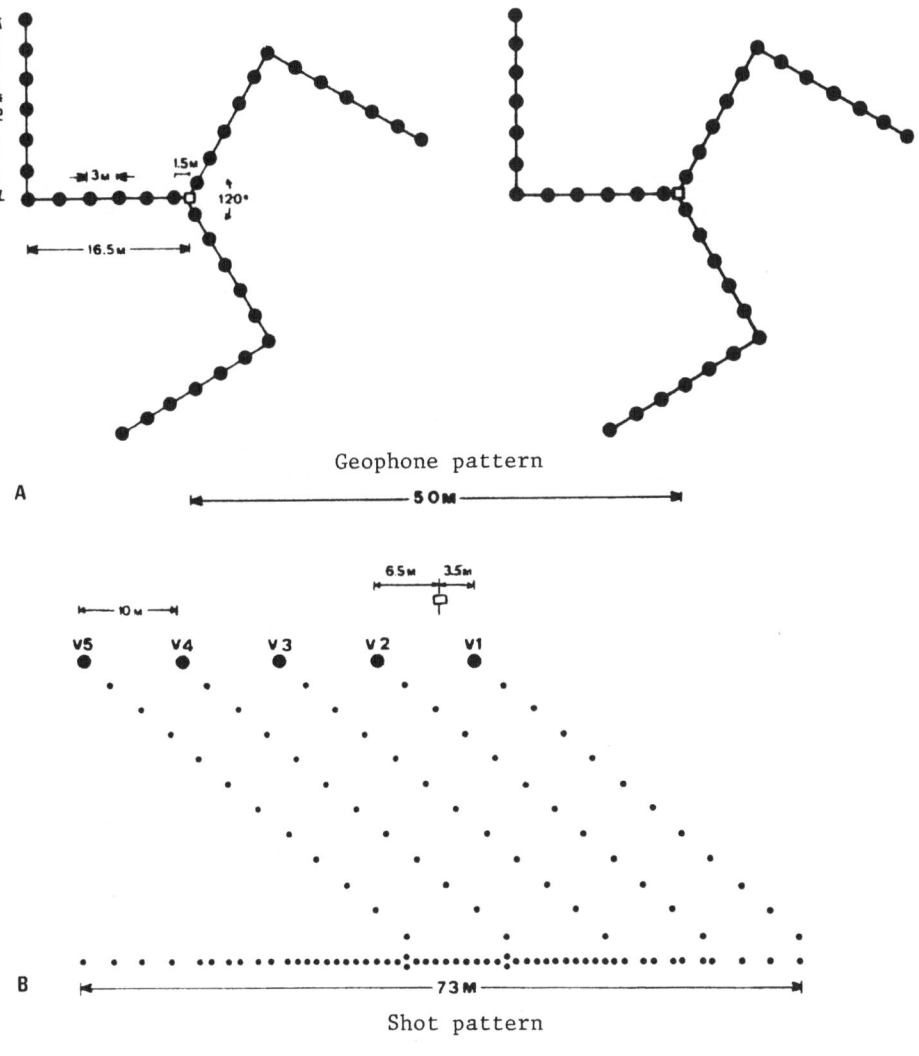

Fig. 31. Geophone Pattern. 3 arm windmill pattern, used for land 3-D surveys.

In marine environment exactly the opposite happens. In fact, the sources can be towed in arrangements allowing the best omni-directionality, whilst the receivers are always arranged along straight lines. A negative characteristic of the straight line is the non-uniform sensitivity in the plane orthogonal to the alignment directions.

The polar response exhibits a series of lobes on this plane, so that the side-swipes are recorded under heavy distortion in the amplitude.

Care should be taken when designing a survey, in order to minimize such negative effects.

In case of multi-source or multi-streamers acquisitions, the problems related to the polar inhomogeneity of the response are partially limited by the interference among the multiple arrays.

2.3.5 Economical considerations in acquisitions. The financial commitments should be considered when designing a 3-D survey, since simple actions can have great relevance on the cost of the project.

Of great importance are the sources allowing the "conjugate shooting". This technique consists of the simultaneous recording of two separated shot-points, and its application saves from 40 to 70% of the acquisition cost.

When using vibrators, if one shot is realized vibrating up-sweep (the vibration frequency increases with time) and the other down-sweep (the vibration frequency decreases from the upper limit down to the lower), it will be possible to extract, through a separate deconvolution of the recorded response with the up-sweep and the down-sweep, two separated records. The operation is equivalent to two separate recordings, but requires half the time to be completed.

There is another technique allowing the separation: the phase control of the vibrators (one group starts vibrating with positive phase while the other start with the opposite).

The two methods can be contemporarily used in order to enhance the separation of the two records, or applied in parallel in order to make possible the separation of four records (up-positive, down-positive, up-negative, down-negative) coming from four groups of vibrators.

The separation is quite satisfactory for all the frequencies of the sweep, except for the frequency common to the up and down sweep. In fact, since the coupling of the baseplate with the ground changes from position to position, there is a narrow band of frequencies not exactly separated.

The problem can be solved:

- by arranging the up and down sweeps in such a way that the crossing frequency can be eliminated by applying a field notch filter,
- by applying the notch during the preprocessor of the data,
- by using non-linear sweeps, so that the crossing frequency falls outside the useful bandwidth of the response.

The residual can then be eliminated by a simple short-period predictive deconvolution.

In marine environment there is the possibility to tow multiple source arrays and multiple streamers, so that it is possible to record more subsurface lines per boat sail.

The reduction in cost is close to proportional to the number of subsurface lines collected.

The cheaper operations are those involving multiple sources with single streamer, since the cost of a streamer is many times the cost of a gun array and to tow additional source arrays requires less power than to tow multiple streamers.

Sometimes some troubles can be encountered while processing the data since the signature of the multiple arrays can be different, and interference or anomalous resonance of one array can be recognized when the other is firing.

In land environment the cost is inversely proportional to the number of recording channels available. In fact, this is true in desert zones, while in areas highly inhabited it is very difficult to operate with more than 240-300 channels.

2.4 Checking

Once the field layout has been found out we must go on and verify if it is possible to carry it out in practice. Looking at the maps of the area to be surveyed, a first check of the influence of the most important natural obstacles can be made, and, where it is necessary, the layout can be modified locally. If the changes to be made are too many or too important, it is more convenient to go back to point b and try to design the block again taking into account the field constraints.

After that it is necessary to record all the exact field positions where shot-points or geophone stations can be located in the field and to assign a field position to, or set up a dead position for, each shot and station provided for by the design. The choice of the coupling between sites and theoretical positions has to be made carefully, taking into account the operative characteristics of sources and recording system utilized to carry on the survey.

2.4.1 Final check. Once the survey planning has been completed, a whole check of design constraints must be carried out satisfactorily . Clearly to verify what happens in the subsurface when some thousands of shots, of its own recording configuration of some hundreds of traces each, are shooted we need a computer simulation. The simulation, usually called shooting on paper, provides the distribution of some parameters in the bins, and the most important are:

- coverage,
- minimum offset,
- maximum offset,
- minimum mute or minimum useful reflection time,
- azimuth distribution.

By means of coverage analysis we check the completeness of the survey bin by bin; by minimum mute the existence of information at the time of the target, while by averaged azimuth we detect the polarization of the energy. The velocity analysis becomes more robust and easy to peak when the azimuths of the traces collected in a bin are scattered in all the directions (average azimuth close to zero), rather than concentrated along one direction only. The easiness increases as the difference between maximum and minimum offset contemporarily present in the bin increases.

The analysis of these distributions of parameters can suggest local alterations of shot-geophone geometry, and/or the addiction of some shots in order to balance the anomalies of the distributions.

Figures 32 to 35 show the resulting distribution of maximum and minimum offset, coverage and azimuth in a theoretical case.

2.5 Marine 3-D Survey Design

All the considerations and formulas given in the former paragraphs are valid for marine surveys too. In fact, since the acquisition geometry if fixed because of the rigid streamer configuration, not all the constraints can easily be satisfied.

Moreover, the marine design is generally simpler since the acquisition is nothing more than a conventional 2-D survey. Things become complicated when a light 3-D has to be designed.

2.5.1 Special marine surveys. The light 3-D are carried out during the early exploratory phase over geologically complicated areas. Available seismic data is generally poor in quantity and in quality as well.

In other words, it is generally very difficult to understand the exact nature of the geophysical problem to solve. With the streamers and recorders available today, it is easy to match any specification in the in-line surveying, while in the cross-line care should be taken about the technique and the line spacing selected. In fact the cross-line sampling is too coarse to allow the running of the migration, so that the data should first be interpolated to a regular, narrower grid.

We will consider the two basic techniques:

- single source-single streamer, and
- multiple source-multiple streamer.

By using the single source-single streamer technique one subsurface line only is collected per boat trip. This provides data less critical to process than the other techniques, but also more limited in handling the

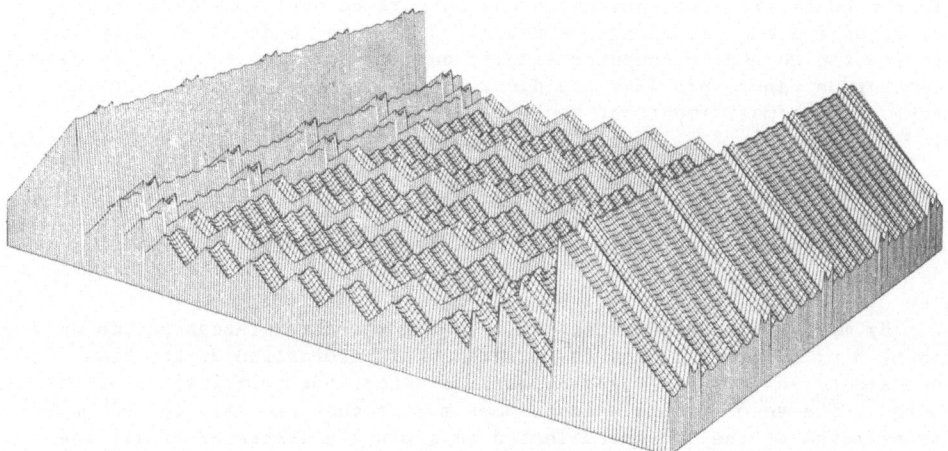

Fig. 32. Theoretical Distribution of Minimum Offset. Perspective view.

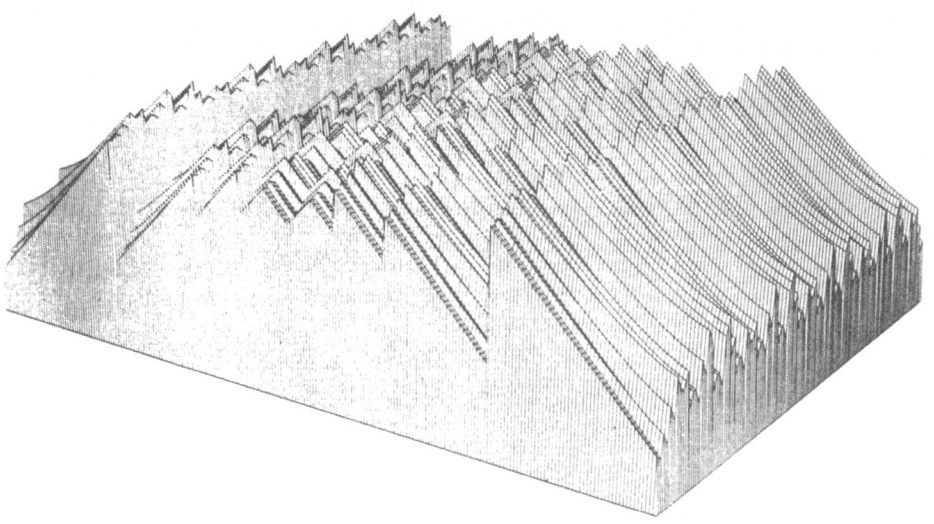

Fig. 33. Theoretical Distribution of Maximum Offset. Perspective view.

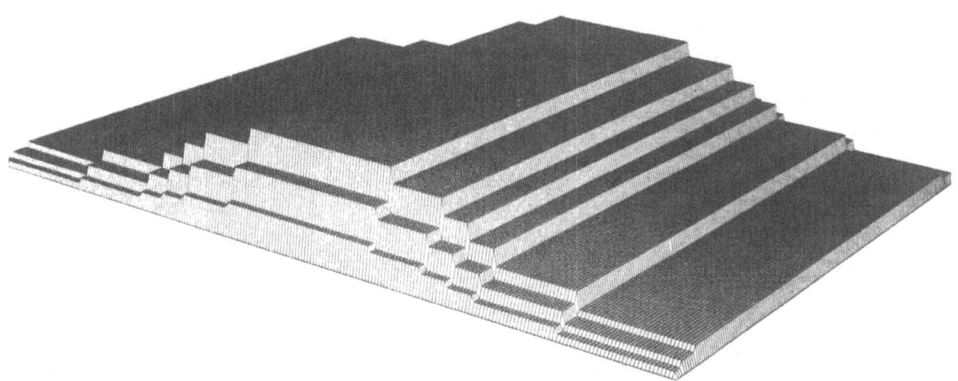

Fig. 34. Theoretical Distribution of Coverage. Perspective view.

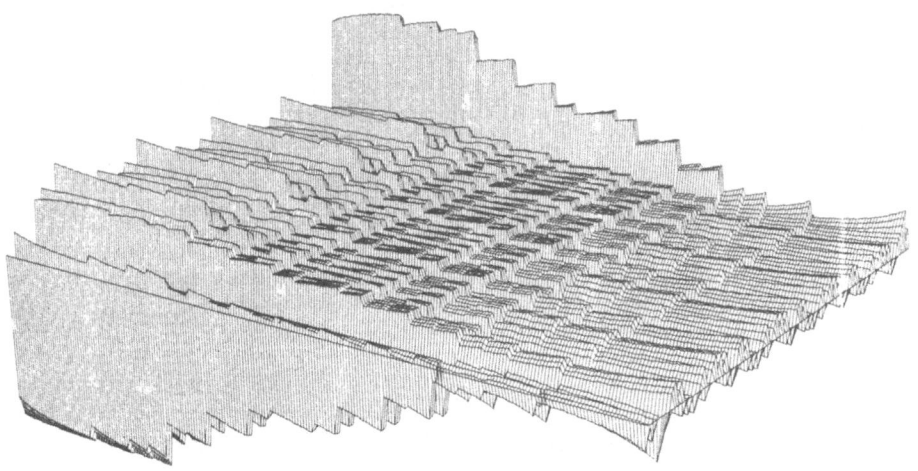

Fig. 35. Theoretical Distribution of Azimuths. Perspective view.

297

higher spatial frequencies in the cross-line direction. In fact
navigation constraints could be less restrictive, allowing large streamer
feathering, but this fact generally leads to the need of binning the data
through the variable bin size technique in order to avoid big differences
in coverage folding and frequency spectrum similarity between adjacent
bins.

Since the lateral extension of the bin changes with the feathering,
the cross-lines are affected by jittering. In order to avoid noise
generation by the cross-line migration, the interpolation schemes applied
before that step should be less sensitive to the higher spatial
frequencies. In other words, this technique can be used for structural
survey only.

By using the multiple source-multiple streamer technique much better
results can be achieved, provided that the acquisition is carried out
respecting narrower constraints than in the previous case.

In single source-single streamer techniques the maximum spatial wave
number that can be handled depends upon the line distance, that is the
boat course. Any interpolation scheme is unable to increase that limit in
spatial resolution.

In multiple source-multiple streamer operations the maximum spatial
wave number physically recordable is associated to the minimum distance
between the subsurface lines collected along the single boat course. This
figure is sometimes higher than the one obtained using the boat course
distance.

Consider the example of Figure 36. Let us imagine we work with two
sources, towed 25 m port and 25 m starboard off the single streamer. For
each boat course two subsurface lines, 25 meters apart, will be recorded.
If the distance between two adjacent boat courses is 150 m and we like to
get a final grid in which the lines are 25 m apart, after acquisition we
will have filled in two lines every six.

Through the interpolation the actually dead lines will be filled in.
The FK spectrum of any section input to the interpolation program shows,
according to the characteristics of the section in terms of dips and
frequencies, a distribution of data point up to 50% in the zone between
wave numbers 0 to 0.004 and for the remaining 50% between 0.004 and 0.020.

The interpolation program should be capable of handling conflicting
dips, also in presence of large differences between the amplitudes and
the frequencies of the events to be interpolated or, in other words,
should be able to follow the weak tail of the diffraction taking it
separated from the seismic horizons.

Hence the design criterium of such a survey is the following: the
distance between the pair of collected lines should match with the
requirements of handling the high dips associated to the tails of
diffraction, while the boat course separation should match with a proper
sampling of the structures.

2.5.1.1 Circle shooting technique. The patented "CIRCLE-SHOOT"
method, trademark of Tensor Geophysical Service Corporation, was designed
with the aim to realize a true marine 3-D by collecting the data in such a
way that into each bin a proper distribution of both azimuths and offsets
be achieved.

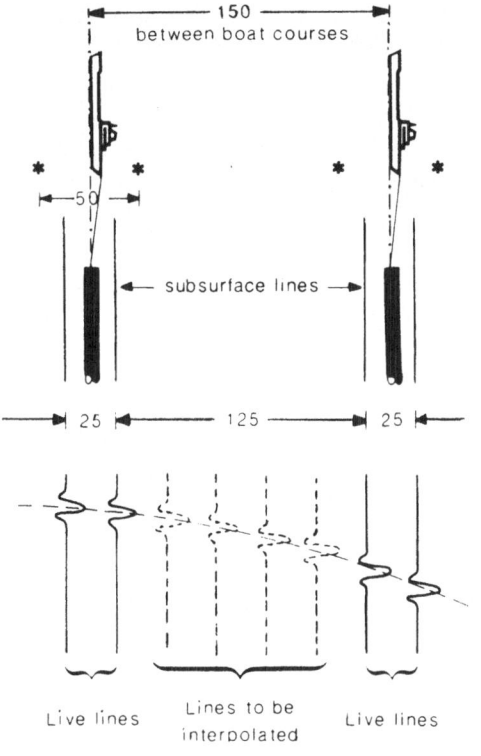

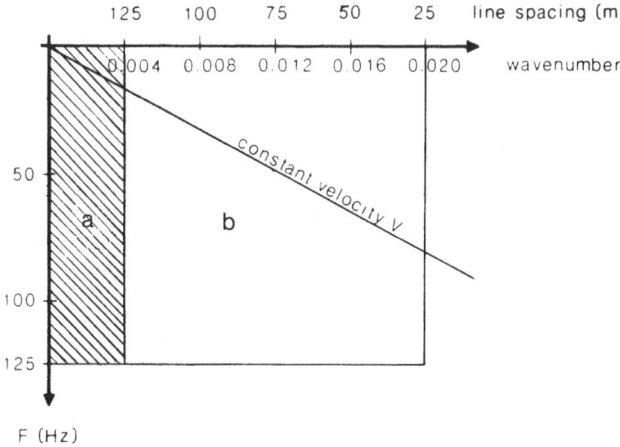

Fig. 36. Multiple Source-Multiple Streamer Operations. Principle of operations with two sources and one streamer when performing a reconnaissance marine 3-D.

The problem was solved by collecting the data along overlapping circles rather than along parallel straight lines. The absence of preferential directions (in-line and cross-line) during the acquisition requires a different processing sequence, through the offset continuation technique, in order to get the proper stack of data into the final bins.

A full 3-D implementation of the offset continuation algorithm is a hard job and Tensor tries to solve it in an approximate way as described in the following sections.

The offset continuation operator, that is the convolutional operator able to remove from the data the effects of CMP dispersion that generates in the presence of dipping layers, depends on:

    h   - Offset (scalar)
    θ   - Propagation azimuth
    x   - CMP theoretical position in the in-line direction
    y   - CMP theoretical position in the cross-line direction
    v   - Velocity (scalar)
    dip - Dip of the horizon in the point of coordinates (t,x,y)

The problem of the 3-D operator synthesis can be simplified to a simpler 2-D synthesis by considering the velocity no longer a scalar value but a vector depending on the offset h, the azimuth θ and the local dip. V be such a vectorial velocity: the synthesis reduces to evaluate an operator, function of x, y and V, exactly like a conventional 2-D survey.

The determination of the 3 components of the velocity is achieved by interpreting, for each point of analysis, three different displays, in a way very similar to the conventional velocity analysis:

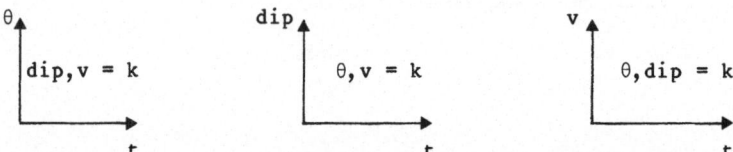

The final product is then assembled in a data base, from which it is possible to extract the vectorial value of the velocity for each point of coordinates x, y and t.

The stacked volume is got by summation, along the offset axis, of the constant offset sections obtained by convolution of the offset continuation operator punctually synthesized with the data collected into the theoretical bins.

Such a algorithm is very useful when used as interpolator (before stack) because of its anti-aliasing characteristic. This means that it can be used also in presence of data recorded in points not regularly spaced and with spatial sampling greater than those commonly accepted in conventional parallel-profiling 3-D surveys.

Apart from these positive aspects the method exhibits the big advantage of drastically reducing the non-producing time during the acquisition phase.

In parallel-profiling 3-D surveys, the percentage of the recording time over the total amount of time required to complete the survey is between 25 and 35%, because of the problems of streamer positioning along a straight line before the beginning of the recording of each collected line.

With the "circle-shooting" technique, the operations are carried on continuously, due to the fact that the vessel passes to the following circular pattern immediately after the former circle has been completed. The useful time percentage increases to 80-85%.

The reduction in the total cost of a survey, taking into account the saving in acquisition and the increase in processing, can be estimated about 50%.

The results of this method are particularly well suited for operations in geologically complicated areas, like the salt domes, where the following problems affect the parallel profilings:

- a large portion of the recorded data is wasted because the salt dome and/or the complex velocity structure in the vicinity of the dome distort one leg of the reflection travel path, thus invalidating the hyperbolic stacking assumption;
- surface structure on the perifery of the salt dome constantly interferes with long-streamer straight-line shooting;
- short hydrophone group lengths are necessary to pass the higher frequency components of steep dip reflections, but these short groups also pass more towing noise;
- the survey is a mixture of dip-lines and strike-lines shooting which varies around the dome;
- a considerable amount of ship time is wasted because shot-lines must be longer than the desired fully fold area and the vessel must go outside the shooting area a significant distance in order to turn and get the streamer straight between successive shot-lines.

On the contrary, the advantages of the circle-shooting are:

- both legs of reflection ray paths propagate through essentially the same sedimentary section and the data satisfy the hyperbolic moveout assumption;
- only the inner circles encounter the surface obstructions, and since they are collecting only shallow data, a very short streamer can be used to minimize problems in this area;
- since the steep dip reflections arrive broadside to the hydrophone groups, longer group lengths can be used to suppress towing noise without affecting the reflections (this results in an increase of the S/N ratio at the acquisition stage);
- the coverage is more uniform and continuous all around the survey and all the shooting area is fully fold;
- the accurate configuration of the streamer produces a swath of subsurface coverage with very small spatial sampling interval in a direction radial to the dome.

Despite all these advantages, some problems affect the method, that can be grouped as follows:

- acquisition noise,
- streamer tracking and positioning very critical,
- low stability for small streamer depth,
- loss of efficiency at smaller target depths.

Acquisition noise. Watching the results exhibited by the contractor, an average noise level of 5 microbars is recognizable, with local peaks up to 25 microbars. By comparing these figures with those coming from conventional in-line survey, we notice that the average figure is of the same value (about 3 microbars), whilst the peaks are much larger.

The higher noise levels are probably generated across the streamer sections that lie on the border of turbulence zone at the stern of the vessel. The final seismic sections do not look particularly noisy, and this is probably because this kind of streamer noise is completely randomized by binning and stacking processes.

The noise level is a problem, but it seems to be less critical than one can expect simply remembering what happens in conventional marine seismics.

Streamer tracking and positioning. The tracking and positioning of the streamer become critical when the acquisition is random-like. At the moment the tracking of the streamer is done through data coming from 12 gyros filtered through a very sophisticated Kalman filter. The actual precision of traces and shots is about a tenth of meters, a figure a little bit high for a very accurate processing.

By mounting some mini-rangers both on the streamer and on the tail buoy it will be possible to provide room for triangulation with the master vessel and a slave vessel fixed, for each circular pattern, more or less at the center of the circle. With this innovative positioning system it is believable to reach 1 meter of precision in the positioning of the subsurface points, thus by-passing the problem.

Streamer stability. A general loss of streamer stability has been detected for radius of the circle less than 800 meters, and a loss of depth stability for nominal depth of the streamer less than 10 meters.

This is due, first of all, to the reduced speed of the vessel (between 3.5 and 4.5 knots) requested in order to keep the noise at relatively low level, and to the circular pattern of the vessel and of the streamer.

The torsion on the depth controller sections of the streamer increases with the curvature of the streamer, so disturbing the correct alignment along the desired pattern. Because of that, the streamer exhibits the tendency to gallop.

Some practical experience show as optimal a curvature radius of 1800-2000 meters when the speed is of 3.5-4.5 knots and the streamer depth of almost 10 meters.

This means that the circle-shooting method is a good tool for structural problems with middle-high depth of the target.

Efficiency. The acquisition along circular patterns provides a means to collect circular strips of data, which width is directly proportional to the streamer length and inversely proportional to the curvature radius.

By considering the effect of the muting, we recognize that the width is also inversely proportional to the recording time.

When the depth of the target is less than 2 seconds, the number of different circles to collect in order to ensure enough coverage at the target level rapidly increases.

The cost of a conventional parallel-profiling equals the circle-shooting at about 1.0 second (two way time).

2.5.1.2 <u>Cross-profiling method</u>. The cross-profiling method was first introduced by <u>AGIP in a 1986</u> 3-D survey. It lies in between the parallel-profiling and the circle-shooting, and was conceived in order to take the maximum advantage from the application of the powerful DMO in its 2-D form during the processing stage also in 3-D environment.

In the basic form it consists of two sets of parallel lines, recorded along orthogonal directions. In practice there are two superimposed parallel-profiling surveys recorded on the same area.

A first advantage comes out from the fact that, where the two sets of lines cross, the bins hold traces with two different azimuths. As previously seen, this means more accurate and reliable velocity analyses as well as improved S/N.

By properly planning the acquisitions it is also possible to reduce the cost of the survey. Figure 37 shows the design criterium applicable when the research vessels has twin sources or twin streamers capability.

For each sail line, two subsurface lines are recorded at a distance equal to an even multiple of the group interval. In the example shown the group interval was 13.33 m and the line distance twice that figure. The sail lines belonging to the two sets cross at a multiple of the recorded line distance. In our case, every twice the distance between lines, that is 106.66 m.

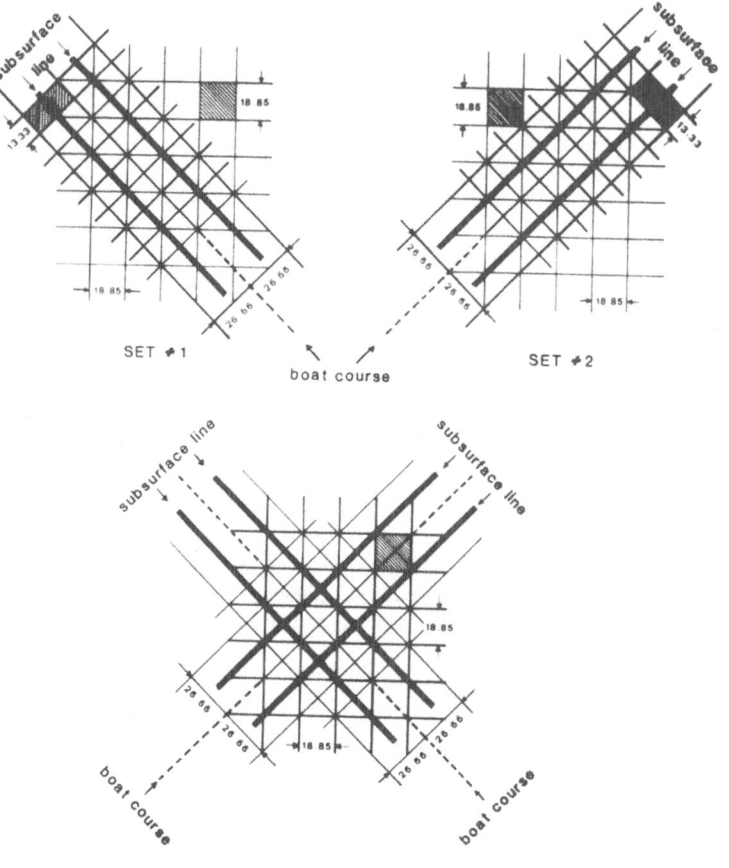

PROCESSING FINAL GRID

Fig. 37. Cross-Profiling Surveying. Principle of operations with two directions.

The actual distances should be determined taking into account the true recording conditions, over all the average figure of the streamer feathering. In our case we expected not less than 5 degrees of feathering, that means, with a streamer 3000 meters long, a lateral displacement of the subsurface traces belonging to the second half of the streamer between 65 and 130 meters.

The theoretical distance between the two nearest lines is 80 meters: we can thus expect to find most of the area in between the two nearest collected lines filled in.

Nevertheless, some bins will anyway be empty or filled with a sub-optimal selection of offset ranges. As in the reconnaissance surveying, an interpolation after stack has to be planned in the processing stage.

In order to help the interpolation, the following processing sequence was designed:

Data are first of all compensated against the dips through the application of DMO, after gathering of data in common offset planes parallel to the acquisition direction. Since data are collected with feathering not exceeding 10 degrees, 2-D DMO works at its best.

After that, the single traces are binned according to a fixed bin scheme. The bin grid, as outlined in Figure 37, is rotated by 45 degrees with respect to the original shooting direction. By this trick the bins, that in acquisition were rectangular, become square, with size midway between the two dimensions of the rectangle. In our case the bins are 18.88 meters square. In order to equalize the frequency spectra among the traces, a partially overlapping binning procedure can be utilized in this step.

Finally, after stacking of the data bin to bin, a spatial inter-polation aiming to fill in the few empty bins is performed. The 3-D migration step closes the sequence.

The quality of the final results is better than a conventional parallel profiling at a cost comparable, sometimes lower despite the higher shooting mileage.

In fact, the boat sailing in the two directions can be arranged to form a double circle, thus saving some unuseful time with respect to the conventional parallel shooting like the circle-shooting. The saving is greatly highlighted when the operative conditions force to operate the "one way profiling", that is to record all the lines in the same sense.

The method has proved powerful in solving the complicated situations associated with thick alloctonous layers.

We examined a cross profiling scheme performed along two directions only, but the method can be generalized to any number of directions. In fact, economical considerations advise against a number of directions greater than three.

3-D PROCESSING

Figure 38 shows a typical flow-chart for marine 3-D processing. Most of the processes are 2-D or 2-D-like, and do not require special

considerations. In fact, since data are recorded so densely, some
benefits derive from the statistical redundancy of information, in
particular at velocity field conditioning level.

In the following we will consider three important processing steps:
DMO, migration and interpolation.

3.1 Dip Move Out Correction

The Dip Move Out technique, also referred to as Offset Continuation,
can be used to improve the quality of the stacked sections and their
spatial resolution (Salvador and Savelli, 1982; Rocca and Ronen, 1984);
moreover, velocity analysis carried out on data continued to zero
offset are better detailed and focused.

In this case the NMO can be applied with roughly approximate velocity
data before or after Offset Continuation without two noticeable
differences.

The water-bottom multiples appear always with water velocity,
simplifying their cancellation with multichannel filters, and diffractions
appear with the velocity of the top of corresponding hyperbola. Their
velocity is thus lowered and the interference with the higher velocity
layers underneath is decreased.

Figure 39 shows the effect of OC on a velocity analysis, while Figure
40 shows the effect of the correction on the CMGs. The aforementioned
effects are quite visible.

In the following the process by which DMO correction reduces the
aliasing noise is explained in the simple case of pure geometrical
acoustics.

Let us consider a single diffraction point (Figure 41) and suppose we
record a single common mid point gather (CMG), that is, a set of traces
with source and receiver symmetrically displaced with respect to the
Common Mid Point (CMP).

The medium is assumed to be homogeneous and we shall neglect as usual
all other energy but primary reflections. We also zero the traces in all
Common Mid Point gathers other than the chosen one. Here we depart from
the response to the physical model of a diffractor point.

Without OC, the corresponding properly stacked trace is identical to
the zero-offset trace, thus presenting a single event. After migration,
the event is smeared over a semi-circle centered in the CMG; therefore the
diffractor could be located anywhere along this migration circle.

On the other end, if we apply OC to non-zero offset traces before
stacking them, we can get more precise information about the actual
position of the scatterer. Each trace of the Common Mid Point Gather
contains a single event that, after OC or pre-stack partial migration,
followed by stack and post-stack migration, results in a different semi-
ellipse (Figure 42).

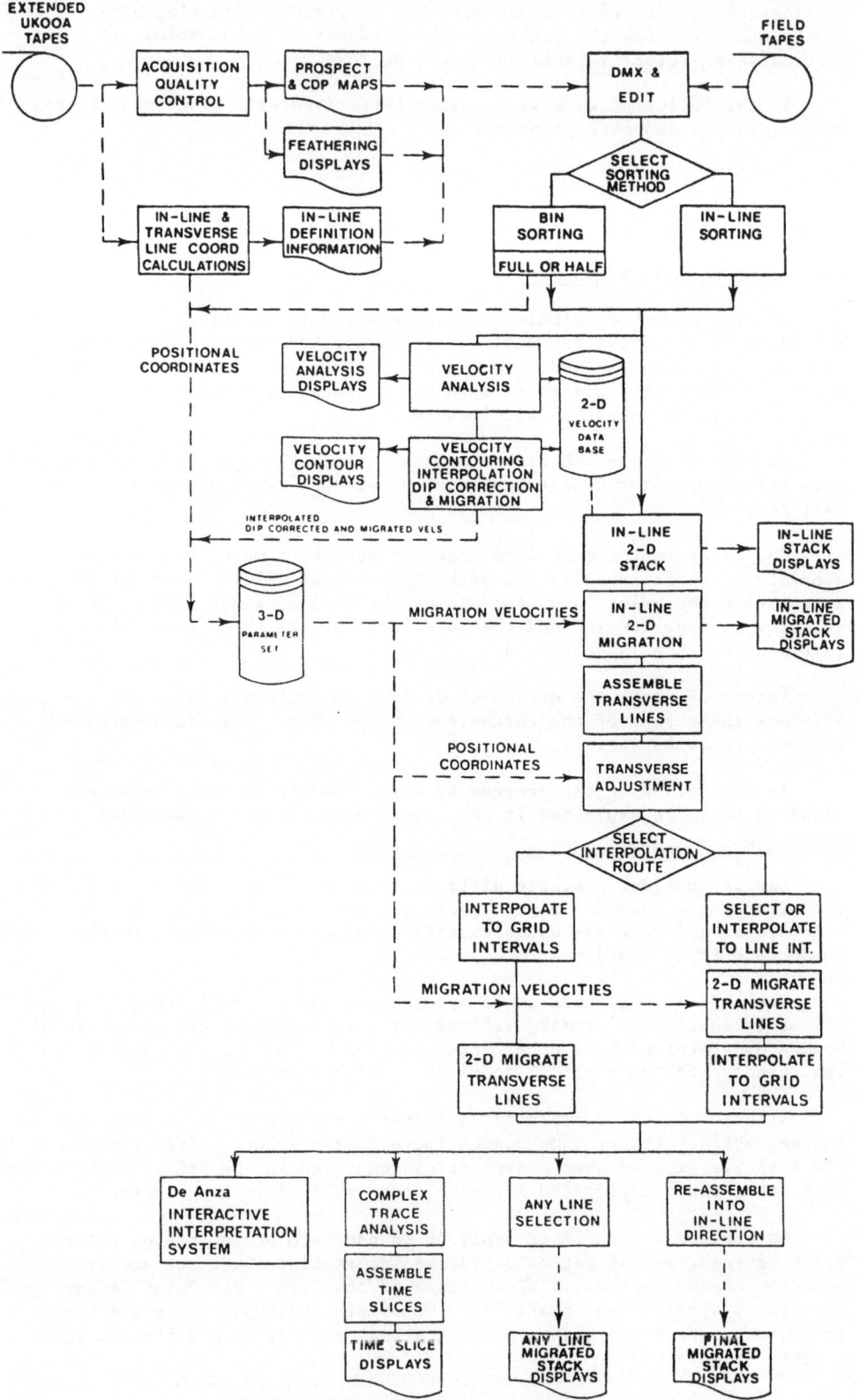

Fig. 38. Marine 3-D Flowchart

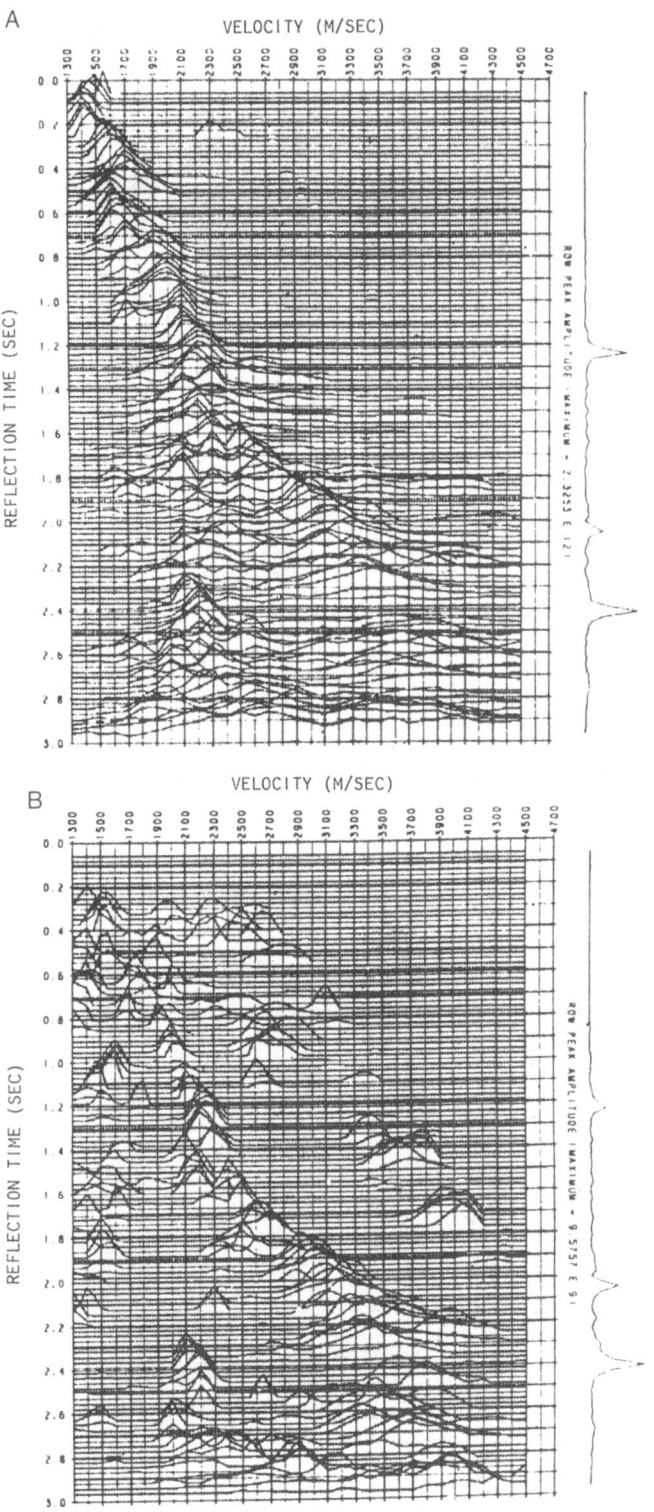

Velocity analysis before and after offset continuation

Fig. 39.   Velan with and without DMO. The right picture exhibits more
concentrated energy due to DMO correction.

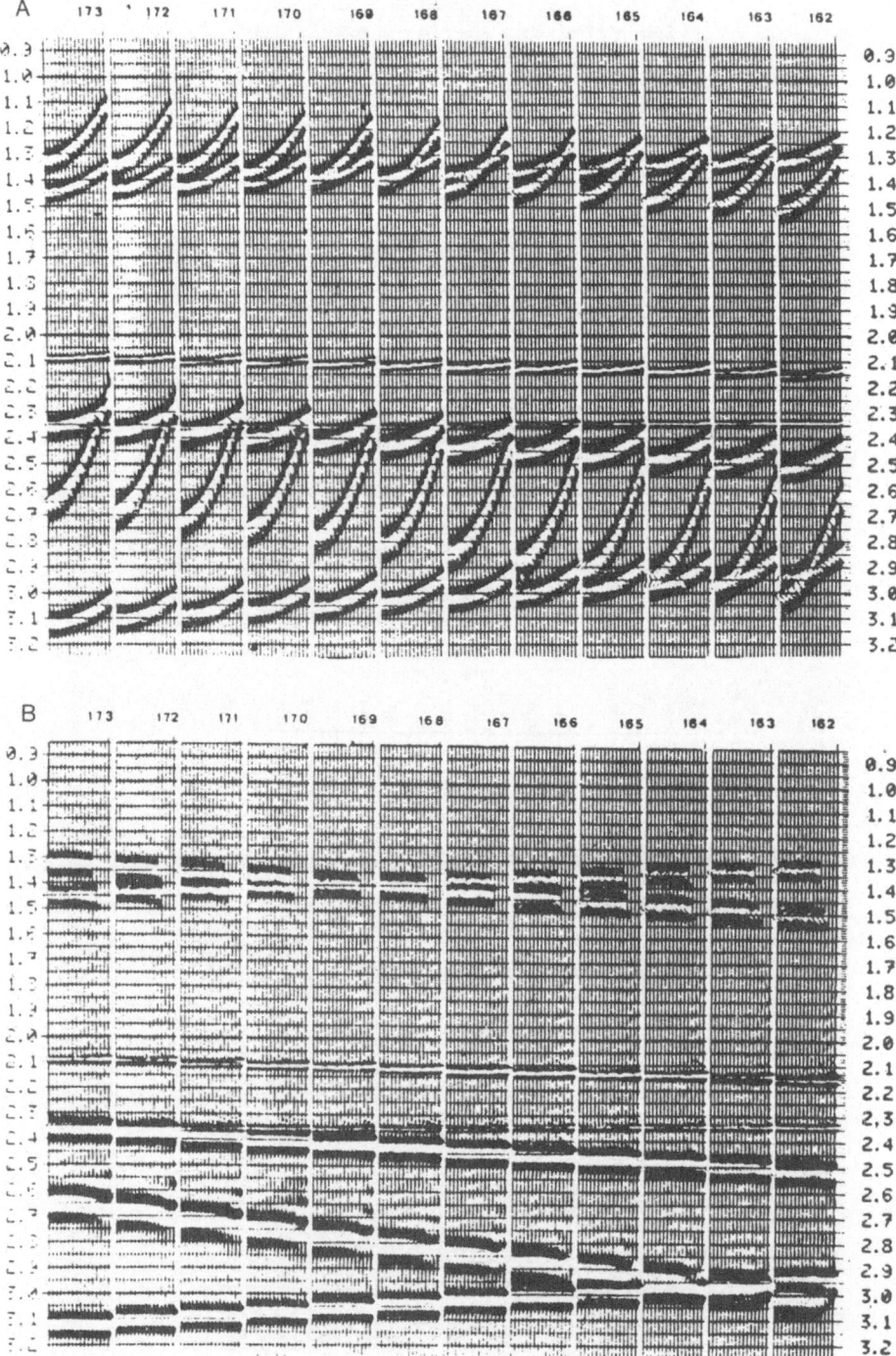

Fig. 40. Dip Move Out on Gather. Effects of DMO correction on CMP
gathers.

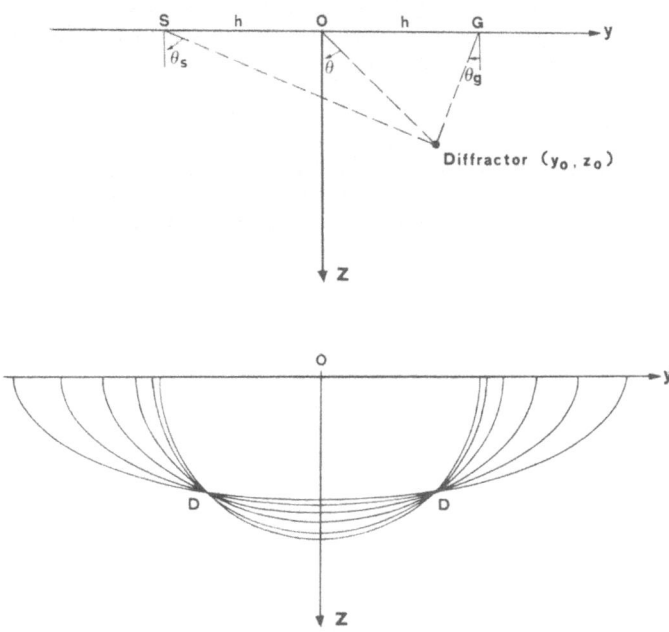

Fig. 41. Single Point Scatterer and Elliptical Mirror.

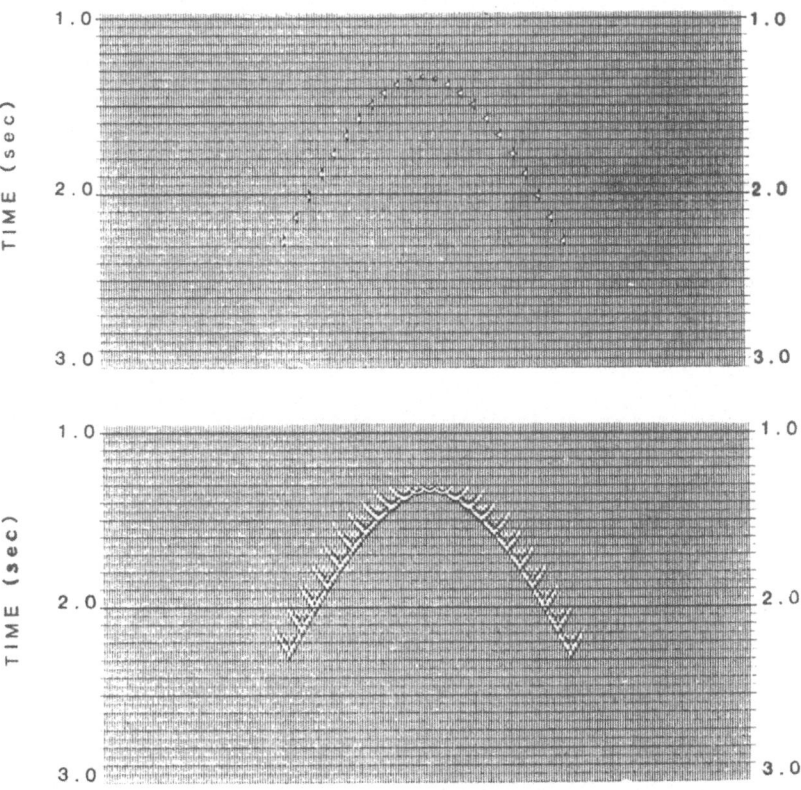

Fig. 42. V-pattern.

The semi-ellipses generated by migration of the traces have different separation of their foci, i.e., the different offsets of the CMG, and they intersect at the location of the diffractor.

The semi-ellipses also intersect in D', symmetrically located with respect to the center of the CMG. Since the semi-ellipses interfere destructively, they cancel each other nearly everywhere but in D and D'. Therefore, the uncertainty about the location of the scatterers is now reduced: instead of the full semi-circle obtained without OC we get only two points D and D'.

Obviously, the more offsets available, the more the actual situation approaches this geometrical limit.

This result shows that the whole CMG indeed contains more information than does any one of its traces; in fact the CMG also holds information about the dip of the recorded events in the cosine-corrected velocity to be used. This information, properly exploited by OC, can decrease the uncertainty about the location of the diffractor. Therefore, the migrated section after OC should be less noisy than the one without OC, since many previously admissible solutions, though actually false, have been eliminated.

In the time domain we have a situation that is dual to that of the depth domain. The zero-offset time section considered is the one corresponding to an initial single point diffractor and to a single CMG, obviously after OC and stack. This time section can be obtained by forward modeling from the combination of the semi-ellipses and the possible locations of the diffractor. Thus we retrieve only parts of the two hyperbolas corresponding to the actual point diffractor D and its alias D'. This happens because the diffraction patterns in D and in D' include only a limited range of semi-ellipses.

The parts of the two hyperbolas form a "V-pattern", which is the basic element of an offset continued stack. Indeed the slope of the V is the slope of the diffraction pattern, the additional information that OC is able to extract from the CMG traces.

An alternative way to derive the V-pattern is to start from the "smiles" (Bolondi et al., 1982), i.e., from the zero-offset time section obtained by the forward modeling of the semi-elliptical distribution of diffractors, corresponding to a single event in a non-zero-offset time section. It is the summation of the smiles from the various offsets of the CMG which produces the V-pattern for the stack.

The smile is another fundamental operator of OC, since by convolution of an NMO-corrected common-offset section with such a time-variant operator it is possible to produce a section that is equivalent to a true zero-offset section.

Figure 43a shows a theoretical zero-offset section corresponding to a single diffractor located at depth z = 1000 m with a velocity v = 1500 m/s. In order to increase aliasing, we zeroed four intermediate traces in every five, thus yielding a strongly aliased diffraction hyperbola, since the distance between adjacent CMP's is 125 m.

We generated also the corresponding 16-fold section, with offset ranging between 100 to 1500 m such that their square was uniformly spaced. The stacked section after Oc is shown in Figure 43b: it exhibits the well-known V-pattern. We see that every event has been substituted by its

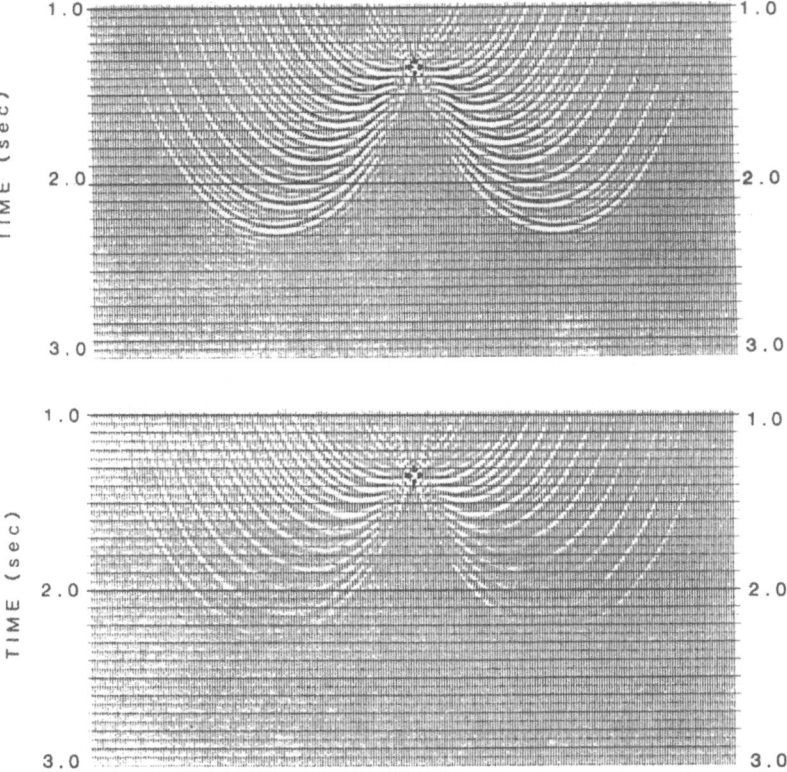

Fig. 43. Aliasing Reduction through DMO.

proper V-operator and that the arms of two contiguous V's touch each
other, thus generating a continuous signal.

The width of the V-pattern increases with the apparent dip angle; one
angle of the V lies on the hyperbola, while the other one points to the
alias diffractor symmetrically to each live Common Mid Point Gather.

Comparison of Figures 44a and b shows the reduction of alias noise
achieved through the OC procedure. The difference becomes even
more evident in the corresponding migrated sections: the migration of
sections in Figures 43a and b (without and with OC) are shown in Figures
44a and b, respectively.

The migrated section without OC exhibits considerable aliasing noise
located along the migration semi-circles.

On the other hand, in the offset continued section the noise is
smeared along semi-circles and semi-ellipses, and the effect is for
signals to cancel each other except at the position of the diffractor and
its aliases (that are symmetric with respect to each live CMG). This is
particularly evident where the dip move out is most noticeable, i.e., for
the lowest parts of the hyperbola. Then the alias noise is strongly
reduced after OC even if the part due to the alias arm of the V is still
present.

It is interesting to notice that with a suitable weighting of common-
offset sections on the stack we could zero the aliasing noise at given
wave numbers, times and frequencies.

OC in 3D

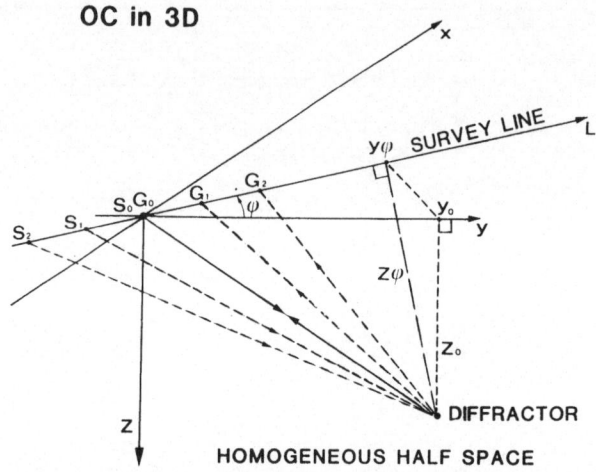

Fig. 44.   3-D DMO

The impact of such a processing technique on data acquisition is evident.

The OC procedure can be extended easily to 3-D data volumes. In fact, each event in a non-zero-offset section is related to a depth distribution of diffractors along a rotational semi-ellipsoid with foci located in the source and receiver position.

By zero-offset forward modeling from the semi-ellipsoidal depth mirror we obtain, as in the 2-D case, a smile that lies completely in a 2-D image plane, bottoms under the mid point between the source and the receiver at the zero-offset arrival time and has the azimuth of the line connecting source and receiver. This implies that during the mapping to zero-offset only the traces with the same azimuth of the source-receiver connection should be mixed.

The advantages achieved by the OC of 3-D data are:

- reduction of alias noise,
- higher spatial resolution,
- independence of velocities from both dip and azimuth,

whilst the main drawback consists of being most useful only if a regular survey is available with all azimuths and all offsets up to cable length.

The characteristic V-pattern also appears in the case of a survey line in a 3-D medium, when a single scatterer is present and a single CMG is considered. In the 3-D case, it can be seen immediately that the width of the V-pattern depends on the relative azimuth of the survey line and of the vertical plane passing through the diffractor and the CMP.

In 2-D the arms of the V-pattern lie on the two hyperbolas corresponding to the diffractor and its alias; in 3-D its arms lie on two rotational hyperboloidal diffraction patterns corresponding to the diffractor and its alias, again symmetrically displaced with respect to the CMP (Figure 46).

Indeed, for every azimuth of the survey line, the experiment can be considered as two-dimensional. With reference to Figure 44 we have, if the azimuth of the survey line is θ:

$$y_\theta = y_o \cos\theta \tag{50}$$

$$z_\theta = \sqrt{y_o^2 \sin^2\theta + z^2} \tag{51}$$

and therefore the zero-offset image of the CMG is a V-pattern bottoming at the azimuth-independent time:

$$t_o = 2(y_o^2 + z_o^2)/v \tag{52}$$

with azimuth-dependent width.

The V is obviously contained in the image plane passing through the CMG, and oriented parallel to the survey line.

Figure 45 shows that the width of the V-pattern is at its maximum when the survey line lies along the direction of the diffractor ($x = 0$), and it is zero when the line is orthogonal to the direction of the diffractor ($y = 0$). Figure 46 shows the hyperboloidal diffraction patterns in three dimensions. The shaded area is the surface described by the V-patterns for azimuths from -90 to +90 degrees. If we omitted OC, the NMO correction would have generated only an event located in correspondence to the CMP.

Theoretically so, we can make good use of the feathering of the streamer or of surveys like squares or loops, whenever we get offsets with different azimuths.

A suitable sampling of the V-pattern area could also be used to define a lower cost 3-D survey.

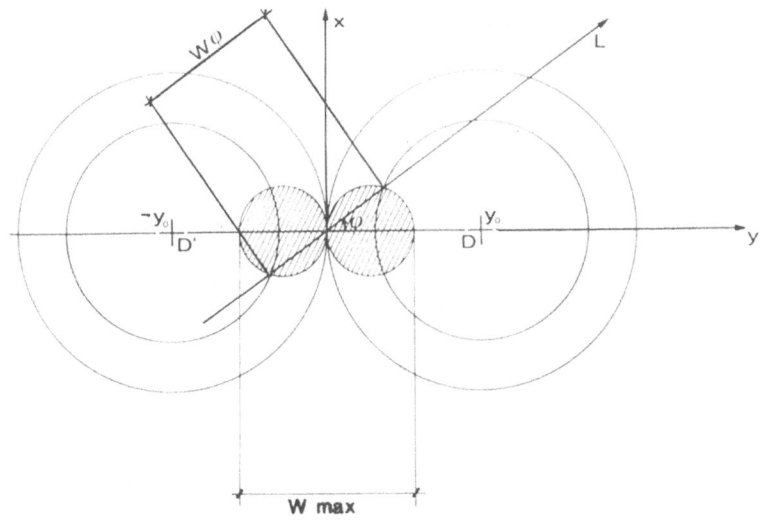

Fig. 45.   3-D DMO.

313

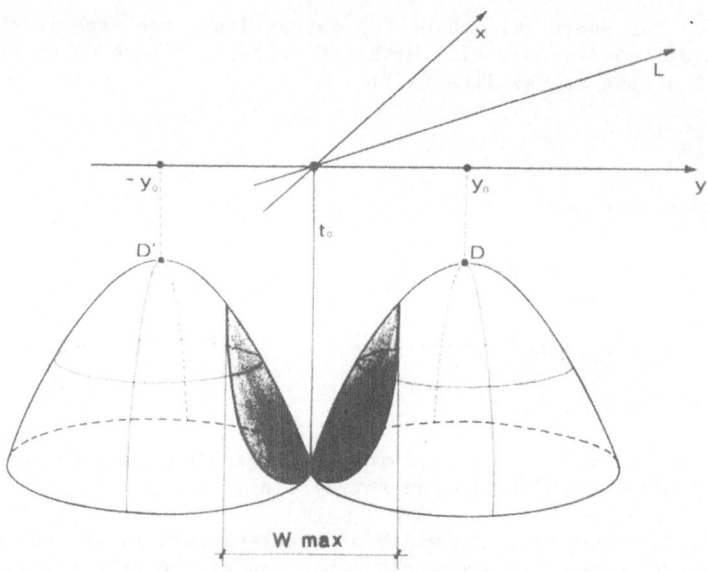

Fig. 46. 3-D DMO

3.2 Migration

The 3-D migration can be performed either in single step (one step migration) or by splitting it into two successive 2-D migrations (two pass method) carried out along two normal directions.

Despite the wide use of the two pass approach, being it simpler and cheaper, it is only approximated.

Let us consider the response of a point scatterer in a constant velocity medium. In 3-D it will be a circular hyperboloid, the vertical sections of which are hyperbolae of "curvature" decreasing with the distance from the axis (Figure 47).

The diffraction hyperbolae satisfy to:

$$\frac{V\, t_o}{2} = \sqrt{S^2 + Z^2} \qquad (53)$$

where S represents the offset. The derivative is:

$$\frac{ds}{dt} = \frac{V^2\, t_o}{4S} \; . \qquad (54)$$

Since:

$$S^2 = x^2 + y^2$$

we get:

$$\frac{ds}{dt} = \frac{V^2\, t_o}{\sqrt{x^2 + y^2}} < \frac{V^2\, t_o}{4x} \; . \qquad (55)$$

This means that the hyperbolae obtained by vertically slicing the hyperboloid look generated at a lower velocity. In fact, if we denote with V' such a velocity, we can write:

$$\frac{V_o^2\, t_o}{\sqrt{x^2 + y^2}} = \frac{V'^2 t_o}{4x} \qquad (56)$$

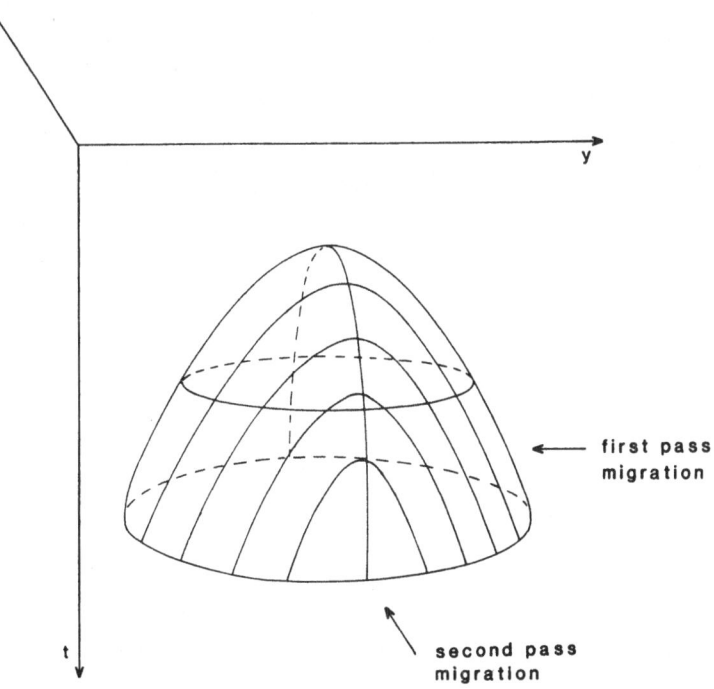

Fig. 47.   3-D Diffraction Hyperboloid.

that can be modified in:

$$\frac{V'}{Vo} = (\frac{x^2}{x^2 + y^2})^{0.25}$$ (57)

that gives the rate between the apparent velocity V' and the true velocity Vo. Figure 48 shows the reduction, in percentage, of the apparent velocity versus the true one.

As a consequence, if we migrate the hyperboloid through a two step 3-D migration by applying the constant velocity of the medium we will produce an overmigrated result.

To avoid the problem, it is necessary to artificially reduce the velocity field to be used for migration. In order to estimate the most suitable reduction, it is better to build up a synthetic model and to migrate it, looking at the artifact produced.

3.3 Interpolation

The interpolation, as seen, is another key process in 3-D environment. In practice there are two methods followed to interpolate data after stack:

- cross-correlation,
- sinc function application.

The first method utilizes the cross-correlation between adjacent traces to identify the most important dip, along which the interpolation is done. The interpolation itself can be linear or more complex, but the method does not change.

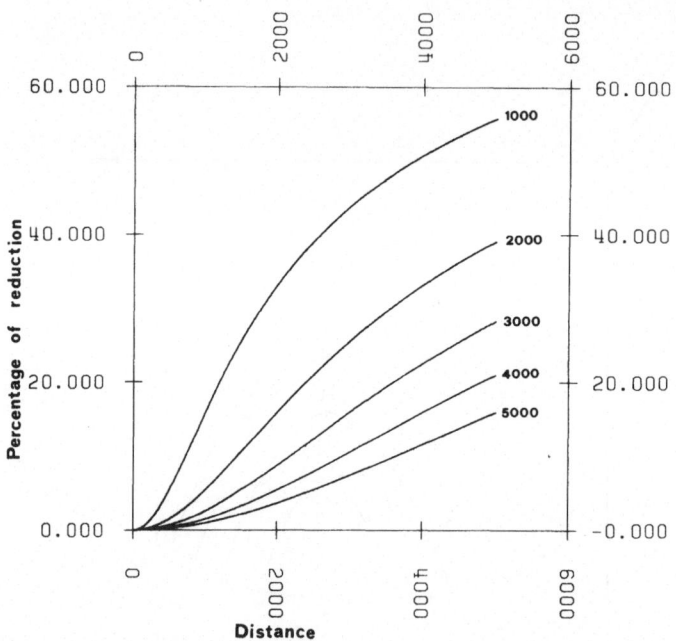

Fig. 48. Velocity Reduction for 2 Step 3-D Migration.

The correlation analysis is carried out looking at the traces falling inside a predetermined fan, and repeating the computations for increasing dips up to the upper limit.

The second is much more sophisticated, and looks the best actually on the market. It operates in two steps. The first step is tracking, that is the identification of the most important dips exactly like the method formerly described. Generally more than one dip is detected and followed along the seismic section.

Once the dips are tracked along the sections, on small square blocks base, the second step starts up. It consists of the application of the sinc function to produce an interpolation.

The sinc function is defined by:

$$\text{sinc } x = \frac{\sin x}{x} .$$ (58)

If T represents the sampling interval, the reconstruction of a signal, that is the extraction of the values in between two samples, is obtained simply by applying (Shannon, 1984):

$$s(t) = \sum_{n=-m}^{m} s(nT)\text{sinc}(t - nT).$$ (59)

In fact, s(t), the continuous signal, is obtained for any value of the continuous variable t simply by summing over a finite number of samples (function of the precision we want), the corresponding values of the sinc function multiplied by the amplitude of the samples.

The interpolation is carried out along "flattened" horizons. After first step completion, the extracted values of dip are averaged along the section in order to provide some consistency with the geological features. Then, for each identified horizon, that is a consistent feature, the

316

section is flattened, and the sinc interpolation applied to that particular horizon.

Shannon's statement is in fact valid only within the band of the signal, that, since we are working in a two-dimensional space, means below the spatial aliasing.

Through the flattening operation the upper bandwidth limit is enlarged to infinity, thus allowing an aliasing free interpolation despite the true dip of the horizon we are dealing with.

After completion of the interpolation for the first horizon, the operation continues on the second, and so on.

Figure 49 represents a seismic stacked line recorded and processed with 6.25 m between the traces.

Figure 50 represents the same section after a trace decimation leaving alive one trace in every five (trace spacing is now 25 m). The decimated line was then interpolated back to 12.5 (Figure 51) and 6.25

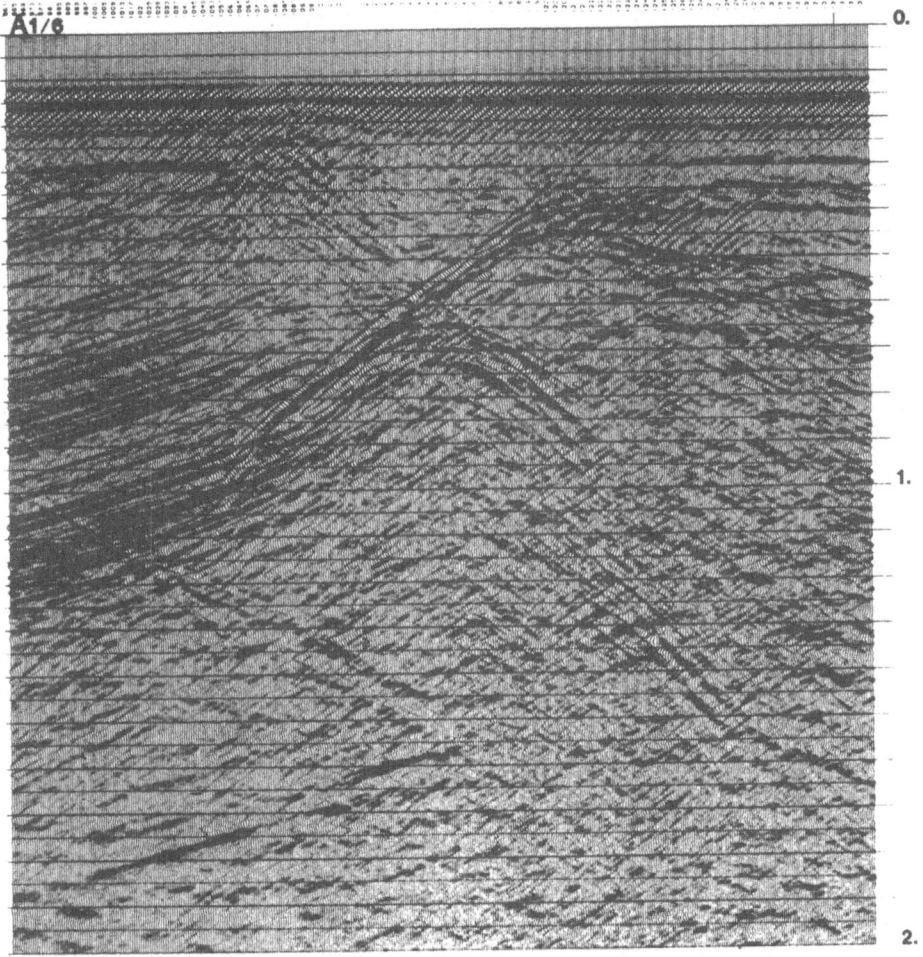

Fig. 49. Seismic Section. Section collected and processed at 6.25 m of trace spacing and to use as reference for interpolation.

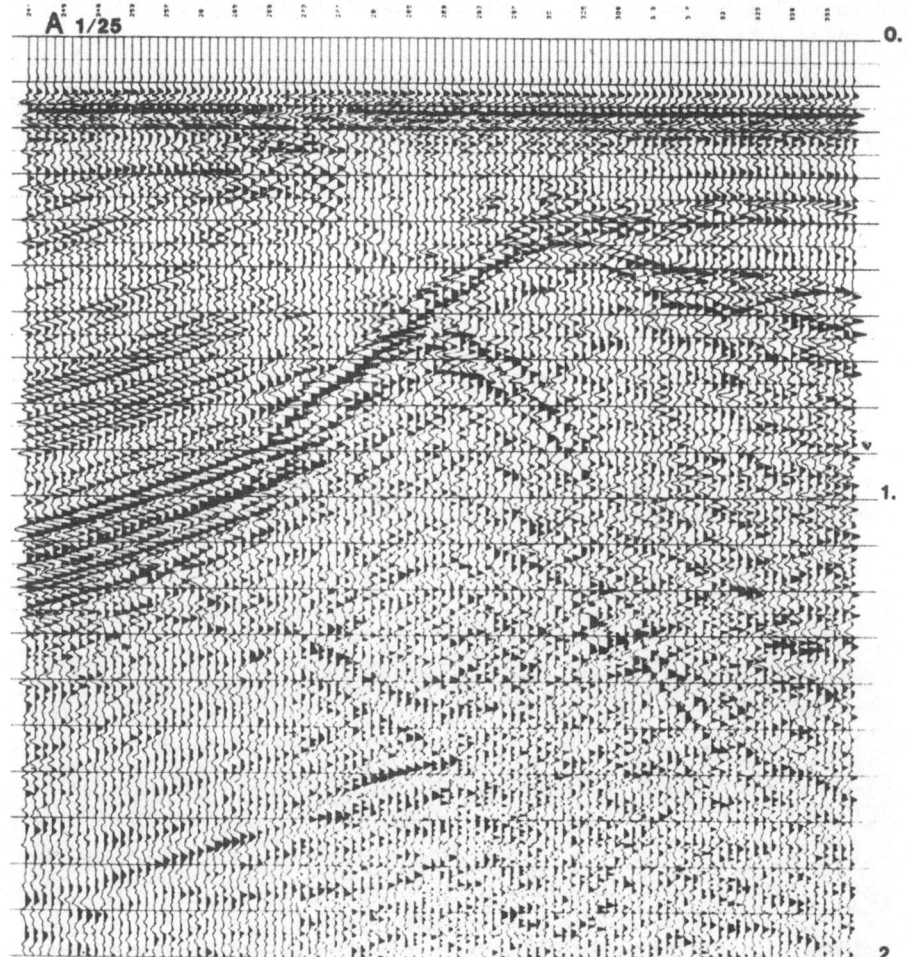

Fig. 50. Decimated Seismic Section. The section of Figure 45 decimated
to produce a section with 25 m of trace interval to be used as
input to the interpolation test.

(Figure 52) meters of trace distance. The method used was the sinc
interpolation.

The section input to interpolation process is clearly aliased in many
places, but the interpolation had the ability to reconstruct in a very
satisfactory way also minor events. Also in presence of conflicting dips
the result is quite satisfactory.

The interpolation programs should provide output sections in which
the amplitudes and the frequency spectra of the traces presented are well
balanced. Otherwise, the following migration will create a lot of
artifact due to non perfect destructive interference among the smiles
generated for each sample.

Generally such an operation is automatically carried out by the
computer programs through an estimation of the average amplitude and
frequency distribution before the interpolation and its subsequent use for
data conditioning.

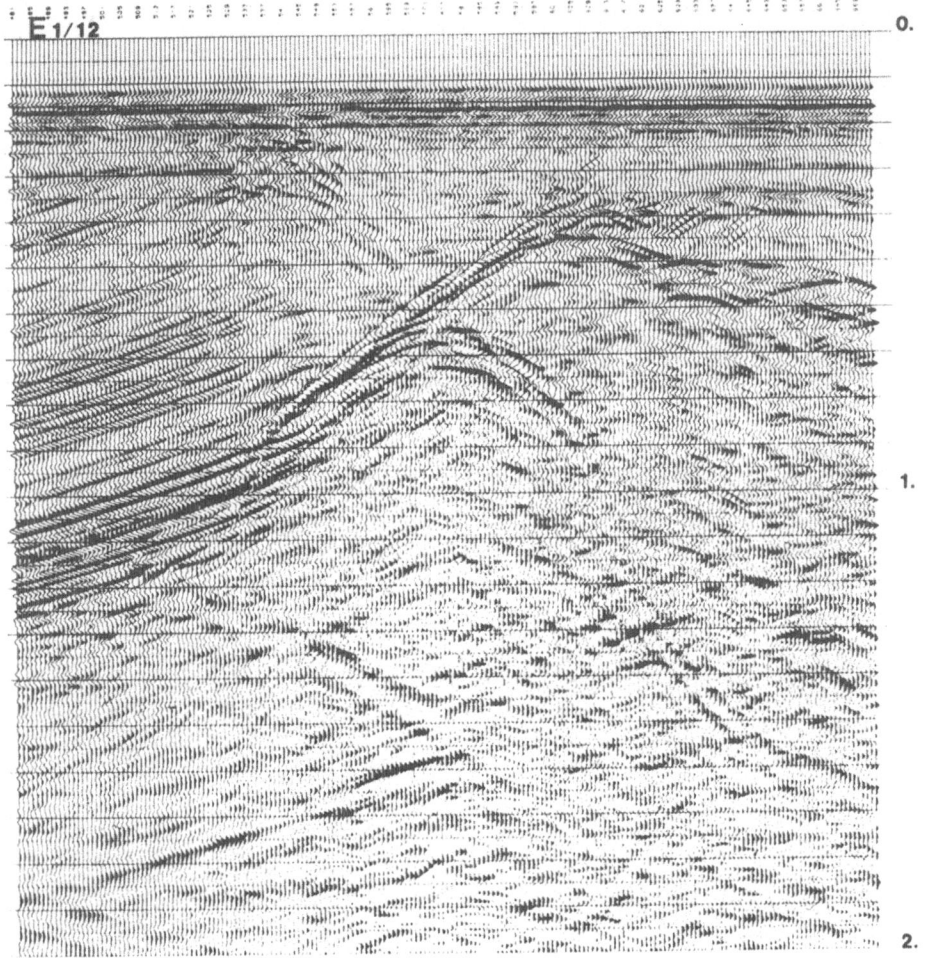

Fig. 51. Interpolation of the Decimated Seismic Section. Result of interpolation back to 12.5 m of trace spacing.

EXAMPLES

In the following two examples are deeply investigated. The first is a full 3-D land survey, while the second is a light marine one.

5.1 Land Example

The survey was shot during the spring of 1984 over the Gaggiano field, located south-west of Milano, in the Po Valley. The target of the survey was the limestones buried at about 5200 meters.

The major dips, about 30 degrees, are recognizable in the E-W direction, along which the envelope of the target structure exhibits an anticline followed by a syncline. The structural dips in the N-S direction are less than 10 degrees. The area of interest was drawn on the existing maps, and it resulted in a rectangle of 10 x 4.5 km, oriented with the longer side parallel to the N-S direction. In order to provide data free from edge effects, the area to be surveyed was enlarged up to 10.5 x 6 km. The in-line axis was chosen parallel to E-W direction, along the dip direction.

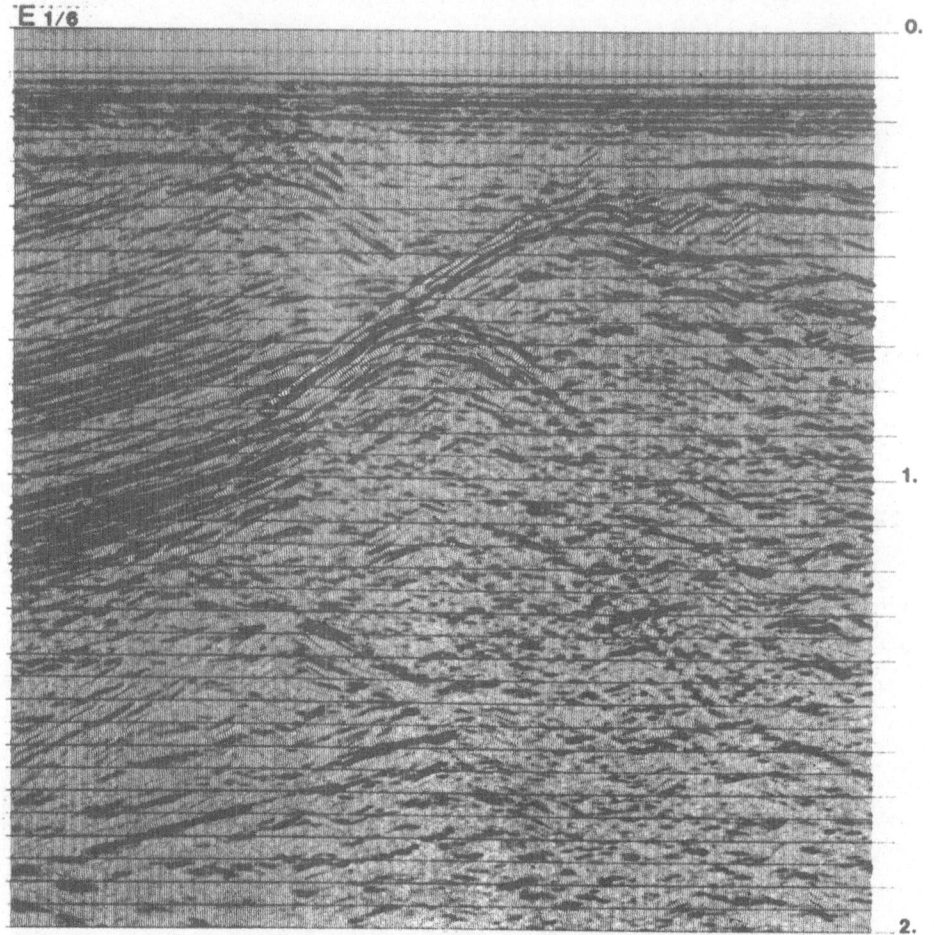

0.

1.

2.

Fig. 52. Interpolation of the Decimated Seismic Section. Result of
interpolation back to 6.25 m of trace spacing.

The acquisition of dips up to 30 degrees in the bandwidth of 40 Hz at
3 sec of TWT is the geophysical problem to solve.

Simple computations show that at target level the Fresnel radius is
greater than 150 meters, thus the choice of the group interval depends
upon the spatial aliasing only. The respect of dip conditions provides 50
meters for the surface in-line group interval and 72 meters for subsurface
line spacing in the cross-line direction.

Considering the uncertainty of the maps used for dip computations the
line spacing was reduced to 60 meters (subsurface distance). Applying the
formulae described at point b, the maximum in-line offset results at 2050
meters and the minimum maximum offset 1470 meters for a multiple bandwidth
of 20 Hz.

Since statics are not problematic, a 3 fold coverage in cross-line
direction was chosen. In-line folding is 6, yielding a total coverage of
18 fold.

The basic block design shows 4 parallel geophone-lines, 360 meters
apart, of 48 groups each. The group interval is 50 m. Each swath

320

consists of 15 blocks, each having its own shooting configuration, that is a relative position of the shot-line with respect to the spreads, since a roll-along move up technique was applied in the in-line direction.

The total number of swaths required to complete the survey is 9, because the cross-line move up is accomplished by shearing one geophone-line between two adjoining swaths. The total number of geophone-lines to deploy in the field is 28, and the number of shot-lines 15.

Figure 53 shows the theoretical survey layout and the shot configurations for each of the 9 swaths. The number and the relative position of the shots along each shot-line in the cross-line direction

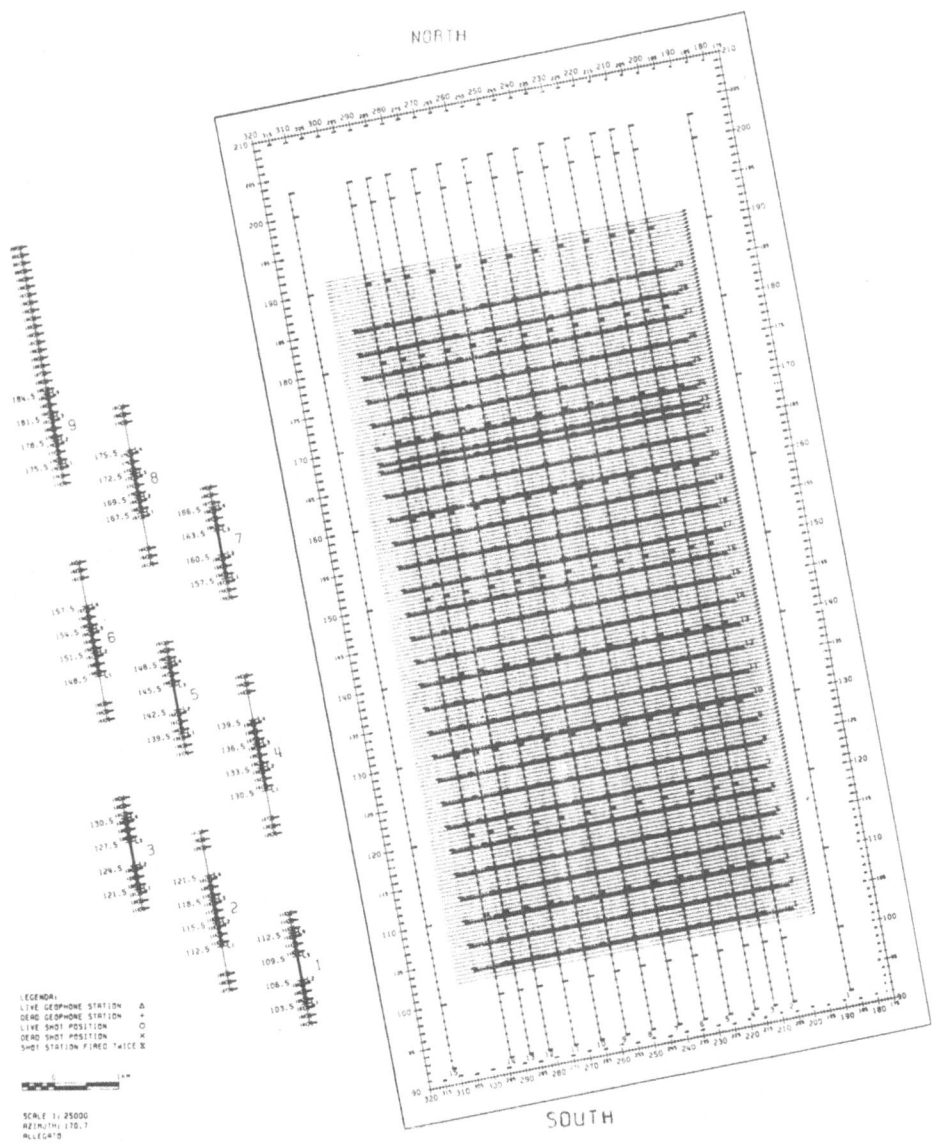

Fig. 53. Land 3-D Example. 3-D theoretical layout and swath configurations. Theoretical 3D Survey Gaggiano

changes with the swath number. For even-numbered swaths there are 15
shots, while for the odd numbered there are only 12 shots. In Figure 53
the small x represents the covered subsurface bins, that are organized in
168 lines, each being 5750 m long. Figure 54 shows the recording
configuration of each of 15 shot-lines, in surface and in the subsurface.

The condition of minimum maximum offset greater than 1470 meters is
satisfied by all of the 105 different recording configurations, so the
multiple rejection has to be considered well suited.

The condition of maximum in-line offset less than 2050 meters is
satisfied for the 5 central shot-lines only (from 6 to 10), whilst the
others exhibit a maximum offset of 2800 m. It is also immediately
recognizable that the shot-lines from 3 to 13 can be recorded using the
same 64 geophone stations.

By using a zig-zag move up technique when passing from one swath to
the next the whole Gaggiano survey was completely recorded in 45 working

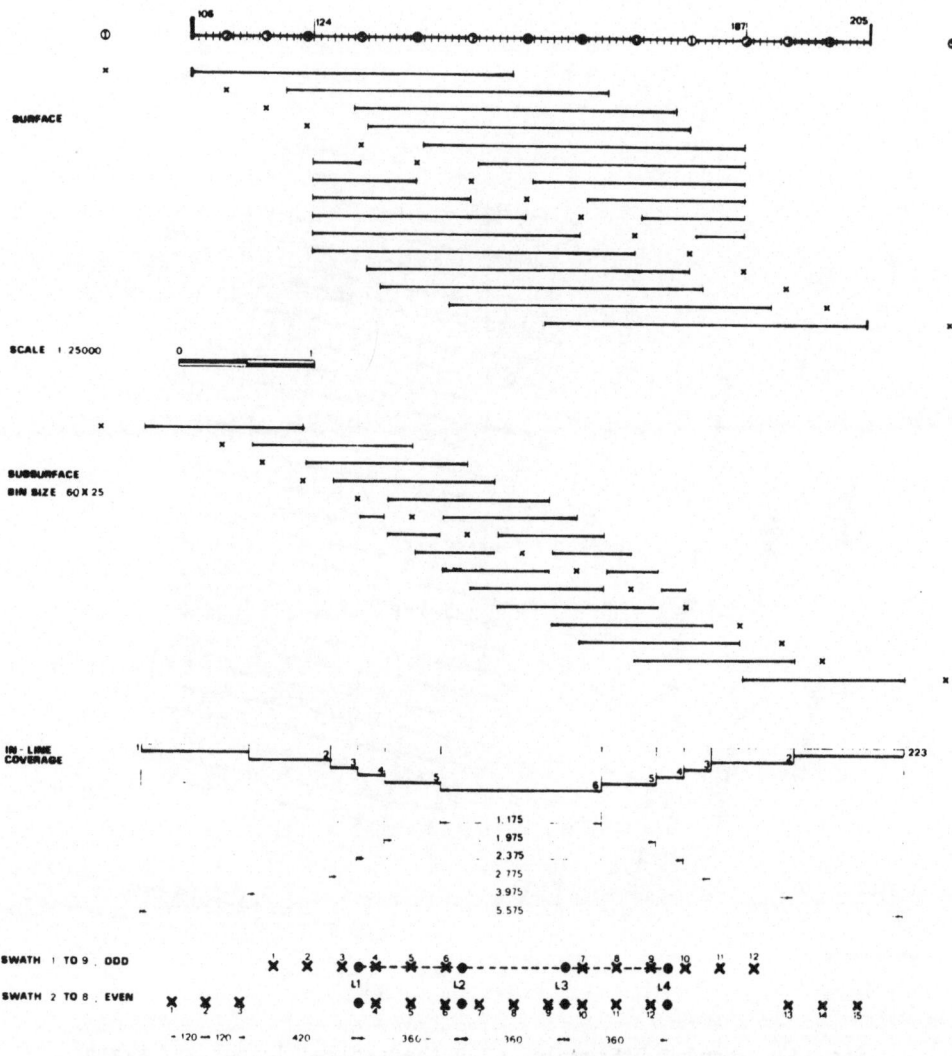

Fig. 54. Land 3-D Example. Shot-line recording configurations.

322

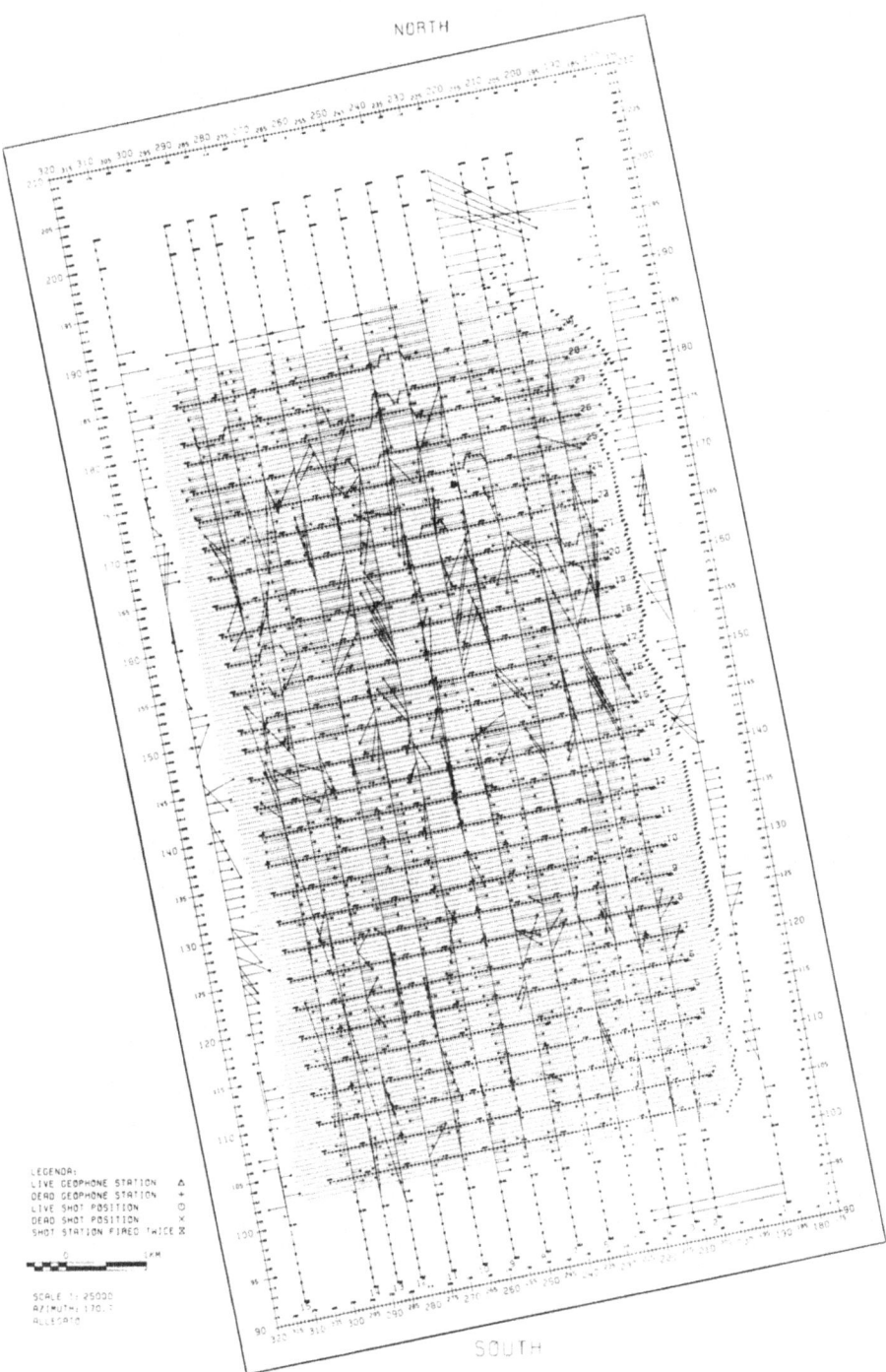

Fig. 55. Land 3-D Example. 3-D actual layout.

days, with the use of only 260 geophone groups. The geophone pattern adopted was the classic "3 arm windmill" of 36 geophones, that is almost omni-directional whilst providing enough rejection (14 db) of noise wavelengths ranging between 3 and 48 meters.

Figure 55 shows the final location of all shot points and geophone stations. The small circle indicates the true shot position, that appears joined with a continuous line to the theoretical position, and the triangles indicate the true receiver positions. The small x indicate again the covered bins.

5.1.1 _Analysis of the results._ Figure 56a shows some in-line vertical sections after the first migration step, carried out along the in-line direction. The diffraction tails look well collapsed and the general appearance of the sections is acceptable.

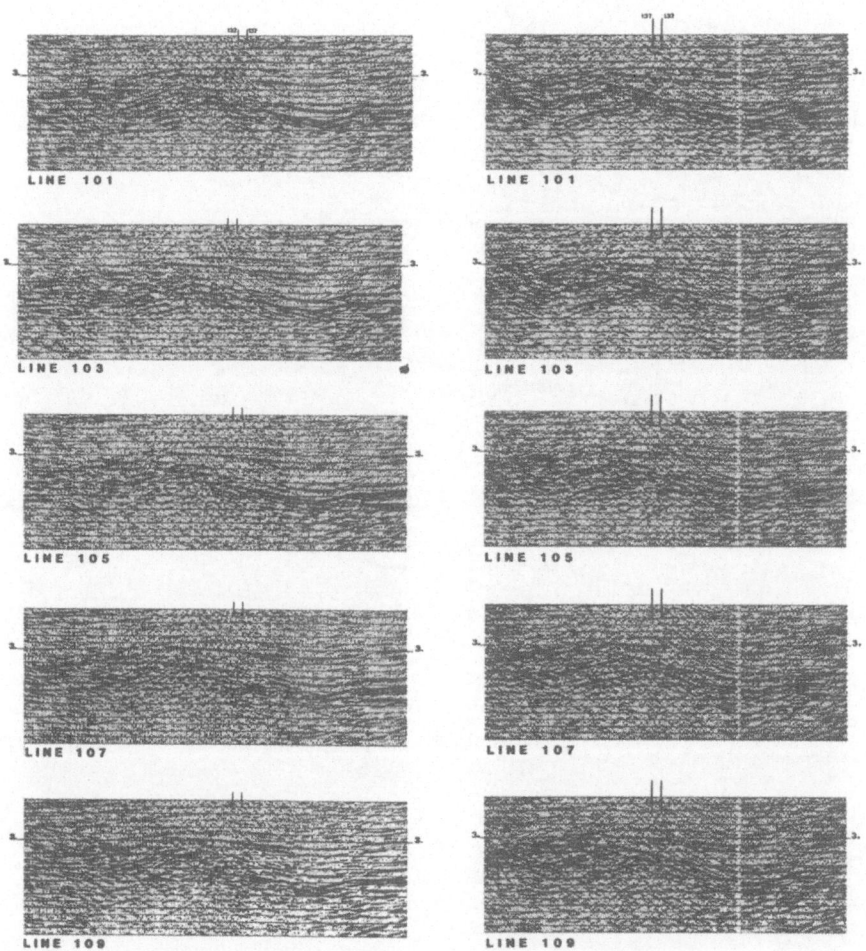

IN-LINE VERTICAL SECTIONS
AFTER IN-LINE-MIGRATION

IN-LINE VERTICAL SECTIONS
FULL 3D MIGRATION

Fig. 56.   Land 3-D Example.
a) Sequence of vertical in-line sections, 60 meters apart, after in-line migration only.
b) Sequence of vertical in-line sections, 60 meters apart, all full 3-D migration.

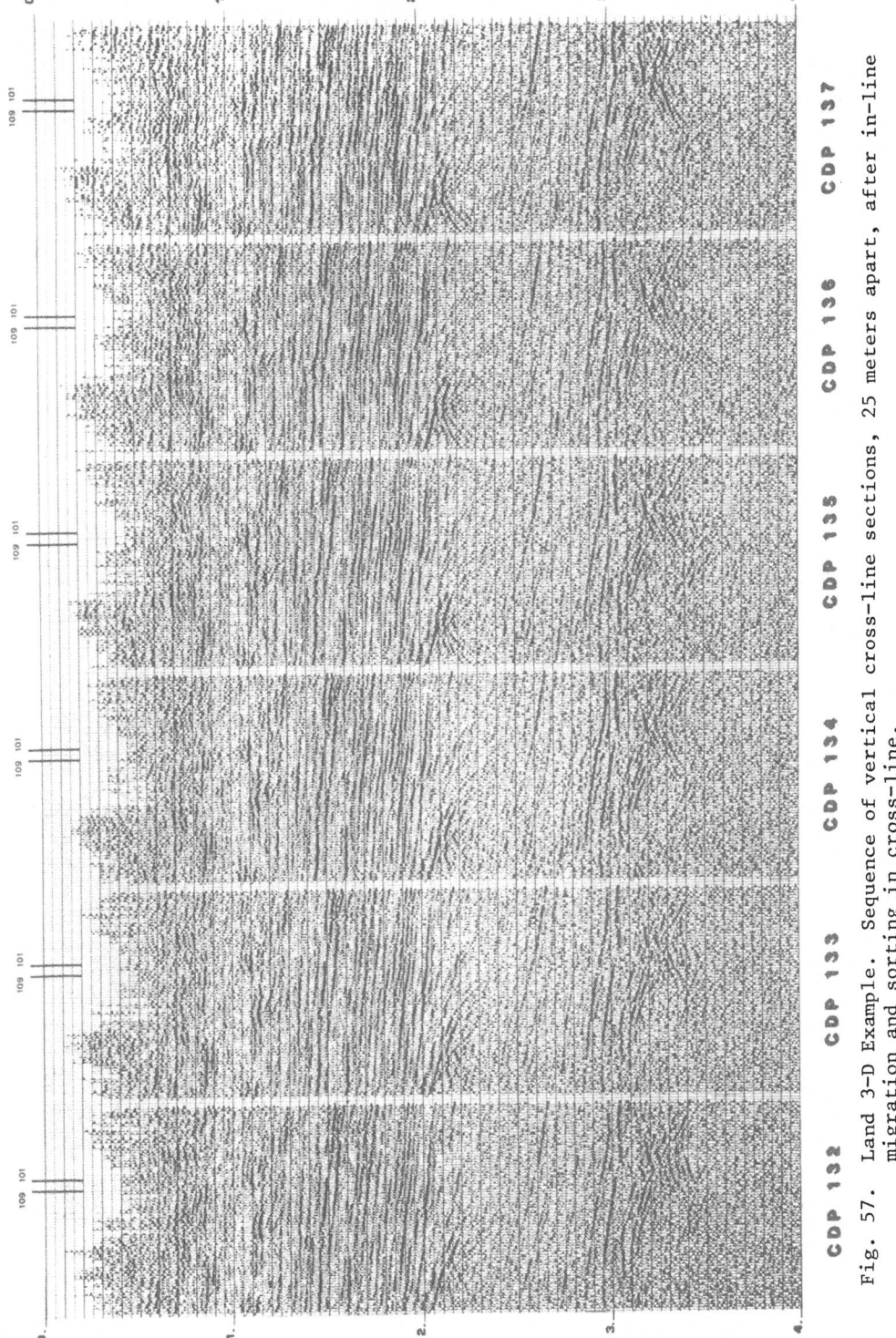

CDP 132   CDP 133   CDP 134   CDP 135   CDP 136   CDP 137

Fig. 57. Land 3-D Example. Sequence of vertical cross-line sections, 25 meters apart, after in-line migration and sorting in cross-line.

Figure 57 represents the same data rearranged in the cross-line direction. The picture clearly explains the need to operate in 3-D. In 2-D operations the processing will stop at this level, and anyone can see the amount of unresolved lateral events that still affects the data. The second migration step is able to clean up these events, and after it has been done we finally get the true layer shape all around the surveyed area. Figure 56b shows the same lines of Figure 56a, but after the completion of the full 3-D migration.

Figure 58 shows a comparison between a 2-D section, shot six months before the 3-D, and vertical sections extracted from the final 3-D volume strictly following the subsurface pattern of the 2-D. In theory the two lines should be equivalent, but the difference can be easily noted. The 2-D line was shot using the same source and recorded using the same recorder as the 3-D, but its coverage was 48 fold, five times higher than the average folding of the 3-D survey.

The use of the multi-azimuthal recording increases the S/N noise ratio by two to three times. Moreover the 3-D migration increases the S/N too since a much larger number of traces is used to collapse the diffracted energy.

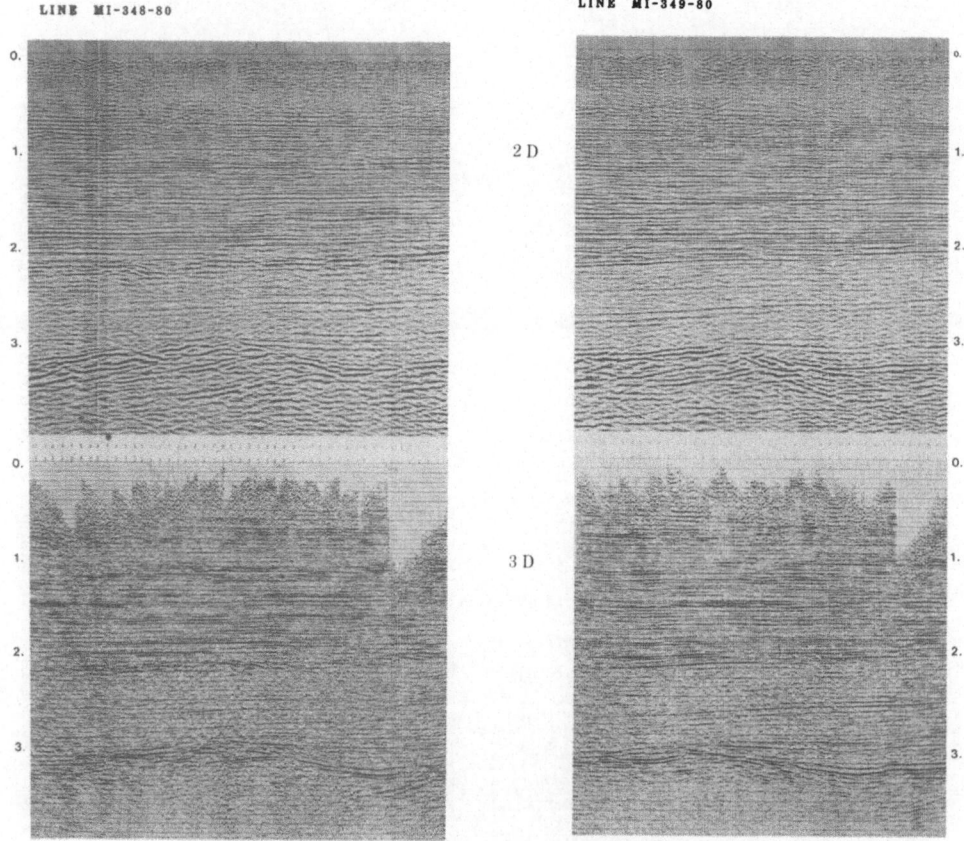

LINE MI-348-80   LINE MI-349-80

2 D

3 D

Fig. 58. Land 3-D Example. Comparison between a 2-D line and the equivalent sorted from 3-D volume after full 3-D migration.

Comparing the two pictures of Figure 58 we can see that the 3-D is simpler to interpret from the structural point of view, since the spurious event has been removed from the section.

Moreover, the result obtained exhibits a larger bandwidth than the 2-D, 60 Hz rather than 45 Hz, and this provides the possibility to analyze the data also from the stratigraphic point of view. It is clear that the vertical resolution at the target level is some 70-80 meters, so that the stratigraphic analyses can be restricted to the macro-zonation of the producing layer, but this represents a big step ahead in petroleum exploration.

The comparison of the seismic results with the high resolution data coming from the wells provides a means to transfer from the wells to the seismic the information about the limit of the most productive layers.

Since the spatial density of the seismic is very high it is possible to spread out such information to all the survey.

In other words, a clear view of the reservoir geometry is achieved, and the possibility of checking the validity of the reservoir model extracted from the well logs and core analyses is also provided.

Figure 59 shows a close up of the comparison between 2-D and 3-D. The upper picture represents the 2-D, the central the traces 3-D processed and extracted following the same trajectory of 2-D, while the lowermost represents again the 3-D output, displayed in instantaneous phase. This presentation provides a means to appreciate the lateral continuity of the signals 3-D processed.

5.2 Marine Example

This offshore survey was recorded in 1985. The target was the sands buried at 2.0-2.2 seconds. The target originating at salt and anhydrites level masks the target and it was impossible to remove it simply by applying a 2-D processing.

The area of the survey was some 800 square kms, and had to be covered by two surveys of the same size but different orientation.

Since the structural dips at target level are moderate it was decided to carry out a "light 3-D" using the twin source-single streamer technique. The field layout is the one depicted in Figure 2, with 150 meters between the boat courses and 50 meters of source separation.

The sources, two air gun arrays, and the streamer, 240 channels with group spacing of 12.5 m, was towed at a depth of 6-7 meters in order to provide broad-band recording.

The processing sampling interval was 2 ms up to stack level, in order to preserve as many high frequencies as possible. The streamer feathering was less than 5 degrees for all the survey, and this made it possible to process the data through the flexible binning technique without affecting the cross-line continuity.

The in-line migration was carried out on the collected lines and data was then sorted into the cross-line. Figure 60a shows three cross-lines as they appear after sorting. The picture was produced keeping the distance constant between the traces, despite the uneven spacing due to

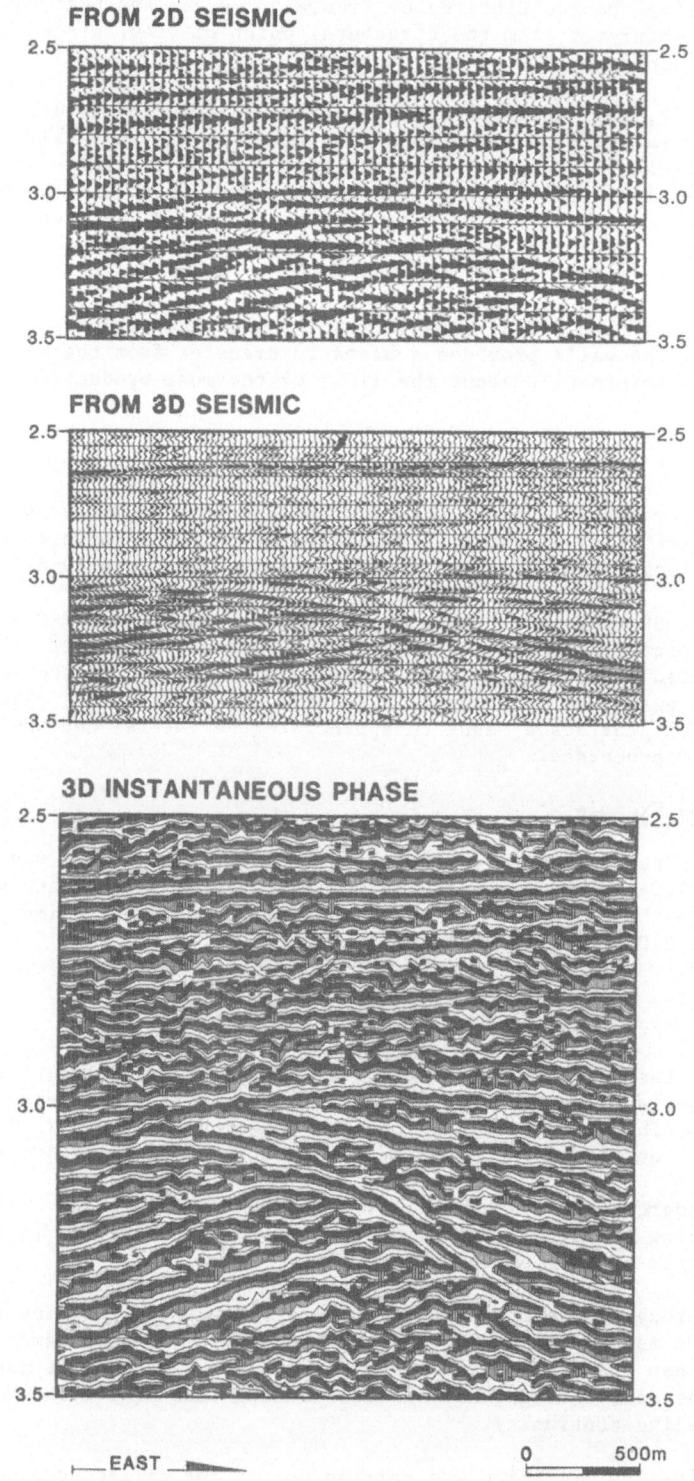

FROM 2D SEISMIC

FROM 3D SEISMIC

3D INSTANTANEOUS PHASE

EAST

0   500m

Fig. 59. Land 3-D Example. Comparison between 2D and 3D processing.

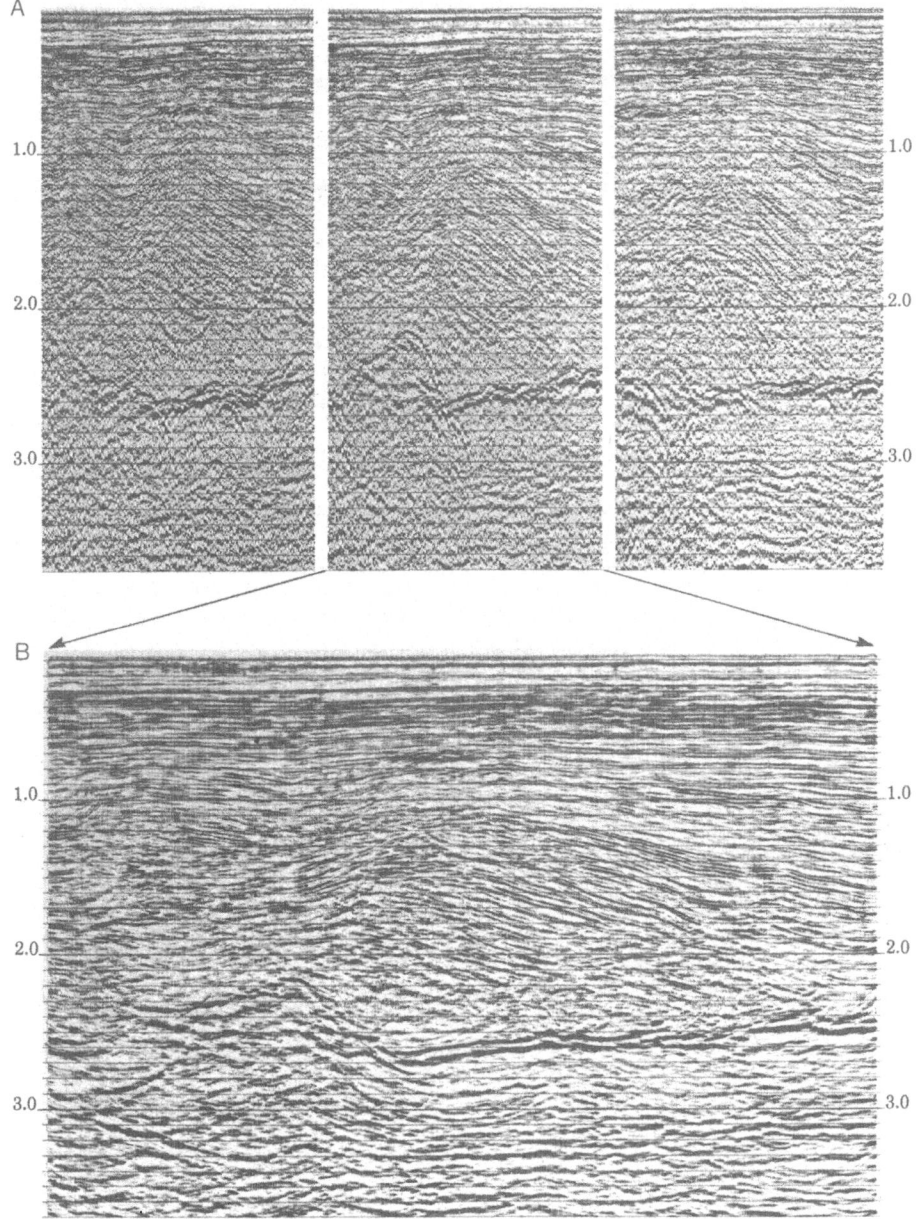

Fig. 60.   Marine 3-D Example.
           a) Cross-line vertical section after sorting.
           b) The same after interpolation and migration.

acquisition technique. Figure 60b represents the upper central line after interpolation to 25 m and migration.

The effect of the applied processes in terms of diffraction collapse is clearly recognizable. In order to get results like that, some rules should be followed during the processing. It is well-known that the migration acts like a high pass filter.

Referring to Figure 36, the zone of the FK spectrum between the wave number axis and the propagation velocity line pertains to the "evanescence zone", so that the energy laying here will be simply filtered out by migration process. The evanescence zone increases as the velocity increases. This means that if the propagation velocity is high most of the lower temporal frequencies associated to steeply dipping events will disappear after migration. In our survey the velocity is 2000 m/s and the temporal frequency limit corresponding to a distance of 25 m is about 40 Hz.

In the diffraction hyperbola the frequency decreases with the angle so that very low frequencies are associated to the tails. Therefore care should be taken in conditioning the data after interpolation and before migration in order to prevent artifact and noise generation by migration, caused by lack of information across the tails of the diffraction.

A spatial filter (reducing lateral local discontinuities) and a temporal filter (enhancing the lower frequencies against the higher) should be applied before migration. Moreover, after migration, a spectral whitening, with the aim of balancing the amplitude spectrum of the seismic traces, is quite often required. Such a sequence provides good results, due to the fact that the character of the signals is preserved as well as the higher frequencies for superior resolution. Since the 3-D migration can only be applied using the two pass approach, the results are only approximated.

In fact if we look at the hyperbola, obtained by vertically slicing the circular hyperboloid that represents the response of a point scatterer in 3-D environment, we can observe that they appear as generated in decreasing velocity media as their distance from the scatterer increases. In other words, if we migrate the hyperboloid using the true velocity by applying a cascade of two 2-D migrations in two orthogonal directions we will drastically overmigrate the data. Since the first step is the most critical, special care should be applied when choosing the reduction coefficient of the supposed true velocity field for the in-line migration step. In our example, since the vertical gradient of the velocity field was very low, and therefore the sensitivity of the migration to the velocity variations very high, 5% of reduction for the in-line was applied, whilst in the cross-line migration only 3% was applied.

ACKNOWLEDGEMENTS

The author wishes to thank the Management of AGIP SpA., Direzione Generale dei Servizi Centrali dell'Esplorazione DES for kind permission to publish this work.

REFERENCES

Bading, R., 1980, How to optimize land survey - some rules for areal data gathering, Festschrift Theodor Krey, Prakla Seismos Gmbh, pp 8-33, Hannover.

Bolondi, G., Rocca, F., and Loinger, E., 1982, Offset continuation of seismic sections, Geophysical Prospecting, 30:813-828.

Bolondi, G., Rocca, F., and Loinger, E., 1984, Offset continuation in theory and practice, Geophysical Prospecting, 32:1045-1073.

Marshall, R., 1984, 3-D Data Acquisition: Which Effort is Needed, Paper presented at Dallas SEG, 9th April 1984.

Rocca, F., and Salvador, L., 1985, Migration in Seismic Processing - Course of Continuous Education for Explorationists Program, held at the 47th EAEG Meeting in Budapest, June 1985.

Rocca, F., and Ronen, S., 1984, Improving Resolution by Dip Move Out, 54th SEG Meeting Proceedings, Atlanta, pp 611-614.

Salvador, L., and Savelli, S., 1982, Offset continuation for seismic stacking, Geophysical Prospecting, 30:820-849.

Salvador, L., December 1984, Integrated approach of design, acquisition and processing for optimum land 3-D seismics, Boll. Geof. Teor. Appl., vol. XXVI, n. 104:178-209.

Salvador, L., 1984, Seismic migration: a review of the practical problems and up to date solutions, Boll. Geof. Teor. Appl., The OGS Silver Anniversary Issue, pp 173-188.

Salvador, L., and Matteini, L., 1986, 3-D Seismic to Solve Deep and Subtle Traps, Proceedings of EGPC Eight Exploration Conference, November 17-23 1986.

Tatham, R. H., Keeney, J., and Massel, W. F., 1981, Spatial Sampling and Realizable 3-D Surveys, 51st SEG Meeting.

REFERENCE MODEL FOR TOMOGRAPHIC IMAGING OF THE UPPER

MANTLE SHEAR VELOCITY STRUCTURE BENEATH EUROPE

A. Zielhuis, W. Spakman and G. Nolet

Institute of Earth Sciences, Department
of Theoretical Geophysics, PO Box 80.021
3508 TA Utrecht, The Netherlands

INTRODUCTION

For several years geophysicists have applied tomographic techniques
to image velocity structures on a global or regional scale or very
detailed structures in exploration seismics. In studies of large areas
they often make use of the delay time data provided by the International
Seismological Centre. Many results have been obtained using P delays.
For instance, Spakman (1986) has used 500,000 ISC P delays to derive a
tomographic model of the Upper Mantle structures beneath Central Europe,
the Mediterranean and the Middle East where the African and Eurasian
plates converge. Less effort has been paid to tomographic imaging of
shear velocity structures using ISC S delays. A tomographic S velocity
model in addition to P model could be of great importance for the
interpretation of P anomalies in terms of mineral or temperature causes.
With the use of S delays instead of P delays some extra problems are to be
expected. Fewer S data are available from the ISC and S delays are
subject to larger errors due to identification problems and inaccuracies
in the readings of the exact onsets. The choice of the starting or
reference model to be used for tomographic inversions should be considered
carefully. An important approximation in delay time tomography is that
the delay time of a certain arrival, i.e. the difference between observed
travel time and travel time according to a theoretical reference model, is
due to velocity anomalies located on the ray path in the reference model.
This assumption is based on Fermat's principle that states that a wave
follows a ray path that is stationary with respect to travel time, so that
a small perturbation of the ray path has only a second order effect on the
travel time. Systematic differences in the velocity structures of the
true earth and the reference model may cause large differences in ray
geometry. In those cases the linear approximation is not justified.
Moreover, even small differences in ray geometry may have unwanted effects
on the solution if the inverse problem is ill-posed (Nolet, 1987).

ISC delays are delay times relative to Jeffreys-Bullen travel times.
Therefore, the use of uncorrected ISC delays for tomography implies that
the Jeffreys-Bullen model is the reference model. The Jeffreys-Bullen
model is a very smooth model without a Low Velocity Zone (LVZ) or
discontinuities except the MOHO discontinuity. Currently several regional
models are available that give a better description of the local velocity
structure of the Upper Mantle. Most of these models show a LVZ between

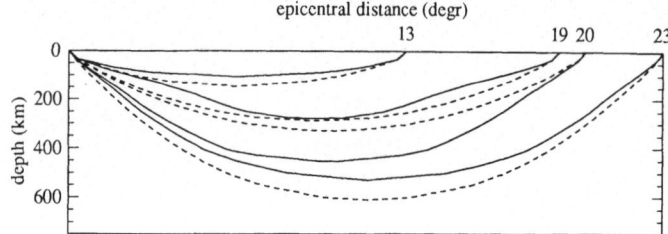

Fig. 1. Comparison of ray paths in Jeffreys-Bullen S velocity model
(dashed lines) and a more realistic Upper Mantle model (solid
lines) with a LVZ and a "400 km" discontinuity.

100 and 200 km and a "400 km" and a "670 km" discontinuity. Since these
features are absent in the Jeffreys-Bullen model it is doubtful if this
model is adequate as a reference model for tomography. Figure 1
illustrates the problem. Four ray paths are calculated from the Jeffrey-
Bullen model and from a more realistic model with a LVZ and a 400 km
discontinuity. Figure 1 shows that at several locations the distance
between rays in the two models arriving at $13°$ and $19°$ is about 50 km.
For rays arriving at $20°$ and $23°$ the difference in depth of the turning
point is 150 km and 90 km, respectively. These differences in ray
geometry are too large for correct tomographic imaging if cells with
dimensions in the order of several hundreds of kilometers or less are
used. Moreover, the attainable accuracy of tomographic results depends
strongly upon the employed ray geometry (Spakman and Nolet, 1987).

This study focuses on the question whether the Jeffreys-Bullen shear
velocity model is appropriate as a reference model for tomographic imaging
of the Lithosphere and Upper Mantle beneath Europe. In this report
"Europe" refers to a large area extending from 10W to 50E and 30N to 80N
(see Figure 2). First, ISC S delays are analyzed to see if observed
travel times differ significantly from Jeffreys-Bullen (J-B) travel times.

ANALYSIS OF ISC S DELAYS

S delays were read from ISC magnetic tapes containing all recordings
of events that occurred during the period 1972-1982. Only arrivals were
selected of which the turning point of the ray path is under the area
described above. In this way the entire or at least the largest part of
the ray path along which the time delay was acquired is located beneath
Europe.

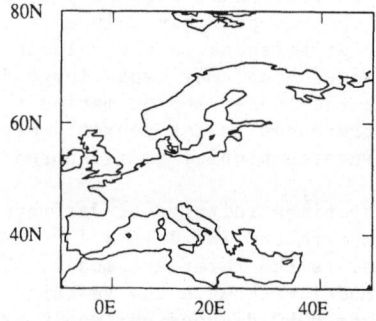

Fig. 2. "Europe" as described in the text, extending from 10W to 50E,
30N to 50N.

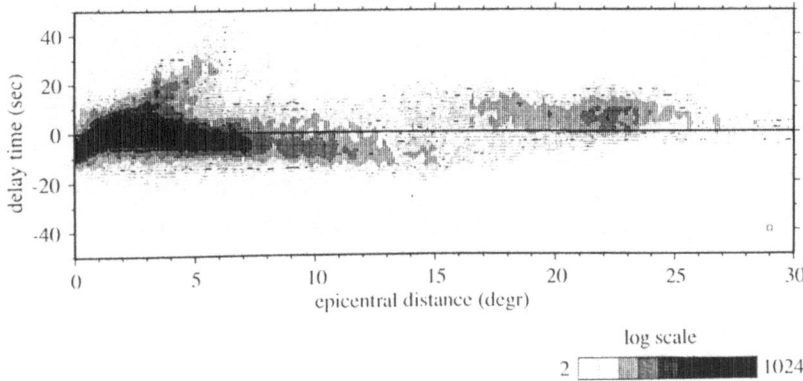

Fig. 3. Density plot of ISC S delays versus epicentral distance. The
number of delays per square area of 0.2° x 2 sec (see box in
lower right corner) is contoured with the intervals shown in the
legend.

Figure 3 shows the relation between the 83,000 selected delay times
and epicentral distance. If the average velocity structure beneath Europe
would be similar to the J-B model, delays would be distributed around the
line of zero delay. This is not the case. From 5° to 27° delays deviate
systematically from 0. Moreover, different branches can be distinguished
in the diagram. There is a clear branch between 0° and 15° with delays
distributed around 0 seconds at short distances, turning negative with
increasing distance to an average of approximately -10 seconds at 15°.
From 15° to 30° delays are mainly positive. Histograms of delay times at
three different intervals of epicentral distance, shown in Figure 4,
indicate the gradual transition from negative to positive delays. Between
8° and 10° the histogram shows a peak around -6 seconds, between 15° and
17° the negative and positive branch are both important, between 19° and
21° the maximum is at 7 seconds and the negative branch has disappeared.
Negative delay times of waves arriving at short distances (< 13°) indicate
that they travelled through layers with velocities higher than velocities
of the J-B model. Positive delays at greater distances indicate that the
"time gain" was lost at greater depth. This can be explained by the
presence of a low velocity layer. There is, however, no clear gap in the
registrations. This is surprising since ray theory predicts a shadow zone
in the travel time curve if a low velocity layer is present in the Earth.
The absence of a shadow zone and the existence of both a positive and a
negative branch between 12° and 18° is presumably due to lateral
heterogeneity. The variation of delay times with epicentral distance was
also investigated for smaller areas within Europe. It turned out that the
distance at which the average value of the delays changes from negative to
positive varies significantly, ranging from 11° in Greece to 17° in
Scandinavia. Therefore, in the diagram containing the total data set
positive and negative branches associated with different areas overlap.
We note that the decrease in the number of reported arrivals between 12°
and 17° cannot be explained by an irregular distribution of station
locations with respect to event locations. This distribution is rather
uniform between 10° and 20°.

A third branch is present between 0° and 7° with values of delays
ranging from -10 to 30 seconds. A large number of arrivals belonging to
this branch are S_g waves but spurious arrivals like S_MS may have been
misinterpreted as first arrivals and thus appear in this branch.

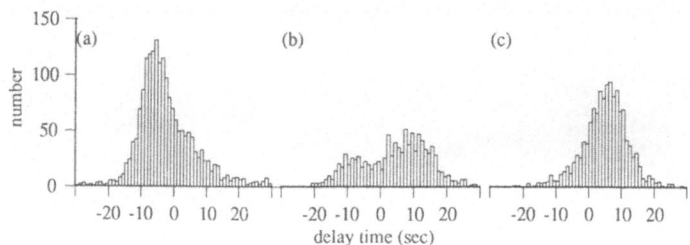

Fig. 4. Histograms of delay times at three different intervals of epicentral distance: 8°-10° (a), 15°-17° (b) and 19°-21° (c).

The delay density plot indicates there is a significant systematic deviation of observed travel times from J-B travel times. This means that the J-B model differs significantly from the average shear velocity structure beneath Europe and, consequently, it is questionable if the J-B model is adequate as the reference model for tomographic imaging of the Upper Mantle. In the following a new reference model is derived that makes the systematic deviation of observed from theoretical (= reference model) travel times as small as possible. First, a delay time curve is derived from the ISC data shown in Figure 3. Next, J-B travel times are added to this curve and the resulting travel times are inverted into a new S velocity model for Europe.

THE TRAVEL TIME CURVE

Derivation of a travel time curve representative for the average velocity structure beneath Europe from the delay time data shown in Figure 3 was not a straightforward procedure. Figure 3 and the histograms in Figure 4 indicate that different branches overlap and that the data set contains a considerable amount of large residuals. Moreover, it is not clear what the extension of the shadow zone corresponding to the average velocity structure is. Fitting one curve to the total data set would not result in a proper separation of the negative and positive branch. In order to estimate the location of the two branches, histograms of delays were made for every degree of epicentral distance, each smoothed over two degrees (so the histogram for 10° contains all data at the interval 9°-11°). For most intervals the histogram resembles a superposition of two Gaussian-like distributions with different amplitude. Therefore, first a distribution consisting of two Gaussian distributions was fit to the histograms. A χ^2-test was performed to see if such a distribution is an adequate description. It turned out that for the histograms belonging

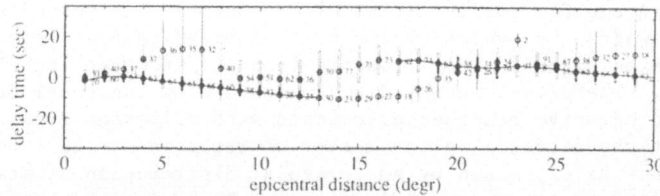

Fig. 5. Parameters of the Gaussian distributions. Circles represent the means, the bars the corresponding standard deviations. An open bar indicates that most data belong to this distribution. The percentage of data belonging to each distribution is placed next to the symbols. The solid line indicates the branches of the delay time curve.

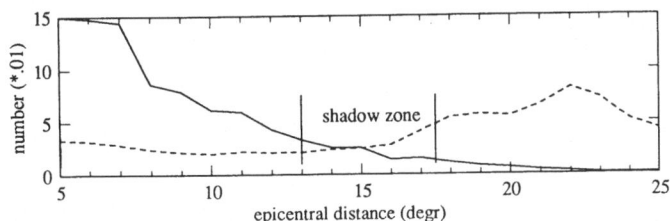

Fig. 6. Number of data within a band of 14 seconds around the negative
branch (solid line) and positive branch (dashed line) as a
function of epicentral distance.

to 9°, 10°, 23°, 24° and 25° a superposition of two Gaussian and one
uniform distribution is a better description because these histograms
contain many large residuals, resulting in heavier tails than can be fit
by Gaussian distributions only. Figure 5 shows the parameters of the
distributions. The location of the negative and positive branches is
determined by the maxima of the two Gaussian distributions. The
percentage of data belonging to each of the two distributions indicates
the relative importance of the branches.

The extension of the shadow zone was estimated in a subjective way.
Around the negative and the positive branches as they appear in Figure 3
two bands with a width of 14 seconds were drawn. Figure 6 displays the
number of reported arrivals within each band as a function of epicentral
distance. The curve corresponding to the positive branch shows that
between 5° and 16° the number of arrivals within the band is low and
approximately constant, from 16° to 18° it increases to a higher level,
indicating that the positive branch becomes important. The gradual
increase demonstrates that, probably due to lateral heterogeneity, the
upper boundary of the shadow zone varies mainly between 16° and 18°. On
the basis of the same argument, the lower boundary varies between 11° and
16°. The respective boundaries of the shadow zone associated with the
average velocity model should be well within these extremes.

Figure 5 shows the final delay time curve. To convert this curve
into a travel time curve J-B travel times were added. J-B travel times
for different focal depths were tested to take care that the model
resulting from the inversions was not biased by the initial choice of
event depth.

INVERSION

Travel times were inverted using the generalized inversion technique
with tapered eigenvector cut off (Wiggins, 1972). In the inversion a
trade-off exists between depth resolution and variance in the model
velocities. The number of eigenvectors determines this trade-off. The
fit of the model to the data is measured by the relative deviation of the
travel time data from the travel times calculated from the model. The
relative deviation corresponds to $\sqrt{\chi^2/N}$, where χ^2 is the statistical
quantity and N the number of data. In the following $\sqrt{\chi^2/N}$ is briefly
denoted by ξ. If ξ is too large the fit of the model is bad. "Too large"
means too large compared with the value to be expected for ξ on the basis
of a χ^2-distribution with N-K degrees of freedom, K being the number of
eigenvectors used in the inversion. The expected value for ξ is
$\sqrt{(N-K-1)/N}$, for this inversion with 26 data and 11 eigenvectors equal to
0.73. A ξ well below this value should be avoided too since this means
the model fit is too good in comparison to the estimated uncertainties in

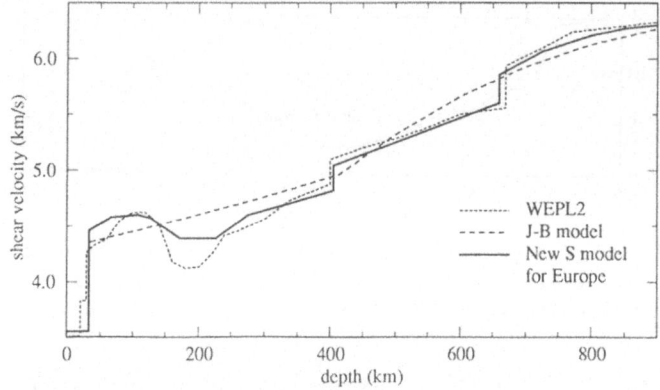

Fig. 7. The new S model for Europe. For comparison, the J-B and WEPL2 models are shown.

the data. Resolution is calculated by doing a separate "inversion", allowing uniform velocity changes over thick layers, thereby forcing the problem to be overdetermined. Inversion then gives the variance in velocity averaged over the layer.

The model resulting from the inversion procedure is influenced by subjective factors. The starting model has an influence on the result, probably due to the fact that the problem is underdetermined for detailed models. Also, inversion of travel times shows highly nonlinear effects. In general, during the inversions the model was changed in such a way that the shadow zone became too large or too small compared with the observations (Figure 6). In those cases the model had to be adjusted "by hand" to see that the requirements concerning the extension of the shadow zone were satisfied.

RESULTS OF TRAVEL TIME INVERSIONS

Several starting models were tested. These inversions resulted in different velocity models fitting the first arrival travel times equally well ($\xi \approx 1$). The preferred model ($\xi = 0.74$) is shown in Figure 7. The travel times calculated from this model, reduced with J-B travel times, best fit the delay time data (Figure 8). The velocities and the resolution are listed in Table 1 and Table 2. The experiments with different starting models provided some information about the uncertainties in the model. To start with, the exact shape of the LVZ is not resolved by the data.

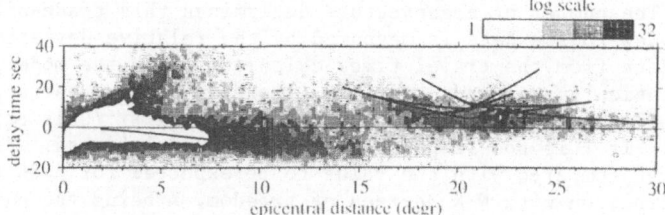

Fig. 8. Travel time curve of the preferred model relative to J-B travel times. The density of data is contoured from 1 to 32 per $0.2° \times 2$ sec.

Table 1. Velocities

depth (km)	v_s (km/s)	depth (km)	v_s (km/s)
0.0	3.550	405.0	4.820
33.0	3.550	405.0	5.041
33.0	4.460	500.0	5.249
67.0	4.578	600.0	5.468
108.0	4.601	660.0	5.600
128.0	4.570	660.0	5.860
171.0	4.390	725.0	6.066
226.0	4.386	800.0	6.211
276.0	4.596	850.0	6.275
340.5	4.708	950.0	6.335

The LVZ of the preferred model has minimum velocity equal to 4.386 km/s, a gradient at 250 km and a discontinuity at 405 km, but two other models with a broad, less pronounced LVZ and minimum velocity of 4.5 km/s, one bounded by a discontinuity at 380 km, the other by a gradient of the same size centered at 325 km, were also permitted by the data. Neither the velocity jumps at the "400 km" nor the "670 km" discontinuity or their precise depths are well constrained. For instance, two models with a jump of about 0.2 km/s at 400 km but discontinuities of 0.09 km/s and 0.4 km/s at 660 km, respectively, fit the data equally well ($\xi \approx 1$).

These features are badly resolved by first arrival data only, whereas the later arrivals - which presumably contain relevant information about these features - cannot easily be identified. The delay density plot, however, provided some constraints on the velocity jumps at the major discontinuities. Between 17° and 26° the shape of the contours of lowest density (see Figure 8) indicates that later arrivals belonging to triplication branches were misidentified as first arrivals. This is likely to happen when the appearance of the first arrival on a seismogram is much weaker than that of later arrivals. In a density plot of P delays for the same area, triplications are more clearly visible since there are many more P registrations. Therefore, with trial and error the velocity jumps of the model resulting from the inversion were adjusted to achieve that the corresponding triplications would match the density plot. Figure 8 shows that the travel time curve of the preferred model fits well to the delay time data. The shadow zone extends between 13° and 18° but the triplication branch corresponding to the 400 km discontinuity folds back to 14°. Hence, apart from the effects of lateral heterogeneity, the absence of a clear shadow zone can also be attributed to the presence of a triplication branch. We note that the general features of the model are in agreement with the WEPL2 model for the West European Platform, recently derived from dispersion measurements of higher modes (Dost, 1987, see Figure 7).

Table 2. Resolution

depth (km)	v_s (km/s) averaged over Δz	st dev (km/s)	vertical resolution Δz (km)
16.5	3.550	0.12	33
50.0	4.519	0.059	34
108.0	4.601	0.061	82
200.0	4.388	0.017	102
328.0	4.687	0.048	154
469.0	5.181	0.048	128
596.5	5.460	0.19	127
745.0	6.105	0.087	170

CONCLUDING REMARKS

The delay density plot (Figure 3) shows that there is a significant systematic deviation of S travel times observed in Europe from the J-B travel times. From the ISC data set a new S velocity model of the Lithosphere and Upper Mantle beneath Europe has been derived. Figure 1 shows the ray paths calculated from the J-B model and this model, in the introduction referred to as "a more realistic model". Comparison of these ray paths indicates that the J-B shear velocity model is inadequate as a reference model for tomographic imaging of the Upper Mantle beneath Europe.

ACKNOWLEDGEMENTS

A. Zielhuis wishes to thank B. Dost for his support with the inversion of the travel time data and R. D. van der Hilst for helping to collect and process the data.

REFERENCES

Dost, B., 1987, The Nars array. A seismic experiment in Western Europe, Ph.D. Thesis, Utrecht, pp 117.

Nolet, G., 1987, Seismic wave propagation and seismic tomography, in: "Seismic Tomography", G. Nolet, ed., Reidel, Dordrecht, 1-24.

Spakman, W., 1986, Subduction beneath Eurasia in connection with the Mesozoic Tethys, Geol. Mijnbouw, 65:145-153.

Spakman, W. and Nolet, G., 1987, Imaging algorithms, accuracy and resolution in delay time tomography, in: "Mathematical Geophysics: A Survey of Recent Developments in Seismology and Geodynamics", N. J. Vlaar et al., eds., Reidel, Dordrecht, 155-188.

Wiggins, R. A., 1972, The general linear inverse problem: implication of surface waves and the free oscillations for Earth structure, Rev. Geophys. Space Phys., 10:251-285.

SYNTHETIC SEISMOGRAMS FOR 3-D INHOMOGENEOUS MEDIA

Reinoud Sleeman

Department of Theoretical Geophysics
PO Box 80.021
3508 TA Utrecht, The Netherlands

INTRODUCTION

Constructing synthetic seismograms for 3-D inhomogeneities seems so far only applicable in homogeneous background models, due to the complexity of the calculations. Strong multiples such as generated at the sea bottom or the basement, are ignored in this approach. These multiples, however, also include information about the model and should not be neglected.

The reflectivity method (Kennett, 1979) is useful in modeling body waves in layered media, but becomes impractical in media with 3-D inhomogeneities. This is because the 3-D body can be considered as a collection of (secondary) point sources, and for each source depth the model response must be calculated again.

This paper presents the use of the locked mode method, in order to calculate the (pulse) response on 3-D inhomogeneities in acoustic, horizontally layered media. The advantage of this method of forward modeling is that all multiples are taken into account.

LOCKED MODE METHOD

Seismic energy will only be confined in a layer (or a system of layers), when the angle of incidence at the boundaries is supercritical, otherwise energy will be transmitted and leak away. In the locked mode approach the energy is forced to stay in the layer (or the system), by introducing perfect reflectors at upper and lower boundaries, so total reflection will occur for all angles of incidence. We consider a system of acoustic, horizontal layers, bounded by a free surface and a rigid bottom boundary (Figure 1).

The problem of solving the homogeneous acoustic wave equation can then be considered as an eigenvalue problem. Solving this problem yields the eigenmodes P_n (which represent the pressure) and its eigenvalues. The eigenmodes must satisfy the boundary conditions $P_n = 0$ at the free surface, and $\partial P_n/\partial z = $ constant at the rigid boundary. These modes can be used to construct the solution of the inhomogeneous acoustic wave equation, in other words, using the orthogonality of the eigenmodes P_n,

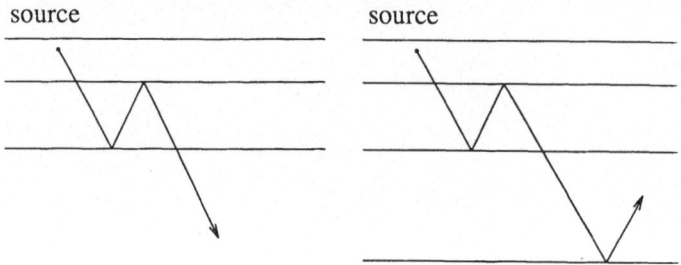

Fig. 1. Locked modes can be used if the energy is forced to stay in the system by introducing a perfect reflector at the deepest boundary.

the Green's function (the medium response to a point source) for the layered model can be expanded into the eigenfunctions of the homogeneous problem.

One advantage of this approach is that all multiples in the layered system are taken into account. Besides, errors in the time signals due to the artificial reflector can be understood and can be corrected. In calculating the disturbed wavefield in a medium where the distortion can be considered as a collection of perturbation points (acting as secondary point sources), the advantage of using the locked mode approach is that once the eigenmodes are known, it is very easy to construct the Green's function for (secondary) sources at different depths (in contrast to the reflectivity method).

DEVIATION OF THE PRESSURE IN AN ACOUSTIC, HORIZONTALLY LAYERED MEDIUM, DUE TO A PERTURBATION IN THE BULK-MODULUS

In this section I will give the expression for the perturbed pressure field (deviation of the pressure field) in an acoustic, horizontally layered medium, due to perturbations in the bulk-modulus. Only first order scattering (Born approximation) will be considered. The system of horizontal layers is bounded by the free surface, where the pressure becomes zero, and a rigid boundary (perfect reflector) where the displacement becomes zero, so the locked mode approach can be applied.

We consider a medium characterized by the density $\rho(r)$ and bulk-modulus $\kappa(r)$. In the frequency domain, the wave equation for acoustic waves in this medium is given by:

$$\nabla \cdot \left[\frac{1}{\rho(r)} \nabla P(r,\omega) \right] + \frac{1}{\kappa(r)} \omega^2 P(r,\omega) = \nabla \cdot \left[\frac{1}{\rho(r)} f(r,\omega) \right] = s(r,\omega) \qquad (1)$$

where $P(r,\omega)$ represents the pressure field and $f(r,\omega)$ the source of the acoustic waves. Suppose we have a point source $s(r,t) = \delta(r-r_s)\delta(t)$ then we can introduce the Green's function $G(r,\omega;r_s)$:

$$\nabla \cdot \left[\frac{1}{\rho(r)} \nabla G(r,\omega;r_s) \right] + \frac{1}{\kappa(r)} \omega^2 G(r,\omega;r_s) = \delta(r - r_s). \qquad (2)$$

For an arbitrary source $s(r,t) = S(r)\delta(t)$ the solution of (1) is then given by:

$$P(r,\omega) = \int G(r,\omega;r_s)S(r_s)dr_s. \qquad (3)$$

Once the solution of (2) is known for the medium specified by $\rho(r)$ and $\kappa(r)$ it is easy to determine the pressure deviation $\delta P(r,\omega)$ due to perturbations $\delta\rho(r)$, $\delta\kappa(r)$ and $\delta f(r,\omega)$ in density, bulk-modulus and source field. We therefore modify Eq. (1) and make use of:

$$\frac{1}{h(r)+\delta h(r)} = \frac{1}{h(r)} - \frac{\delta h(r)}{h^2(r)} + O(\delta h^2).$$

Neglecting second order terms (this is the Born approximation), Eq. (1) then reduces to:

$$\nabla\cdot\left[\frac{1}{\rho(r)}\nabla\delta P(r,\omega)\right] + \omega^2 \frac{1}{\kappa(r)}\delta P(r,\omega) = \nabla\cdot\left[\frac{\delta\rho(r)}{\rho^2(r)}\nabla P(r,\omega)\right] +$$

$$+ \omega^2 \frac{\delta\kappa(r)}{\kappa^2(r)} P(r,\omega) + \nabla\cdot\left[\frac{1}{\rho(r)}\delta f(r,\omega) - \frac{\delta\rho(r)}{\rho^2(r)} f(r)\right]. \tag{4}$$

The terms on the right hand side of this equation can be seen as the secondary source $\delta S(r,\omega)$ (which is considered as a collection of scattering points acting as point sources) in the homogeneous medium $\rho(r)$, $\kappa(r)$. These scattering points cause the pressure field $\delta P(r,\omega)$. The Born approximation implies that the scattering points only disturb the primary wave field and not the scattered field. In general $\delta\rho$ is small, compared to $\delta\kappa$. If the primary source is a point source $\delta(r-r_s)$, then:

$$\delta S(r,\omega) = \omega^2 \frac{\delta\kappa(r)}{\kappa^2(r)} G(r,\omega;r_s)$$

and the solution of Eq. (4) is given by (according to (3)):

$$\delta P(r,\omega) = \iiint G(r,\omega;r')\omega^2 \frac{\delta\kappa(r')}{\kappa^2(r')} G(r',\omega;r_s)dx'dy'dz' \tag{5}$$

where $r = (x,y,z)$.

This equation can easily be understood (see Figure 2). We see that the primary wave field travels from the source located in r_s, to a perturbation point in r', described by the Green's function $G(r',\omega;r_s)$. Each perturbation point acts as a secondary point source, and causes a secondary wave field propagating to a receiver in r, described by $G(r,\omega;r')$. The strength of the secondary source is given by $\omega^2\delta\kappa(r')/\kappa^2(r')$. Summation over all scattering points gives the total disturbed field, as expressed by Eq. (5).

According to Nolet et al. (1987) the Green's function in a horizontally layered medium, determined with the locked mode summation, is:

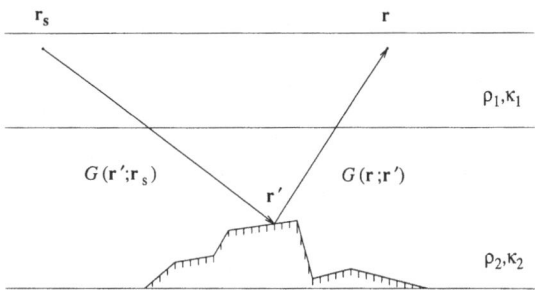

Fig. 2. 3-D inhomogeneous medium in a layered system. $G(r';r_s)$ describes the wave propagation from source to a perturbation point in r'; $G(r;r')$ describes the propagation from this point to a receiver in r.

$$G(r',\omega;r_s) = -\frac{1}{2}\pi i \sum_{n=0}^{\infty} P_n(z')P_n(z_s)H_o^{(1)}(k_n R_s) \tag{6}$$

with $R_s = |(x'-x_s)^2 + (y'-y_s)^2|^{1/2}$. P_n are the eigenfunctions and k_n the eigenvalues of the layered medium at frequency ω. $H_o^{(1)}$ is the Hankel function of the first kind. Substituting (6) in (5) gives the expression for the disturbance in the pressure field:

$$\delta P(r,\omega) = -\frac{1}{4}\pi^2 \iiint \sum_{n=0}^{\infty} P_n(z)P_n(z')H_o^{(1)}(k_n R') \times$$

$$\times \sum_{m=0}^{\infty} P_m(z')P_m(z_s)H_o^{(1)}(k_m R_s)\omega^2 \frac{\delta\kappa(r')}{\kappa^2(r')}\,dx'dy'dz' \tag{7}$$

with $R' = |(x-x')^2 + (y-y')^2|^{1/2}$.

NUMERICAL RESULTS

In order to check the validity of the locked mode approximation, in calculating the perturbed wave field, some numerical tests have been made. First, let us consider a very simple model consisting of only one homogeneous layer, with the following parameters:

P velocity: 1500 m/s
density: 1000 kg/m^3
quality factor: 100
depth: 200 m

The material quality factor Q, is related to the attenuation by means of a complex bulk-modulus $\kappa = \kappa_r + i\kappa_i$ so that $Q = \kappa_r/\kappa_i$.

In the following examples the source is a point source located at a depth of 20 m, and the receivers are located at 400, 500, 600, 700 (and 800) m, also at 20 m depth. Receivers are indicated as REC. A,B,C,D (and E). All calculations have been made for frequencies from 1 to 75 Hz, with sample interval of 0.25 Hz.

The Green's function for the one layer model is shown in Figure 3a. Due to the limited number of frequencies, the source pulse is smeared out and as a consequence of this, the direct wave and its ghost almost cancel each other. Also multiples and their ghosts will partly cancel each other. In Figure 3b the ray paths are to be seen, corresponding with the five numbered pulses in the time section of the first receiver.

In all numerical examples the multiple reflections would continue for infinite time and slowly decrease in amplitude, due to geometrical spreading and attenuation. Since sampling is done in the frequency domain, multiple reflections will wrap around in the time domain, to contaminate the entire seismogram. However, because the total length of the calculated time signal is 4 seconds, the amplitudes of these multiples will be small and hardly to be seen (see arrows in Figure 3a).

We now consider a plate of only 6 m thick at a depth of 180 m with horizontal dimensions of 200 m (Figure 4). P velocity in the plate is increased with 3% (also in further examples). Calculations were done for 4 receivers (at 400, 500, 600 and 700 m). Figure 5 shows the results.

Figure 6a shows the time signals due to a "hidden pyramid" in the water layer (Figure 6b), for 4 receivers. The horizontal dimensions of

344

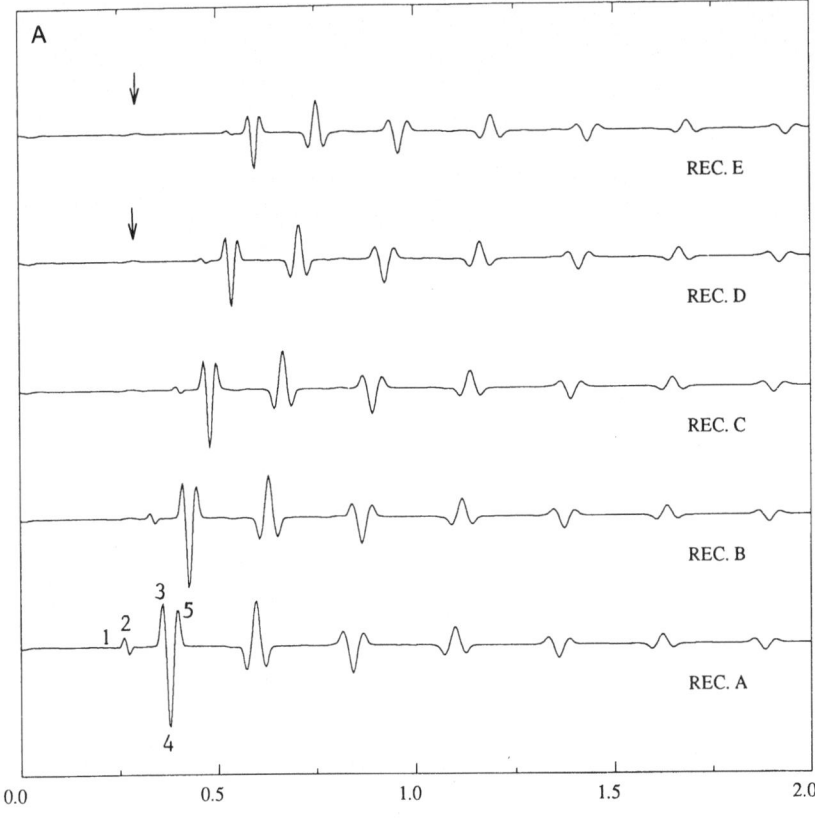

TIME IN SECONDS

Fig. 3a.   Green's function for the one layer model.
  3b.   Five different ray paths, corresponding with the five numbered
        pulses in Figure 3a.

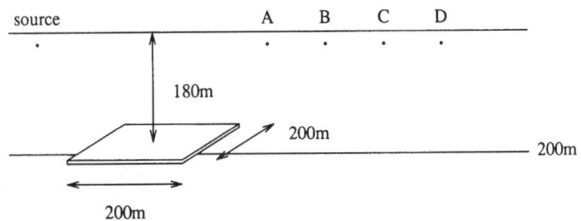

Fig. 4.   Horizontal plate at 180 m and horizontal dimensions of 200 m
         (100 < x < 300, -100 < y < 100, and 180 < z < 186).

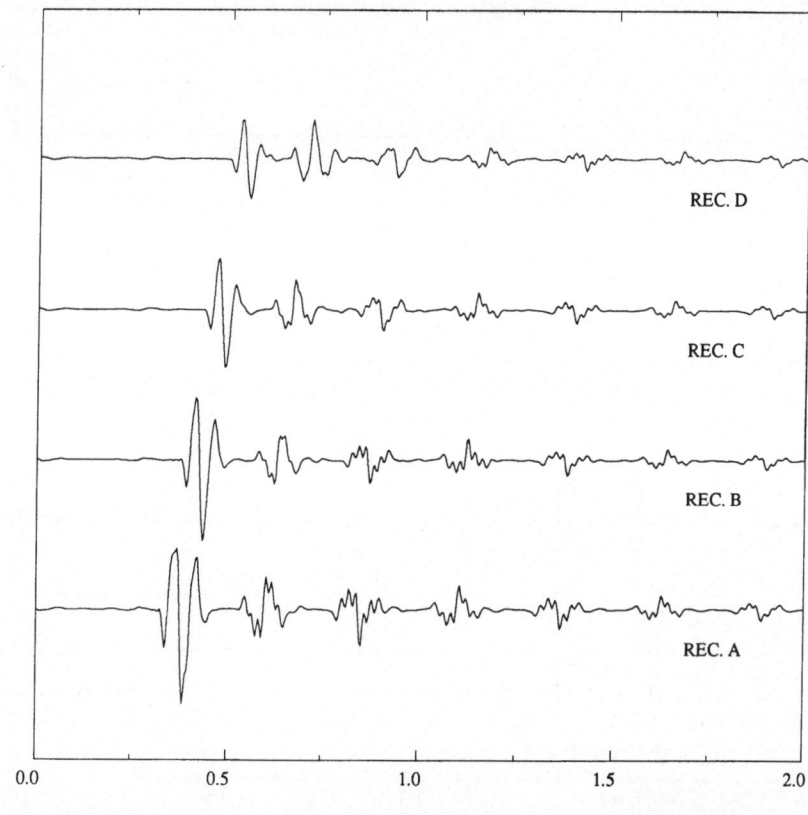

REC. D

REC. C

REC. B

REC. A

| 0.0 | 0.5 | 1.0 | 1.5 | 2.0 |

TIME IN SECONDS

Fig. 5. Disturbed wave field in the one layer model, due to a horizontal plate at 180 m depth. Scale factor (with respect to Figure 3a) is 0.15.

the bottom plate (at 180 m) are 200 x 200 m, while the top is placed at a depth of 140 m. Comparing the Green's function with the disturbed time signals in Figures 5 and 6a, it is clear that also the multiples contain information about the model.

We now consider a two layer model, for instance a water layer and a basalt layer, with parameters:

    Layer 1
            P velocity: 1500 m/s
            density: 1000 kg/m^3
            quality factor: 100
            depth: 100 m
    Layer 2
            P velocity: 2000 m/s
            density: 3000 kg/m^3
            quality factor: 50
            depth: 300 m

Figure 7 gives the Green's function for 10 different receivers, located at 20 m depth and offsets of 300,325,...,525 m.

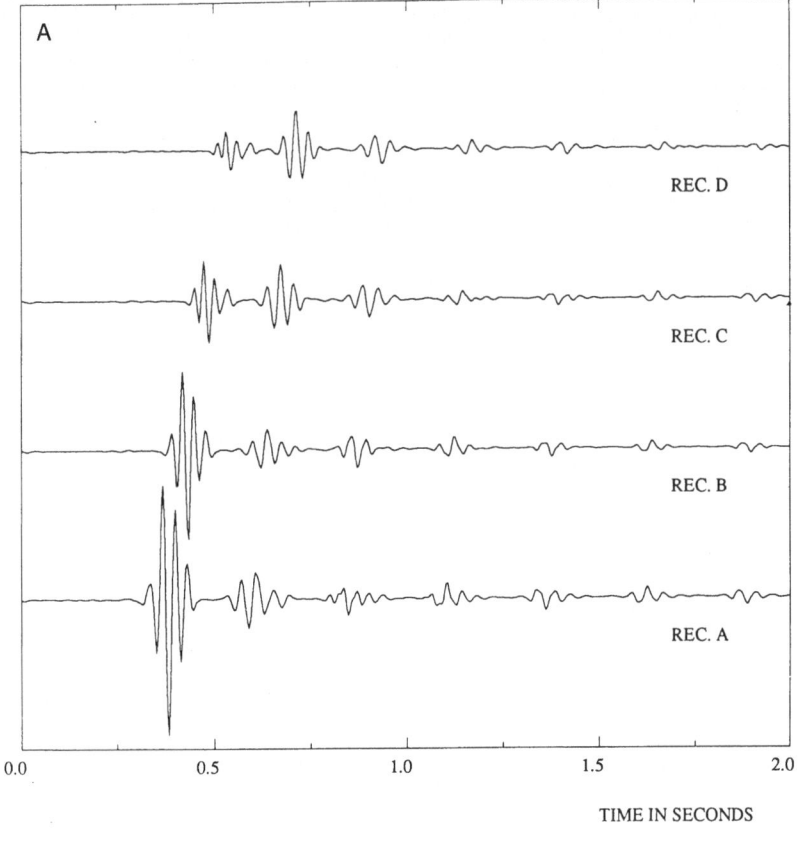

TIME IN SECONDS

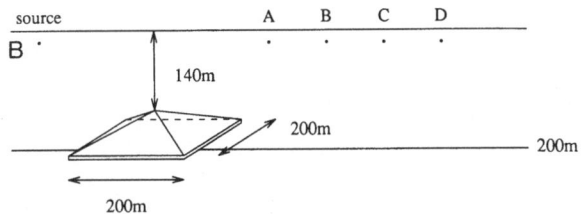

Fig. 6a. Disturbed wave field due to a pyramidal structure in the one
layer model. Scale factor (with respect to Figure 3a) is 1.0.
6b. Structure of the pyramid. The bottom plate is located at 180 m
depth with the same horizontal dimensions of the plate in Figure
8. The top of the pyramid is located at 140 m depth.

    In Figure 8 the response on a pyramid at 280 m depth (horizontal
dimensions 200 x 200 m, top is at 220 m and has an offset of 150 m of the
source-receiver line) is given. Again the latter parts of the signals
contain also information about the inhomogeneity.

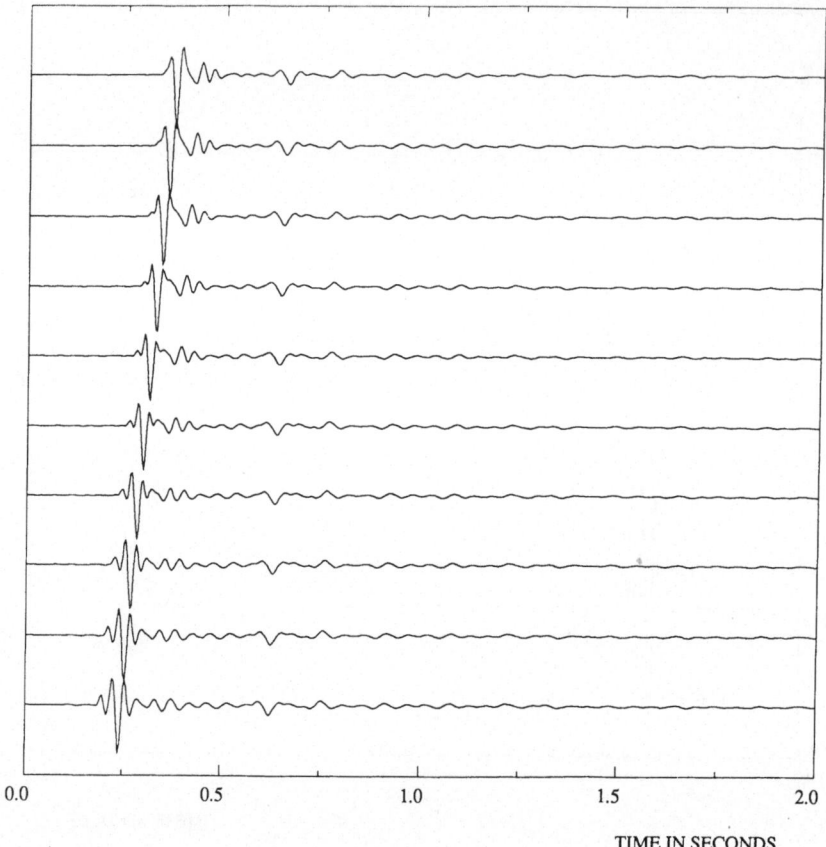

TIME IN SECONDS

Fig. 7. Green's function in the two layer model.

CONCLUSIONS

Considering the numerical examples given in this paper it seems justified to conclude that the locked mode approach is useful in calculating the perturbated wave field in an acoustic, horizontally layered medium. The seismograms include all information about the model, including information about the 3-D inhomogeneity in the multiples. At this stage the numerical implementation of the method is not fast enough to be used in interactive work on minicomputers.

ACKNOWLEDGEMENTS

I would like to thank Vincent Nijhof for his practical help during this research. Also I thank Guust Nolet and Roel Snieder for giving me helpful suggestions in the beginning of this research. Besides, I am grateful to Aly Brandenburg and Wim Groenewoud for their mental support and humoritistic input during moments of despair.

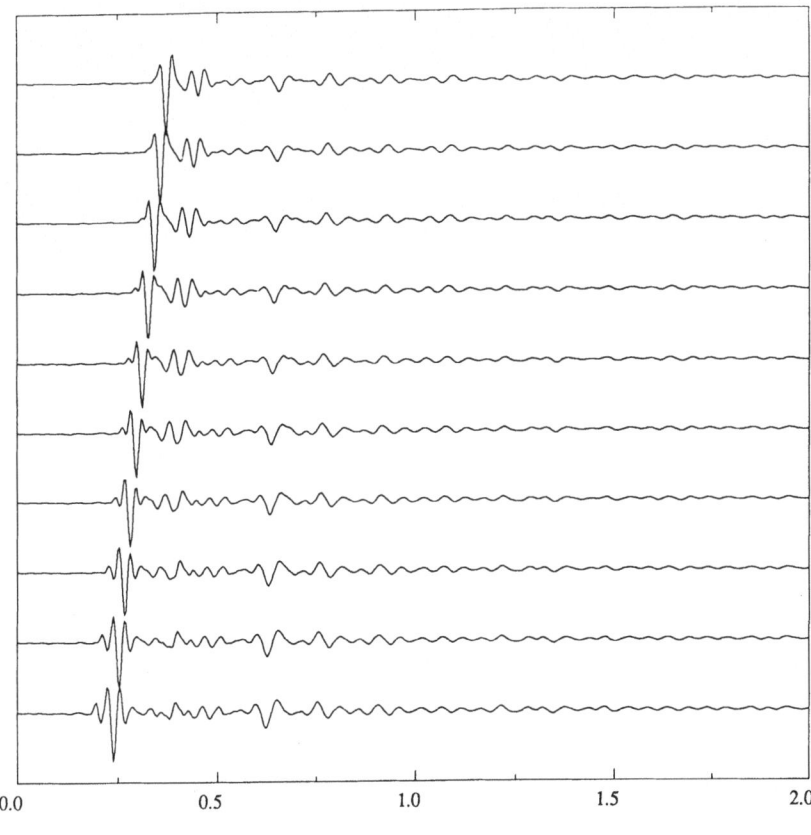

TIME IN SECONDS

Fig. 8. Total response (disturbed plus undisturbed field) on a pyramid at
         280 m depth in the two layer model. Horizontal dimensions are
         the same as in Figure 6b.

REFERENCES

Kennett, B. L. N., 1979, Theoretical reflection seismograms for elastic
        media, Geophys. Prospect., 27:301-321.
Nolet, G., Nijhof, V., Sleeman, R., and Kennett, B. L. N., 1987, Synthetic
        Reflection Seismograms in 3D by a Locked Mode Approximation,
        submitted to Geophysics.

BROAD-BAND P-WAVE SIGNALS AND

SPECTRA FROM DIGITAL STATIONS

M. Bezzeghoud, A. Deschamps and R. Madariaga

Laboratoire de Sismologie, UA CNRS No 195
Institut du Globe de Paris and
UER Sciences de la Terre, Université Paris VII
4 Place Jussieu, 75252 Paris Cedex 05, France

1. INTRODUCTION

The first determination of the seismic moment was made by Aki (1966) for the Niigata earthquake of 1964. Since then, it has become the single most important source parameter used to characterize the strength of earthquake sources. Several methods are currently used in order to compute seismic moments from long and short period P- and S-waves (see, for example, Kikuchi and Kanamori, 1982; Ruff and Kanamori, 1983; Houston and Kanamori, 1986) and from long period surface waves (see, for example, Dziewonski et al., 1982; Monfret and Romanowicz, 1986). The observed high frequency waveforms from large earthquakes are very complex and it is sometimes difficult to determine the seismic moment from time domain studies. Furthermore, because of the limited bandpass of most of the standard instruments it is easy to miss low frequency components that would contribute significantly to the moment of the earthquake. For this reason, the spectral method in which the observed spectra are corrected for instrument response is probably a more reliable method for estimating seismic moments. Of course, if the instrument response of the seismograph does not include the corner frequency in its bandpass the recovered moment will be wrong. This difficulty has been invoked several times in order to explain differences between seismic moments determined from surface wave analyses, and from body wave studies. A striking case of this difference is the Valparaiso earthquake of March 3, 1985 which had a surface wave moment of $1.5 \ 10^{28}$ dyne.cm (Romanowicz and Monfret, 1986) and reported body wave moments of only 4 to $6 \ 10^{27}$ (Korrat and Madariaga, 1986; Christensen and Ruff, 1986). Recently, Houston and Kanamori (1986) used Global Digital Seismograph Network (GDSN) data to compute average source spectra of several large earthquakes. They compared their results to the standard "ω-square" model and a modified model proposed by Gusev (1983). Their study shows that neither of these models agree completely with spectra data. We used their method on the Chilean earthquake and found again a discrepancy between surface wave and body wave moments.

The availability of broad-band digital data from several networks (RSTN, GEOSCOPE, GDSN) may help in resolving these ambiguities. In this paper we will study the spectra and body wave signals from the Chilean earthquake in detail in order to understand the apparent difference between body and surface wave moments. This study is made possible by the

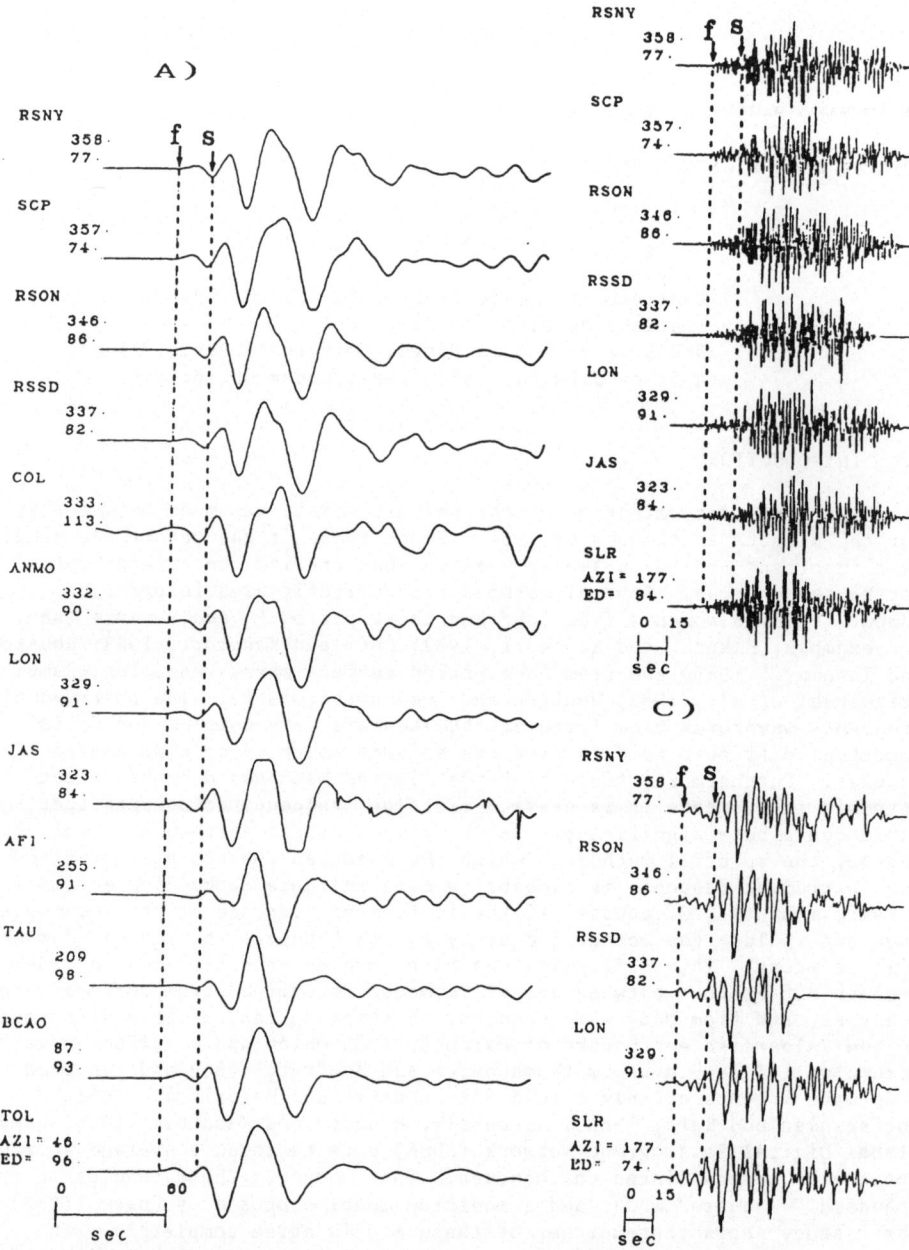

Fig. 1. Long (A), intermediate (C) and short period (B) GDSN records of
the March 03, 1985, Valparaiso earthquake, f and s denote,
respectively, foreshock and main shock. ED and AZI denote,
respectively, epicentral distance and azimuth.

availability of broad-band recordings from several networks. Using techniques introduced by Choy and Boatwright (1984) and Bezzeghoud et al. (1986) we deconvolve recorded P-wave pulses in order to obtain very broad band records that are then processed by standard techniques in order to extract very long period (VLP) body wave seismic moments, and to study seismic moment release rate as a function of time and frequency. The difference between body wave and surface wave moments will be shown to be due to problems with the correction of body wave spectra for the effect of reflected depth phases.

2.   DATA

In the last years a number of broad-band (BRB) digital networks with high-dynamic range have been installed at several sites around the world. We intend to show, briefly, the possibilities that are open for the study of earthquake source processes by the availability of BRB data. These possibilities are illustrated by the analysis of the recordings of the Valparaiso, Chile, earthquake of March 1985 from four digital networks (GDSN, GRF, NARS and GEOSCOPE). The instrumental response of narrow-band systems limits the information that can be retrieved about the details of the source process for shallow earthquakes. To illustrate the interest of broad-band data, in Figure 1 we show the (A) long, (B) short and (C) intermediate period teleseismic GDSN records of the earthquake of March 3, 1985 near Valparaiso, Chile. Comparison of these three records shows how the information contained in the narrow-band systems (LP and SP) is reduced with respect to that of a broad-band system (IP) for large earthquakes. The P-wave is divided into two clearly different phases in the intermediate period records (Figure 1C). The first relatively weak phase (denoted f in the·figure) is a precursor to the main shock, the second (s) phase corresponds to the beginning of the main shock. Let us look now at the long and short period records, the precursor is difficult to identify in these seismograms. Similarly, the third phase is not present on the narrow-band records.

Several authors (Choy and Boatwright, 1981; Kind and Seidl, 1982; Eydogan et al., 1985; Bezzeghoud et al., 1986) have proposed to increase the resolution of narrow-band digital records by deconvolving the instrumental response and reshaping the seismic signal with a high pass filter with an appropriate cut-off period. On Figure 2, we show a series of high pass filtered ground velocity records obtained from the short period GDSN record of the 1985 main shock at station LON. The frequency contents of the original short period record (bottom) is broadened by (1) deconvolving the response of the standard short period instrument and (2) applying a high pass Butterworth filter to obtain the broad-band ground velocity signal. Several cut-off periods for the Butterworth filter were tried in order to study the source complexity in different frequency bands. The T = 10 s record deconvolved from the short period GDSN record at LON is practically identical to the intermediate period (IP) recording at the same station. The latter is shown at the bottom of Figure 3. In this figure we show a set of numerically calculated high pass filtered records obtained by deconvolution of the IP record at LON. Let us remark that the two reshaped records at T = 25 s, deconvolved from the short period GDSN response in Figure 2 and from intermediate period in Figure 3 are identical. This is a proof of the numerical stability of the instrument reshaping that is applied to the records.

From Figure 3 we observe that it is possible to recover the ground velocity up to 200 s from the standard intermediate period GDSN channel. The signal at 400 s shows sign of amplification of long period noise. Finally, in Figure 4 we show a similar deconvolution of the broad-band

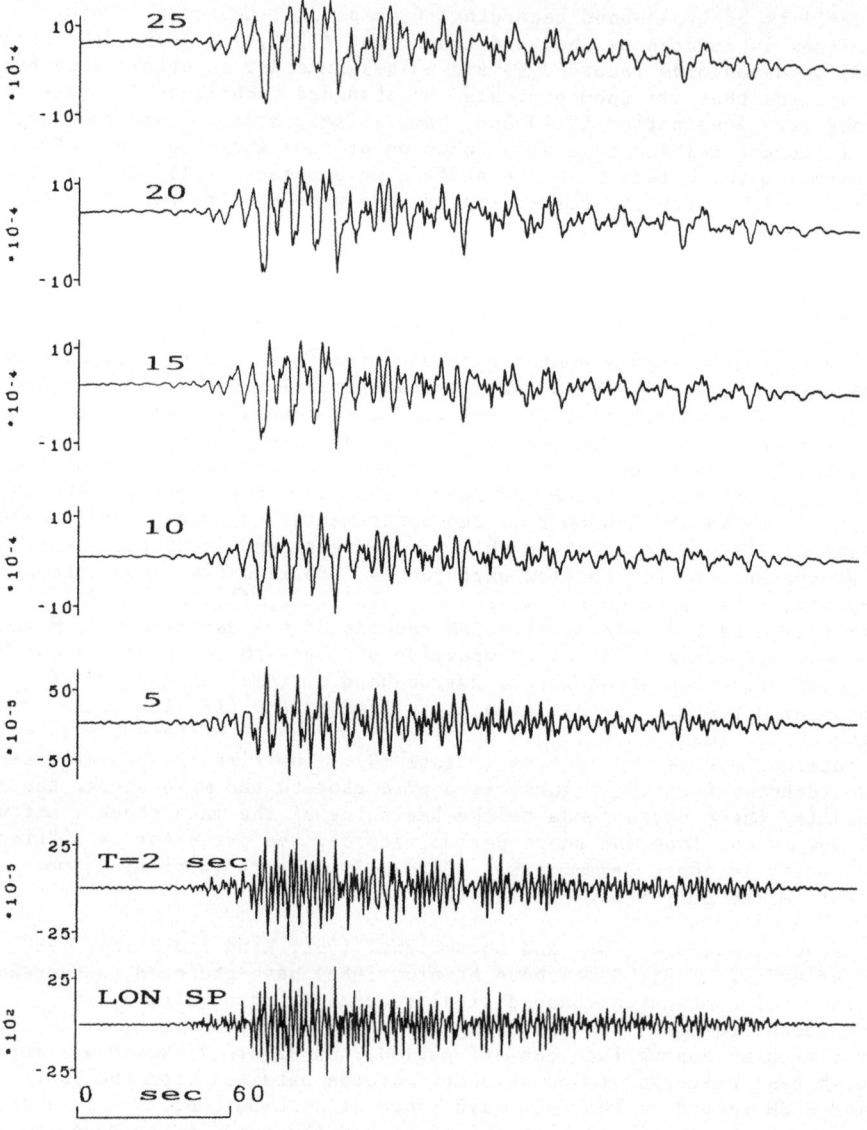

Fig. 2.   A series of high pass filtered ground velocity records obtained
          from short period GDSN records of the March 3, 1985, Valparaiso
          earthquake at station LON.

record obtained at station SSB of the GEOSCOPE network. The data from
this station which is located in central France may be stably deconvolved
until 300 s. Thus, by appropriate deconvolution of the digital data we
can successfully generate the seismograms that would have been recorded by
broader band instrument than those actually installed. This is the great
advantage of digital recording.

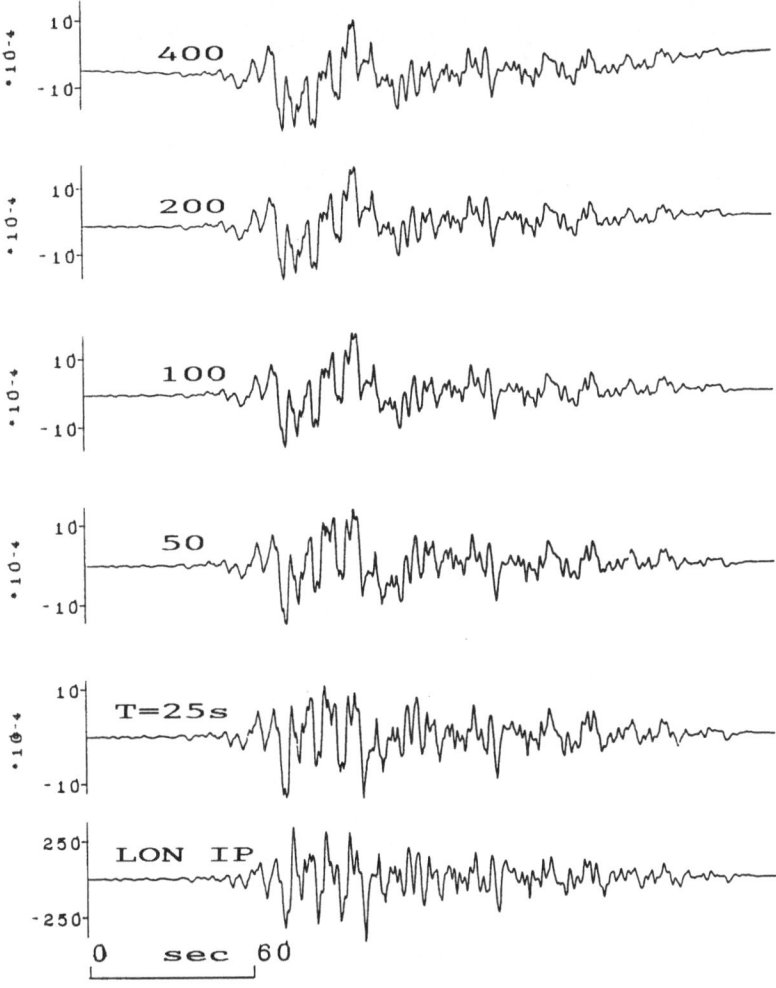

Fig. 3. A series of high pass filtered ground velocity records obtained
from the intermediate period GDSN recording of the March 3,
1985, Valparaiso earthquake at station LON.

Table 1. Location of the Studied Earthquakes

Date	Time	LAT(S°)	LON(W°)	m_b	M_s
1985 March 03	22:47	33.14	71.87	6.9	7.8
1985 March 17	10:41	32.63	71.55	5.9	6.6
1985 March 19	04:01	33.20	71.65	5.9	6.6
1985 April 09	01:56	34.13	71.62	6.3	7.2

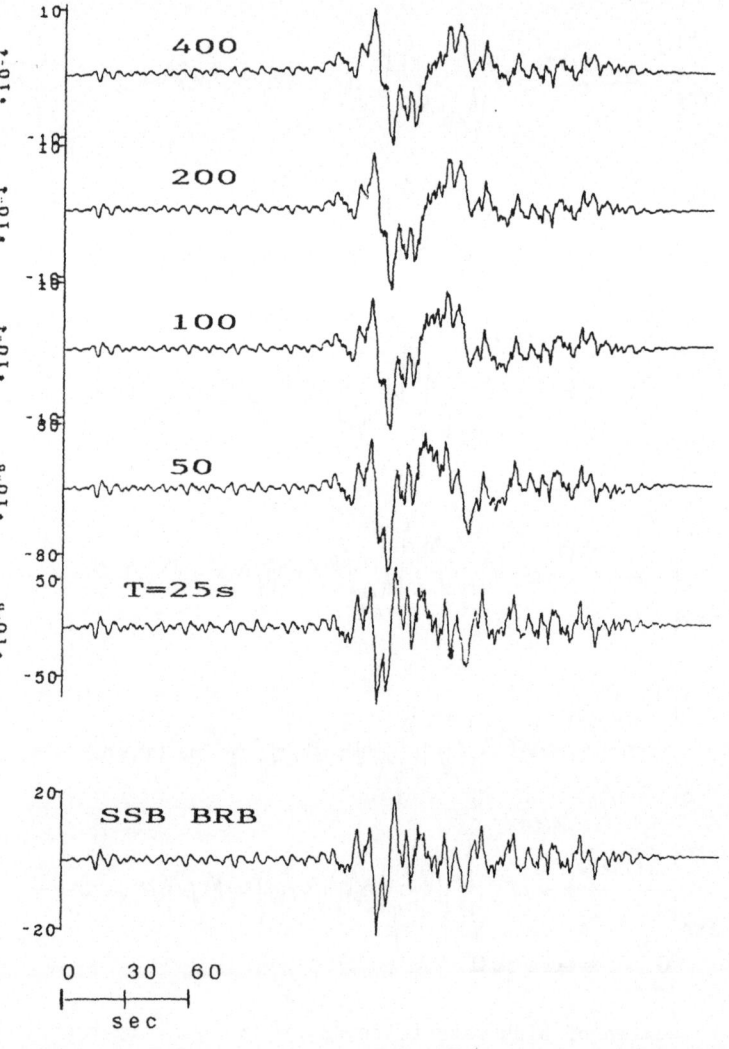

Fig. 4. A series of high pass filter ground velocity records obtained
         from the broad-band GEOSCOPE record of the March 3, 1985,
         Valparaiso earthquake at station SSB.

        For the Valparaiso earthquake and three of its largest aftershocks
(see Table 1), we examined 51 digital records from GDSN long, intermediate
and short period vertical instruments at teleseismic distances from 30° to
90°. The records used in this study belong to RSTN and DWWSSN stations.
Inversion of P-waves will be performed on both long period and broad-band
records obtained from intermediate period recordings by the deconvolution
procedure explained above. Spectral analysis will be applied directly to
the original records and the spectra will be corrected for instrument
response. We complete our analysis of the main shock with the BRB record
of the GEOSCOPE station SSB. Detailed description about this global
network is contained in the paper of Romanowicz et al. (1983).

3. CHARACTERISTICS OF THE 1985 VALPARAISO EARTHQUAKE

The subduction process along the western margin of the South America is very complex. Several studies of the seismicity (Baranzangi and Isacks, 1976) of the Peru-Chile trench indicate that the oceanic Nazca plate is divided into five segments of variable dip as it descends beneath the continental South American plate. The earthquakes studied in this paper are located in central Chile (Figure 5) just to the South of the boundary between the central Chile shallow subduction zone and the normal subduction zone to the South of 32°. The exact nature of the transition between the two zones is not known, but it is likely that the northern edge of the Valparaiso earthquake zone is limited by the transition between the two segments of the subduction zone. The large earthquake of March 3, 1985 occurred in a well-known gap, identified by several authors (Kelleher, 1972; McCann et al., 1972; Nishenko, 1985). It is well established that this region had been the site of several major earthquakes in 1647, 1730, 1822 and 1906 (see for instance Comte et al., 1986).

As proposed by Korrat and Madariaga (1986) the rupture of the Valparaiso gap started with the large M_s = 7.8 earthquake of July 9, 1971 in the northern part of the gap (see Figure 5). This event was followed by an M_s = 7.3 event on October 5, 1973 that occurred near the southern end of the 1971 event. A few smaller shocks (M_s < 6) occurred near the latitude of 34°S in October 1981. Thus, before March 1985 about one half of the Valparaiso gap had already been broken. The earthquake of March 3 was preceded by an intense foreshock activity beginning on February 21 (Comte et al., 1985), about 13 days before the main shock. Thus, we proposed that the rupture process of the Valparaiso gap started on July 9, 1971 and probably finished by the sequence of earthquakes that occurred on March and April 1985 (Korrat and Madariaga, 1986). Whether this series of events is equivalent to the previous events of 1821 or 1906 is difficult to establish and will be the subject of further study. Most of the major earthquakes (between 1965-1985) in the Valparaiso area take place at two distinct depths, either inside the subducted plate or at the plate interface. Modeling of P-waves (Korrat, 1986) shows that most plate interface events have the shallow thrust faulting mechanism, typical of subduction zone interplate earthquakes. Their fault planes dip at about 22°-24° under the South American continent.

4. SPECTRAL ANALYSIS

4.1 Method

Modeling and inversion of body waves are frequently used to compute earthquake source parameters. In our study, we applied two different methods to extract the source time function and the seismic moment of the Chilean earthquakes of March and April 1985: time domain inversion and spectral analysis of body waves. We present very briefly in the following the spectral method used in our study.

The body wave signal at a far field station is given by Kanamori and Stewart (1976) or Deschamps et al. (1980) as:

$$W(t) = I(t)*Q(t)*U(t). \tag{1}$$

Where $Q(t)$ is the Futterman (1962) attenuation operator, $I(t)$ is the instrumental response and $U(t)$ is the far field displacement. The vertical displacement $U(t)$ of P-waves can be represented by:

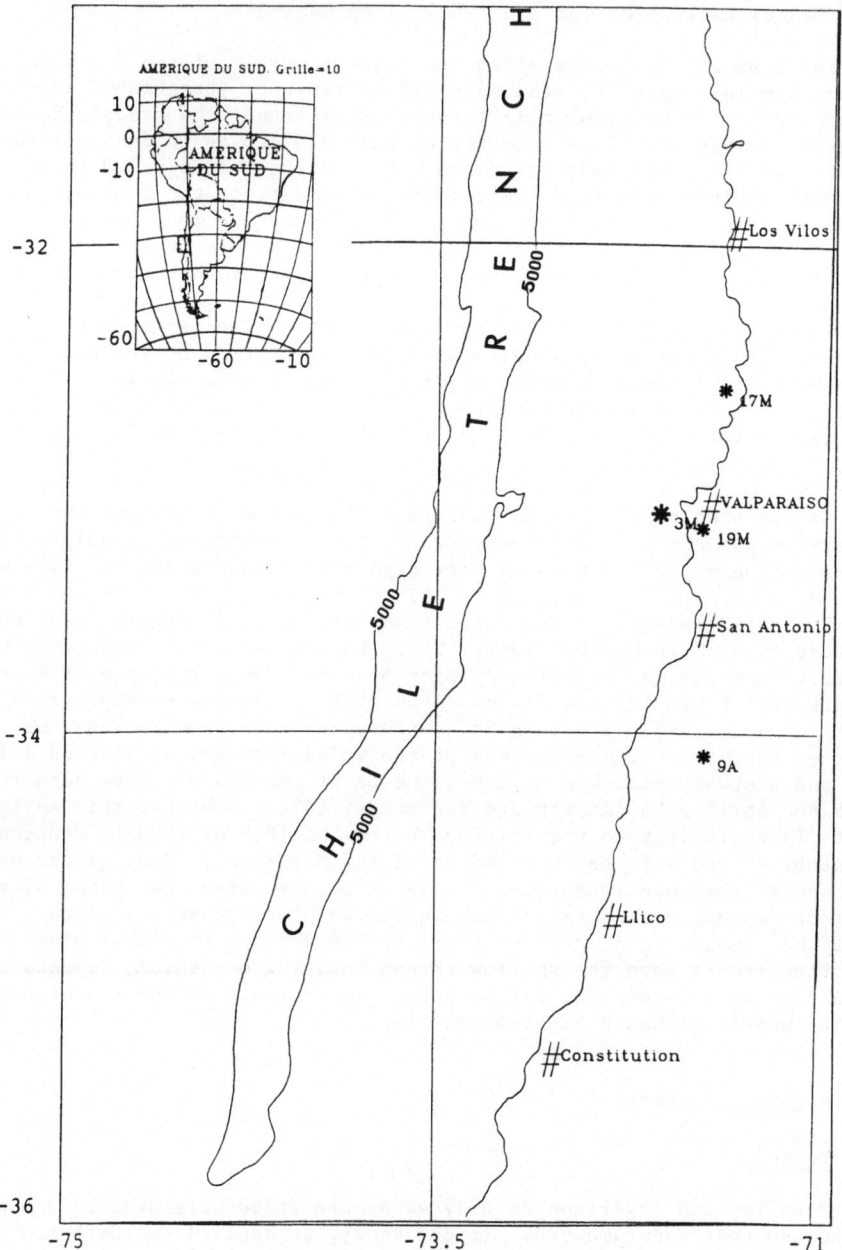

Fig. 5. Location of the March 3, 1985, Valparaiso earthquake and its three largest aftershocks (March 17, 19 and April 9).

$$U(t) = \frac{1}{4\pi\rho_h\alpha_h^3} \left[\frac{g(\Delta,h)}{a}\right] C(i_o)S_{rad}(t)*\dot{M}(t)$$

where * : convolution operator; a : earth's radius; $g(\Delta,h)$: geometrical spreading factor; $C(i_o)$: free surface response at the receiver station; ρ_h, α_h: density and velocity at the source; $\dot{M}(t)$: seismic moment rate; and

$$S_{rad}(t) = R^P(\theta,\phi)\delta(t) + R^{pP}(\theta,\phi)\delta(t-\Delta t_{pP}) + R^{sP}(\theta,\phi)\delta(t-\Delta t_{sP}) \quad (2)$$

is the combined effect of the radiation pattern and the near-source surface reflexions. $\Delta t_{pP} = t_{pP} - t_P$ and $\Delta t_{sP} - t_P$ are the time delays between the reflected and direct P phase. $R^i(\theta,\phi)$ is the radiation pattern for the phase i. $\delta(t)$ is Dirac's delta function.

After Fourier transforming equation (1), we obtain the moment rate spectral density $\dot{M}(\omega)$:

$$\dot{M}(\omega) = \frac{4\pi\rho\alpha^3 a}{g(\Delta,h)C(i_i)} \left[\frac{Q(\omega)}{I(\omega)} \frac{W(\omega)}{S_{rad}(\omega)} \right] \tag{3}$$

where ω is the circular frequency.

The real part of the attenuation operator is $Q(\omega) = \exp(-\omega t^*/2)$, where t^* is the attenuation parameter defined by:

$$t^* = \int \frac{ds}{Q(s)\alpha(s)} \tag{4}$$

and $\alpha(s)$ is the P-wave velocity. Many authors discuss the dependence of t^* on different parameters. Kanamori (1967) established a relation between t^* and station distance, Der and Lees (1985) discussed the variation of t^* with tectonic structure in the Central and South Western United States. In our case, no studies of attenuation along the source station paths of Chilean earthquakes is available. For this reason we chose a constant $t^* = 0.7$ second. However, reasonable changes of this parameter are significant only at high frequencies and they should not affect the spectral amplitude at intermediate and long periods.

Fourier transforming equation (2) we obtain the effective radiation pattern $S(\omega)$ as:

$$S_{rad}(\omega) = R^P(\theta,\phi) + R^{pP}(\theta,\phi)e^{i\omega\Delta t_{pP}} + R^{sP}(\theta,\phi)e^{i\omega\Delta t_{sP}} \tag{5}$$

which becomes at very low frequency,

$$S_{rad}(0) \to R^P(\theta,\phi) + R^{pP}(\theta,\phi) + R^{sP}(\theta,\phi). \tag{6}$$

Moment rate spectra from body waves may be obtained from (3). The static seismic moment $M_o = \dot{M}(0)$ may also be obtained from (3) using (6) for $S_{rad}(0)$:

$$M_o = \frac{4\pi\rho\alpha^3 a}{g(\Delta,h)C(i_o)} \left[\frac{W(0)}{I(0) * S_{rad}(0)} \right].$$

Unfortunately, for shallow thrust earthquakes a very good knowledge of the fault plane is required because the different terms in (6) almost cancel each other. For this reason, the estimation of seismic moment becomes very unstable. Houston and Kanamori (1986) discussed this problem and proposed to replace the frequency dependent function $S(\omega)$ by an average radiation pattern function that does not depend on frequency. From the statistical average of the spectra of eight thrust earthquakes with fault planes dipping around $30°$ and observed at several stations, they proposed to replace $S(\omega)$ by the root mean squared average scalar radiation pattern defined by:

$$S^{RMS} \simeq \sqrt{(R^P)^2 + (R^{pP})^2 + (R^{sP})^2}. \tag{7}$$

The seismic moments for the Chilean earthquakes of March and April, 1985 computed using (7) are systematically smaller than those found using the exact frequency dependent function $S_{rad}(\omega)$. Since we have a relatively good knowledge of the source mechanism for these earthquakes we decided to explore the problem of estimating the seismic moment in more detail.

As noted above, the two expressions (6) and (7) for the correction of observed spectra for radiation pattern and depth of the source may give significantly different estimates of the seismic moment, especially if one takes into account errors in depth and source time function. This is particularly true for Chilean earthquakes recorded at northern hemisphere stations. The reason is that the function $S(\omega)$ has a minimum at $\omega = 0$ and several peaks and troughs at higher frequencies. Its first peak is at a frequency ω_1 that is inversely proportional to the time delay of the depth phases with respect to the direct P. Its amplitude depends on the relative values of the radiation patterns for P, pP and sP. Let us designate by S_{max} this peak. Thus, theoretically at least, the observed spectra should present a peak at low frequencies. This peak is not observed in practice because of noise: spectra of the P-wave train for the Valparaiso earthquake look just like Brune spectra with a flat low frequency level. Therefore the problem of estimating low frequency seismic moments from body wave trains reduces to that of extrapolating observed body wave spectra to low frequencies.

In order to illustrate this problem we show on Figure 6 the theoretical body wave spectra calculated for the Chilean earthquakes of March 3, 1985 and April 9, 1985 as they would have been observed at station SCP in the absence of noise. For the smaller aftershock the predicted spectrum presents a clear peak at low frequencies (see Figure 6b). In the figure ω_1 represents the peak of $S(\omega)$ and ω_0 the corner frequency. It is clear that for earthquakes such that $\omega_1 < \omega_0$ the spectral peak is well developed. This will occur for small events such that ω_0 is large, and for deeper events such that the depth phase delay is large and, consequently, Δt_{pP} is relatively large. If the decreasing part of the spectrum at low frequencies is hidden by noise, the spectrum should be corrected by S_{max} in order to obtain M_0. For small or deeper earthquakes such that $\omega_1 < \omega_0$, the correction S^{RMS} proposed by Houston and Kanamori (1986) is a good approximation to S_{max}. Thus, their method of correction is correct for events with duration shorter than the depth phase delay.

For larger or shallower earthquakes, like that of March 3, 1985, in Valparaiso, the two frequencies ω_1 and ω_0 are close to each other and the theoretical spectrum shown in Figure 6a presents a much flatter shape. The peak S_{max} of $S(\omega)$ is now much lower than S^{RMS}. In this case the appropriate correction is between (6) and (7) as can be observed in the figure.

In this study we used Houston and Kanamori's (1986) correction (7) in order to invert the moment of the three larger Valparaiso aftershocks (17, 19 March and 9 April) whose source time duration is smaller than the delay time of the pP phase and the low frequency correction (Eq. (6)) for the Valparaiso main shock where the source time duration is longer than the delay time of pP phase.

4.2 Results of Spectral Analysis of P-Waves

Seismic moment rate spectra were calculated for all available digital recordings of the four events under study. The P-wave trains in the original digital records were windowed, detrended and tapered before performing a fast Fourier transformation. Then the corrections of

360

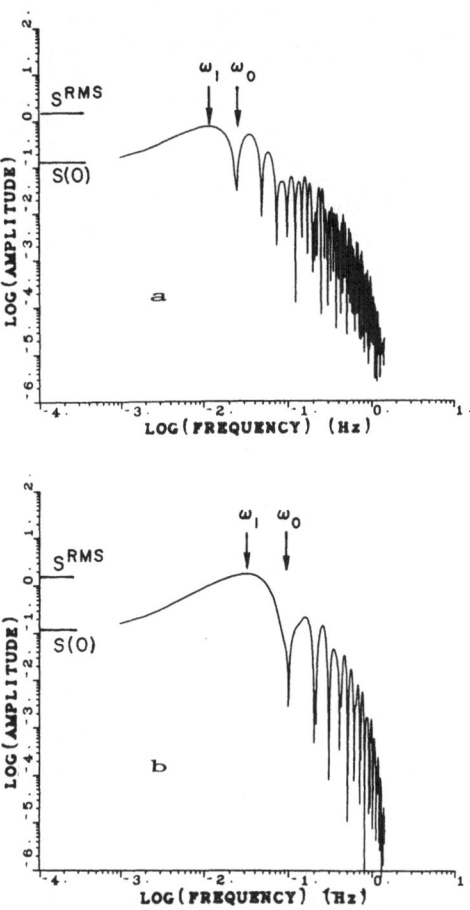

Fig. 6. Synthetic spectra calculated for two earthquakes in the region of
Valparaiso observed at station LON. (a) For a large event with a
source duration of 40 s. (b) For a small event of duration of
5 s. The spectra of the synthetic data were calculated taking
into account only the terms $\dot{M}(\omega)$ and $S(\omega)$ in (1). ω_1 is the peak
of $S(\omega)$ and ω_0 is the corner frequency.

equation (4) were applied. We studied in this way the spectra of
intermediate and long period teleseismic GDSN records of the four
earthquakes and we used, in addition, the broad-band GEOSCOPE recording
obtained at station SSB in France. All the spectra were corrected for the
radiation pattern using the same fault plane orientation for all the
events (see Table 2). We used the fault plane solution determined by
Korrat and Madariaga (1986) from long period body waves. These thrust
mechanisms are also close to the CMT (Centroid Moment Tensor) solution
listed in the PDE report. The depth reported on Table 2 for each event
were determined by modeling (Korrat and Madariaga, 1986) and inversion
(this study) of P-waves. The depth found by the two methods are similar.
From simultaneous determination of depth and source time function,
Christensen and Ruff (1985) found that the depth of the major shock
extended between 10 and 40 km. The depths used in the study of each
event are within this interval and agree with that of other underthrusting
events in this region (Malgrange and Madariaga, 1983). The crustal

Table 2. Mechanism and Moment of the Studied Earthquakes

Date	ϕ	λ	δ	H	M_o (inv)		M_o (spe)		M_o (mod)
	°	°	°	Km	LP & BRB		LP & IZ		LP
							$\times 10^{26}$ dyn.cm		
1985 March 03	0	24	100	25	62	55	180		40
1985 March 17	0	24	100	25	0.4	0.4	0.4	0.5	-
1985 March 19	0	24	100	25	0.65	-	0.6	0.8	0.5
1985 April 09	0	24	100	35	2.5	2.4	3.0	4.0	4.0

inv : from inversion of P-waves
spe : from spectral method
mod : from modeling (Korrat, 1986)

velocity and density used for the computation of the P-wave train are 6.5
km/s and 3 g/cm^3 respectively. The geometric spreading g(Δ,h) is taken
from the table by Kanamori and Stewart (1976).

Figure 7 shows a comparison between the spectral amplitude of the
seismic signal (top) and that of the noise (bottom) preceding the P
arrival. The signal to noise ratio is greater than 1000 for frequencies
above 0.18 and 0.14 Hz for stations RSON and RSNY, respectively. This
example shows that the signal level is considerably higher than the noise
level near the peak frequency. So that the corner frequency and seismic
moment determined from the data are well defined.

The main shock. On Figure 8 we plot the corrected moment rate
spectrum determined from long and intermediate period recordings obtained
at three stations (RSON, RSNY and LON). We verify that the spectra from
long period records (dashed line) is comparable to that from the
intermediate periods (solid line) for the three stations. Obviously the
intermediate period record gives broader frequency range information
compared to that from long period seismograms. At frequencies greater
than 0.3 Hz, the moment rate spectrum obtained from long period records
becomes lower than the noise level and may not be used. We conclude that
the seismograms recorded by broad-band instruments are the most suitable
for the study of the seismic source. The arrow in each of the spectra on
Figure 8 indicates the maximum spectral peak that was identified and the
dashed lines indicate the extrapolated flat level and the high frequency
slope. In practice, we estimate the flat level from the maximum of the
moment rate spectrum. From the flat level we determine the value of the
seismic moment (M_o) using the correction S^{RMS} for the smaller events and
$S_{rad}(0)$ for the main shock. These choices were discussed in the previous
section. The source time function duration (T) was calculated from the
corner frequency using the simple expression $T = f_o^{-1}/2$. There is a
significant variation of the values of M_o and T from station to station.
The variation in estimated moment (0.9 to 3.0 10^{28} dyne.cm) is probably
due to errors in the radiation pattern.

On Figure 9a, we plot the average moment rate spectrum of the main
shock. We used separately 5 long and 5 intermediate period records in
order to compute these averages. We deduced from these spectra two
average parameters: $M_o \simeq 1.8 \ 10^{28}$ dyne.cm and $T \simeq 50$ s. The same
technique applied to the broad-band GEOSCOPE record at SSB (see Figure 10)
gives $M_o \simeq 2.0 \ 10^{28}$ dyne.cm and $T \simeq 60$ seconds. These values are
comparable to those gotten from spectra of GDSN records.

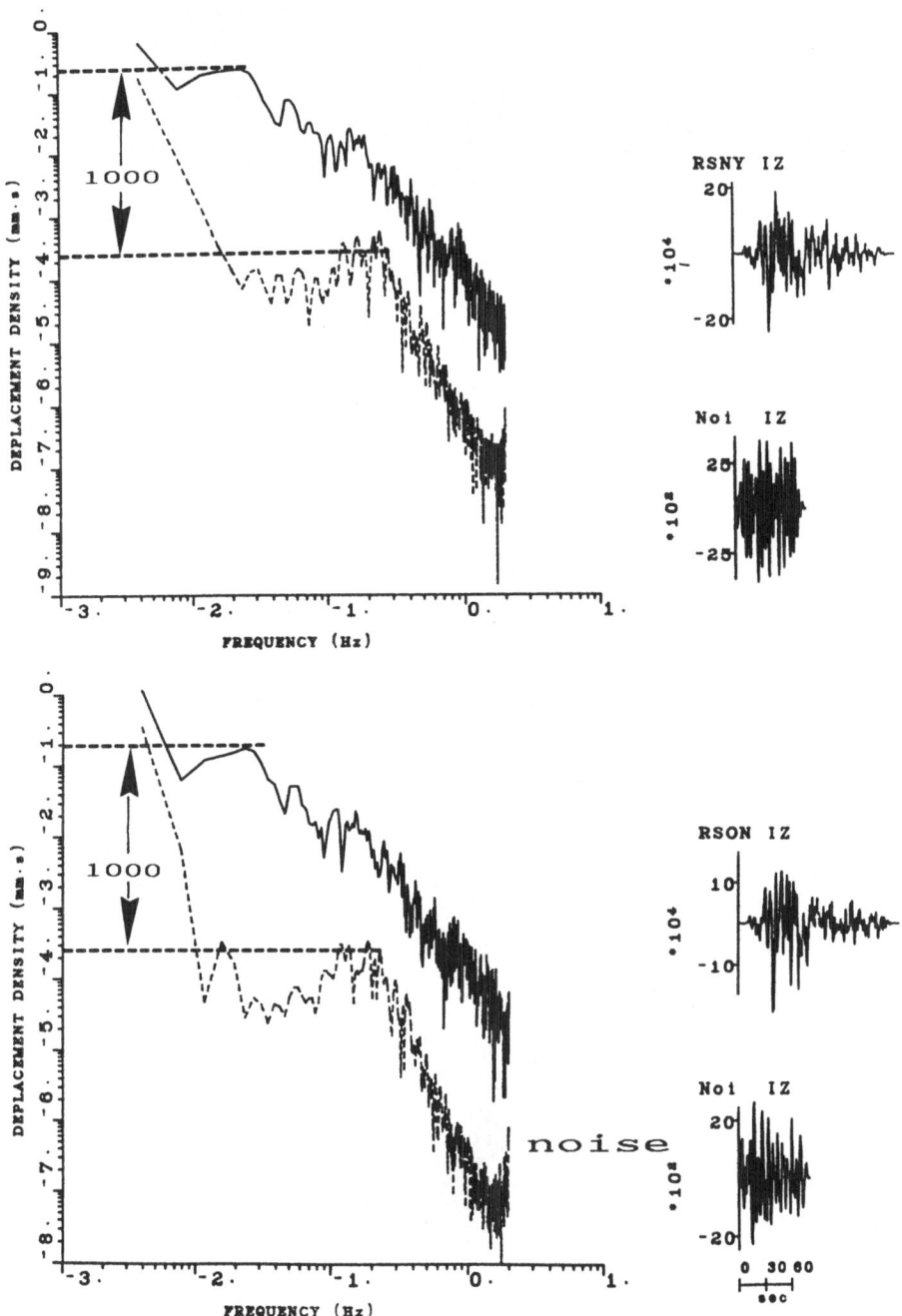

Fig. 7.   Comparison between spectral amplitude of the seismic signal of
          March 3, 1985, Valparaiso earthquake (top) and that of the noise
          (bottom) preceding the P arrival. Let us remark that the signal
          to noise ratio is greater than 1000.

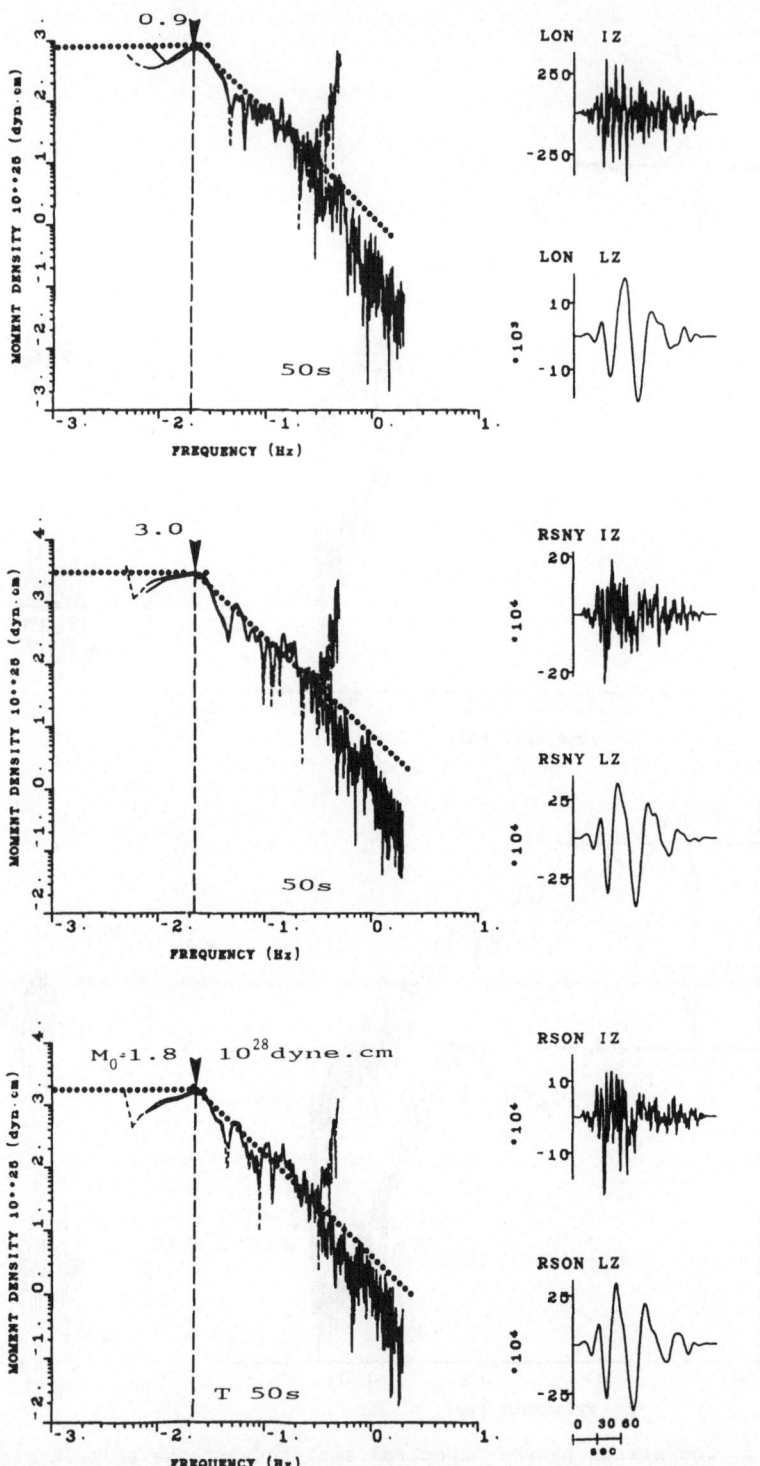

Fig. 8. The corrected moment rate spectrum of March 3, 1985, Valparaiso
earthquake determined from Long (dashed line) and Intermediate
period GDSN (continuous line) period recordings shown on the
right side of each spectrum. Spectral shape (slope and flat) and
corner frequency are denoted, respectively, by the dotted line
and arrow.

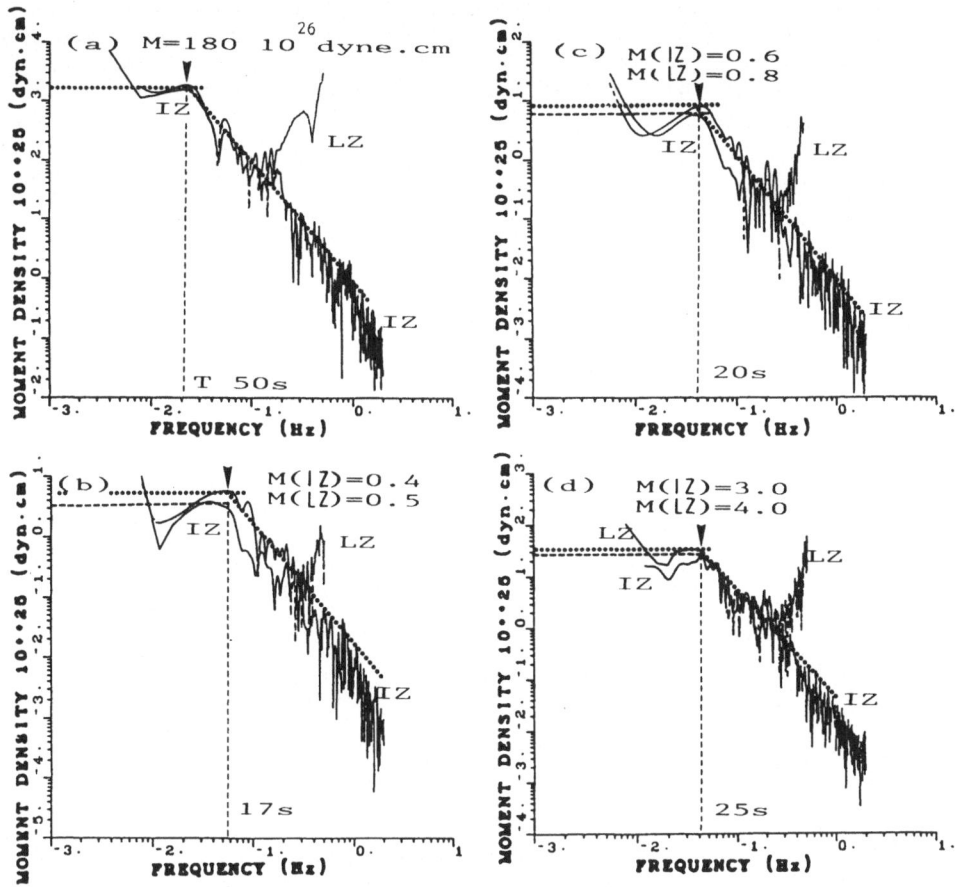

Fig. 9. Average moment rate spectrum of the 1985, (a) March 3, (b) March
17, (c) March 19 and (d) April 9 Valparaiso earthquakes. Each
figure presents a comparison between long (dashed line) and
intermediate (continuous line) period. For notation, see
Figure 8.

The large aftershocks. The spectra from the large aftershocks of
March 17, 19 and April 9 are shown in Figures 9b, 9c and 9d. From 5 long
and 4 intermediate periods records of the 9 April aftershock we deduced
the averages values (see Figure 10d) of M_O and T which are, respectively,
$3-4 \ 10^{26}$ dyne.cm and 25 seconds. The two smaller aftershocks of March 17
and 19, 1985 are very similar and so are their average moment rate spectra
(given from 3 LP and 3 IP). On Figure 9b and 9c we can see that the
seismic moments are 4-5 and 6-8 10^{25} dyne.cm for the earthquakes of March
17 and 19 respectively. The source time duration is ($T \simeq$) 17 and 20
seconds for these two aftershocks, respectively.

4.3 Interpretation of Results and Validity of the Spectral Method

A seismic moment of $1.2 \ 10^{28}$ dyne.cm was determined for the main
shock from surface waves recorded at VLP (Very Long Period) GEOSCOPE
stations (Monfret and Romanowicz, 1986). Independently, from the
aftershock area and average slip, Comte et al. (1986) estimated that the
seismic moment was greater than 10^{28} dyne.cm. These two values are close
to that computed above by spectral analysis of P-waves ($M_O \simeq 1.8 \ 10^{28}$
dyne.cm). The corner frequency extracted from spectral shape gives a

characteristic source length in the range 150 to 180 km for rupture velocities in range 2.5 to 3 km.s^{-1}. The coherence of the moment and size of the main shock obtained by different methods, demonstrates that the spectral method yields good estimates of the seismic moment and the source size.

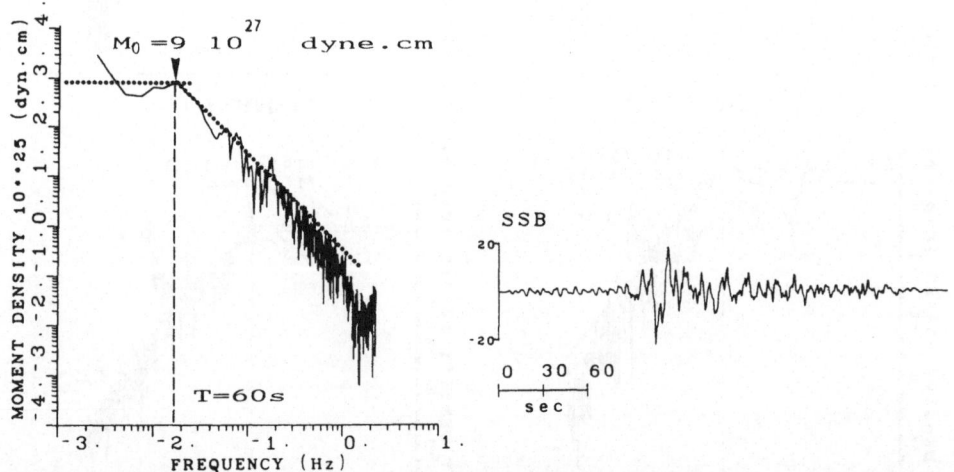

Fig. 10. Moment rate spectrum of the March 3, 1985, Valparaiso earthquake obtained from the broad-band GEOSCOPE record at SSB (Saint Sauveur in France). For notation, see Figure 8.

It is well established that the surface wave magnitude M_s saturates for rupture lengths greater than about 100 km (Kanamori and Anderson, 1975). In order to avoid this saturation, Kanamori (1977) introduced the moment magnitude M_W. From the seismic moment obtained by the spectral method, we determined M_W for the four Chilean events. We obtained M_W = 8.1 (M_s = 7.8) for the main shock of March 3, 1985, M_W = 6.9 - 7.0 (M_s = 7.2) for the aftershock of April 9, 1985, and M_W = 6.3 - 6.4 (M_s = 6.6) and M_W = 6.4 - 6.5 (M_s = 6.6), respectively, for the two smaller aftershocks of 17 and 19 March 1985. The three aftershocks, in particular those of March, had fault lengths smaller than 100 km. The moment magnitude of these aftershocks agrees with the 20 seconds surface wave magnitude M_s determined by USGS.

Finally, we observe, that none of the events studied here had the complex spectral shape proposed by Gusev (1983). Their spectra do not present the decrease in aspectral amplitude in the frequencies range from 0.2 to 10 Hz suggested by Gusev (1983). In fact the spectra look extremely simple and as can be observed from Figure 9 for the main shock they can be modeled with the simple spectral shape proposed by Brune (1970).

366

5. TIME DOMAIN INVERSION OF P-WAVES

5.1 Method

The details of the method of inversion used in this study have been described by numerous authors. The algorithm of inversion is described by Kikuchi and Kanamori (1982) but the version used here was presented recently by Bezzeghoud et al. (1986). The observed far-field displacement (1) can be rewritten in the form $W(t) = \omega(t)*\dot{M}(t)$. Where $\omega(t)$ is the Green function and $\dot{M}(t)$ the moment rate function for an equivalent point source. With this definition the Green function $\omega(t)$ contains three spikes at the arrival times of P, pP and sP waves. In order to make a stable estimate of $\dot{M}(t)$ we have to use a Green function with finite duration. For this purpose we write the moment rate in the form:

$$\dot{M}(t) = \sum_i M_i s(t - T_i)$$

where $s(t)$ is an elementary sampling function. The source model is represented by the discrete set of values (M_i, T_i). Inserting this model of $\dot{M}$ in (1) we obtain the following form for the synthetic:

$$W(t) = \sum_i M_i g(t - T_i)$$

where $g(i)$ is the elementary seismogram calculated using(1) with a sampling function $s(t)$ as the source time function. In our inversions we used a simple trapezoidal function for $s(t)$. This function has two time constants: (1) the dislocation rise time (r) and (2) the elementary rupture time t_r. Each point source is extracted by a least square method iteratively from the correlation between the observed record and the Green function $g(t)$ (see Kikuchi and Kanamori for further discussion). For a fixed number of iterations, the convergence rate depends on the choice of the combination of (r, t_r). Kikuchi and Fukao (1985, Figure 5) show an example in which they studied three combinations of (r, t_r). In our inversions we used two sets of values $(r = 4$ seconds, $t_r = 10$ seconds) and $(3, 4$ seconds) for the main shock. For the large aftershock of April 9 and the two smaller aftershocks of March 17 and 19 we used a single set of values $(r = 3$ seconds, $t_r = 4$ seconds). The focal mechanism, depth, P-wave velocity and density used to compute the elementary far field seismogram are the same as those used in the section on spectral analysis.

5.2 Results

In the following we discuss the results of inversion of the source time function from long period and deconvolved broad-band velocity records of the March and April, 1985 earthquakes. The broad-band records were obtained using a low frequency cut-off of 0.04 Hz (T = 25 seconds) for all the Intermediate Period records available.

The main shock. On the left hand side of Figure 11a, we compare synthetic (dashed line) and broad-band data (solid line) of the 1985 Valparaiso main shock. The source time functions obtained by inversion are shown in the right hand side of this figure. The results of the inversion of the long period records is shown at the bottom of Figure 11. The source time functions are simple and coherent whether they are inverted from long period (LP) or from deconvolved broad-band (BRB) records. All the source time functions, except AFI at LP and TOL at BRB, contain three pulses. Firstly, a small precursor of a duration of about 15 seconds followed by a very strong pulse which we identify as the main shock which has a total duration of about 40 seconds. The third and last

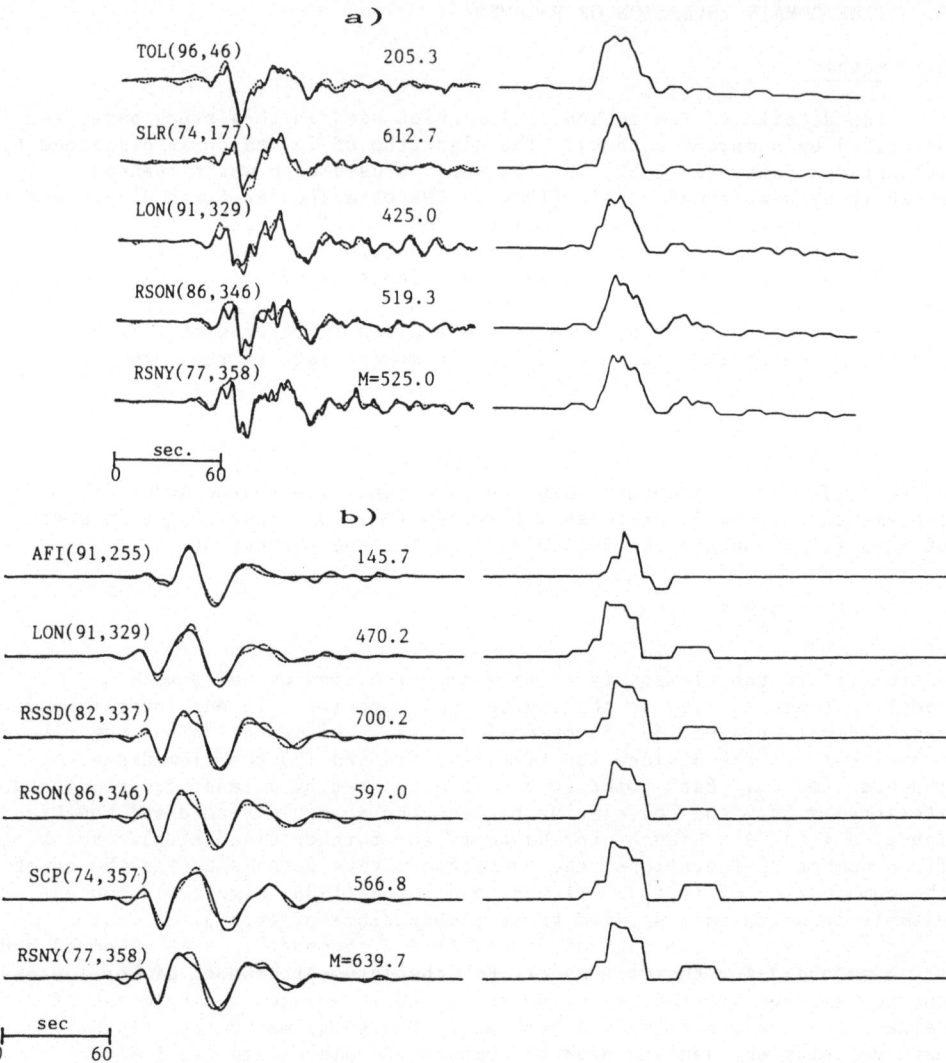

Fig. 11. The March 3, 1985, Valparaiso earthquake waveform data together with the best fit synthetics for (a) broad-band and (b) long period records. On the right hand side of each signal is plotted the source time function obtained by iterative inversion. M_o, A and D denote, respectively, seismic moment value (x 10^{25} given in dyne.cm), station azimuth and distance (given in degrees).

pulse has a time delay of about 58 seconds after the precursor. It is very clear in the inversion of long period data but it is less obvious in the inversion of broad-band data. The total duration of the source time function is in the range from 77 to 83 seconds. Comparison between the data and synthetics shows that the best fit is obtained from broad-band records (Figure 11a). In particular, the broad-band data give the best fit for the precursor. This is not surprising, because the deconvolved broad-band record has a better resolution at high frequencies than the long period one. The form of the source time function agrees with those found by other authors who studied the main shock (Korrat and Madariaga,

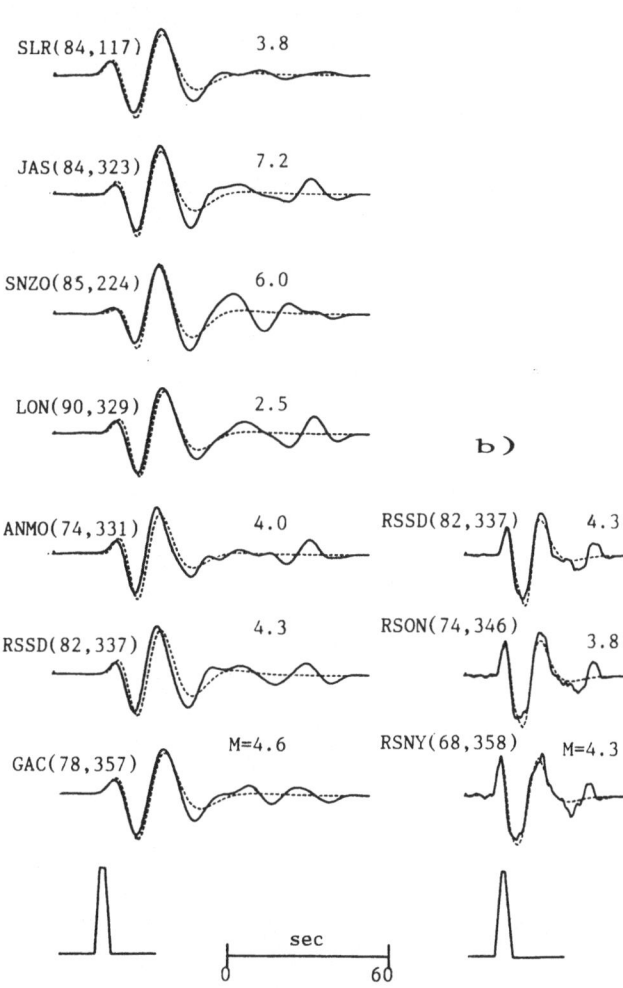

Fig. 12. The March 17, 1985, Valparaiso earthquake waveform data together with the best fit synthetics for (a) long and (b) broad-band period records. The source time function for each bandpass is plotted at the bottom. For notation, see Figure 11.

1986; Christensen and Ruff, 1986). The seismic moment (including the precursor) ranges from 4.7 to 7.0 10^{27} dyne.cm (except at AFI which is too far from the epicenter) for LP and from 4.2 to 6.1 10^{27} dyne.cm (except TOL which is diffracted) for BRB records. The average seismic moments computed separately from LP (6.2 10^{27} dyne.cm) and BRB (5.5 10^{27} dyne.cm) are remarkably coherent between them. The moment computed from long period P-wave modeling by Korrat and Madariaga (1986) ($M_o \simeq 4\ 10^{27}$ dyne.cm) and by deconvolution of the source function from P-waves by Christensen and Ruff (1986) ($M_o \simeq 7\ 10^{27}$ dyne.cm) are very close to our estimate.

The large aftershocks. Figures 12, 13 and 14 shows the aftershock waveform data together with the best fitting synthetics for the long period (a) and broad-band records (b). The slight time shift that is

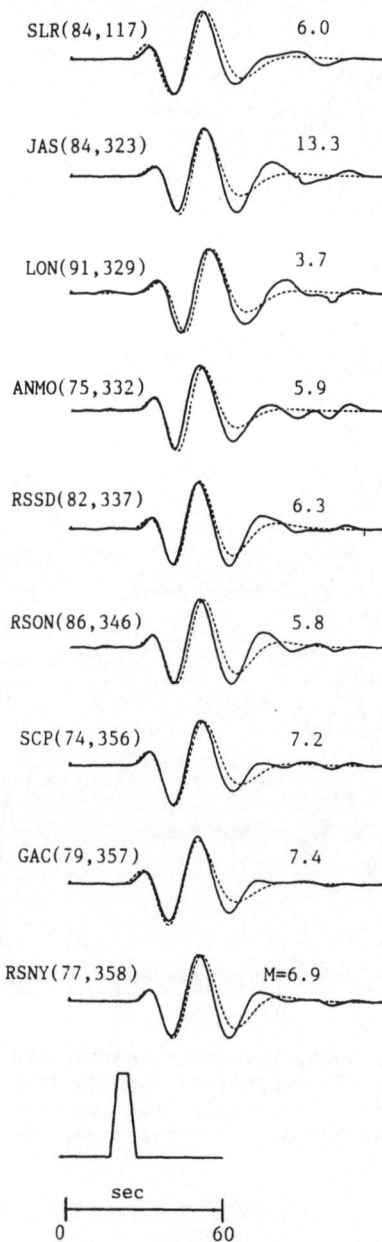

Fig. 13. The March 19, 1985, Valparaiso earthquake. See Figure 11.

observed between synthetics and observed records is due mainly to a poor modeling of the PcP phase. Most of the station used to model the aftershocks had epicentral distances greater than 84°. In this case the PcP phase arrives less than 5 seconds after the onset of the P-wave. We observe clearly this phase on the broad-band records of the aftershock of April 9, 1985 (Figure 14b). The source time function found by iterative deconvolution from LP and BRB records is very simple for the three aftershocks. Nevertheless the duration is about 6 seconds for the March 17 event and about 9 seconds for the two others (March 19 and April 9); they are simple trapezoids as seen in the bottom of Figures 12, 13 and 14.

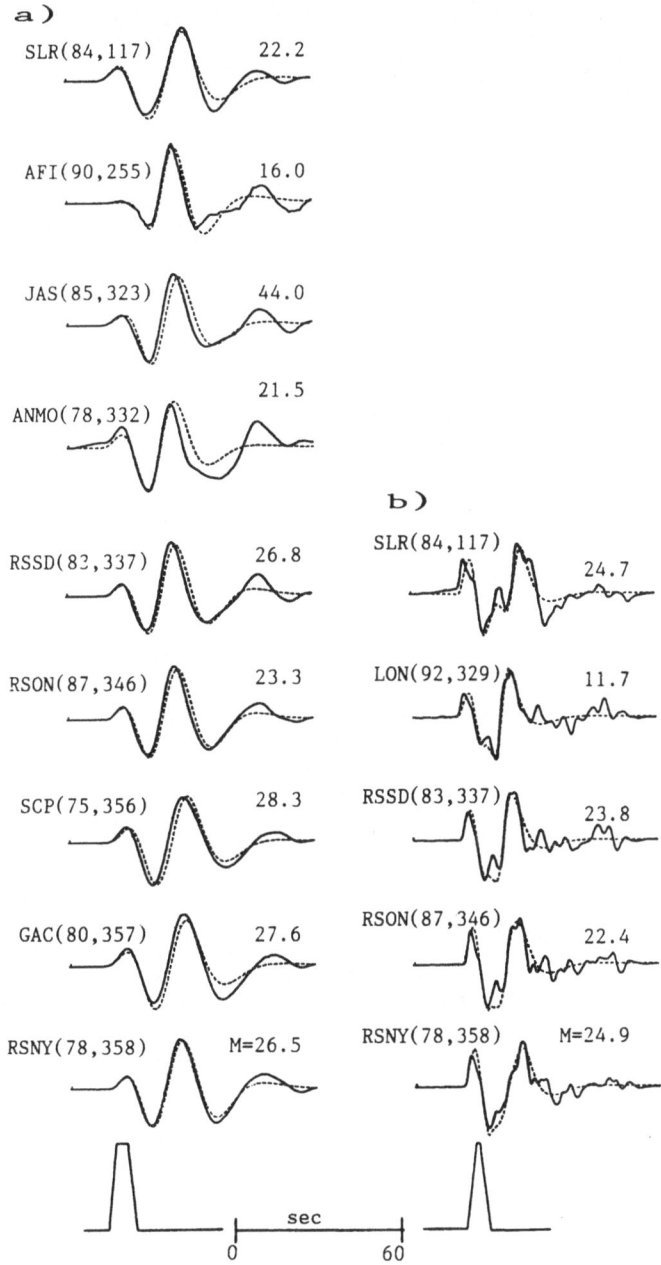

Fig. 14. The April 9, 1985, Valparaiso earthquake. See Figure 11.

On the other hand the sP reflected phase is clearly identified on the broad-band records of the April 9, 1985 aftershock. We used this observation to show that this aftershock is deeper than the two aftershocks of March 1985. The PDE report also estimates that the events of March 19 and April 9 are deeper than those of March 3 and 17. However, because we lack broad-band data we can not determine precisely the depth of the earthquake of March 19. The seismic moment estimated from different stations are very similar and coherent. The average seismic moment of the April 9 event is close to 2.4 10^{26} dyne.cm (2.5 from LP and

2.4 from BRB). Those of the aftershocks of March 17 and 19 are $4.0 \ 10^{25}$ and $6.5 \ 10^{25}$ dyne.cm, respectively. Finally, let us remark that the seismic moment released by the main shock is larger by a factor of about 30 with respect to the event of April 9 and of about 150 with respect to the March aftershocks.

6. DISCUSSION

We showed that it is now possible to recover the seismic moment and the time history of moment rate release at the source of large earthquakes from the digital recordings obtained by the GDSN stations and the broad-band data from a GEOSCOPE station (SSB). From intermediate and broad-band data we are able to reconstruct the source spectrum over a very broad frequency band, from about 0.01 Hz to 1 Hz. From these data we can estimate two source parameters: (1) source time duration and (2) the seismic moment.

Our results show that the seismic moment of the main shock inverted from spectral data is three times larger than the moment obtained by time domain modeling or inversion of body waves. The spectral seismic moment ($1.8 \ 10^{28}$ dyne.cm) is quite close to that determined from surface wave excitation ($1.2 \ 10^{28}$ dyne.cm, Monfret and Romanowicz, 1986) and the CMTS method ($M_O = 1.0 \ 10^{28}$ dyne.cm, PDE report). On the other hand, for the three largest aftershocks we found that both methods (spectral and inversion) give similar seismic moments. Thus, we believe the source parameters estimated from the spectral method for shallow events with source time durations less than 20 s are quite acceptable. For larger earthquakes, like the Valparaiso main shock, the main problem for obtaining the total seismic moment from body wave spectra lies in the appropriate estimation of the effective radiation pattern $S(\omega)$ of equation (6). For shallow thrust mechanisms, $S(\omega)$ may vary by an order of magnitude at low frequencies presenting a minimum at zero frequency and a maximum at a finite frequency inversely proportional to the delay between the direct and the surface reflected phases. Correcting the observed spectra for radiation and source depth requires a very good knowledge of the source mechanism. In addition, there is a strong interference between the spectral peak due to the depth phases and the corner frequency. We propose that this interference between the depth phases and the source duration is at the origin of the difference usually observed between surface wave and body wave moments.

We did not observe in any of our spectra the variation in slope from 0.1 to 10 Hz proposed by Gusev (1983). According to Aki (1983), Gusev's bump should be generated by the heterogeneity of fault plane characterized by barriers and asperities. It is natural to think that the fault complexity of the seismic source may produce an inflection in the spectrum. However, the spectra of the Chilean earthquakes are simple and do not show the characteristic frequencies predicted by Gusev (1983). Houston and Kanamori (1986) reached a similar conclusion from the analysis of the spectra of seven large earthquakes recorded at GDSN stations and from hand-digitized WWSSN records. Spectra are certainly more complex than the simple model proposed by Brune (1970), but they do not show systematic behavior like that proposed by Gusev (1983).

REFERENCES

Aki, K., 1966, Generation and propagation of G waves from Niigata earthquake of June 16, 1964. Part 2. Estimation of earthquake moment, from the G wave spectrum, Bull. Earthquake Res. Ins. Univ., Tokyo, 44:73-88.

Aki, K., 1983, "Strong-motion Seismology", Proceedings of the International School of Physics, Enrico Fermi, Earthquakes Observation, Theory and Interpretation, H. Kanamori and E. Boschi, eds.

Barazangi, M., and Isacks, B., 1976, Spatial distribution of earthquakes and subduction of the Nazca plate beneath South America, Geology, 4:686-692.

Bezzeghoud, M., Deschamps, A., and Madariaga, R., 1986, Broad-band modeling of the Corinth Greece earthquakes of February and March 1981, Anns. Geophys., B3:295-304.

Bezzeghoud, M., 1987, Inversion et analyse spectrale des ondes P. Potentialité des données numériques large bande. Applications à des séismes Méditerranéens et Chiliens, Thèse de doctorat, Université Paris VII.

Brune, J., 1970, Tectonic stress and the spectra of seismic shear waves from earthquakes, J. Geophys. Res., 5:4997-5009.

Choy, G. L., and Boatwright, J., 1981, The rupture characteristics of two deep earthquakes inferred from broad-band GDSN data, Bull. Seismol. Soc. Am., 1:691-711.

Christensen, D. H., and Ruff, L. J., 1986, Rupture process of the March 3, 1985 Chilean earthquake, Geophys. Res. Lett., 3:721-724.

Comte, D., Eisenberg, A., Lorca, E., Pardo, M., Ponce, L., Saragonie, R., Singh, S. K., and Suarez, G., 1986, The central Chile earthquake of 3 March 1985: a repeat of previous great earthquake in the region?, Science, 33:449-452.

Deschamps, A., Lyon Caen, H., and Madariaga, R., 1980, Mise au point des méthodes de calcul de sismogrammes synthétiques de longue période, Ann. Geophys., 6:167-178.

Dziewonski, A. M., Chou, T. A., and Woodhouse, J. H., 1981, Determination of earthquake source parameters from waveform data for studies of global regional seismicity, J. Geophys. Res., 6:2825-2852.

Eyidogan, H., Nabelek, J., and Toksoz, M. N., 1985, The Gazli, USSR, 19 March 1984 earthquake: the mechanism and tectonic implication, Bull. Seism. Soc. Am., 5:661-675.

Gusev, A. A., 1983, Descriptive statistical model of earthquake source radiation and its application to an estimation of short-period strong motion, Geophys. J. R. Astron. Soc., 4:787-808.

Houston, H., Kanamori, H., 1986, Source spectra of great earthquake: teleseismic constraints on rupture process and strong motion, Bull. Seism. Soc. Am., 6:19-42.

Kanamori, H., and Anderson, D. L., 1975, Theoretical basis of some empirical relation in seismologie, Bull. Seism. Soc. Am., 5:1073-1095.

Kanamori, H., and Stewart, G., 1976, Mode of strain release along the Gibbs fracture zone, Mid-Atlantic Ridge, Phys. Earth Planet. Interiors, 1:312-332.

Kanamori, H., 1977, The energy release in great earthquakes, J. Geophys. Res., 2:2981-2987.

Kelleher, J., 1972, Rupture zones of large South American earthquakes and some predictions, J. Geophys. Res., 7:2087-2103.

Kikuchi, M., and Kanamori, H., 1982, Inversion of complex body waves, Bull. Seism. Soc. Am., 2:491-506.

Kind, R., and Seidl, D., 1982, Analysis of broad-band seismograms from the Chile-Peru area, Bull. Seism. Soc. Am., 2:2131-2145.

Korrat, I., 1986, Mécanisme et distribution spatiale des épicentres en
    relation avec la rupture de la lacune de Valparaison (Chili) en
    Mars 1985, Thèse de doctorat, Université de Paris VII.
Korrat, I., and Madariaga, R., 1986, Rupture of the Valparaiso (Chile) gap
    from 1971 to 1985, in: "Earthquake Source Mechanics", Maurice Ewing
    volume 6, S. Das, J. Boatwright and C. H. Scholz, eds., pp 247-258,
    American Geophysical Union, Washington DC.
Malgrange, M., and Madariaga, R., 1983, Complex distribution of large
    thrust and normal fault earthquakes in Chilean subduction zone,
    Geophys. J. R. Astron. Soc., 3:489-505.
McCann, W., Nishenko, S., Sykes, L., and Krause, J., 1979, Seismic gaps
    and plate tectonics: seismic potential for major boundaries,
    Pageoph, 17:1082-1147.
Monfret, T., and Romanowicz, B., 1986, Importance of on scale observation
    of first arriving rayleigh wave trains for source studies: example
    of the Chilean event of March 3, 1985, observed on the GEOSCOPE and
    IDA networks, Geophys. Res. Lett., 3:1015-1018.
Nishenko, S., 1985, Seismic potential for large and great interplate
    earthquakes along the Chilean and Southern Peruvian margins of
    South America: a quantitative re-appraisal, J. Geophys. Res.,
    0:3589-3615.
Romanowicz, B., Cara, M., Fels, J. F., and Rouland, D., 1984, GEOSCOPE: a
    French initiative in long-period three component global seismic
    network, EOS, Trans. AGU, 2:753-754.
Ruff, L. J., Kanamori, H., 1983, The rupture process and asperity
    distribution of three great earthquakes from long-period diffracted
    P-waves, Phys. Earth Planet. Inter., 1:202-230.

SITE RESPONSE USING DIGITAL TECHNIQUES:

AN APPLICATION TO THE ABRUZZO AREA - CENTRAL ITALY

Edoardo Del Pezzo(*) and Marcello Martini

Osservatorio Vesuviano
80056 Ercolano, Italy

INTRODUCTION

The geological characteristics of the upper layers beneath the
seismological station affect the spectral shape of the recorded seismic
waves. This effect is particularly known in the strong motion seismology
and earthquake engineering.

The effects of a strong earthquake are dependent on the nature of the
ground on which the building is constructed. In particular the so-called
"hard rock" as for instance granite, do not present particular
amplifications in determined frequency bands of the seismic spectrum,
while, on the contrary, loose materials, especially when they are filled
by water, amplificate seismic motion in frequency bands that can be close
to the self-oscillation frequency of the buildings. These effects are
called "site effects". Their knowledge is very useful in seismic source
studies too, when it is necessary to remove the anomalous peaks from the
spectra for fitting them to the source models.

The attenuation along the ray-path from hypocenter to the recording
station can produce a similar effect on the spectrum of the incoming
waveform, so it is in principle a quite difficult task to separate both
effects if the attenuation is unknown.

The spectrum of the recorded seismic phase (the P phase, for example)
can be written (Bath, 1974) as

$$R(w) = S(w) \cdot R(\phi,\theta) \cdot I(w) \cdot Si(w,\phi,\theta) \cdot G \cdot A(w) \qquad (1)$$

where $S(w)$ is the source spectrum, $R(\phi,\theta)$ is the radiation pattern
function, $I(w)$ is the instrument transfer function, G is the geometrical
spreading coefficient and $A(w)$ is the attenuation along the ray-path.
$Si(w,\phi,\theta)$ is the site response function. As it is shown there is an
angular dependence in the site response function $Si(w,\phi,\theta)$, which takes
into account the effects due to the possible variation of the local
geology with the azimuth and incidence of the incoming wavefront.

(*) Present address: Istituto di Scienze della Terra, University of
    Catania, Italy.

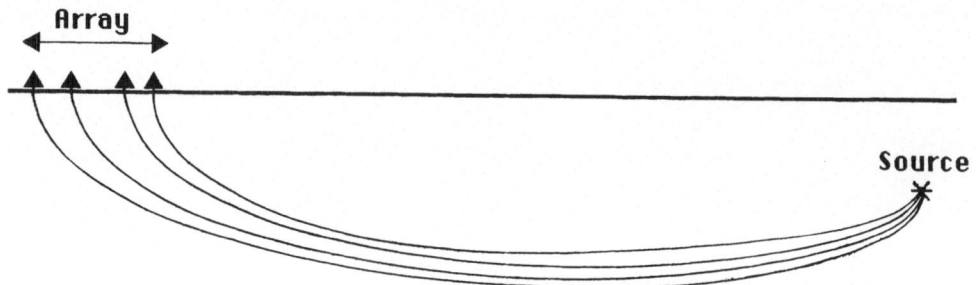

Fig. 1. Scheme in which an array of stations records an earthquake located outside the array itself. Ray-paths share an almost common zone outside the array and sample different portions of earth crust only beneath the stations.

In Figure 1 is shown a scheme in which an array of stations records a wavefront coming from an earthquake located outside the array. Because the travel paths hypocenter – stations share a common zone for long distances compared to the size of the array – the modification of the spectra of the seismic pulse travelling along these paths can reasonably occur beneath the array. It is clear that, for example, if the distance among the stations is of the order of some tenths of kilometers, the modification at the site occurs approximately in a volume with linear dimension of the same order of magnitude. This means that "site effect" can assume different meanings, depending on the size of the array and hypocentral distance.

In small dimension arrays (local networks) with hypocentral distances of the same order of magnitude of the array extension, it is in principle difficult to separate the path effects from site effects. This is because each seismic ray samples a different portion of the earth's crust: in this case differences in the shape of the spectra among different stations can be due to differences in the attenuation and/or site effects.

REVIEW OF TECHNIQUES AND RESULTS

Direct Wave Techniques

Site response function measurements are traditionally based on data from spectra of direct (P and S) phases. In these type of studies it is necessary that almost one station of the network is located on "hard rock". The site response for other stations is obtained relatively, dividing each spectrum for the spectrum of the reference station. In this way if attenuation is uniform under the array and if the calibration curve of each instrument is well-known, it is possible to infer from (1) the site transfer function. Many problems arise if an insufficient azimuth coverage is available: in this case it is not possible to average out these azimuth effects and the estimate of the site response results are biased. Moreover, it is necessary that the radiation pattern for each source is known. To obtain a set of data with such characteristics is quite a difficult task especially when it is necessary to use strong motion data. Despite these difficulties results on site effect were obtained using direct phases and small arrays.

Recently Tucker et al. (1984), King and Tucker (1984) and Tucker and King (1984) studied the differences between the site effects on "hard rock" and on sediments. They found that significative variation can exist in the relative amplification of sediments with respect to hard rock sites.

For example in the Chusal valley (USSR) the motion of sediments can be as much as ten times the motion on nearby rock sites. In only one case (Runo valley) of the three examined no amplification was observed. Dependence of the site amplification from the amplitude of the recorded wave was studied by the same authors for the Chusal valley in the range from 10^{-5} to 0.2 g. No dependence on the amplitude of the signals was observed in the site amplification. Frankel (1982) shows that an amplification at low frequencies of the seismic spectrum exists in the Caribbean area. All the referenced authors use direct phases and restrict to the low frequency band of the seismic spectrum.

Coda Wave Method

Coda waves are the tail part of the seismograms of the local earthquakes. These waves have so relevant experimental properties that they become a power resource in many fields of the short period seismology. A very extensive review of coda wave studies is made by Herraiz and Espinosa (1987). Phillips (1985) reports an intensive study of coda waves from local earthquakes in California. In both these papers it is possible to find most of the theory and the description of the methods. In the present paper are reported very shortly the most important features of the single scattering theory for coda wave generation.

It is possible to express the amplitude spectrum of the coda waves evaluated around a time value, t, chosen along the coda wave train as:

$$A(w|t) = S(w) \cdot C(w|t) \tag{2}$$

where w is the angular frequency. $S(w)$ includes source and site effects while $C(w|t)$ represent the effect of scattering in a volume whose linear dimensions are very large if compared with the source-station distance. $C(w|t)$ is common to all the events of the same seismogenetic area (if their locations are close together) and is also defined as a regional coda decay term. The analytical expression for $C(w|t)$ is

$$C(w|t) = \exp(-\pi f t/Qc)$$

in the case of single scattering. Taking the geometrical spreading into account the complete expression for $A(w|t)$ becomes the following

$$A(w|t) = S(w)t^{-1}\exp(-\pi f t/Qc). \tag{3}$$

This model is obtained assuming that coda waves are backscattered from randomly distributed inhomogeneities in the earth. Observational support on this model is given by several authors, as for example Aki (1980). Coda waves are naturally averaged over azimuth and incident angle at the receiver since they are composed by the total contribution of singly scattered wavelets coming from every direction. This type of "natural smoothing" is very useful because it reduces the influence of radiation pattern, which instead strongly affects the spectra of primary waves.

Considering two stations, 1 and 2, we can write

$$A_1(w|t)/A_2(w|t) = Si_1(w)/Si_2(w) \tag{4}$$

where the two different stations record the same event. The ratio at the right hand side of this expression is equal to the ratio of the site response functions for the two stations. It is clear that coda amplitude spectral ratio between two stations, one of which is the station chosen as reference, gives the site response function for the waves that compose the coda. If we make use of the S-wave or P-wave spectra, we may add more uncertainties due to azimuthal and source depth effects.

In the initial work on coda waves by Aki (1969) site effect was identified in several stations in the Parkfield-Cholame area (California). These stations presented the same pattern of coda amplitude decay but different levels of excitation. Later Tsujura (1978) compared the site effect calculated with direct S- and P-waves with that obtained with coda waves. He obtained more stable results working with coda waves and pointed out that coda site effect function is very similar to the average site function obtained with S-waves. P-wave site effect results were different. Del Pezzo et al. (1985.a) estimated the site effect in a small area surrounding the city of Ancona (Central Italy) showing an amplification of approximately ten in the frequency band centered at 1.5 Hz. More recently Phillips and Aki (1986) developed an array technique to evaluate the site effect using digital data from coda waves recorded in California. They found that young sediments amplificate ground motion more than 8 times with respect to hard rock sites in the low frequency band centered at 1.5 Hz. This result was obtained with a large number of data; their paper (see also Phillips, 1985) can be considered as a good review on the state of art on site effect measurements and interpretation.

DATA ANALYSIS AND RESULTS

Digital Network and Data Collection

A short period (1 Hz) three component digital network was set up in Abruzzo region (Central Italy) two days after the shock of May 7, 1984 (Local Magnitude 5.8). The network is composed of 7 stations with local triggering and PCM recording. Triggering is controlled for each station by the ratio of a short term average of the signal over a long term average. When this ratio becomes greater than a pre-fixed value (set up by the console of the microprocessor system) the triggering is over and the seismic event is recorded. Dynamic range of this type of instrument is more than 100 dB. Sampling rate was set up at 100 Hz.

The data set consists of good quality records of small aftershocks of the mainshock of May 7. P onsets are generally very clear, and the uncertainty in their time picking is less than twice the sampling interval. S onset are also very clear and easily readable on the three component seismogram.

Data were pre-processed (time picking) with a semi-automatic computer-interactive procedure at the seismic laboratory of the University of Wisconsin in Madison, and re-formatted for data exchange. Hypocenter location is described in Del Pezzo et al. (1985.b). A map of the seismic stations and of the events analyzed in the present paper is reported in Figure 2.

Coda Wave Analysis for the Site Response

Earthquakes recorded at more than six stations with a good quality location were selected for this analysis. Data were processed in the following steps.

1. A numerical time window slides along the entire seismogram (vertical component) in steps of 1.5 seconds. Duration of this window is 3 seconds. Amplitude Fourier spectra are evaluated in each step using a discrete Fourier Transform (semi-fast) algorithm. In this way the left hand side of the relationship (2), $A(w|t)$, is numerically estimated at discrete time values, of 1.5 seconds distance of each other.

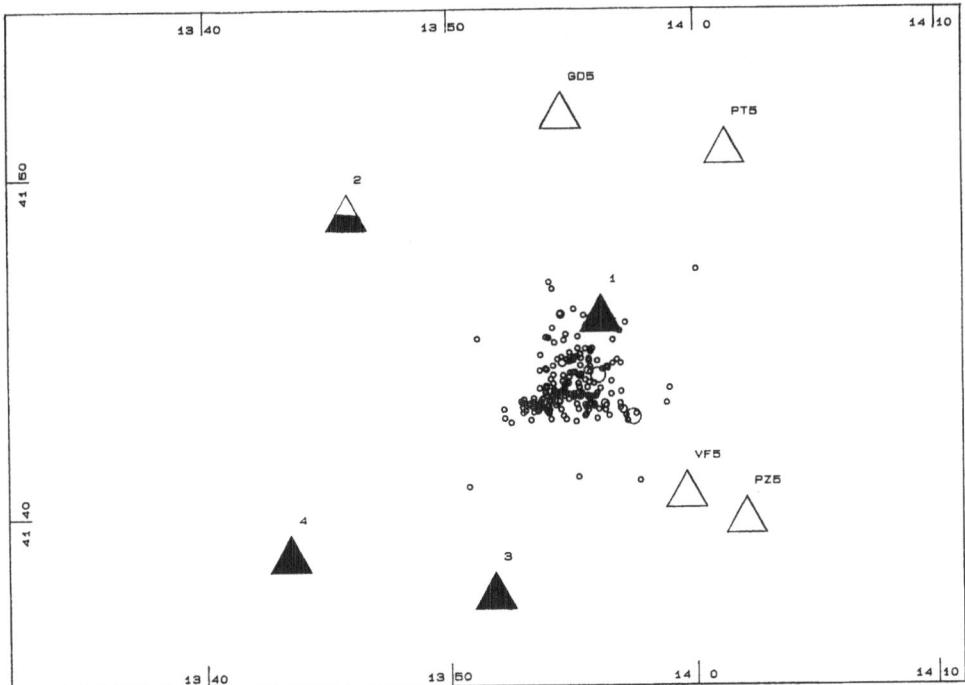

Fig. 2. Map of selected epicenters used for the site effect analysis. Triangles are the locations of the digital seismic stations and the solid triangles indicate the stations used for site effect calculations. Half filled triangle is the reference station.

2. This quantity, A(w|t), for each value of the time, is numerically integrated in the frequency bands centered at 1.7, 3, 6, 12, 16 Hz and with 60 percent of the central value.

3. Linearizing the expression (3) it is possible to fit data with a straight line from the beginning to the end of the coda. Beginning of coda is the point in which seismogram begins to decay regularly and end of coda is the point in which signal becomes of the same order of magnitude as the noise. These two time values are automatically estimated from data. It is noteworthy that the slope of this line is proportional to the quality factor, Q_c, of coda waves. An estimate of Q_c for the same area is reported in Del Pezzo et al. (1986). Correlation coefficient is then evaluated on the line. Events with a correlation coefficient less than 0.7 are not considered for site response analysis.

4. The value of the central point in the time interval from the beginning to the end of the coda envelope for the reference station is chosen as the reference time, denoted as t*. the value of A(w|t*) is evaluated from the straight line fitted by data (see step 3) for each station. Then the ratios of Rel. (4) are estimated. In this way the values of the ratios between the spectra present a very stable distribution around their average. Finally the arithmetic average over all the events is taken as the estimate of the site response.

Results

Distributions of the site response for different frequency bands and for different stations are shown in Figure 3. Reference station is the station number 2. Because no information is available about the geology

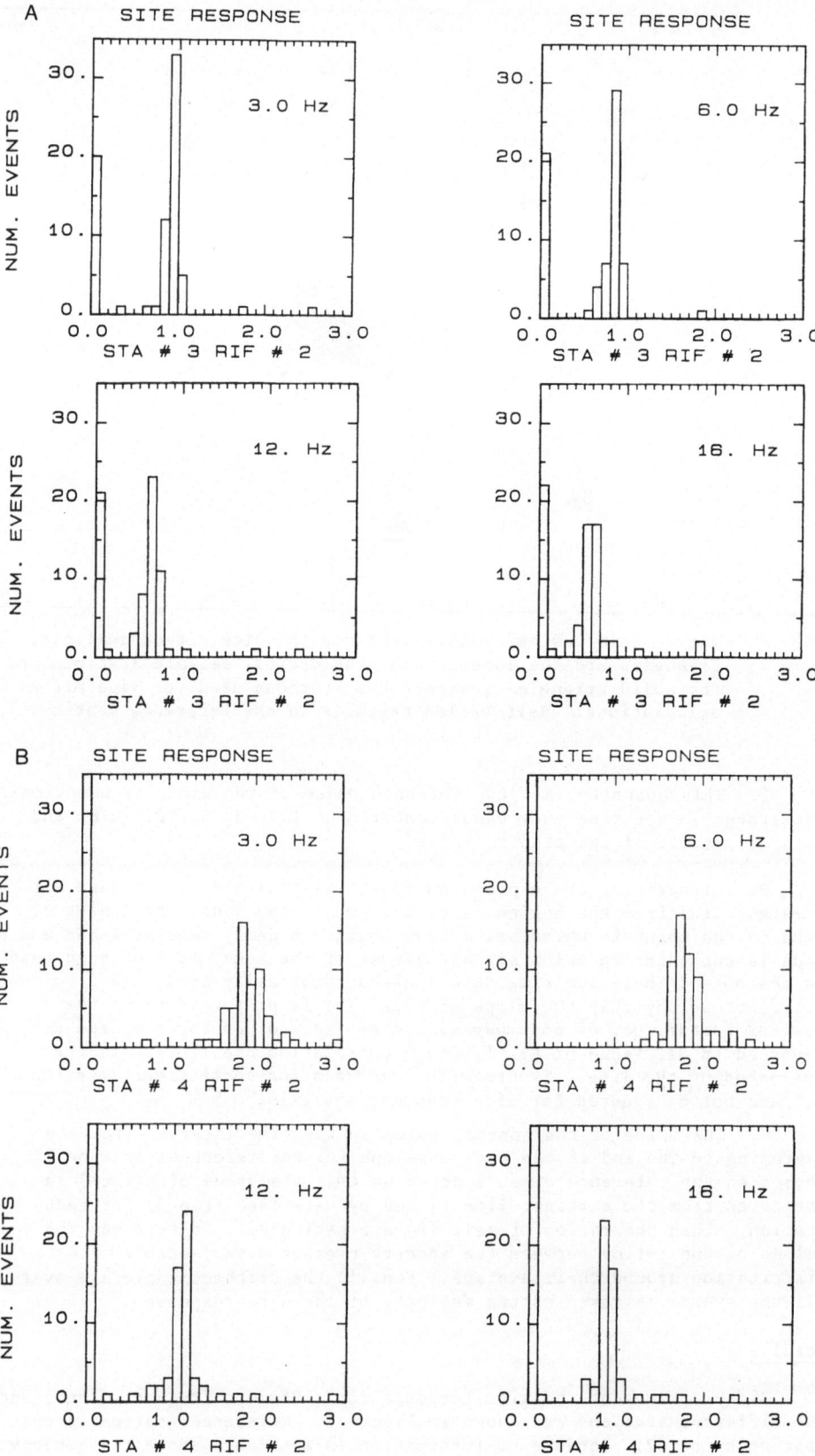

380

of the sites which have been examined, the choice of the station 2 as reference is purely numerical, in the sense that this station had the most continuous operating period.

The peaks in the histograms around zero are the events which have been rejected for analysis, due to the low quality of signal to noise ratio (see step 3). It is quite clear from these plots that station 4 and station 1 present amplification at low frequencies. Station 3 has an opposite effect. The site response for all the examined stations is reported in Figure 4.

DISCUSSIONS AND CONCLUSIONS

The site effect through the use of coda waves has been calculated for a local short period seismic network, with stations of less than 8 km distance of each other and located in sites with apparently the same geology.

The method is very stable and there is an advantage in its use with respect to the direct wave technique. The coda waves give a naturally

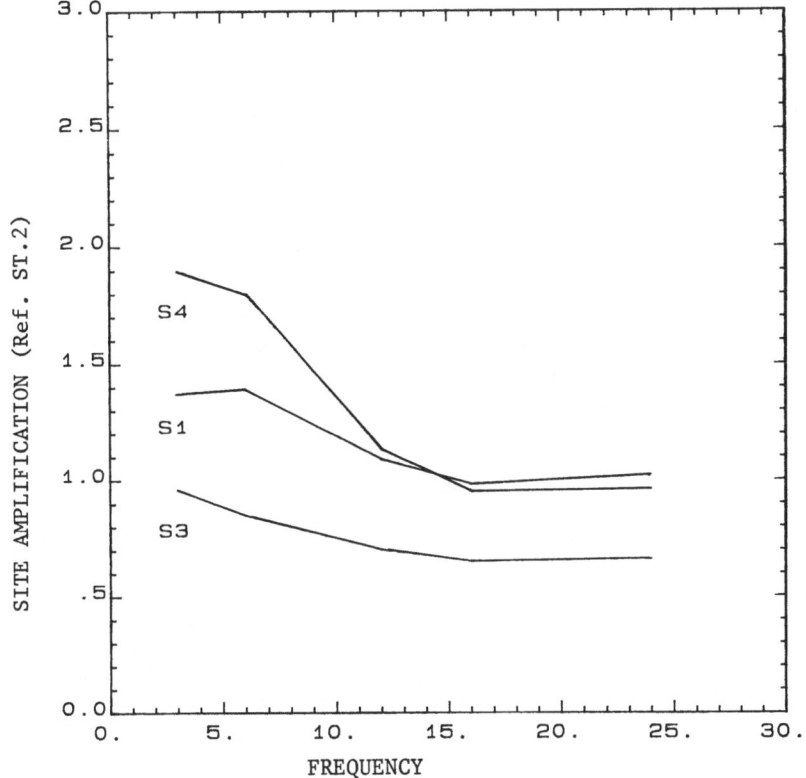

Fig. 4. Site effect estimates for the stations 1, 3 and 4 as a function of the frequency. The reference station is the station 2 (see Figure 2).

Fig. 3. Histograms showing the distributions of the amplitude ratios with respect to the reference station for (a) station 3 and (b) station 4 of Figure 2 at different frequency bands.

smoothed value of the spectra because their amplitude does not depend on the position of the hypocenter. Moreover, the use of coda allows to separate the site from the path term. Our results show that there is a variation of the site effect inside a relatively small area. These variations are as high as 2 in the low frequency band (around 5Hz). Unfortunately the low quality of coda signal to noise ratio in the 1.5 Hz frequency band does not permit to extend this analysis to lower frequencies and to make a comparison with other data in that frequency band. For example, Phillips reports an amplitude ratio as high as 20 at 1.5Hz for stations located in sites with different geology. A value around 10 was found by Del Pezzo et al. (1985) in Central Italy (Ancona area) in 1.5 Hz band for a station located on soft sediments.

Our data show that some variations in site response can be found also for stations located on apparently the same geological units. Care must be spent in the analysis of source parameters and magnitude even if a good signal to noise ratio and hard rock sites are available.

REFERENCES

Aki, K., 1969, Analysis of the seismic coda of local earthquakes as scattered waves, J. Geophys. Res., 74:615-631.
Aki, K., 1980, Scattering and attenuation of the shear waves in the lithosphere, J. Geophys. Res., 85:6496-6504.
Bath, M., 1974, "Spectral Analysis in Geophysics", Elsevier, Amsterdam.
Del Pezzo, E., Rovelli, A., and Zonno, G., 1985.a, Seismic Q and site effects on seismograms of local earthquakes in the Ancona region (Central Italy), Ann. Geophysicae, 3,5:629-636.
Del Pezzo, E., De Natale, G., Iannaccone, G., Martini, M., Scarpa, R., and Zollo, A., 1985.b, Analisi preliminare della sequenza sismica dell' Abruzzo mediante i dati di una rete sismica digitale, Atti del Convegno GNGTS, Roma.
Frankel, A., 1982, The effects of attenuation and site response on the spectra of micro-earthquakes in the North-Eastern Caribbean, Bull. Seism. Soc. Am., 72:1379-1402.
Herraiz, M., and Espinosa, A. F., 1987, Coda waves: a review, PAGEOPH, vol. 125, n. 4.
King, J. L., and Tucker, B. E., 1984, Observed variations of earthquake motions across a sediment filled valley, Bull. Seism. Soc. Am., 74:137-151.
Phillips, W. S., 1985, The separation of source, path and site effects on high frequency seismic waves. An analysis using coda wave techniques, Ph.D. Thesis, MIT, pp 195.
Phillips, W. S., and Aki, K., 1986, Site amplification of coda waves from local earthquakes in central California, BSSA, 76:627-648.
Tsujiura, M., 1978, Spectral analysis of the coda waves from local earthquakes, Bull. Earthquake Res. Inst., Tokyo Univ., 53:1-48.
Tucker, B. E., King, J. C., Hatzfeld, D., and Nersesov, I. L., 1984, Observation of hard rock site effects, Bull. Seism. Soc. Am., 74:121-136.
Tucker, B. E., and King, J. L., 1984, Dependence of sediment filled valley response on input amplitude and valley properties, Bull. Seism. Soc. Am., 74:153-165.

A SIMPLE LINEARIZED METHOD FOR INVERSION OF TRAVEL

TIME DATA IN TWO-DIMENSIONAL HETEROGENEOUS MEDIA

J. A. Madrid

Centro de Investigacion Cientifica y Educacion
Superior de Ensenda

I. INTRODUCTION

The broad complexity of geological situations in regions of economical
or geotechnical interest requires the development and implementation of
numerical and computational techniques that allow to handle many, if not
all, these situations adequately. The analysis of data may be carried by
solving either the direct or the inverse problem, or both.

One of the most efficient algorithms for the direct problem of ray
tracing (c.f. Aric et al, 1980; Gebrande, 1976; Madrid, 1985) is the so-
called "circular approximation". This is a fast, analytical method where
the model space is divided in triangular regions wherein the wave velocity
is linear in the coordinates (x,z). The rays are arcs of circles and the
travel time is very simple. All other quantities, like slownesses, the τ
(TAU) function, etc., are easily evaluated too.

The (inverse) problem of estimating the velocity structure of a medium
from travel time observations has received great attention in the last two
decades. Some procedures have been developed that are quasi-linear (Aki
and Lee, 1976; Aki et al, 1977). Their method involved the determination
of velocity perturbations in individual cells. This kind of parameterization generally yields an ill-conditioned matrix system. The instability
due to errors in the data can be controlled by methods like Singular Value
Decomposition when the problem involves relatively few parameters. For
very large systems of equations, the errors generated by the numerical
calculations add to those coming from the data.

Other attempts have been made to solve the inverse problem in
continuous media, like in Firbas (1981), or to parameterize geological
features analytically, and include the source (c.f. Julian and Gubbins,
1977).

The problem of an adequate parameterization is essential to the
inverse problem. Very probably, the result of an inversion is influenced
by the type of parameterization used. The parameterization is practically
determined by the available ray-tracing method. In general, it is clear
that the result of an inversion is ambiguous if the number of parameters
used is small, whereas redundancy and instability will develop if too many

parameters are used. In this case, the computer time may also increase excessively.

In this paper, a method to compute travel time derivatives with respect to the velocity defined in a point, and a correction to produce travel time residuals (Madrid, 1986) are reformulated and illustrated with synthetic as well as with real data.

In the next section, a brief account of the relevant theory is given. Questions such as effects on the boundaries, optimum discretization of the model, and the presence of noise in the observations are discussed in section III. In section IV an example using real data is exposed.

II. PARTIAL DERIVATIVES AND RESIDUALS

In this section, formulae for computing the derivative of the travel time with respect to a velocity-point, and the residuals at the points of observation, are developed. Since these formulae are based in the circular approximation, we start by giving a short account of this method.

Ray tracing is governed by a system of equations that can be written in different ways. In the following paragraphs, one form is adopted (Cerveny, 1985, 1986) that is suitable for the problem:

$$\frac{dx_i}{d\sigma} = p_i, \quad \frac{dp_i}{d\sigma} = \frac{1}{2} \frac{\partial}{\partial x_i} \left(\frac{1}{v^2} \right), \tag{1}$$

where $d\sigma = vdS$, and S is the arc length. We will use $x_1 = x$, $x_2 = z$, $p_1 = p$, $p_2 = q$ indistinctly. Due to the linear character of the velocity

$$v = a_o + \sum_i a_i x_i, \quad i = 1,2 \tag{2}$$

the second equation of (1) is

$$\frac{dp_i}{d\sigma} = -\frac{a_i}{v^3}. \tag{3}$$

Equations (1) hold in any system of coordinates. Solving (1), it is easy to show that, in a medium where the velocity increases linearly with depth, ($a_1 = 0$ in (2)) a ray is an arc of a circle of radius:

$$R = -\frac{1}{pa}$$

where $a = |grad(v)|$, p is the horizontal component of the slowness vector $\underline{u} = (p,q)$. The center of the circle is easily found from the initial conditions (x_o, z_o, p_o, q_o).

If the velocity varies linearly with x ($a_1 \neq 0$ in (2)), we have a tilted gradient of velocity, but a simple rotation of the axes of coordinates reduces the case to the previous one. The rotated quantities are:

$$p' = p \cos r + q \sin r$$

$$q' = -p \sin r + q \cos r \tag{4}$$

with similar expressions for the coordinates (x,z). In the rotated system, hereafter called "local system" the horizontal component of the slowness (p') is constant through a whole ray, since $V = V_{(z')}$.

In the circular approximation, a grid of points (x,z) is used to generate a set of triangular regions, inside every one of which, the velocity is given by (2). A ray is generated by establishing an origin (source), and solving for the circular path with the appropriate side of the triangle where the ray is currently located (Figure 1b). The final point (pf,qf,xf,zf) is used as an initial condition in the new triangle. The travel time is the sum of all the local travel times along the path:

$$T = \sum_j \Delta T_j \tag{5}$$

and the local travel time is

$$\Delta T = \frac{1}{|\underline{a}|} \left[\ln\left(\frac{1 + qv}{pv} \right) \right]_o^f . \tag{6}$$

Figure 1a shows an example of a ray in a medium divided in triangular regions.

If the velocity indicated as vk (vk(xk,zk)) is perturbed, the ray shown will be distorted, and the travel time curve will be different. A perturbation of vk will affect all rays passing through the regions marked as A,B,C,D,E,F, although one ray does not need to pass through all of them.

The inverse problem can be stated as follows: Given a set of travel time observations, and an initial model, how must we perturbate the model in order that

$$T_c \rightarrow T_c + \delta T = T_o \tag{7}$$

where Tc = travel time computed from an initial model, To = observed time, and δT is the perturbation in travel time. Now, if Tc = Tc $(v_1, v_2, \ldots, v_N)$, we have, after expanding in a Taylor series and keeping only the first order term:

$$\delta T_i = \sum_k \frac{\partial T_i}{\partial V_k} \delta V_k \tag{8}$$

(i = 1,2...,M).

The travel time perturbation δT can be identified with the residuals:

$$r = \delta T = T_o - T_c \tag{9}$$

then we have the well known expression

$$y = A\underline{x} \tag{10}$$

where $Y = \underline{r}$ and $\underline{x} = \delta\underline{v}$

Techniques for solving (10) are well known and will not be discussed here. Now, we must determine $\frac{\partial T_i}{\partial V_k}$ and $\underline{r}$ in (8). To do this, let's consider a triangular region like the one in Figure 1b. Let's assume that vl, where l can be 1, 2 or 3, is subject to a slight perturbation while the other two velocities are kept fixed, then

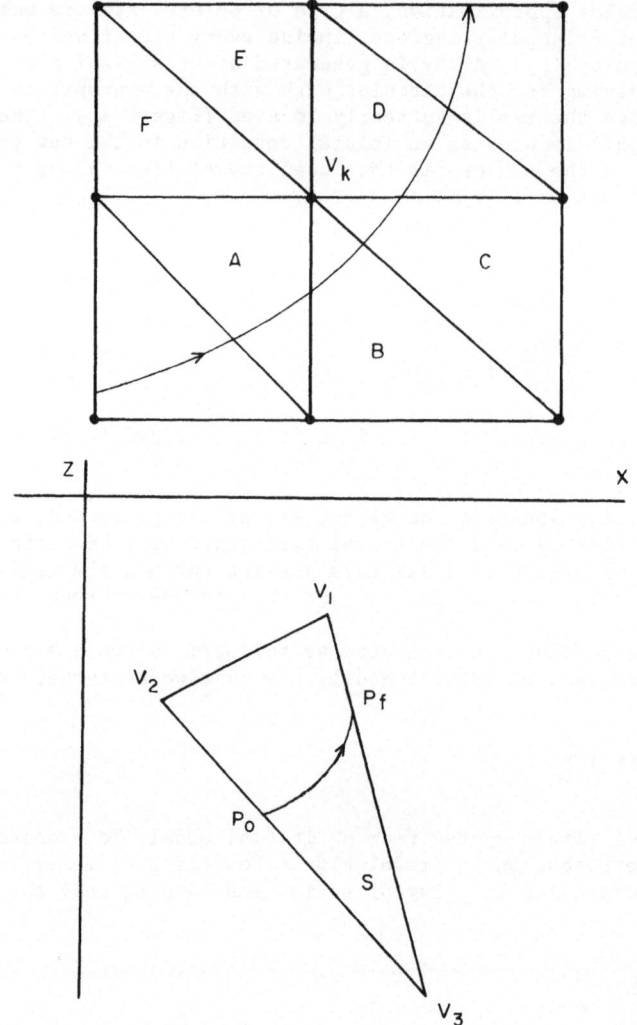

Fig. 1. (a) Regions A,B,C,D,E and F are affected by the value of the velocity at point (X_k, Z_k). (b) A circular ray segment in a region with velocity $V = a_0 + a_1 X + a_2 Z$.

$$\delta V_i = \delta V_i \delta_{i1}, \quad i,1 = 1,2,3 \tag{11}$$

using the Kramer's rule, it is easy to solve for δa_0, δa_1, δa_2; dividing by $\delta v1$ and taking the limit as $\delta v1 \rightarrow 0$, one gets:

$$\frac{\partial a_o}{\partial V_1} = \frac{1}{\Delta} \varepsilon_{1jk} x_j z_k \equiv A_{o1}$$

$$\frac{\partial a_o}{\partial V_1} = \frac{1}{\Delta} \varepsilon_{i1k} e_1 z_k \equiv A_{11} \tag{12}$$

$$\frac{\partial a_2}{\partial V_1} = \frac{1}{\Delta} \varepsilon_{ij1} e_i x_k \equiv A_{21}$$

where Δ is the determinant of the system, $\varepsilon_{ijk} = \pm 1$ if ijk is a cyclic or non-cyclic permutation of the numbers 1,2,3, and $\varepsilon_{ijk} = 0$ if any two indices coincide, and $e_i = 1$ for $i = 1,2,3$.

The travel time inside the region is

$$\Delta T = \int_o^f \frac{dS}{V} . \tag{13}$$

After $\delta v1 \rightarrow v1 + \delta v1$, $\Delta T \rightarrow \Delta T + \delta(\Delta T)$, so that

$$\delta(\Delta T) = - \delta V_1 \int_o^f \frac{\partial V}{\partial V_1} \frac{dS}{V^2} \tag{14}$$

using (12), dividing by $\delta v1$ and taking the limit as $\delta v1 \rightarrow 0$:

$$\frac{\partial(\Delta T)}{\partial V_1} = - \int_o^f (A_{o1} + A_{11}x + A_{21}z) \frac{dS}{V^2} . \tag{15}$$

In the local system (15) is

$$\frac{\partial(\Delta T)}{\partial V_1} = - \int_o^f (A_{o1} + A_{11}'x' + A_{21}'z') \frac{dS}{V^2} \tag{16}$$

to integrate (16) we use (3) as expressed in the local system:

$$\frac{dp_i'}{a_i'} = \frac{dS}{V^2}$$

substituting in (16):

$$\frac{\partial(\Delta T)}{\partial V_1} = \frac{1}{a_2'} \int (A_{o1} + A_{11}'x' + A_{21}'z')dq' . \tag{17}$$

Integrating by parts the second term we get

$$\frac{\partial(\Delta T)}{\partial V_1} = \left[\frac{Aq'}{a_2'} \right]_o^f - \frac{1}{a_2'} \int_o^f q'(A_{11}'dx' + A_{21}'dz')$$

Where $A = A_{o1} + A_{11}'\, x' + A_{21}'\, Z'$. Now, using $T = \int p'\, dx' + q'\, dz'$ and $\frac{dx'}{dz'} = \frac{p'}{q'}$,

$$\frac{\partial(\Delta T)}{\partial V_1} = \left[\frac{Aq'}{a_2'} \right]_o^f - \frac{A_{11}'}{a_2'} \int_o^f p'dz' - \frac{A_{21}'}{a_2'} \left[\Delta T - \int_o^f p'dx' \right]$$

so that finally

$$\frac{\partial(\Delta T)}{\partial V_1} = \left[\frac{Aq'}{a_2'} \right]_o^f - \frac{p'}{a_2'} \left[A_{11}'z' - A_{21}'x' \right]_o^f - \frac{A_{21}'}{a_2'} \Delta T. \tag{18}$$

In the example shown in Figure 1, the ray is affected in the regions marked A,B,C,D. From (5):

$$\frac{\partial T}{\partial V_k} = \sum_i \frac{\partial(\Delta Ti)}{\partial V_k} = \sum_{j=A,B,C,D} \frac{\partial(\Delta Tj)}{\partial V_k}.$$ (19)

The calculation of the residual requires that the travel time be evaluated in the observation point (Xobs). Usually, this is accomplished by the so-called "fixed ends" (bending) method (c.f. Julian and Gubbins, 1977), and more recently by dynamic ray tracing (Cerveny, 1985). Here, we implement a simple correction that makes a bending method unnecessary.

The time difference between the ray emerging at Xp and the one emerging at Xobs is

$$Tx = \int_{Xp}^{Xobs} p(X)dX$$ (20)

here, $p(X) = p(x=X,z=0)$ is the horizontal component of the slowness for rays emerging between Xp and Xobs. Since an analytical expression for $p(X)$ is not available, a Taylor expansion at first order may give a good approximation, whenever Xp and Xobs are close to each other. This is readily achieved if, for example, interval ray tracing is used (Cerveny, 1985). We have:

$$p(X) = p(Xp) + \left(\frac{\partial p}{\partial x}\right)_{Xp} (X - Xp);$$ (21)

$\frac{\partial p}{\partial x}$ is given by

$$\frac{\partial p}{\partial x} = \frac{a_1^2}{q'V^3 a} .$$ (22)

An analogous expression is easily obtained for $\frac{\partial q}{\partial z}$

$$\frac{\partial q}{\partial z} = \frac{a_2^2}{q'V^3 a}$$ (22')

(Madrid, 1986).

Using (22') and (22) in (20) we obtain

$$Tx = p(Xobs - Xp) + \left(\frac{a_1^2}{2q'V^3 a}\right) (Xobs - Xp)^2.$$ (23)

This expression holds for receivers aligned with the x-axis. If, as we shall see later, we have receivers in the z direction (Figure 2), we will require the corresponding expression, i.e.:

$$Tz = q(Zobs - Zq) + \left(\frac{a_2^2}{2q'V^3 a}\right) (Zobs - Zq)^2.$$ (23')

At the point of observation (x-direction):

$$Tc = Tp + Tx$$ (24)

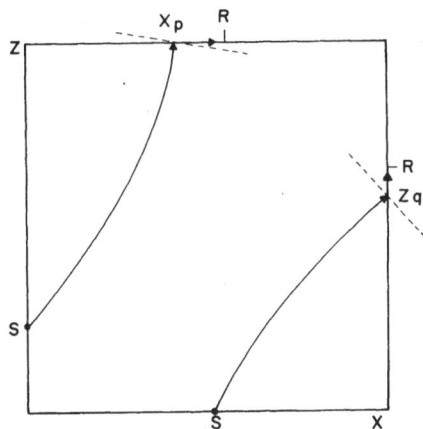

Fig. 2. Example of a couple of rays from different sources arriving at Xp
and Zq. The receiver for the ray on the left is located at Xobs,
the receiver for the ray on the right is located at Zobs. To
reach Xobs, the intercept of the wavefront (dotted line) with the
receiver line z = const. must go from Xp to Xobs (R).

and

$$r = To - Tc = To - Tp - Tx$$

Since we do not really have a ray emerging at Xobs, the actual point
of observation is Xp. The last expression may be rewritten as

$$r\,(Xp) = (To - Tx) - Tp \tag{26}$$

that is, applying the correction to To instead than to Tp we generate a new
"observed time" at Xp. Then the residual is actually associate to Xp
rather than to Xobs.

Rigid Discretization vs. "Flexible" Discretization

In Figures 3a and 3b, the same ray is shown drawn on two different
systems of coordinates. In Figure 3a, the coordinate axis are x and z; in
Figure 3b, the axis are (x,v). We can pass from one to the other by just
drawing the isovelocity lines superposed to the (x,z) axes, and then
removing these axes. This is more easily seen if v is a function of z
only, because the isovelocity lines coincide with the lines z = const. If
the linear law holds, then the transformation $(x,z) \rightarrow (x,v)$ is given by just

$$z = -\frac{a_o}{a_2} - \frac{a_1}{a_2} x + \frac{V}{a_2} .$$

If the velocity is non-linear, a one-to-one relation is required.

The inversion procedure, such as it has been presented here, consists
of perturbating the values of the velocity at nodes (x,z). But suppose we
describe the procedure in the transformed system (x,v). Here, we would be
changing the value of the vertical coordinate (z) at nodes (x,v). If the
initial model is designed using isovelocity lines, the deformation of
isovelocities may be directly seen after an inversion step.

389

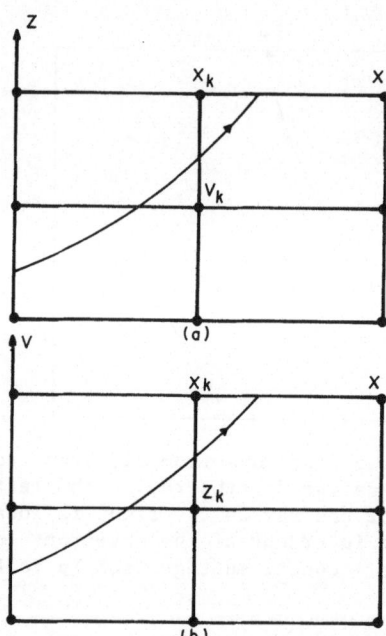

Fig. 3. (a) A ray in the usual coordinate system (x,z). Here the isolines
correspond to equal velocity. (b) The same ray depicted in the
(x,v) system, where the isolines correspond to equal z-coordinate.

We will see that, when representing this change in the (x,z) space,
the mesh becomes adaptive in the z-direction i.e., the nodes move
vertically.

From (2) and (19), we obtain

$$\frac{\partial(\Delta T_j)}{\partial v_k} = \frac{1}{a_{2j}} \frac{\partial(\Delta T_j)}{\partial v_k}$$

so that

$$\frac{\partial T}{\partial z_k} = \sum_j \frac{\partial(\Delta T_j)}{\partial z_k} = \sum_j a_{2j} \frac{\partial(\Delta T_j)}{\partial v_k} \tag{27}$$

then, in (x,v) we can solve the problem

$$\sum_k \frac{\partial T_i}{\partial z_k} \delta z_k = r_i \tag{28}$$

instead of (8).

The Inversion

In the following examples, the data, either synthetic or real, were
subject to a minimization of the functional (Frez, 1985):

$$\| \underline{Y} - AX \| Cy^{-1} + \alpha \| X_I - X_{I-1} \| \tag{29}$$

which corresponds to the sum of the quadratic errors plus the norm of the solution. In (29), α is a global stabilizing factor that controls the change in the parameters. Cx is the covariance matrix of the initial model and Cy is the covariance matrix for the errors.

Solving for the minimum of (29), one finds

$$Xi = (A^TCy^{-1}A + C_{xI-1}^{-1})^{-1}A^TCy^{-1}Yi. \tag{30}$$

Calling $A' = U^T\Lambda V$, where A' is normalized (Frez, 1985), we have

$$X_i = \frac{\lambda_j V_{ij} U_{ki}}{\lambda_j^2 + \alpha} r_i \tag{31}$$

here N is the number of non-zero eigenvalues and M the number of observations. The best value of α is settled by trial and error.

Equation (31) represent the standard least squares solution multiplied by a filter

$$F(\alpha) = \frac{\lambda^2}{\lambda^2 + \alpha} = \frac{\lambda^2/\alpha}{1 + \lambda^2/\alpha} \tag{32}$$

(Figure 7). The value of α establishes a cutoff in the value of the eigenvalues to be used in an iteration. It is easily seen that, for a given α, the eigenvalues $\lambda j < \alpha^{\frac{1}{2}}$ are more effectively attenuated than the largest ones.

III. EXAMPLES

A variety of numerical experiments were performed to check the validity of the method. The experiments were as follows: first, a laterally heterogeneous ("actual") model is established, and a number of observed travel times with their corresponding observation distances are obtained through exact ray tracing. Then, an initial model is assumed and sampled by direct ray tracing to determine adequate limits for the shooting angle. The initial model does not have to be laterally homogeneous. After inversion, the resulting model is used as initial model in a new iteration.

The first set of experiments was run with no noise included. In all cases the convergence was practically perfect (final residuals null). The model shown in Figure 4b includes discontinuities at depths -3.0 and -5.0 u. and a low velocity zone around point 10 (x=10, z=-5 u.). About seven iterations were required for a very satisfactory convergence. The initial model is shown in Figure 4a.

The discretization used for this model is exactly the same as the one used to generate the data. In Figure 5a, an example is given where the "actual" and the initial models were designed using different (non-coincident) number of points. Here, solid lines represent the "actual" model. Broken thin lines are isovelocities of the resulting model, heavy horizontal lines are isovelocities of the initial model. (Figure 5b) shows the initial (dots) and final (crosses) residuals. The oscillating part is due to the different discretization between "actual" and inverted models.

Figures 6a and 6b show the difference in results when the data are inverted in the rigid (a) and flexible (b) modes. The "actual" model

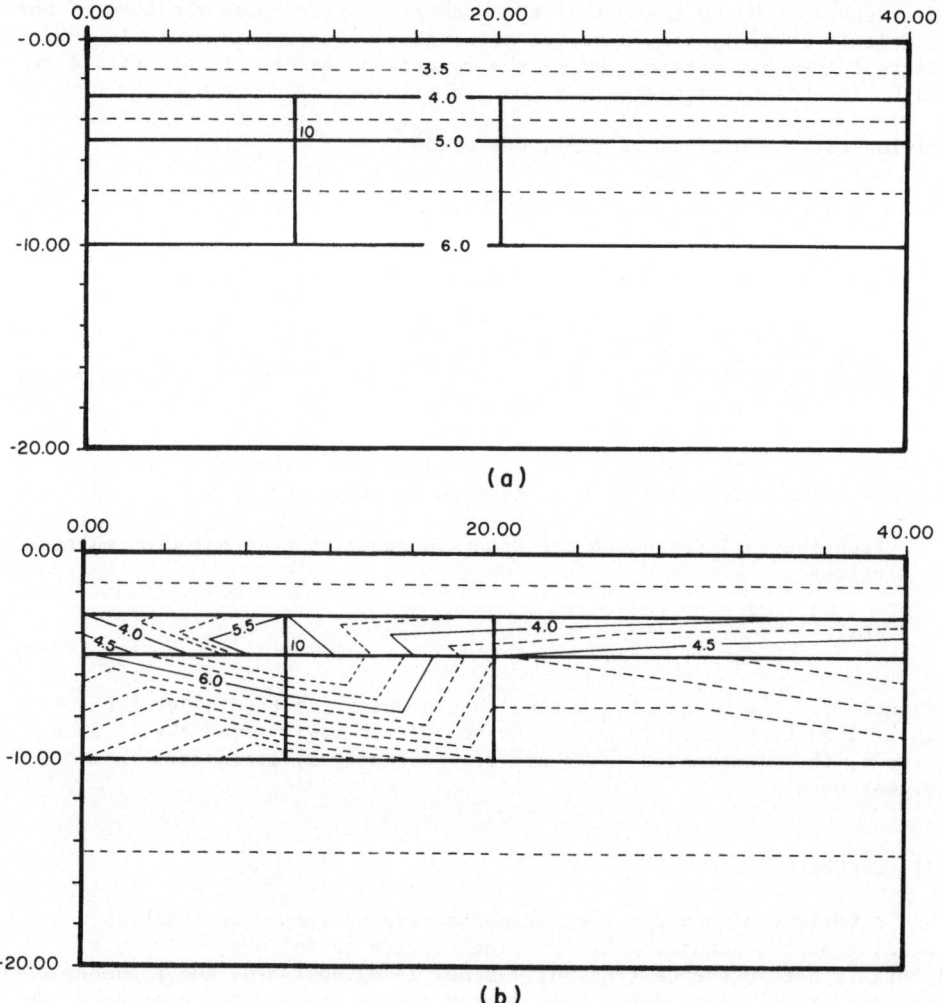

Fig. 4. Example of the inversion in a complex model. Two discontinuities
are included, at z = -3.0 and z = 5.0 μ. Also included is a LVZ
around point 10.

corresponds to Figure 6b. It is clearly seen that, in this case, the
result of the flexible discretization is better than that of the rigid one.
In the flexible discretization the grid becomes adaptive, at least in the
vertical direction. In the rigid discretization the isolines of velocity
are forced to end at the (fixed) sides of triangles.

Many synthetic examples (Traslosheros, 1987) were run with no noise,
with the purpose of checking the effects of boundaries, the quantity of
information and of different number of parameters.

The results may be resumed as follows: Parameters in the center of
the model are resolved more accurately. Peripheral parameters are
generally poorly resolved. This is because the central parameters are
sampled by more rays than peripheral parameters (Figure 8). Using more
data for the inversion procedure improves the estimation, whenever this
information is correct.

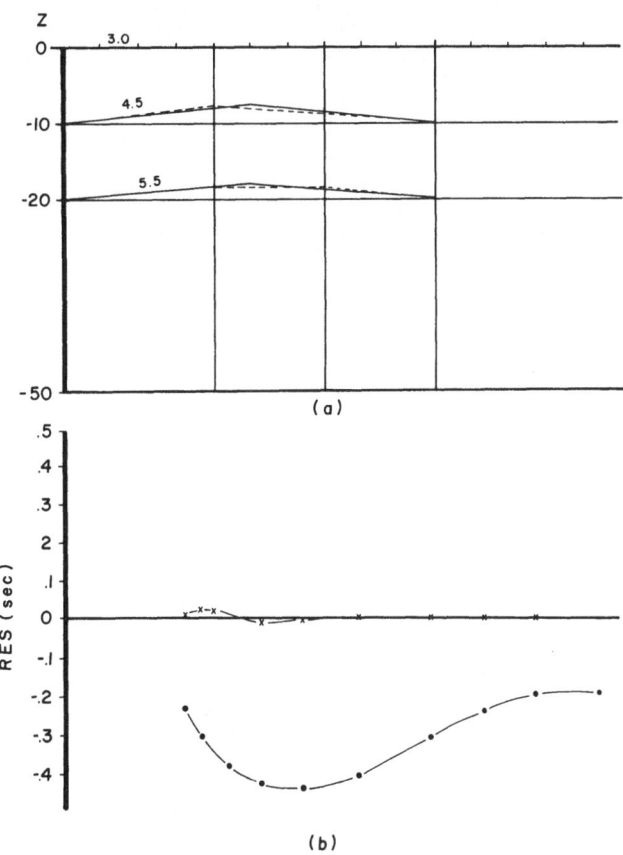

Fig. 5. An example of incompatibility between the discretization of the "actual" and initial models.

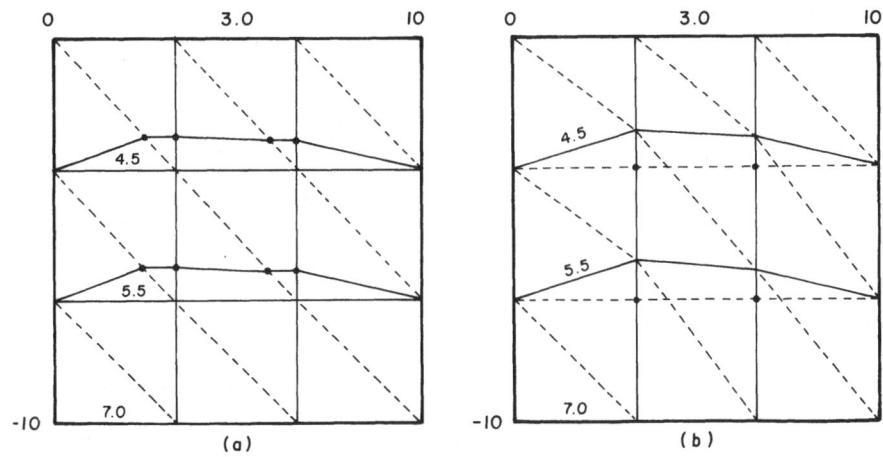

Fig. 6. Example showing the difference in the results of an inversion (a) in the (x,z) space, (b) in the (x,V) space. The actual model is exactly (b). In both cases the initial model consisted of horizontal isovelocity lines of 3.0, 4.5, 5.5 and 7.0 u/sec. The resulting isovelocity lines in (b) are the lower sides of triangles.

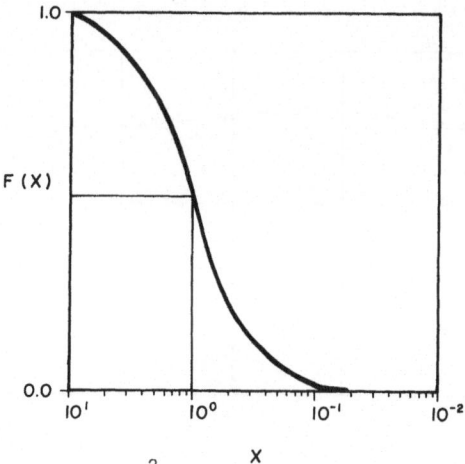

Fig. 7.   The filter $F(X) = \dfrac{x^2}{1 + x^2}$ $x^2 = \lambda^2/\alpha$, is here plotted in the direction of decreasing λ. At $\lambda^2 = \alpha$, $F(X) = \frac{1}{2}$. From here on, the eigenvalues $\lambda_j^2 < \alpha$ are more effectively attenuated than the larger $\lambda_j^2 > \alpha$ (left side).

The number of parameters used has an effect on the resolution of the final model. From numerical experiments where the number of parameters used was varied, it was concluded that, when the number of parameters is very small, the fine detail of the results is lost. If, on the contrary, the number of parameters if very large, the model is oversampled and there will be redundancy in the matrix of partial derivatives. The redundancy produces instabilities which result in spurious fine details in the solution. A useful criterion (Newmann, 1971) is to choose the spatial separation between nodes of the order of the wavelength dominant in the signal.

In the last set of experiments, the data were contaminated with gaussian noise of standard deviation 1%, and 5% of the mean error. Due to the presence of noise, the problem becomes unstable. For the first figure, it was possible to find a value of α that stabilizes the problem. This value is obtained by trial and error. At 5%<r>, the problem becomes extremely unstable. The resulting criteria were used in the analysis of a set of real seismic sections obtained using sources distributed along the left and lower sides of a square area on the surface, and receivers along the other sides, so that the ray tracing steps are performed in a 2-D medium. In this case, the triangle division means that the 3-D medium is composed of vertical triangular prisms, and the surface area is assumed representative of the stratigraphic column at depth.

IV. AN EXAMPLE WITH REAL DATA

The theory developed was applied to a set of seismic sections provided by the Mexican National Electricity Agency (Comision Federal de Electricidad). All sections were obtained using sources and receivers as shown in Figure 9. Each source is recorded in each receiver, thus the sampling of the area is dense. The signal was detected using geophones "MARK" with a central frequency of 2Hz.

Previous to the inversion procedures, it was found convenient to perform an analysis of the frequency content in the signals recorded, to have an idea of the dominant wavelength. A value of about 6. m was found

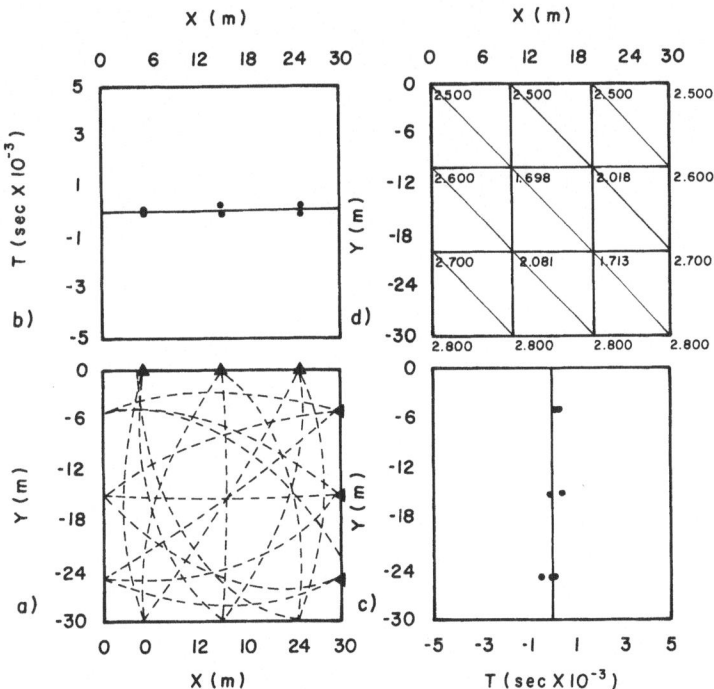

Fig. 8. Rays coming from multiple sources sample the internal parameters more times than the external parameters. The model is shown in the right upper corner. The residuals in the left upper corner correspond to the receivers (triangles) along z = 0. The residuals in the lower right correspond to the receivers along X = 30 μ.

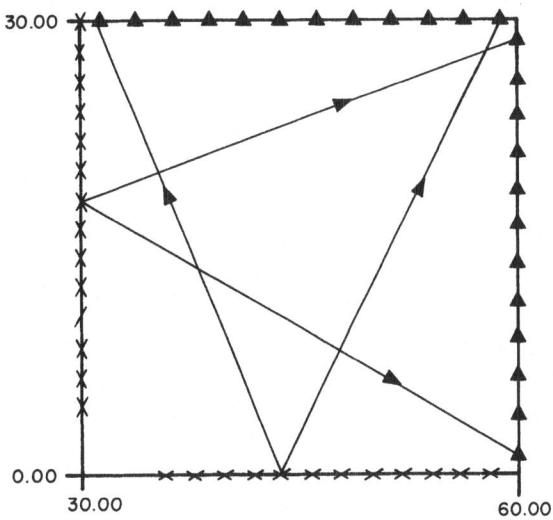

Fig. 9. Configuration of sources (crosses) and receivers (triangles) for a typical inverse experiment. In total, 15 sources and 12 receivers were placed in the sides of the square shown. Typical dimensions are 30 x 30 m.

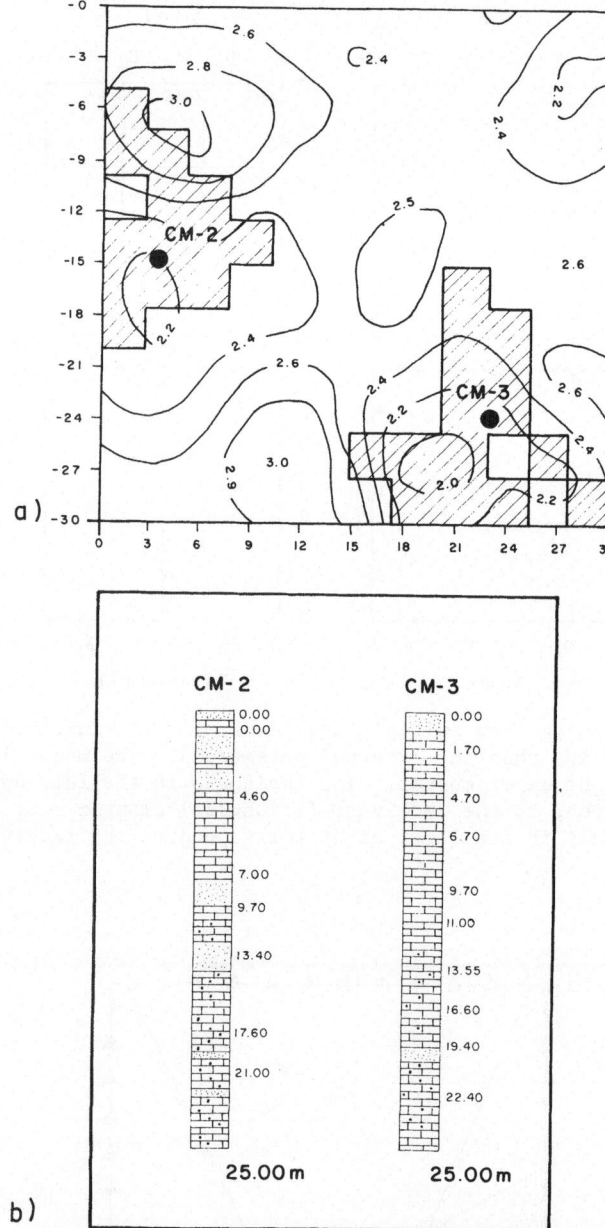

Fig. 10. A typical result of inverting the data collected in an experiment similar to that in Figure 9. The shaded areas are the result of inverting using the method of Aki and Lee (1977). It can be seen that instabilities are present within the shaded areas. More than twenty iterations were required. The isovelocity lines represent our result. No more than three iterations were required. (b) The stratrigraphic columns obtained at points CM-1 and CM-2 by drilling.

to be appropriate. With this value, we could expect to recover features a little larger (say ~10 m), and avoid the over discretization of the medium, as well as a deficient discretization. Also, a search for the best laterally homogeneous initial model was done by trial and error.

In Figure 10a the results of inverting one of the sections is shown. All other sections were found to behave in an analogous way. The shaded zones (low velocity) result from applying an inverse procedure (Aki and Lee, 1977) where the medium is divided in homogeneous blocks. The iso-velocity lines correspond to the present method. The resulting values of the velocity coincide quite well. In Figure 10b the lithological columns obtained by drilling are illustrated.

It is interesting to notice that both columns have low velocity material immediately on the surface, so that the ray paths and travel times are truly affected by this feature. It is also important to outline that, whereas the result obtained with blocks required about twenty iterations or more, our result required at most five or six iterations.

REFERENCES

Aki, K., Christofferson, A., and Husebye, E. S., 1977, Determination of the three dimensional seismic structure of the lithosphere, J.Geophys. Res., 82:277-296.

Aki, K. and Lee, W. H. K., 1976, Determination of three dimensional velocity anomalies under a seismic array using first P arrival times from local earthquakes, 1. A homogeneous initial model, J.Geophys. Res., 81:4381-4399.

Aric, K., Gutdeatch, R., and Sailer, A., 1980, Computation of travel times and rays in a medium of two-dimensional velocity distribution, Pure and Applied Geophysics, 118:796-805.

Cerveny, V., 1985, The application of ray tracing to the numerical modeling of seismic wave fields in complex structures, in: "Handbook of Geophysical Exploration, Section I: Seismic Exploration", Helbig and S. Treitel, eds., vol. 15 A: Seismic shear waves, Part A: Theory, G. Dohr, ed., Geophysical Press, London, 1-124.

Cerveny, V., 1985, Ray synthetic seismograms for complex two-dimensional and three-dimensional structures, J.Geophys., 58:2-26.

Cerveny, V., 1986, Lecture notes for the Autumn Course on Seismology, ICTP, Trieste.

Firbas, A., 1981, Inversion of travel time data for laterally heterogeneous velocity structure-linearization approach, Geophys.J.R.Astr.Soc., 67:189-198.

Frez, J., Notas de Clases: "Teoria de Inversion de datos geofisicos", 1985, CICESE, Ensenada, BC, Mexico.

Gebrande, H., 1976, A seismic-ray tracing method for two-dimensional inhomogeneous media, in: "Explosion Seismology in Central Europe: data and results", P. Giese and C. Prodehls Skin, eds., Springer-Verlag, Berlin, 162-167.

Julian, B. R. and Gubbins, D., 1977, Travel time inversion for simultaneous earthquake location and velocity structure determination in laterally varying media, 1980, Geophys.J.R.Astr.Soc., 63:95-116.

Madrid, J., 1985, Travel times and ray paths for continuous media, Geof. Int., 24:439-458.

Madrid, J., 1986, New formulae for linear travel time inversion in 2-D heterogeneous media. Theory and results, Geof.Int., 25:361-382.

Newmann, G., Determination of lateral inhomogeneities in reflection seismics inversion of travel time residuals, 1981, Geophys.Prosp., 29:161-177.

Traslosheros, J.C., Un esquema no lineal de inversion de datos sismicos con referencia a medios bi-dimensionales lateralmente heterogeneos, 1987, Tesis de Maestria, CICESE, Ensenada, BC, Mexico.

EARTHQUAKE FOCAL MECHANISMS FROM INVERSION OF

FIRST P AND S WAVE MOTIONS

Giuseppe De Natale and Aldo Zollo

Osservatorio Vesuviano
N. Manzoni, 249
80123 Napoli Italia

1. INTRODUCTION

The advent of digital recording of seismic signals has greatly improved the capability to study the details of earthquake processes. This is mainly due to the large dynamic range of digital instruments allowing high quality recording of ground motion over a wide amplitude interval. Consequently, a new generation of techniques applied to time and frequency domain digital data has been developed which have the objective to increase the resolution on the estimate of source and medium parameters.

The study of earthquake processes from seismic data, requires a good knowledge of elastic and anelastic medium properties which affect the shape of seismic signals along the wave path.

Several techniques have been developed to study the source mechanisms of moderate-to-large earthquakes which are based on the distribution of first P-wave polarities over the focal sphere, waveform modelling, amplitude ratios of main seismic phases, moment tensor inversion (Kasahara, 1981; Sipkin, 1982; Pearce, 1977; Dziewonsky and Woodhouse, 1983).

Most of them have been applied to long period data (teleseismic and surface waves) in order to avoid the influence of small scale heterogeneities using only the well known large scale velocity models of the Earth crust. Furthermore, at long periods, small scale source complexities are averaged out so that the overall rupture process is seen like a point source. However, even using long period waves, there is evidence of complex rupture processes for large earthquakes, which is reflected in the computed source mechanisms as apparent large deviations from the double couple (Sipkin, 1986).

The detailed analysis of data from small size events (microearthquakes) ($M_o < 10^{22}$ dyne x cm, $M_L < 4$) using high frequency ($f > 1$ Hz) body waves represents a different approach to the study of rupture phenomena in active tectonic or volcanic regions.

In tectonic areas which are characterized by active faulting processes with the periodic occurrence of large earthquakes, microearthquakes can be used to study the geometrical features of the faulting zone, the actual stress regime, fracturing properties of the rocks and the efficiency of the medium to transmit the seismic energy radiated by the sources.

In geothermal and volcanic areas, microearthquakes can provide useful information on the physical processes involved like magma intrusion, gas emission or hydrothermal and fluid circulation activity.

For instance, in volcanic areas, a detailed analysis of the source mechanisms is very important in view of possible deviations from the double couple model and their relation with volcanic explosions or intrusive phenomena.

The major problem which arises when estimating source mechanisms from microearthquake data is the effect on signals of unknown small scale het- ereogeneities of the medium. This problem generally prevents the clear identification and interpretation of different wave trains. It can be partially overcome at short source-receiver distances by using only direct arrivals when they are clearly readible as first arrivals.

At short distances, in fact, direct arrivals carry on the most infor- mation about the radiating source, since they are the least affected by propagation effects.

This paper describes the theory and an application of a method to determine focal mechanisms by inversion of amplitude and polarity data of first P and S pulses. This method avoids any assumption about the shape of the slip time function, provided that, for the selected wavelengths, the source is effectively point-like and the far-field approximation is valid.

Moreover, it is shown an application to a microearthquake sequence recorded at the volcanic region of Campi Flegrei by a short period 3-com- ponent digital network during the recent episode of ground uplift (1982-1984) (De Natale et al, 1987).

SEISMIC MOMENT TENSOR

The seismic radiation recorded at the Earth surface carries on a combined information about the source excitation mechanism and filtering effects due to the wave propagation in generally inhomogeneous media.

The representation theorem allows one to separate source and path terms relating the displacement at any point of the medium to the discon- tinuities across a buried fault surface through a convolution with a linear operator (the Green tensor) which accounts for propagation in elastic media.

According to the representation theorem the displacement u at a point x of the medium and at a time t is given by (Aki & Richards, 1980):

$$u_n(\underline{x},t) = \iint_\Sigma m_{pq} * G_{np,q}(\underline{x},t;\underline{\xi},\tau) d\Sigma \tag{1}$$

(*denotes convolution and ',' denotes space variable derivative) where

$$m_{pq} = \Delta u_i(\underline{\xi},t) \nu_i c_{ijpq} \tag{2}$$

is the moment tensor density (moment per unit area), Δu is the slip func- tion, ν is the direction normal to the fault, ξ is the coordinate of points on the fault and c_{ijpq} is the 4-th order tensor of elastic constants. G is

400

the 2-th order elasto-dynamic Green tensor which is a function of the source and receiver relative positions as well of the properties of the medium encompassed by the seismic waves. The representation theorem states that motions at the earth surface can be constructed by summing up the contributions of couples uniformly distributed over the fault surface whose strength is given by m_{pq}. In the case of an isotropic medium c_{ijpq} is a function of the only Lame's constants λ and μ, for which

$$m_{pq} = \lambda \nu_k \Delta u_k \delta_{pq} + \mu(\nu_p \Delta u_q + \nu_q \Delta u_p).$$

At periods much greater than the rupture times the single elements of the fault are seen to radiate in phase and the whole of the fault can be considered as a point source, i.e. a system of couples located at the center of Σ with moment tensor:

$$M_{pq} = \iint_\Sigma m_{pq} d\Sigma.$$

In this case the representation theorem (1) can be written in the following compact form:

$$u_n = M_{pq} *G_{np,q}. \tag{3}$$

Relation (3) has been largely used to investigate the source mechanisms of moderate to large earthquakes mostly using seismic records at teleseismic distances. There are no a priori restrictions to apply the representation (3) to short period data provided that the proper frequency band is considered and the details of the velocity structure are known at the scale of the observed minimum wavelength.

The Green's function in (3) represents the complete wave field including both near and far field terms. Several methodologies have been recently developed to compute complete and high frequency approximate Green's functions even in quite complex velocity structures (Bouchon, 1979; Cerveny and Hron, 1980). Stump and Johnson (1977) used complete Green's function formulation for inverting the moment tensor components by using body wave seismograms.

At large distances from the source the far- field terms dominate over the near-field ones because they propagate a stronger slip discontinuity and are undergone to a weaker attenuation with distance (Aki and Richards, 1980).

In practice, it can be proved that at distances which are few wavelengths away from the source, Green's functions which include the only far-field terms are sufficient to describe the main features of the radiated displacement field (Bernard and Madariaga, 1985).

If we assume for simplcitity a homogeneous, unbounded and isotropic medium, and take the only far-field terms of the related Green's function, the displacement at $\underline{x}$ can be analytically determined carrying out the integral convolution in (3) (Aki and Richards, 1980):

$$u_n(\underline{x},t) = (4\pi\rho\alpha^3 r)^{-1}(\gamma_n\gamma_p\gamma_q)\dot{M}_{pq}(t - r/\alpha) +$$
$$- (4\pi\rho\beta^3 r)^{-1}(\gamma_n\gamma_p - \delta_{np})\gamma_q\dot{M}_{pq}(t - r/\beta) \tag{4}$$

where γ is the direction cosine of the vector $\underline{x}$, r is the source-to-receiver distance, α and β are the P and S wave velocity, ρ is the density and σ_{mp} is the Kronecker symbol.

Turning the expression (4) from the Cartesian coordinate system to the ray system and separating the radial and transverse wave motions, the far-field displacement components can be written in a matrix form:

$$\underline{u}^P(\underline{x},t) = (4\pi\rho\alpha^3 r)^{-1}(\underline{\gamma}M(t - r/\alpha)\underline{\gamma})\underline{1}$$

$$\underline{u}^{SV}(\underline{x},t) = (4\pi\rho\beta^3 r)^{-1}(\underline{p}M(t - r/\beta)\underline{\gamma})\underline{p} \qquad\qquad (5)$$

$$\underline{u}^{SH}(\underline{x},t) = (4\pi\rho\beta^3 r)^{-1}(\underline{\phi}M(t - r/\beta)\underline{\gamma})\underline{\phi}$$

The unit vectors $\underline{\ell}$, $\underline{p}$ and $\underline{\phi}$ give respectively the directions of radial and transverse motions as shown in Fig. 1.

Relations (5) can be easily generalized to the case of a spherically symmetric medium replacing ρ by $\rho^{1/2}(\underline{x})\rho^{1/2}(\underline{\xi})$, α by $\alpha^{5/2}(\underline{\xi})\ \alpha^{1/2}(\underline{x})$, β by $\beta^{5/2}(\underline{\xi})\beta^{1/2}(\underline{x})$ and r by the geometrical spreading:

$$R(\underline{x},\underline{\xi}) = |\underline{x}||\underline{\xi}|c(\underline{\xi})^{-1}(p^{-1}\cos i_x \cos i_\xi \sin\Delta|\partial\Delta/\partial p|)^{1/2}$$

where i_x and i_ξ are respectively the ray incidence angle at the station and take-off angle at the source, Δ is the epicentral distance, c is the wave velocity and p is the ray parameter.

Formulae (5) can also be used in the case of a layered medium once the proper transmission-reflection coefficients at the interfaces are taken into account. The amplification effect due to the traction-free surface of the Earth which is not considered in (5) can be taken into account by computing the correction factor in the plane wave approximation given the incidence angle i_x (Aki and Richards, 1980).

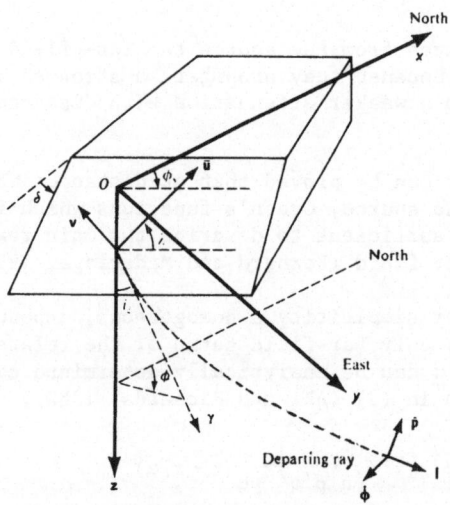

Fig. 1. Geometry of the seismic ray (Aki and Richards, 1980).

Radiation patterns in (5) can be expressed in terms of the take-off angle i_ξ (assumed to be the same for P and S waves), azimuth ϕ and moment tensor components:

$$\underline{\gamma \dot{M}\gamma} = \sin^2 i_\xi (\cos^2\phi \dot{M}_{11} + \sin2\phi \dot{M}_{12} + \sin^2\phi \dot{M}_{22} - \dot{M}_{33}) +$$
$$+ 2\sin i_\xi \cos i_\xi (\cos\phi \dot{M}_{13} + \sin\phi \dot{M}_{23}) + \dot{M}_{33}$$

$$\underline{p \dot{M}\gamma} = \sin i_\xi \cos i_\xi (\cos^2\phi \dot{M}_{11} + \sin2\phi \dot{M}_{12} + \sin^2\phi \dot{M}_{22} - \dot{M}_{33}) +$$
$$+ (1 - 2\sin^2 i_\xi)(\cos\phi \dot{M}_{13} + \sin\phi \dot{M}_{23})$$

$$\underline{\phi \dot{M}\gamma} = \sin i_\xi (1/2\sin2\phi (\dot{M}_{22} - \dot{M}_{11}) + \cos2\phi \dot{M}_{21})\cos i_\xi (\cos\phi \dot{M}_{23} - \sin\phi \dot{M}_{13}).$$

$$(6)$$

As shown in (5) and (6) the relationship between moment tensor and far-field displacement can be arranged in the form of a linear system:

$$A \underline{m} = \underline{u}$$

where $\underline{m} = (m_1, m_2, m_3, m_4, m_5, m_6) = (M_{11}, M_{12}, M_{22}, M_{13}, M_{23}, M_{33})$, M_{ij} being the 6 independent components of moment tensor; $\underline{u}$ is the vector of the recorded displacement amplitudes and A is a matrix which contains the Green's function and the radiation pattern coefficients.

Assuming that the source mechanism does not change during the fracture process we can separate the temporal part and write:

$$\dot{M}(t) = M \dot{s}(t) \qquad (7)$$

where s(t) is the slip time function. This assumption is reasonable in case of small size events whereas it seems to be too strong for complex fracture mechanisms occurring during large earthquakes (Magnitude > 6) as observed for Mammouth Lakes (1980), Coalinga (1983) and Yemen (1982) earthquakes (Sipkin, 1986). Given a set of observations $\underline{u}(t)$, relations (5) and (6) allow one to estimate by least square inversion techniques the 6 component of the moment tensor and the related source time function.

In case of time independent moment tensor components (rel.(7)) the amplitudes of P and S wave motions at a given time t_o (for instance the time corresponding to the maximum first P and S pulse amplitude) can be used to determine the relative amplitudes of moment tensor elements, being in this case $s(t_o)$ just a scaling factor. In this case no information can be achieved about the source strength (scalar seismic moment) unless to assume an analytical form for $s(t_o)$. The eigenvectors of the moment tensor give the directions of the principal stress axes and the eigenvalues give their magnitude. The relative magnitudes of the principal stress axes and their directions determine the radiation pattern, i.e. the distribution over the focal sphere of the normalized amplitude of direct P and S waves. This method will be exstensively described in the next section.

Equivalently, taking the Fourier Transform of eq. 5, it can be shown that the low frequency levels of the P and S displacement spectra can also be used to determine the moment tensor components provided that the first motion polarity is known (Oncescu, 1986).

As shown in the previous section, equations (5) and (6) can be written in the form of a linear system:

$$A \underline{m} = \underline{u} \tag{10}$$

Solving (10) is straightforward by ordinary methods of matrix inversion (see for example Lawson and Hanson, 1974). Due to the linearity of (10), the solution will be always univoquely determined when the number of data is greater than or equal to 6 (number of model parameters), provided that at least one P or SV amplitude is present in the data set.

Moreover, assuming gaussian errors in the data set, the likelihood of the model $\underline{m}$ is gaussian and it can be exactly specified by mean value m and associated covariance matrix (Jackson and Matsu'ura, 1985).

Given the covariance matrix of the data: $C_u = W^T W$, where W is a diagonal matrix (because the data are uncorrelated), the maximum likelihood estimate of $\underline{m}$ can be obtained by the least squares solution of:

$$A_w \underline{m} = \underline{u}_w \tag{11}$$

with $\quad A_w = WA \; ; \; \underline{u}_w = W\underline{u}$

In the framework of our algorithm, (11) is solved by the Singular Value Decomposition (Lawson and Hanson, 1974), giving:

$$\underline{m} = VS^{-1}U^T \underline{u}_w \tag{12}$$

where V is a 6x6 orthogonal matrix formed by the eigenvectors of $A_w^T A_w$, U (dimension m x m, m = number of data), is formed by the eigenvectors of $A_w A_w^T$, S^{-1} is a diagonal matrix with elements $1/s_i$, s_i being the eigenvalues of $A^T A$.

The model co variance matrix is computed as:

$$C_m = V^T S^{-2} V \quad \text{(Lawson and Hanson, 1974).} \tag{13}$$

Equations (11) through (13) allow to get from data all the information about a generic 6 parameter moment tensor. However, for many practical purposes, we can impose some a-priori constraints on the form of the moment tensor. For instance, the assumption of double couple mechanisms for tectonic earthquakes is often supported by other geophysical and geological observations. In other cases, we know that the isotropic component of the moment tensor must vanish, because volume changes cannot occur at considerable depth due to the lithostatic pressure.

Moreover, we would simply to test the suitness of the commonly used models (e.g. double couple) against others. In all these situations, we have to compute particular kinds of solutions, and possibly to compare them in order to find the most satisfactory in a statisticl sense. The most adopted source mechanism model is the double couple one. It is represented by a moment tensor having zero trace (no isotropic component) and zero determinant.

The constraint of zero trace is linear. It can be imposed by the method of Lagrange multipliers (Claerbout, 1976). The system to be solved is:

$$
\begin{bmatrix} A_w^{\,T} A_w & V^T \\ V & 0 \end{bmatrix} \begin{bmatrix} \underline{\delta m} \\ \lambda \end{bmatrix} = \begin{bmatrix} A^T \underline{\delta u} \\ -\mathrm{Tr}\,(\underline{m}) \end{bmatrix}
\tag{14}
$$

with V = [101001], λ = Lagrange multiplier, Tr = trace

Inverting (14) instead of (11), a zero trace moment tensor is obtained, corresponding to a mechanism with no isotropic component.

The constraint of zero determinant is non linear; a method to impose this constraint is to linearize and to solve it iteratively for the constrained solution (Oncescu, 1986).

So doing, the constrained system can be written as:

$$
\begin{bmatrix} A_w^{\,T} A_w & D^T \\ D & 0 \end{bmatrix} \begin{bmatrix} \underline{m} \\ \underline{L} \end{bmatrix} = \begin{bmatrix} A^T \underline{u} \\ \underline{C}_d \end{bmatrix}
\tag{15}
$$

with $D = \begin{bmatrix} 1 & 0 & 1 & 0 & 0 & 1 \\ R_1 & R_2 & R_3 & R_4 & R_5 & R_6 \end{bmatrix}$ and

$R_1 = m_3 m_6 - m_5^{\,2}$; $R_2 = 2(m_4 m_5 - m_2 m_6)$; $R_3 = m_1 m_6 - m_4^{\,2}$; $R_4 = 2(m_2 m_5 - m_3 m_4)$;
$R_5 = 2(m_2 m_4 - m_1 m_5)$; $R_6 = m_1 m_3 - m_2^{\,2}$; $\underline{L} = \lambda_1, \lambda_2$ (Lagrande multipliers),
$\underline{C}_d = -\mathrm{Tr}(\underline{m}), -\mathrm{Det}(\underline{m})$ (Det = determinant).

Solving (14) iteratively, up to convergence to a stable minimum is reached, a double couple solution is found. The covariance matrix for the model parameters is obtained as the inverse of the coefficient matrix in (14).

It is important to note, however, that, due to the intrinsic non-linearity of the zero-determinant constraint, the probability density function on the model parameters is no longer gaussian. This means that (15) is only an approximation of the true covariance matrix, and it is often called 'asymptotic covariance matrix'. It represents a gaussian approximation of the likelihood in the neighboring of the obtained solution. Exact error estimates could be obtained by combining the constraints of zero-trace and zero-determinant with the equation (10) and determining the likelihood in the whole space of interest (Tarantola and Valette, 1980; Jackson and Matsu-ura, 1985). This method, however, would be much more cumbersome, giving markedly different results only for complicate multimodal likelihood functions (Jackson and Matsu'ura, 1985). In all the applications of this method, however, both for sinthetic tests and real data, we have always found that the solution of (14) was well determined and a stable minimum reached, without needing for damping or other stabilizing methods (see Lawson and Hanson, 1974). This means that for practical purposes, the likelihood function has only one maximum so that the asymptotic covariance matrix (15) is a reasonable approximation of the true one.

Once different kinds of solutions have been computed from the data, it is possible to compare their performance by statistical tests of hypothesis. Since the models having more free parameters always give a better

fit to the data, we must decide if the improvement in the fit is significant, with respect to simpler models (having less free parameters). If it
is not, we have no reason to prefer a more complex model (supposing we have
no a priori information about the kind of mechanism involved). We use a
statistical F test on variances in order to decide if a models with more
parameters is significantly better than a simple one. This is made by
comparing the variance-ratio between the competitive models with different
degrees of freedom, with the 95% probability level for an F distribution.
If the variance-ratio is greater than the threshold we must reject the
hypothesis that variances belong to the same distribution, and choose the
best fitting model as more appropriated

DATA PROCESSING FOR MOMENT TENSOR ANALYSIS

A set of high quality 3 component ground motion records is needed to
perform a moment tensor inversion using first P and S time domain pulse
amplitudes. In fact, such a kind of analysis requires the use of seismograms with high signal to noise ratio, clear P and S phase identification,
simple first pulse waveforms.

General requirements to apply source mechanism inversion techniques
are common to most of the adopted methods:

1) good data sampling of the focal sphere,
2) knowledge of the details of velocity structure at the scale of the
   observed maximum wavelength,
3) estimate of anelastic attenuation and instrument filtering effects on
   waveforms,
4) accurate locations of the earthquakes.

The processing of seismic records is performed in several steps as
detailed in the following.

1. Graphic display of the seismic signals on high resolution videographic
   computer terminal and selection of the P and S phases. Wave motions
   which are recorded in the N-E-Vert reference system are rotated in the
   ray system (P-radial, SH and SV). P wave signals are rotated into the
   radial component according to the polarization angle estimated from
   the vectorial composition of the first P pulse amplitude.
   S wave signals are rotated into the SH and SV components using the
   azimuth from the epicenter location.

2. Correction for the instrument response can be performed in frequency
   domain by division of the complex Fourier transform of the recorded
   signal by the instrument complex response curve. Short period velocity response curves can be generally parameterized by the damping and
   natural frequency of the geophones for the high pass filter and the
   cut-off frequency of the low pass anti-aliasing filter (Blair, 1982).
   Instrument corrected time signals are obtained by inverse Fourier
   transform.

3. Waveform distorsion due to the anelastic properties of the rocks can
   be removed provided that the attenuation of seismic waves in the
   propagation medium has been independently estimated. Several analytical expressions of the anelastic attenuation filter have been
   proposed which assume a nearly constant quality factor Q in the
   seismic frequency band (Aki and Richards, 1980) and parameterized by a
   cut-off parameter which is related in some way to the constant Q
   frequency range.

Correction of seismic records for the attenuation effect can again be performed in frequency domain as for the instrument filter using the analytical form of attenuation filter.

4. Digital band pass filtering of seismic records may be needed since the frequency band for the analysis must be within the range of validity of both far-field and point source approximations as required by the theory (re.(5)). Band-pass filtering is also useful to enhance the signal-to-noise ratio in case of low and high frequency noise is present on the seismic records.

5. Integration of velocity records to get displacements and readings of first P and S pulse amplitudes are the final steps of data processing. Integration can be equivalently performed in frequency or time domain. Since the validity of (7) readings of maximum first P and S pulse amplitude on velocity records can equivalently be used as input data for moment tensor inversion according to (5), (6) and (7). In fact, in this case we have:

$$\dot{\underline{u}}(\underline{x},t) \propto M\ddot{s}(t) \tag{9}$$

where the moment tensor elements are now (at a given time t_o) linearly related to the far-field velocity field with the scaling factor being represented by the acceleration of the slip time function computed at t_o.

Robustness of the method

In order to assess the robustness of the inverse method with respect to errors in the data, we can distinguish between two cases:

1) Random errors in the amplitude data set
2) Random and systematic errors in hypocenter and velocity model.

Errors of the first kind, are naturally taken into account by the inverse procedure, and their influence on the moment tensor estimates can be correctly inferred by the covariance matrix, due to the linearity of the problem. Also for costrained double couple solutions, the asymptotic covariance matrix generally represents a good approximation to the true covariance.

A good indicator of the robustness of an inverse method is represented by the condition number C; it is defined like the ratio of the larger to the smaller singular value in the Singular Value Decomposition of the matrix A in the inverse problem. It represents a scale factor for the propagation of the relative errors on the data into the relative errors on the model parameters. For a linear inverse problem we have:

$$\delta\underline{x}/\underline{x} < C\delta\underline{y}/\underline{y} \quad \text{(Forsythe and Moler, 1967)}.$$

Typical values obtained, with several tests, involving both real and simulated situations range from few units to few tenths, indicating the method is very robust. The influence of kind 2)-errors cannot be analyzed in a general way, but it requires to consider the specific problem involved, being related to the accuracy of hypocenter locations and the knowledge of structural complexities and possible departures from the assumed models. The robustness of the method, as inferred from the condition number is of course indicative of the method capability to give correct estimates even

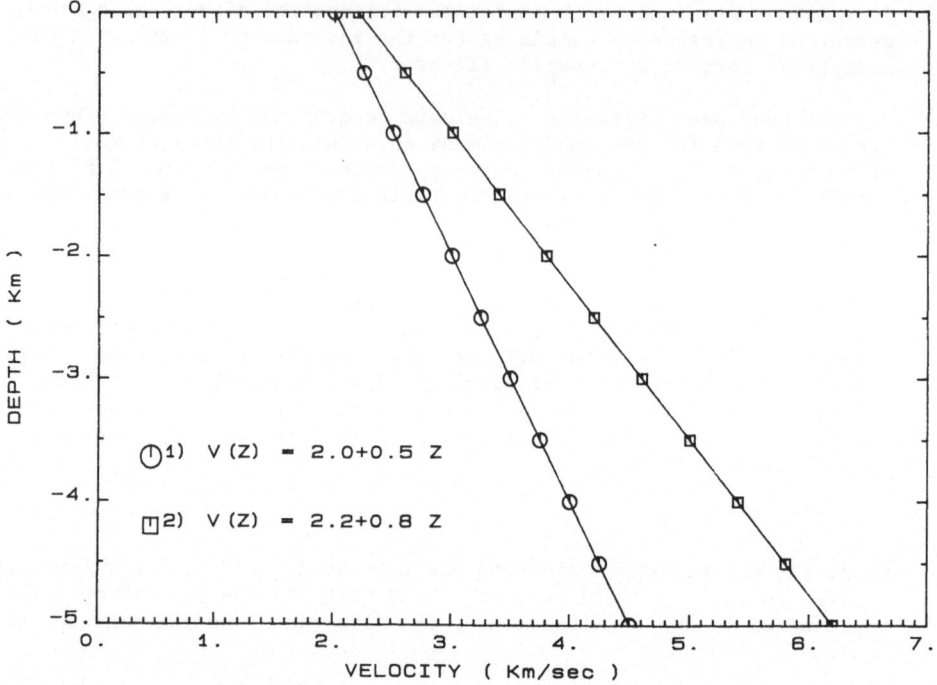

Fig. 2. Velocity models used for radiation pattern simulations.

when some spourious effects are present. However it cannot give statistical confidence limits for the estimated solution due to errors in the theoretical model. The only way to assess in a quantitative way the robustness of the inverse method against errors in hypocenter location and velocity model, is to perform synthetic tests, in which the true model is exactly known. Such a kind of synthetic tests, obviously, cannot account for all the possible situations, but they must be suitable to simulate the actual problem that one is working out.

In view of the application of moment inversion method to microearthquakes recorded at Campi Flegrei volcanic area, we have simulated similar conditions for earthquake location and velocity model.

Radiation patterns for an earthquake with location given in Table 1, recorded by 12 stations (see Table 1), have been computed.

The correct velocity model used for simulations is named model 1) in Fig. 2. The focal mechanism has been then recomputed by moment tensor inversion of P-radial and SH wave amplitudes, using a perturbed hypocenter location (see Table 1) and a different velocity model (model 1 in Fig. 2).

Results are shown in Table 1 and Fig. 3 and 4. The true focal mechanism for test 1 was left lateral pure strike- slip with planes dipping 90°, striking respectively N-S and N90°E (Fig. 3).

For test 2, a pure normal faulting mechanism with planes dipping 45°, oriented N-S was the true solution (Fig. 3b). As it is shown, for the strike slip faulting, the solution is practically unaffected by the wrong velocity model, and it is only slightly influenced by hypocentral errors. This is the reason because for the second test only the combined effect of both kinds of errors is shown (Fig. 4). Also in this case, the bias on the solution is minor.

Table 1

TRUE HYPOCENTER

Latitude	Longitude	Depth
40° 49'	14° 7'	3 Km

PERTURBED HYPOCENTER

Latitude	Longitude	Depth
40° 49.4'	14° 6.7'	2.5 Km

TEST N.1 (Pure Strike-Slip Fault)
TRUE MECHANISM

PRINCIPAL STRESS AXES

EIGENVALUES			Maximum A	P	Intermediate A	P	Minimum A	P
DC 1.	0.	−1.	45	0	−	90	135	0.

MECHANISM COMPUTED WITH PERTURBED HYPOCENTER

PRINCIPAL STRESS AXES

EIGENVALUES			Maximum A	P	Intermediate A	P	Minimum A	P
FS 1.	−0.02	−.98	46.	4	126.	71	138.	19
TRO 1.	−0.02	−.98	46.	4	126.	71	138.	19
DC 1.	0.	− 1.	46.	4	126.	71	138.	19

TEST N.2 (Pure normal fault)

TRUE MECHANISM

PRINCIPAL STRESS AXES

EIGENVALUES			Maximum A	P	Intermediate A	P	Minimum A	P
DC 1.	0.	−1.	90	0.	0.	0.	−	90.

MECHANISM COMPUTED WITH PERTURBED HYPOCENTER AND VELOCITY MODEL

PRINCIPAL STRESS AXES

EIGENVALUES			Maximum A	P	Intermediate A	P	Minimum A	P
FS. .87	0.03	−1.	92.	6.	1.	8.	39.	80.
TRO .95	0.05	−1.	92.	7.	1.	9.	39.	78.
DC 1.	0.	−1.	92.	8.	1.	9.	43.	79.

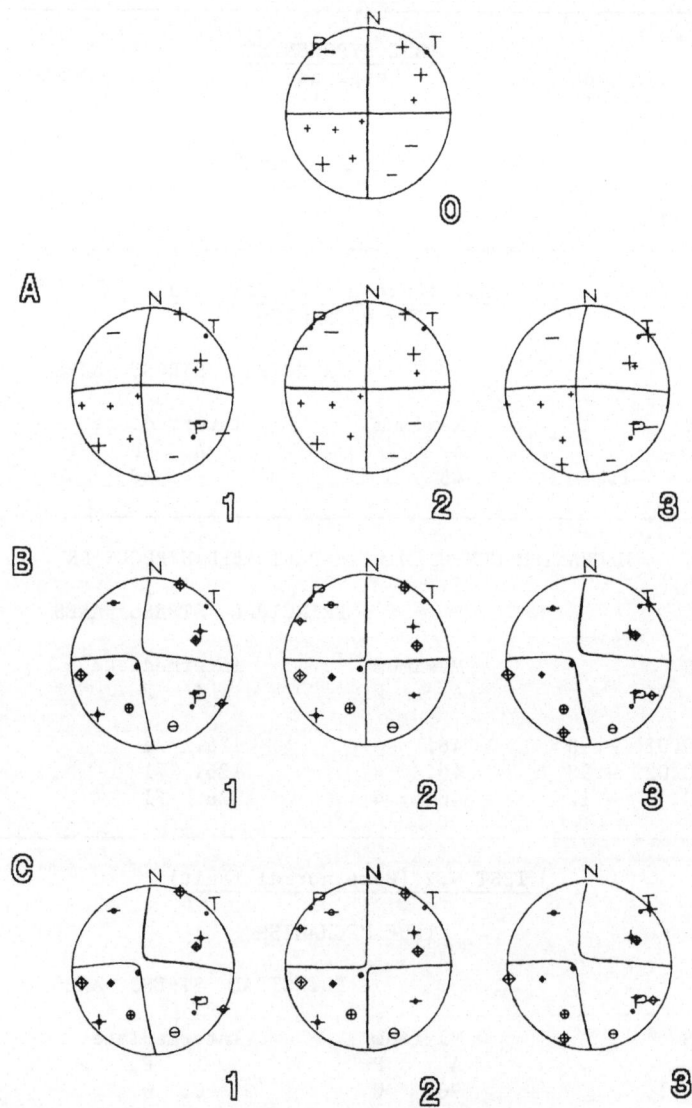

Fig. 3. Results of synthetic tests for a strike-slip fault (test n.1. in
Table 1). 0 = true model; 1 = solution computed after perturbing
hypocenter location (see Table 1); 2 = solution computed after
perturbing the velocity model (using model 2 in Fig. 2); 3 =
solution computed after perturbing both hypocenter location and
velocity model as in 1 and 2. A) double couple solutions; B)
zero-trace solutions; C) generic 6-parameter free solutions. Plus
and minus indicate positive and negative P amplitude; circle and
rhombus are for SH waves. Symbol size is proportional to the
amplitude.

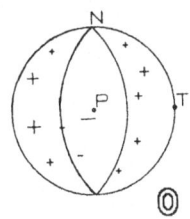

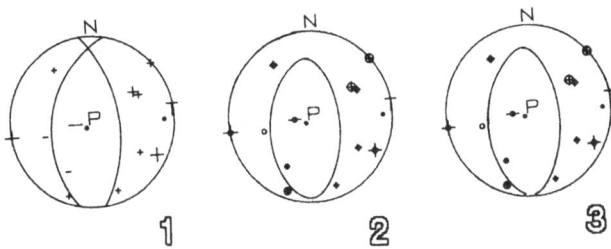

Fig. 4. Results of synthetic tests for a normal fault (test n.2. in Table 1). 0 = true model; 1 = double couple solution computed after perturbing both hypocenter location and velocity model as done in test n.1; 2 = zero-trace solution computed as in 1; 3 = generic 6-parameter free solution computed as in 1 and 2. Symbols are as in Fig. 3.

It is important to stress that the simulated hypocentral errors are larger than the average location errors in the area, for the best located events (De Natale and Zollo, 1986). Also the errors on the velocity model represent an overestimate of the possible errors.

Obviously, the key to understand why the misinterpretation of the velocity model plays a minor role lies in the assumed linearity of velocity with depth. If more complicated structures, for example layers with strongly varying wave velocities would have been unmodeled, results would have been surely worst. However, in Campi Flegrei area, such kind of structural complexities have not been evidenced, after detailed analyses (Aster and al, 1987), and are hardly hypothesizable within the first 3-4 Km of the crust. So, we believe the tests performed are effective to assess the robustness of the method in the real application.

Focal Mechanisms of Microearthquakes at Campi Flegrei Volcanic Area

During the period 1982-1984 a large ground uplift episode occurred in this region of Southern Italy 10 km west of the city of Naples. More than 10000 earthquakes were recorded ranging from M_L = 0. to M_L = 4..

A digital 3-component seismic network owned by the University of Wisconsin, Madison was installed since January to May 1984. A comprehensive analysis of seismic activity with the aim to investigate both source and medium properties is exstensively described in Aster et al, (1987). A map with digital station locations and a representative epicenter distribution is shown in Fig. 5.

In the following we report an application of moment tensor inversion method to digital recordings of 18 Campi Flegrei microearthquakes.

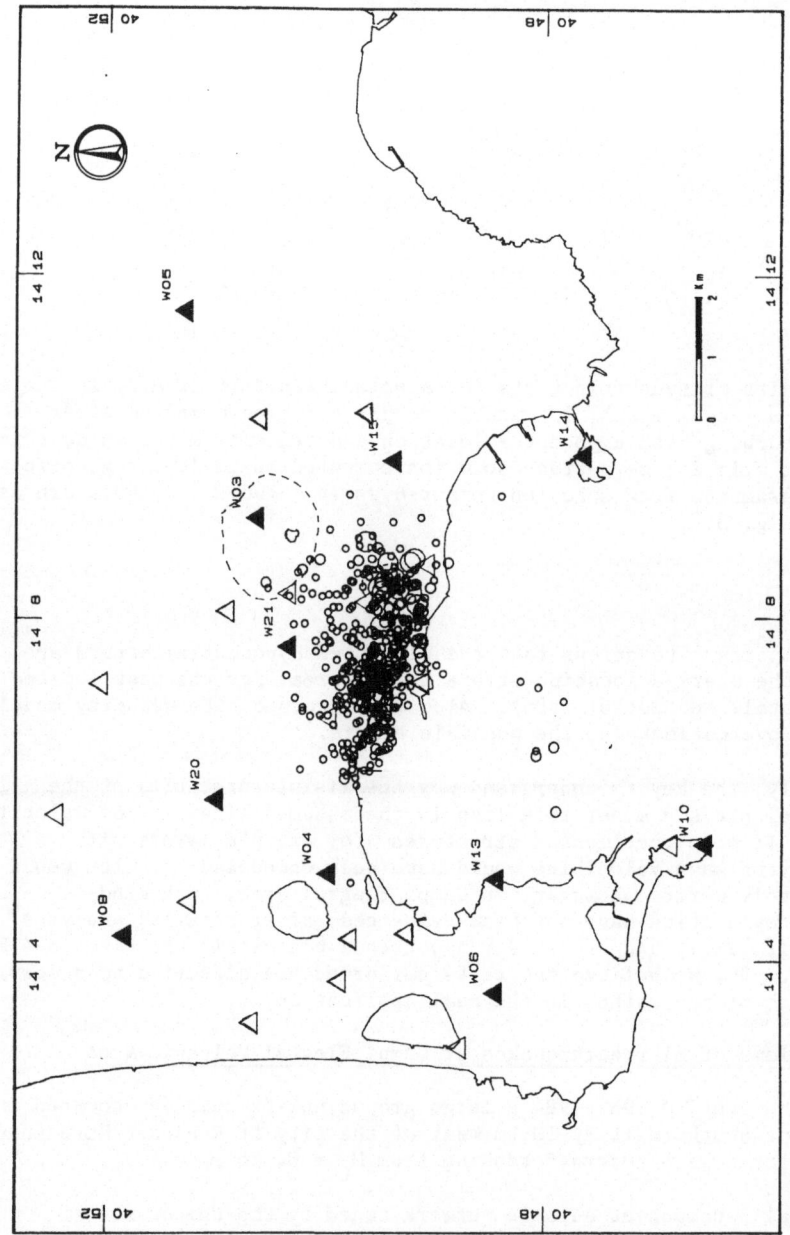

Fig. 5. Map with station locations and epicenter distribution at Campi Flegrei in the period 1983-1984. Solid triangles are the University of Wisconsin portable digital stations, the open ones are the Osservatorio Vesuviano survey analog stations. A representative plot of seismicity is also shown with 450 selected earthquakes (De Natale and Zollo, 1986).

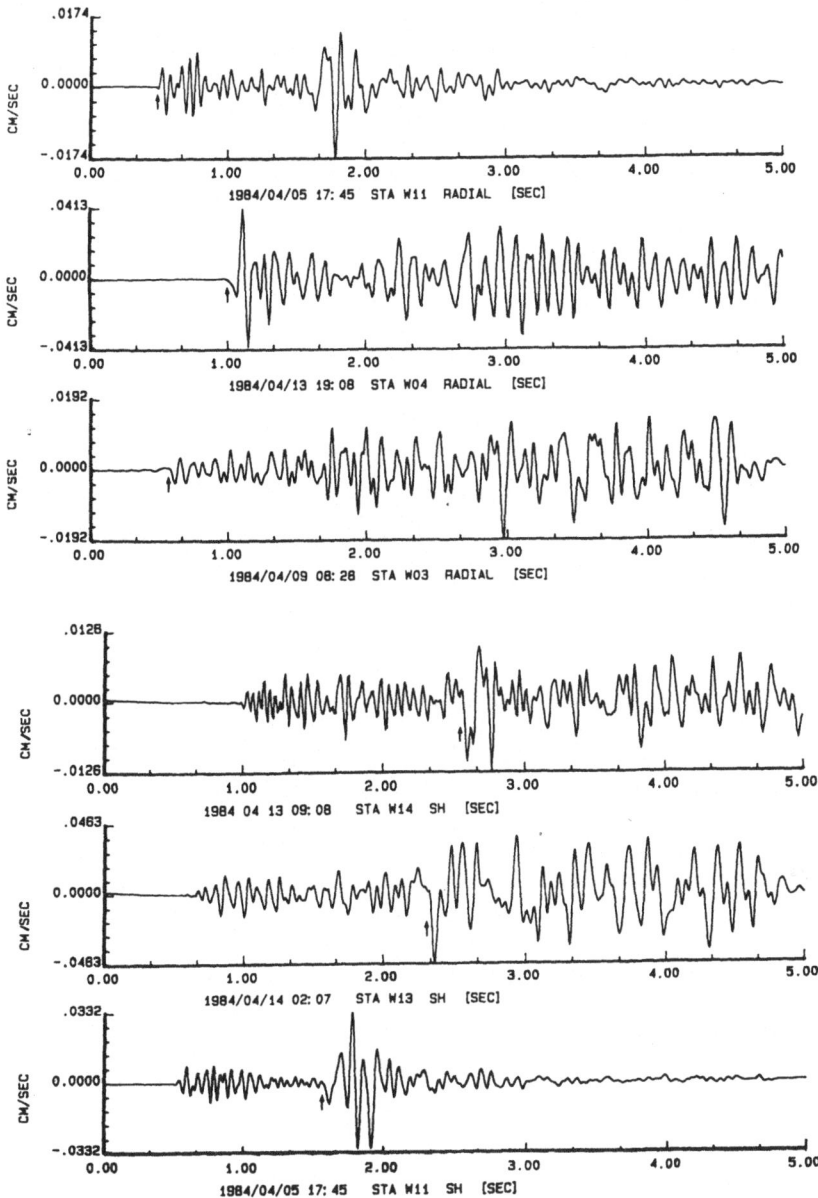

Fig. 6. A sample of Radial and SH seismograms recorded at Campi Flegrei.
Arrows indicate first P and S pulses.

Table 2. Summary of the moment tensor analyses. Principal axes.

#	D A T A	TYPE	EIGENVALUES			maximum azimuth	plunge	intermediate azimuth	plunge	minimum azimuth	plunge
1	28.01.84 23:51	FS	1.00	-.15	-.19	293.2+-16.3	73.4+-10.5	82.4+-89.4	14.4+-13.3	174.6+-84.7	8.1+-25.9
		Tr0	1.00	-.24	-.76	298.3+- 5.2	55.4+-14.9	57.0+-13.1	18.3+- 8.0	157.3+-13.4	28.2+-15.2
		DC	1.00	.00	-1.00	297.9+- 4.1	56.7+-25.0	74.2+-16.9	25.4+-15.3	174.2+- 2.5	20.0+-17.7
2	10.02.84 01:07	FS	1.00	.59	-.00	179.2+-46.1	18.3+-15.2	273.4+-41.6	12.5+-33.3	35.9+-61.6	67.6+-24.5
		Tr0	.68	.32	-1.00	121.9+- 0.4	36.1+- 2.3	319.6+- 1.7	52.6+- 2.2	218.0+- 0.4	8.6+- 0.7
		DC	1.00	.00	-1.00	105.0+- 0.3	44.5+- 0.8	322.0+- 1.1	39.1+- 0.5	215.0+- 0.4	19.5+- 0.6
3	10.02.84 02:22	FS	1.00	.21	-.80	6.3+-31.4	9.2+-30.8	119.7+-74.9	67.7+-37.3	272.9+-22.7	20.1+-32.6
		Tr0	.65	.35	-1.00	172.7+- 2.0	8.3+- 5.7	52.0+-20.3	74.1+- 2.4	264.7+- 0.7	13.5+- 0.8
		DC	1.00	.00	-1.00	358.7+- 0.7	2.7+- 0.8	96.5+- 2.4	70.9+- 1.2	267.7+- 0.4	18.9+- 1.1
4	10.02.84 05:48	FS	1.00	.53	.09	32.4+-39.3	36.1+-34.6	301.1+-43.5	1.8+-23.3	209.0+-62.7	53.8+-34.8
		Tr0	.57	.43	-1.00	291.5+- 8.4	1.9+- 4.3	22.2+-10.2	22.2+- 4.2	197.0+- 2.4	67.7+- 4.4
		DC	1.00	.00	-1.00	299.1+- 3.2	5.8+- 4.0	34.3+- 6.8	42.2+- 3.0	203.0+- 1.4	47.2+- 3.0
5	12.02.84 11:01	FS	.48	.32	-1.00	168.2+-98.5	49.4+-17.6	270.6+-44.8	10.4+-49.8	9.1+- 4.3	38.6+- 2.7
		Tr0	.70	.30	-1.00	132.9+- 3.4	36.7+- 1.6	247.6+- 2.9	29.3+- 1.9	5.1+- 0.7	39.4+- 1.4
		DC	1.00	.00	-1.00	120.7+- 0.6	31.9+- 0.2	235.8+- 0.6	34.3+- 0.6	359.9+- 0.5	39.4+- 0.5
6	15.02.84 01:59	FS	1.00	-.12	-.89	316.8+-23.7	39.4+- 5.7	226.3+-19.5	.7+- 9.2	135.4+-18.0	50.6+- 5.6
		Tr0	1.00	-.11	-.89	317.6+- 1.4	39.4+- 0.5	226.8+- 1.2	1.0+- 0.4	135.6+- 1.0	50.6+- 0.4
		DC	1.00	.00	-1.00	321.9+- 0.8	42.8+- 0.5	52.2+- 0.5	.3+- 0.4	142.5+- 0.4	47.2+- 0.3
7	17.02.84 20:41	FS	.25	-.08	-1.00	224.4+-14.9	42.3+- 4.2	127.5+-13.5	7.6+- 4.9	29.4+-13.2	46.8+- 4.1
		Tr0	1.00	.00	-1.00	122.7+- 2.0	41.2+- 0.5	242.6+- 2.1	29.7+- 1.8	355.7+- 1.9	34.5+- 1.7
		DC	1.00	.00	-1.00	122.5+- 1.5	41.2+- 0.4	242.4+- 1.5	29.6+- 1.5	355.5+- 1.4	34.5+- 1.4
8	19.02.84 00:39	FS	.93	.05	-1.00	19.5+-52.7	71.0+-11.2	285.4+-18.3	1.4+-15.2	195.0+-17.9	19.0+-11.5
		Tr0	.95	.05	-1.00	19.9+- 1.0	71.2+- 0.2	285.7+- 0.3	1.4+- 0.3	195.0+- 0.3	18.8+- 0.2
		DC	1.00	.00	-1.00	25.7+- 0.7	72.1+- 0.1	286.6+- 0.3	2.9+- 0.3	196.0+- 0.3	17.6+- 0.2
9	20.02.84 01:11	FS	1.00	.15	-.94	31.2+-24.0	40.5+- 8.2	293.7+-22.2	8.7+- 8.1	194.0+-19.6	48.2+- 8.0
		Tr0	1.00	.90	.10	36.7+- 0.6	38.9+- 0.3	299.1+- 0.6	9.3+- 0.2	198.0+- 0.6	49.6+- 0.3
		DC	1.00	.00	-1.00	21.9+- 0.4	41.5+- 0.3	288.3+- 0.6	4.0+- 0.3	194.0+- 0.7	48.2+- 0.3
10	22.02.84 17:20	FS	1.00	.55	.17	334.4+-66.0	52.5+-83.7	156.7+-121.	37.4+-82.9	65.9+-88.9	1.1+-50.9
		Tr0	.88	.12	-1.00	110.7+- 4.0	33.0+- 5.2	259.6+-10.8	52.8+- 4.4	10.5+- 4.0	15.2+- 5.4
		DC	1.00	.00	-1.00	110.9+- 3.8	32.2+- 3.6	255.7+- 8.2	52.4+- 3.6	9.6+- 4.1	17.3+- 4.0
11	07.04.84 01:11	FS	1.00	.01	-.19	215.0+- 4.3	45.6+- 2.5	104.6+- 6.2	18.8+- 4.6	359.0+- 8.8	38.3+- 3.2
		Tr0	.74	.26	-1.00	164.5+- 1.7	27.4+- 0.6	69.8+- 1.5	8.9+- 0.7	323.4+- 1.2	61.0+- 0.6
		DC	1.00	-.08	-0.92	155.4+- 0.8	31.1+- 0.5	60.6+- 0.8	7.9+- 0.8	317.9+- 1.8	57.7+- 0.5
12	08.04.84 05:04	FS	1.00	.13	-.45	181.6+- 3.1	14.7+- 4.1	291.4+-11.0	52.3+- 7.5	81.5+- 4.0	33.9+- 8.7
		Tr0	.95	.05	-1.00	192.7+- 3.7	33.9+- 7.8	327.9+-14.8	46.5+- 3.7	85.5+- 3.9	23.8+- 6.2
		DC	1.00	-.00	-1.00	192.7+- 3.1	33.2+- 6.4	326.0+-12.6	46.3+- 2.6	85.1+- 3.6	24.9+- 5.3
13	09.04.84 08:28	FS	1.00	.45	-.86	170.9+-29.0	26.0+-68.5	32.2+-86.7	56.9+-57.2	270.5+- 9.1	18.9+-13.9
		Tr0	.83	.17	-1.00	177.0+- 7.3	5.2+-13.0	77.0+-31.3	62.1+- 2.7	269.7+- 2.0	27.3+- 2.3
		DC	1.00	-.00	-1.00	.4+- 1.9	2.6+- 2.5	95.3+- 5.4	62.6+- 2.7	269.1+- 1.9	27.3+- 2.5
14	13.04.84 19:08	FS	0.41	.13	-1.00	187.0+- 9.3	3.7+- 7.3	94.9+-13.0	29.6+- 2.9	283.4+- 6.6	60.1+- 2.6
		Tr0	1.00	-.09	-.91	8.0+- 0.6	6.5+- 0.8	101.0+- 0.8	24.5+- 1.8	264.1+- 2.5	64.5+- 1.7
		DC	1.00	-.00	-1.00	8.5+- 0.5	6.0+- 0.7	101.5+- 0.8	26.2+- 0.9	266.6+- 1.2	63.0+- 0.9
15	14.04.84 02:19	FS	1.00	-.22	-0.25	289.5+-29.8	51.1+-21.9	124.7+-474.	37.9+-75.8	28.8+-303.	7.5+-235.
		Tr0	1.00	-.28	-.72	284.8+- 0.8	19.9+- 0.7	18.5+- 1.7	10.0+- 2.2	133.7+- 5.7	67.5+- 1.5
		DC	1.00	-.00	-1.00	291.7+- 0.6	21.8+- 0.6	27.3+- 0.7	13.8+- 0.3	147.3+- 0.5	63.8+- 0.5
16	14.04.84 07:30	FS	1.00	.18	-0.27	160.0+-16.7	53.6+-14.0	19.1+- 8.7	29.8+-15.4	277.7+- 8.0	19.0+- 5.6
		Tr0	1.00	-.09	-.91	178.6+- 2.0	8.1+- 0.9	271.1+- 2.5	17.1+- 1.9	64.1+- 2.6	71.0+- 2.1
		DC	1.00	-.00	-1.00	178.0+- 1.8	7.5+- 0.7	270.1+- 2.2	15.4+- 1.1	62.9+- 1.9	72.8+- 1.3
17	14.04.84 21:45	FS	1.00	.58	-0.78	135.2+-52.7	34.7+-21.5	29.9+-41.8	20.9+-32.7	275.0+-21.9	47.8+-12.1
		Tr0	1.00	-.11	-.89	140.0+- 1.6	32.5+- 1.0	24.4+- 3.2	34.2+- 2.6	261.1+- 2.9	39.0+- 2.5
		DC	1.00	-.00	-1.00	139.1+- 1.7	33.0+- 1.0	23.7+- 2.7	33.5+- 2.0	261.0+- 2.2	39.2+- 1.9
18	14.04.84 22:14	FS	0.24	.04	-1.00	15.8+-14.1	8.2+- 1.9	105.9+-14.5	0.7+- 3.2	201.0+-33.7	81.7+- 2.0
		Tr0	0.81	0.19	-1.00	76.9+- 3.4	34.7+- 1.2	179.8+- 2.9	17.9+- 2.5	292.2+- 2.7	49.7+- 1.3
		DC	1.00	-.00	-1.00	31.8+- 2.9	10.9+- 4.0	132.2+- 4.6	43.1+- 6.5	290.8+- 3.3	44.9+- 5.1

(continued...)

Data processing and inverse procedure have already been described in the previous sections of this paper. Examples of P-radial and SH velocity time series after data processing are shown in Fig. 6.

Fig. (7 a and b) show, respectively, zero-trace and double couple focal mechanisms obtained for the 18 analyzed microearthquakes. Fig. 8 shows theoretical (lower trace) and observed first P-radial and SH pulses relative to the earthquake n.6 in Table 2. Theoretical amplitudes are

Table 2. (continued)

#	Variance Ratio $\sigma^2_{DC}/\sigma^2_{FS}$	Degrees of Freedom DC	Degrees of Freedom FS	Variance Ratio $\sigma^2_{DC}/\sigma^2_{TrO}$	Degrees of Freedom DC	Degrees of Freedom TrO	Variance Ratio $\sigma^2_{TrO}/\sigma^2_{FS}$	Degrees of Freedom TrO	Degrees of Freedom FS
1	1.9	5	3	1.1	5	4	1.7	4	3
2	4.5	4	2	1.1	4	3	4.2	3	2
3	1.1	4	2	1.0	4	3	1.0	3	2
4	3.5	7	5	1.0	7	6	3.4	6	5
5	2.2	4	2	1.3	4	3	1.7	3	2
6	1.0	4	2	1.0	4	3	1.0	3	2
7	1.5	3	1	1.0	3	2	1.5	2	1
8	1.0	5	3	1.0	5	4	1.0	4	3
9	1.3	10	8	1.3	10	9	1.0	9	8
10	2.1	6	4	1.0	6	5	2.1	5	4
11	3.2	5	3	1.2	5	4	2.8	4	3
12	10.6	7	5	1.0	7	6	10.5	6	5
13	1.4	6	4	1.0	6	5	1.3	5	4
14	3.1	7	5	1.0	7	6	3.0	6	5
15	12.0	5	3	1.0	5	4	1.1	4	3
16	9.1	4	2	1.0	4	3	8.6	3	2
17	1.3	7	5	1.0	7	6	1.3	6	5
18	8.0	4	2	1.3	4	3	6.3	3	2

FS = Free solution, TrO = Zero trace solution, DC = Double couple solution

computed from the double couple radiation pattern as inferred from moment tensor inversion and scalar seismic moment (M_o) and rise-time of the slip function $(\tau_{1/2})$ as obtained by De Natale et al. (1987).

Values and statistics of the parameters are reported in Table 2, also for generic non constrained solutions. The fit is excellent specially considering that average values of M_o and $\tau_{1/2}$ have been used.

Stress axes orientations are almost stable over the three kinds of solutions. Moreover, zero-trace solutions look very similar to the double couple ones (see solutions n. 6,7,8,9,10,12,13,14,16,17). The variance-ratio between double couple and zero-trace solutions is nearly one for almost all the microearthquakes (see Table 2). This means that, on the basis of an F test on variances, zero-trace solutions give not a significant improvement in fit to the experimental data, so that there is no need for introducing a further degree of freedom, namely free determinant, with respect to the double couple. On the other hand, the variance-ratios of free solutions (6 free parameters) to double couple are generally higher (see Table 2), except for the solutions n. 3, 6, 8, 9, 13, 17. However, on the basis of a 95% confidence level F-test on variances, the improvements in variance with respect to the double couple are not significant, so we cannot reject the hypothesis of double couple mechanism for any of the 18 earthquakes. The focal mechanisms show a complex pattern, with no predominant kind of mechanism. This is probably due to the following factors:

1) The small magnitude (around 2) of the focal mechanisms, so that the fracture follows local stress heterogeneities rather than the dominant stress pattern at large scale.

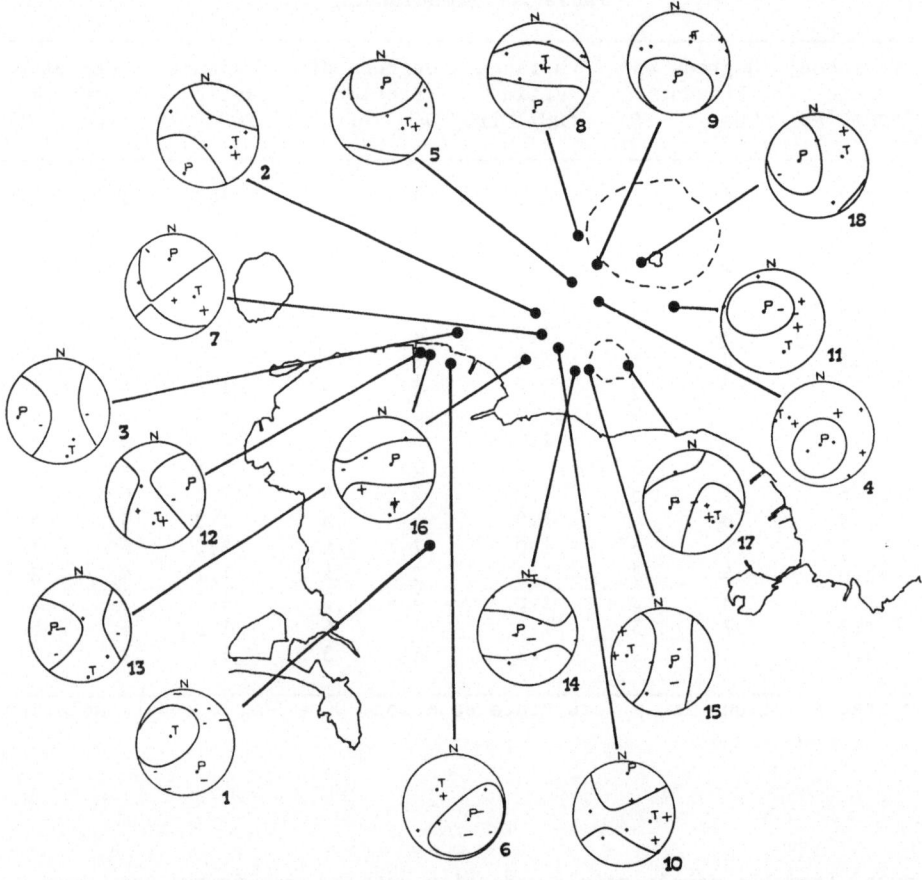

Fig. 7. Lower emisphere projections of focal mechanism from moment tensor
inversion. (a) zero-trace constrained solutions; (b) double
couple constrained solutions.

(continued...)

2) The probably complex stress pattern resulting from the superposition
   of a regional tectonic stress and a local one, due to the volcanic and
   geothermal activity causing the ground deformation.

CONCLUSIONS

The far-field representation theory has been applied to estimate focal
mechanisms of earthquakes by moment tensor inversion of first P and S wave
amplitudes. This method can be successfully applied to the study of source
properties of microearthquakes provided that the velocity structure is
known at the wavelength scale of interest.

However, when only direct arrivals are considered there are many
realistic situations in which the detailed knowledge of the velocity model
doesn't represent a major problem.

The main advantage of this method with respect to the standard techniques based on first P pulse polarities is the possibility to use both P and S waves and the amplitude information. Moreover the method allows one to get confidence limits on the estimated solutions as well as to compare the likelihood of different kinds of fracture mechanisms.

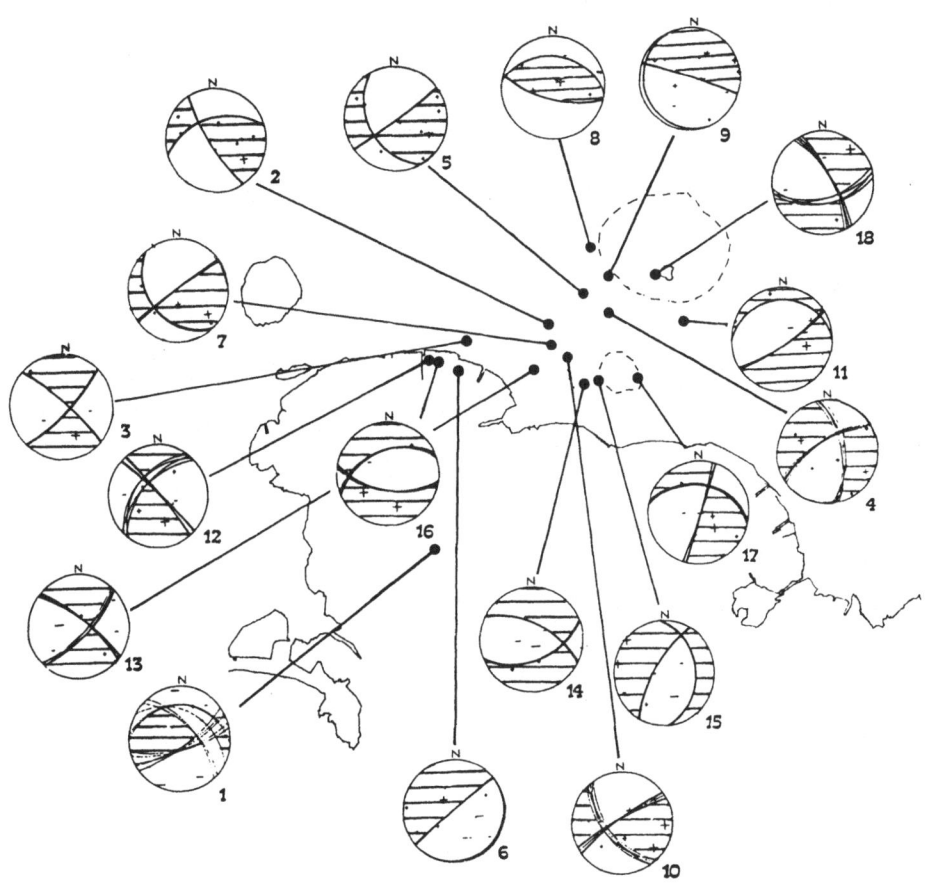

Fig. 7.   (continued)

The application of moment tensor investion technique to several microearthquakes occurred at Campi Flegrei volcanic area shows that double couple mechanisms are by far more significant than other general fracture mechanisms with no isotropic components (e.g. Compensated Linear Vector Dipole, Knopoff and Randall, 1970). On the other hand, although more general fracture mechanisms with isotropic components better fit the data the improvement in variance is not statistically significant. So there is no solid evidence for mechanisms different from double couple in the area.

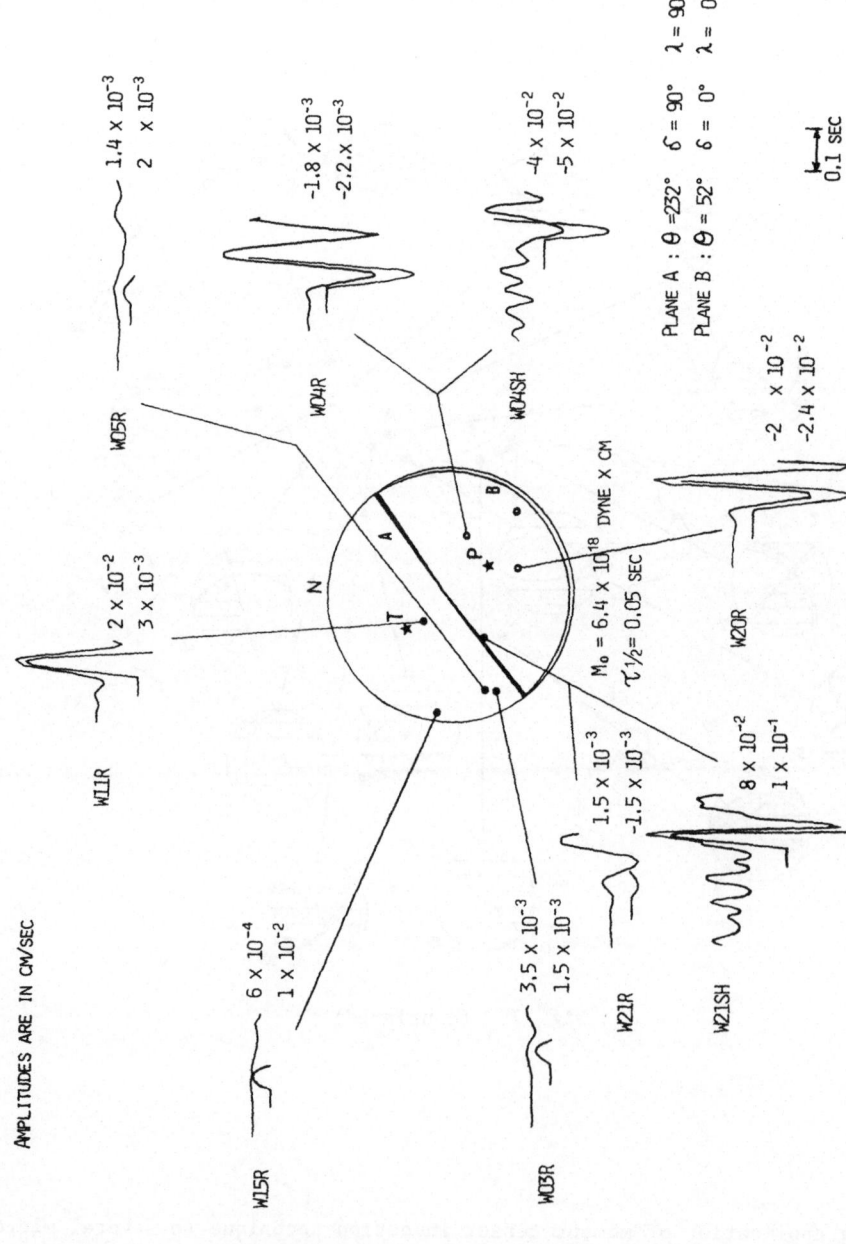

Fig. 8. Theoretical amplitudes of first P-radial and Sh pulses (lower trace) compared with observed ones for the earthquake n.6 in Table 2 (see the text).

REFERENCES

Aki, K., and Richards, P. G., 1980, "Quantitative Seismology," W.H. Freeman, S. Francisco, Calfiornia.

Aster, R., Meyer, R. P., De Natale, G., Zollo, A. Del Pezzo, E., Martini, M., Scarpa, R., Iannaccone, G., 1987, Seismic investigation of the Campi Flegrei Caldera, submitted to Journ.Geophys.Res.

Bernard, P., and Madariaga, R., 1985, Ray theroetical strong motion synthesis, J.Geophys., 58:73-81.

Bouchon, M., 1979, Discrete wave number representation of elastic waves in three-space dimensions, Journ.Geophys.Res., 84:3609-3614.

Cerveny, V., Horn, F., 1980, The ray series method in the dynamic ray tracing in 3-D inhomogeneous media, Bull.Seism.Soc.Am., 70:47-77.

Claerbout, J., 1976, "Fundamentals of geophysical data processing," McGraw Hill, S. Francisco.

De Natale, G., and Zollo, A., 1986, Statistical analysis and clustering features of the Phlegraean Fields earthquake sequence (May 1983 - May 1984) Bull.Seism.Soc.Am., 76:801-814.

Dziewonsky, A. M., and Woodhouse, J. H., 1983, An experiment in systematic study of global seismicity: centroid-moment tensor solutions for 201 moderate and large earthquakes of 1981, Journ.Geophys.Res., 86:2825-2852.

Forsythe, G. E., and Moler, C. B., 1967, "Computed Solution of Linear Algebraic Systems," Prentice Hall Inc., Englewood Cliffs, N.J.

Jackson, D. D., and Matsu'ura, N., 1985, A Bayesian approach to non linear inversion Journ.Geophys.Res., 90:581-591.

Kasahara, K., 1981, "Earthquake Mechanics", Cambridge Un. Press.

Knopoff, L., and Randall, M. J., 1970, The compensated linear vector dipole, Journ.Geophys.Res., 75:4957-63.

Lawson, C. L., and Hanson, R. J., 1974, "Solving Least Squares Problems", Prentice Hall Inc., Englewood Cliffs, N.J.

Oncescu, M. C., 1986, Relative seismic moment tensor determination for Vrancea intermediate depth earthquakes, Pure Appl.Geoph., in press.

Pearce, R. G., 1976, Fault plane solution using relative amplitudes of P and pP, Geophys.J.R. Astr.Soc., 50:381-94.

Sipkin, S. A., 1982, Estimation of earthquake source parameters by the inversion of waveform data: synthetic waveform, Phys.Earth Pla.Int., 30:242-59.

Sipkin, S. A., 1986, Interpretation of non-double couple earthquake mechanisms derived from moment tensor inversion, Journ.Geophys.Res., 91:531-47.

Stump, B. W., and Johnson, L. R., 1977, The determination of source properties by linear inversion of seismograms, Bull.Seism.Soc.Am., 1489-1502.

Tarantola, A., and Valette, B., 1982, Inverse problems = quest for information, J.Geophys., 50:159-170.